SCIENCE OF EVERYDAY THINGS

SCIENCE OF EVERYDAY THINGS

VOLUME 3: REAL-LIFE BIOLOGY

EDITED BY NEIL SCHLAGER
WRITTEN BY JUDSON KNIGHT

A SCHLAGER INFORMATION GROUP BOOK

Detroit • New York • San Diego • San Francisco • Cleveland • New Haven, Conn. • Waterville, Maine • London • Munich

Science of Everyday Things
Volume 3: Real-Life Biology

A Schlager Information Group Book
Neil Schlager, Editor
Written by Judson Knight

Project Editor
Kimberley A. McGrath

Editorial
Mark Springer

Permissions
Lori Hines

Imaging and Multimedia
Robert Duncan, Leitha Etheridge-Sims, Mary K. Grimes, Lezlie Light, Dan Newell, David G. Oblender, Robyn V. Young

Product Design
Michelle DiMercurio, Michael Logusz

Manufacturing
Evi Seoud, Rhonda Williams

For more information contact
The Gale Group, Inc.
27500 Drake Rd.
Farmington Hills, MI 48331-3535
Or you can visit our Internet site at
http://www.gale.com

LIBRARY OF CONGRESS CATALOG-IN-PUBLICATION DATA

Knight, Judson.
Science of everyday things / written by Judson Knight, Neil Schlager, editor.
p. cm.
Includes bibliographical references and indexes.
Contents: v. 1. Real-life chemistry – v. 2 Real-life physics.
SBN 0-7876-5631-3 (set : hardcover) – ISBN 0-7876-5632-1 (v. 1) – ISBN 0-7876-5633-X (v. 2)
1. Science–Popular works. I. Schlager, Neil, 1966-II. Title.

Q162.K678 2001
500–dc21 2001050121

ISBN 0-7876-5631-3 (set), 0-7876-5632-1 (vol. 1), 0-7876-5633-X (vol. 2), 0-7876-5634-8 (vol. 3), 0-7876-5635-6 (vol. 4)

Printed in the United States of America
10 9 8 7 6 5 4 3 2 1

CONTENTS

INTRODUCTION v
ADVISORY BOARD vii

BIOCHEMISTRY

Carbohydrates. 3
Amino Acids. 11
Proteins. 18
Enzymes 24

METABOLISM

Metabolism 33
Digestion 44
Respiration. 55

NUTRITION

Food Webs 67
Nutrients and Nutrition 77
Vitamins. 87

GENETICS

Genetics 99
Heredity. 110
Genetic Engineering 117
Mutation 126

REPRODUCTION AND BIRTH

Reproduction 135
Sexual Reproduction. 142
Pregnancy and Birth. 151

EVOLUTION

Evolution. 161
Paleontology 176

BIODIVERSITY AND TAXONOMY

Taxonomy 191
Species. 204
Speciation 215

DISEASE

Disease. 229
Noninfectious Diseases. 236
Infectious Diseases 244

IMMUNITY

Immunity and Immunology 255
The Immune System. 262

INFECTION

Parasites and Parasitology 273
Infection 283

BRAIN AND BODY

Chemoreception 295
Biological Rhythms. 306

LEARNING AND BEHAVIOR

Behavior. 319
Instinct and Learning 327
Migration and Navigation 335

THE BIOSPHERE AND ECOSYSTEMS

The Biosphere 345
Ecosystems and Ecology. 360
Biomes. 370

BIOLOGICAL COMMUNITIES

Symbiosis. 383
Biological Communities. 391
Succession and Climax. 400
GENERAL SUBJECT INDEX 411

INTRODUCTION

Overview of the Series

Welcome to *Science of Everyday Things.* Our aim is to explain how scientific phenomena can be understood by observing common, real-world events. From luminescence to echolocation to buoyancy, the series will illustrate the chief principles that underlay these phenomena and explore their application in everyday life. To encourage cross-disciplinary study, the entries will draw on applications from a wide variety of fields and endeavors.

Science of Everyday Things initially comprises four volumes:

Volume 1: *Real-Life Chemistry*
Volume 2: *Real-Life Physics*
Volume 3: *Real-Life Biology*
Volume 4: *Real-Life Earth Science*

Future supplements to the series will expand coverage of these four areas and explore new areas, such as mathematics.

Arrangement of Real-Life Biology

This volume contains 40 entries, each covering a different scientific phenomenon or principle. The entries are grouped together under common categories, with the categories arranged, in general, from the most basic to the most complex. Readers searching for a specific topic should consult the table of contents or the general subject index.

Within each entry, readers will find the following rubrics:

- **Concept:** Defines the scientific principle or theory around which the entry is focused.
- **How It Works:** Explains the principle or theory in straightforward, step-by-step language.
- **Real-Life Applications:** Describes how the phenomenon can be seen in everyday life.
- **Where to Learn More:** Includes books, articles, and Internet sites that contain further information about the topic.

In addition, each entry includes a "Key Terms" section that defines important concepts discussed in the text. Finally, each volume includes many illustrations and photographs throughout.

In addition, readers will find the comprehensive general subject index valuable in accessing the data.

About the Editor, Author, and Advisory Board

Neil Schlager and Judson Knight would like to thank the members of the advisory board for their assistance with this volume. The advisors were instrumental in defining the list of topics, and reviewed each entry in the volume for scientific accuracy and reading level. The advisors include university-level academics as well as high school teachers; their names and affiliations are listed elsewhere in the volume.

Neil Schlager is the president of Schlager Information Group Inc., an editorial services company. Among his publications are *When Technology Fails* (Gale, 1994); *How Products Are Made* (Gale, 1994); the *St. James Press Gay and Lesbian Almanac* (St. James Press, 1998); *Best Literature By and About Blacks* (Gale, 2000); *Contemporary Novelists*, 7th ed. (St. James Press,

2000); *Science and Its Times* (7 vols., Gale, 2000-2001); and *Science in Dispute* (Gale, 2002). His publications have won numerous awards, including three RUSA awards from the American Library Association, two Reference Books Bulletin/Booklist Editors' Choice awards, two New York Public Library Outstanding Reference awards, and a *CHOICE* award for best academic book.

Judson Knight is a freelance writer, and author of numerous books on subjects ranging from science to history to music. His work on science includes *Science, Technology, and Society, 2000 B.C.-A.D.* 1799 (U•X•L, 2002), as well as extensive contributions to Gale's seven-volume *Science and Its Times* (2000-2001). As a writer on history, Knight has published *Middle Ages Reference Library* (2000), *Ancient Civilizations* (1999), and a volume in U•X•L's *African American Biography* series (1998). Knight's publications in the realm of music include *Parents Aren't Supposed to Like It* (2001), an overview of contemporary performers and genres, as well as *Abbey Road to Zapple Records: A Beatles Encyclopedia* (Taylor, 1999).

COMMENTS AND SUGGESTIONS

Your comments on this series and suggestions for future editions are welcome. Please write: The Editor, *Science of Everyday Things,* Gale Group, 27500 Drake Road, Farmington Hills, MI 48331-3535.

ADVISORY BOARD

SCIENCE OF EVERYDAY THINGS

REAL-LIFE BIOLOGY

BIOCHEMISTRY

CARBOHYDRATES

AMINO ACIDS

PROTEINS

ENZYMES

CARBOHYDRATES

CONCEPT

Carbohydrates are nutrients, along with proteins and other types of chemical compounds, but they are much more than that. In addition to sugars, of which there are many more varieties than ordinary sucrose, or table sugar, carbohydrates appear in the form of starches and cellulose. As such, they are the structural materials of which plants are made. Carbohydrates are produced by one of the most complex, vital, and amazing processes in the physical world: photosynthesis. Because they are an integral part of plant life, it is no wonder that carbohydrates are in most fruits and vegetables. And though they are not a dietary requirement in the way that vitamins or essential amino acids are, it is difficult to eat without ingesting some carbohydrates, which are excellent sources of quick-burning energy. Not all carbohydrates are of equal nutritional value, however: in general, the ones created by nature are good for the body, whereas those produced by human intervention—some forms of pasta and most varieties of bread, white rice, crackers, cookies,and so forth—are much less beneficial.

HOW IT WORKS

What Carbohydrates Are

Carbohydrates are naturally occurring compounds that consist of carbon, hydrogen, and oxygen, and are produced by green plants in the process of undergoing photosynthesis. In simple terms, photosynthesis is the biological conversion of light energy (that is, electromagnetic energy) from the Sun to chemical energy in plants. It is an extremely complex process, and a thorough treatment of it involves a great deal of technical terminology. Although we discuss the fundamentals of photosynthesis later in this essay, we do so only in the most cursory fashion.

Photosynthesis involves the conversion of carbon dioxide and water to sugars, which, along with starches and cellulose, are some of the more well known varieties of carbohydrate. Sugars can be defined as any of a number of water-soluble compounds, of varying sweetness. (What we think of as sugar—that is, table sugar—is actually sucrose, discussed later.) Starches are complex carbohydrates without taste or odor, which are granular or powdery in physical form. Cellulose is a polysaccharide, made from units of glucose, that constitutes the principal part of the cell walls of plants and is found naturally in fibrous materials, such as cotton. Commercially, it is a raw material for such manufactured goods as paper, cellophane, and rayon.

MONOSACCHARIDES. The preceding definitions contain several words that also must be defined. Carbohydrates are made up of building blocks called monosaccharides, the simplest type of carbohydrate. Found in grapes and other fruits and also in honey, they can be broken down chemically into their constituent elements, but there is no carbohydrate more chemically simple than a monosaccharide. Hence, they are also known as simple sugars or simple carbohydrates.

Examples of simple sugars include glucose, which is sweet, colorless, and water-soluble and appears widely in nature. Glucose, also known as dextrose, grape sugar, and corn sugar, is the principal form in which carbohydrates are assimilated, or taken in, by animals. Other monosaccha-

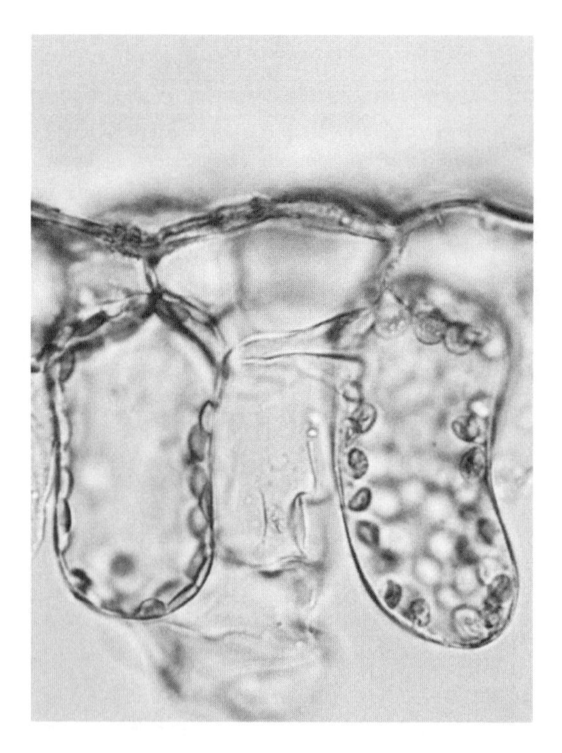

MICROGRAPH OF PLANT CELL CHLOROPLASTS, WHERE PHOTOSYNTHESIS, THE BIOLOGICAL CONVERSION OF LIGHT FROM THE SUN INTO CHEMICAL ENERGY, TAKES PLACE. HIGHER PLANTS HAVE THESE STRUCTURES, WHICH CONTAIN A CHEMICAL KNOWN AS CHLOROPHYLL THAT ABSORBS LIGHT AND SPEEDS UP THE PROCESS OF PHOTOSYNTHESIS. *(© Science Pictures Limited/Corbis. Reproduced by permission.)*

rides include fructose, or fruit sugar, and galactose, which is less soluble and sweet than glucose and usually appears in combination with other simple sugars rather than by itself. Glucose, fructose, and galactose are isomers, meaning that they have the same chemical formula ($C_6H_{12}O_6$), but different chemical structures and therefore different chemical properties.

DISACCHARIDES. When two monosaccharide molecules chemically bond with each other, the result is one of three general types of complex sugar: a disaccharide, oligosaccharide, or polysaccharide. Disaccharides, or double sugars, are composed of two monosaccharides. By far the most well known example of a disaccharide is sucrose, or table sugar, which is formed from the bonding of a glucose molecule with a molecule of fructose. Sugar beets and cane sugar provide the principal natural sources of sucrose, which the average American is most likely to encounter in refined form as white, brown, or powdered sugar.

Another disaccharide is lactose, or milk sugar, the only type of sugar that is produced from animal (i.e., mammal) rather than vegetable sources. Maltose, a fermentable sugar typically formed from starch by the action of the enzyme amylase, is also a disaccharide. Sucrose, lactose, and maltose are all isomers, with the formula $C_{12}H_{22}O_{11}$.

OLIGOSACCHARIDES AND POLYSACCHARIDES. The definitions of oligosaccharide and polysaccharide are so close as to be confusing. An oligosaccharide is sometimes defined as a carbohydrate containing a known, small number of monosaccharide units, while a polysaccharide is a carbohydrate composed of two or more monosaccharides. In theory, this means practically the same thing, but in practice, an oligosaccharide contains 3-6 monosaccharide units, whereas a polysaccharide is composed of more than six.

Oligosaccharides are found rarely in nature, though a few plant forms have been discovered. Far more common are polysaccharides ("many sugars"), which account for the vast majority of carbohydrate types found in nature. (See Where to Learn More for the Nomenclature of Carbohydrates Web site, operated by the Department of Chemistry at Queen Mary College, University of London. A glance at the site will suggest something about the many, many varieties of carbohydrates.)

Polysaccharides may be very large, consisting of as many as 10,000 monosaccharide units strung together. Given this vast range of sizes, it should not be surprising that there are hundreds of polysaccharide types, which differ from one another in terms of size, complexity, and chemical makeup. Cellulose itself is a polysaccharide, the most common variety known, composed of numerous glucose units joined to one another. Starch and glycogen are also glucose polysaccharides. The first of these polysaccharides is found primarily in the stems, roots, and seeds of plants. As for glycogen, this is the most common form in which carbohydrates are stored in animal tissues, particularly muscle and liver tissues.

PHOTOSYNTHESIS

Photosynthesis, as we noted earlier, is the biological conversion of light or electromagnetic energy from the Sun into chemical energy. It occurs in green plants, algae, and some types of bacteria and requires a series of biochemical reactions. Higher plants have structures called chloroplasts, which contain a dark green or blue-black chemical known as chlorophyll. Light absorption by chlorophyll catalyzes, or speeds up, the process of photosynthesis. (A catalyst is a substance that accelerates a chemical reaction without participating in it.)

In photosynthesis, carbon dioxide and water react with each other in the presence of light and chlorophyll to produce a simple carbohydrate and oxygen. This is one of those statements in the realm of science that at first glance sounds a bit dry and boring but which, in fact, encompasses one of life's great mysteries—a concept far more captivating than any number of imaginary, fantastic, or pseudoscientific ideas one could concoct. Photosynthesis is one of the most essential life-sustaining processes, making possible the nutrition of all things and the respiration of animals and other oxygen-breathing organisms.

In photosynthesis, plants take a waste product of human and animal respiration and, through a series of chemical reactions, produce both food and oxygen. The food gives nourishment to the plant, which, unlike an animal, is capable of producing its own nutrition from its own body with the aid only of sunlight and a few chemical compounds. Later, when the plant is eaten by an animal or when it dies and is consumed by bacteria and other decomposers, it will pass on its carbohydrate content to other creatures. (See Food Webs for more about plants as autotrophs and the relationships among primary producers, consumers, and decomposers.)

A carbohydrate is not the only useful product of the photosynthetic reaction. The reaction produces an extremely important waste by-product—waste, that is, from the viewpoint of the plant, which has no need of oxygen. Yet the oxygen it generates in photosynthesis makes life possible for animals and many single-cell life-forms, which depend on oxygen for respiration.

THE PHOTOSYNTHESIS EQUATION. The photosynthesis reaction can be represented thus as a chemical equation:

$$6CO_2 + 6H_2O \xrightarrow{\text{Light}} C_6H_{12}O_6 + 6O_2$$

Note that the arrow indicates that a chemical reaction has taken place with the assistance of light and chlorophyll. In the same way, heat from a Bunsen burner may be required to initiate some other chemical reaction, without actually being part of the reactants to the left of the arrow. In the present equation, neither the added energy nor the catalyst appears on the left side, because they are not actual physical participants consumed in the reaction, as the carbon dioxide and water are. The catalyst does not participate in the reaction, whereas the energy, while it is consumed in the reaction, is not a material or physical participant—that is, it is energy, not matter.

One might also wonder why the equation shows six molecules of carbon dioxide and six of water. Why not one of each, for the sake of simplicity? To produce a balanced chemical equation, in which the same number of atoms appears on either side of the arrow, it is necessary to show six carbon dioxide molecules reacting with six water molecules to produce six oxygen molecules and a single glucose molecule. Thus, both sides contain six atoms of carbon, 12 of hydrogen, and 18 of oxygen.

The equation gives the impression that photosynthesis is a simple, one-step process, but nothing could be further from the truth. In fact, the process occurs one small step at a time. It also involves many, many intricacies and aspects that require the introduction of scores of new terms and ideas. Such a discussion is beyond the scope of the present essay, and therefore the reader is encouraged to consult a reliable textbook for further information on the details of photosynthesis.

REAL-LIFE APPLICATIONS

FRUITS AND VEGETABLES

One of the principal ways in which people obtain carbohydrates from their diets is through fruits and vegetables. The distinctions between these two are based not on science but on custom. Traditionally, vegetables are plant tissues (which may be sweet, but usually are not), that are eaten as a substantial part of a meal's main course. By contrast, fruits are almost always sweet and are eaten as desserts or snacks. It so happens, too, that people are much more likely to cook vegetables than they are fruits, though vegetables are nutritionally best when eaten raw.

Fruits and vegetables are heavy in carbohydrate content, in the form of edible sugars and starches but also inedible cellulose, whose role in the diet will be examined later. In a fresh veg-

CELLULOSE, SOMETIMES CALLED FIBER, IS AN IMPORTANT DIETARY COMPONENT THAT AIDS IN DIGESTION. IT IS ABUNDANT IN FRUITS AND VEGETABLES, YET HUMANS LACK THE ENZYME NECESSARY TO DIGEST IT. WITH THE HELP OF MICROBES IN THEIR GUT, TERMITES CAN DIGEST CELLULOSE. *(© George D.Lepp/Corbis. Reproduced by permission.)*

etable, for instance, water may account for about 70% of the volume, and proteins, fat, vitamins, and minerals may make up a little more than 5%, with nearly 25% taken up either by edible sugars and starches or by inedible cellulose fiber.

THE EXAMPLE OF THE ARTICHOKE. Every fruit or vegetable one could conceivably eat—and there are hundreds—contains both edible carbohydrates, which are a good source of energy, and inedible ones, which provide fiber. An excellent example of this edible-inedible mixture is the globe, or French, artichoke—*Cynara scolymus,* a member of the family Asteraceae, which includes the sunflower. The globe artichoke (not to be confused with the Jerusalem artichoke, or *Helianthus tuberosus*) appears in the form of an inflorescence, or a cluster of flowers. This vegetable usually is steamed, and the bracts, or leaves, are dipped in butter or another sauce.

Not nearly all of the bract is edible, however; to consume the starchy "meat" of the artichoke, which has a distinctive, nutty flavor, one must draw the leaves between the teeth. Most of the artichoke's best parts are thus hidden away, and the best part of all—the tender and fully edible "heart"—is enclosed beneath an intimidating shield of slender thistles. Whoever first discovered that an artichoke could be eaten must have been a brave person indeed, and whoever ascertained *how* to eat it was a wise one. Thanks to these adventurous souls, the world's cuisine has an unforgettable delicacy.

THE CARBOHYDRATE CONTENT OF VEGETABLES. In terms of edible carbohydrate content, the artichoke has a low percentage. A few vegetables have a smaller percentage of carbohydrates, whereas others have vastly higher percentages, as the list shown here illustrates. In general, it seems that the carbohydrate content of vegetables (and in each of these cases we are talking about *edible* carbohydrates, not cellulose) is in the range of about 5–10%, somewhere around 20%, or a very high 60–80%. There does not seem to be a great deal of variation in these ranges.

Water, Protein, and Carbohydrate Content of Selected Vegetables:

- Artichoke: 85% water, 2.9% protein, 10.6% carbohydrate
- Beets, red: 87.3% water, 1.6% protein, 9.9% carbohydrate
- Celery: 94.1% water, 0.9% protein, 3.9% carbohydrate
- Corn: 13.8% water, 8.9% protein, 72.2% carbohydrate
- Lima bean: 10.3% water, 20.4% protein, 64% carbohydrate
- Potato: 79.8% water, 2.1% protein, 17.1% carbohydrate
- Red pepper: 74.3% water, 3.7% protein, 18.8% carbohydrate
- Summer squash: 94% water, 1.1% protein, 4.2% carbohydrate

STARCHES

Not all the carbohydrates in these vegetables are the same. Some carbohydrates appear in the form of sugar and others in the form of inedible cellulose, discussed in the next section. In addition, some vegetables are high in starch content. As we noted earlier, starch is white and granular, and, unlike sugars, starches cannot be dissolved in cold water, alcohol, or other liquids that normally act as solvents.

Manufactured in plants' leaves, starch is the product of excess glucose produced during photosynthesis, and it provides the plant with an emergency food supply stored in the chloroplasts. Vegetables high in starch content are products of plants whose starchy portions happen to be the portions we eat. For example, there is the tuber, or underground bulb, of the potato as well as the seeds of corn, wheat, and rice. Thus, all of these vegetables, and foods derived from them, are heavy in the starch form of carbohydrate.

In addition to their role in the human diet, starches from corn, wheat, tapioca, and potatoes are put to numerous commercial uses. Because of its ability to thicken liquids and harden solids, starch is applied in products (e.g., cornstarch) that act as thickening agents, both for foods and nonfood items. Starch also is utilized heavily in various phases of the garment and garment-care industries to impart stiffness to fabrics. In the manufacture of paper, starch is used to increase the paper's strength. It also is employed in the production of cardboard and paper bags.

NINETEENTH-CENTURY ADVERTISEMENT FOR STARCH. IN ADDITION TO THEIR ROLE IN THE HUMAN DIET, STARCHES ARE PUT TO NUMEROUS COMMERCIAL USES, FOR EXAMPLE, AS THICKENING AGENTS FOR FOOD, IN THE PRODUCTION OF CARDBOARD, AND IN VARIOUS PHASES OF THE GARMENT INDUSTRY TO IMPART STIFFNESS TO FABRICS. *(© BettmannCorbis. Reproduced by permission.)*

CELLULOSE

One of the aspects of fruits and vegetables to which we have alluded several times is the high content of inedible material, or cellulose. (Actually, it is edible—just not digestible.) A substance found in the cell walls of plants, cellulose is chemically like starch but even more rigid, and this property makes it an excellent substance for imparting strength to plant bodies. Animals do not have rigid, walled cells, but plants do. The heavy cellulose content in plants' cell walls gives them their erect, rigid form; in other words, without cellulose, plants might be limp and partly formless. Like human bone, plant cell walls are composed of fibrils (small filaments or fibers) that include numerous polysaccharides and proteins. One of these polysaccharides in cell walls is pectin, a substance that, when heated, forms a gel and is used by cooks in making jellies and jams. Some trees have a secondary cell wall over the primary one, containing yet another polysaccharide called *lignin.* Lignin makes the tree even more rigid, penetrable only with sharp axes.

CELLULOSE IN DIGESTION. As we have noted, cellulose is abundant in fruits and vegetables, yet humans lack the enzyme necessary to digest it. Termites, cows, koalas, and horses all digest cellulose, but even these animals and insect do not have an enzyme that digests this material. Instead, they harbor microbes in their guts that can do the digesting for them. (This is an example of symbiotic mutualism, a mutually beneficial relationship between organisms, discussed in Symbiosis.)

Cows are ruminants, or animals that chew their cud—that is, food regurgitated to be chewed again. Ruminants have several stomachs, or several stomach compartments, that break down plant material with the help of enzymes and bacteria. The partially digested material then is regurgitated into the mouth, where it is chewed to break the material down even further. (If you have ever watched cows in a pasture, you have probably observed them calmly chewing their cud.) The digestion of cellulose by bacteria in the stomachs of ruminants is anaerobic, meaning that the process does not require oxygen. One of the by-products of this anaerobic process is methane gas, which is foul smelling, flammable, and toxic. Ruminants give off large amounts of methane daily, which has some environmentalists alarmed, since cow-borne methane may con-

KEY TERMS

CARBOHYDRATES: Naturally occurring compounds, consisting of carbon, hydrogen, and oxygen, whose primary function in the body is to supply energy. Included in the carbohydrate group are sugars, starches, cellulose, and various other substances. Most carbohydrates are produced by green plants in the process of undergoing photosynthesis.

CATALYST: A substance that speeds up a chemical reaction without participating in it. Catalysts, of which enzymes are a good example, thus are not consumed in the reaction.

CELLULOSE: A polysaccharide, made from units of glucose, that is the principal material in the cell walls of plants. Cellulose also is found in natural fibers, such as cotton, and is used as a raw material in manufacturing such products as paper.

COMPLEX CARBOHYDRATE: A disaccharide, polysaccharide, or oligosaccharide. Also called a *complex sugar.*

DEXTROSE: Another name for glucose.

DISACCHARIDE: A double sugar, composed of two monosaccharides. Examples of disaccharides include the isomers sucrose, maltose, and lactose.

ENZYME: A protein material that speeds up chemical reactions in the bodies of plants and animals.

FRUCTOSE: Fruit sugar, a monosaccharide that is an isomer of glucose.

GALACTOSE: A monosaccharide and isomer of glucose. Less soluble and sweet than glucose, galactose usually appears in combination with other simple sugars rather than by itself.

GLUCOSE: A monosaccharide that occurs widely in nature and is the form in which animals usually receive carbohydrates. Also known as dextrose, grape sugar, and corn sugar.

GLYCOGEN: A white polysaccharide that is the most common form in which carbohydrates are stored in animal tissues, particularly muscle and liver tissues.

GUT: A term that refers to all or part of the alimentary canal, through which foods pass from the mouth to the intestines and wastes move from the intestines to the anus. Although the word is considered a bit crude in everyday life, physicians and biological scientists concerned with this part of the anatomy use it regularly.

ISOMERS: Two substances that have the same chemical formula but differ in

tribute to the destruction of the ozone high in Earth's stratosphere.

Alhough cellulose is indigestible by humans, it is an important dietary component in that it aids in digestion. Sometimes called fiber or roughage, cellulose helps give food bulk as it moves through the digestive system and aids the body in pushing out foods and wastes. This is particularly important inasmuch as it helps make possible regular bowel movements, thus ridding the body of wastes and lowering the risk of colon cancer. (See Digestion for more about the digestive and excretory processes.)

OVERALL CARBOHYDRATE NUTRITION

A diet high in cellulose content can be beneficial for the reasons we have noted. Likewise, a healthy diet includes carbohydrate nutrients, but only under certain conditions. First of all, it should be

KEY TERMS CONTINUED

chemical structure and therefore in chemical properties.

LACTOSE: Milk sugar. A disaccharide isomer of sucrose and maltose, lactose is the only major type of sugar that is produced from animal (i.e., mammal) rather than vegetable sources.

MALTOSE: A fermentable sugar generally formed from starch by the action of the enzyme amylase. Maltose is a disaccharide isomer of sucrose and lactose.

MONOSACCHARIDE: The simplest type of carbohydrate. Monosaccharides, which cannot be broken down chemically into simpler carbohydrates, also are known as simple sugars. Examples of monosaccharides include the isomers glucose, fructose, and galactose.

OLIGOSACCHARIDE: A carbohydrate containing a known, small number of monosaccharide units, typically between three and six. Compare with *polysaccharide.*

PHOTOSYNTHESIS: The biological conversion of light energy (that is, electromagnetic energy) from the Sun to chemical energy in plants. In this process, carbon dioxide and water are converted to carbohydrates and oxygen.

POLYSACCHARIDE: A carbohydrate composed of more than six monosaccharides. A polysaccharide sometimes is defined as containing two or more monosaccharides, but this definition does little to distinguish it from an *oligosaccharide.*

SACCHARIDE: A sugar.

SIMPLE SUGAR: A monosaccharide, or simple carbohydrate.

STARCHES: Complex carbohydrates, without taste or odor, which are granular or powdery in physical form.

SUCROSE: Common table sugar ($C_{12}H_{22}O_{11}$), a disaccharide formed from the bonding of a glucose molecule with a molecule of fructose. Sugar beets and cane sugar provide the principal natural sources of sucrose, which the average American is most likely to encounter in refined form as white, brown, or powdered sugar.

SUGARS: One of the three principal types of carbohydrate, along with starches and cellulose. Sugars can be defined as any of various water-soluble carbohydrates of varying sweetness. What we think of as "sugar" (i.e., table sugar) is actually sucrose.

understood that the human body does not have an essential need for carbohydrates in and of themselves—in other words, there are no "essential" carbohydrates, as there are essential amino acids or fatty acids.

On the other hand, it is very important to eat fresh fruits and vegetables, which, as we have seen, are heavy in carbohydrate content. Their importance has little do with their nutritional carbohydrate content, but rather with the vitamins, minerals, proteins, and dietary fiber that they contain. For these healthy carbohydrates, it is best to eat them in as natural a form as possible: for example, eat the whole orange, rather than just squeezing out the juice and throwing away the pulp. Also, raw spinach and other vegetables contain far more vitamins and minerals than the cooked versions.

SUGAR HIGHS AND FAT STORAGE. Carbohydrates can give people a short burst of energy, and this is why athletes may "bulk up on carbs" right before competition. But if the carbohydrates are not quickly burned off, they eventually will be stored as fat. This is the case even with healthy carbohydrates, but the situation is much worse with junk-food carbohydrates, which offer only empty calories stripped of vitamin and mineral content. One example is a particular brand of candy bar that, over the years, has been promoted in commercials as a means of obtaining a quick burst of energy. In fact, this and all other white-sugar-based candies give only a quick "sugar high," followed almost immediately by a much lower energy "low"—and in the long run by the accumulation of fat.

Fat is the only form in which the body can store carbohydrates for the long haul, meaning that the "fat-free" stickers on many a package of cookies or cakes in the supermarket are as meaningless as the calories themselves are empty. Carbohydrate consumption is one of the main reasons why the average American is so overweight. With an inactive lifestyle, as is typical of most adults in modern life, all those French fries, cookies, dinner rolls, and so on have no place to go but to the fat-storage centers in the abdomen, buttocks, and thighs. Of all carbohydrate-containing foods, the least fattening, of course, are natural nonstarches, such as fruits and vegetables (assuming they are not cooked in fat). Next on the least-fattening list are starchy natural foods, such as potatoes, and most fattening of all are processed starches, whether they come in the form of rice, wheat, or potato products.

WHY YOU CAN EAT MORE CARBOHYDRATES THAN PROTEINS. One of the biggest problems with starches is that the body can consume so many of them compared with proteins and fats. How many times have you eaten a huge plate of mashed potatoes or rice, mountains of fries, or piece after piece of bread? All of us have done it: with carbohydrates, and particularly starches, it seems we can never get enough. But how many times have you eaten a huge plate of nothing but chicken, steak, or eggs? Probably not very often, and if you have tried to eat too much of these protein-heavy foods at one time, you most likely started to get sick.

The reason is that when you eat protein or fat, it triggers the release of a hormone called *cholecystokinin* (CCK) in the small intestine. CCK tells the brain, in effect, that the body is getting fed, and if enough CCK is released, it signals the brain that the body has received enough food. If one continues to consume proteins or fats beyond that point, nausea is likely to follow. Carbohydrates, on the other hand, do not cause a release of CCK; only when they enter the bloodstream do they finally send a signal to the brain that the body is satisfied. By then, most of us have piled on more mashed potatoes, which are destined to take their place in the body as fat stores.

WHERE TO LEARN MORE

Carbohydrates. Hardy Research Group, Department of Chemistry, University of Akron (Web site). <http://ull.chemistry.uakron.edu/genobc/Chapter_17/>.

Dey, P. M., and R. A. Dixon. *Biochemistry of Storage Carbohydrates in Green Plants.* Orlando, FL: Academic Press, 1985.

Carpi, Anthony. "Food Chemistry: Carbohydrates." Visionlearning.com (Web site). <http://www.visionlearning.com/library/science/chemistry-2/CHE2.5-carbohydrates.htm>.

Food Resource, Oregon State University (Web site). <http://food.orst.edu/>.

Kennedy, Ron. "Carbohydrates in Nutrition." *The Doctors' Medical Library* (Web site). <http://www.medical-library.net/sites/carbohydrates_in_nutrition.html>.

"Nomenclature of Carbohydrates." Queen Mary College, University of London, Department of Chemistry (Web site). <http://www.chem.qmw.ac.uk/iupac/2carb/>.

Snyder, Carl H. *The Extraordinary Chemistry of Ordinary Things.* New York: John Wiley and Sons, 1998.

Spallholz, Julian E. *Nutrition, Chemistry, and Biology.* Englewood Cliffs, NJ: Prentice-Hall, 1989.

Wiley, T. S., and Bent Formby. *Lights Out: Sleep, Sugar, and Survival.* New York: Pocket Books, 2000.

AMINO ACIDS

CONCEPT

Amino acids are organic compounds made of carbon, hydrogen, oxygen, nitrogen, and (in some cases) sulfur bonded in characteristic formations. Strings of amino acids make up proteins, of which there are countless varieties. Of the 20 amino acids required for manufacturing the proteins the human body needs, the body itself produces only 12, meaning that we have to meet our requirements for the other eight through nutrition. This is just one example of the importance of amino acids in the functioning of life. Another cautionary illustration of amino acids' power is the gamut of diseases (most notably, sickle cell anemia) that impair or claim the lives of those whose amino acids are out of sequence or malfunctioning. Once used in dating objects from the distant past, amino acids have existed on Earth for at least three billion years—long before the appearance of the first true organisms.

HOW IT WORKS

A "MAP" OF AMINO ACIDS

Amino acids are organic compounds, meaning that they contain carbon and hydrogen bonded to each other. In addition to those two elements, they include nitrogen, oxygen, and, in a few cases, sulfur. The basic structure of an amino-acid molecule consists of a carbon atom bonded to an amino group ($-NH_2$), a carboxyl group (-COOH), a hydrogen atom, and a fourth group that differs from one amino acid to another and often is referred to as the -*R* group or the side chain. The -*R* group, which can vary widely, is responsible for the differences in chemical properties.

This explanation sounds a bit technical and requires a background in chemistry that is beyond the scope of this essay, but let us simplify it somewhat. Imagine that the amino-acid molecule is like the face of a compass, with a carbon atom at the center. Raying out from the center, in the four directions of the compass, are lines representing chemical bonds to other atoms or groups of atoms. These directions are based on models that typically are used to represent amino-acid molecules, though north, south, east, and west, as used in the following illustration, are simply terms to make the molecule easier to visualize.

To the south of the carbon atom (C) is a hydrogen atom (H), which, like all the other atoms or groups, is joined to the carbon center by a chemical bond. To the north of the carbon center is what is known as an amino group ($-NH_2$). The hyphen at the beginning indicates that such a group does not usually stand alone but normally is attached to some other atom or group. To the east is a carboxyl group, represented as -COOH. In the amino group, two hydrogen atoms are bonded to each other and then to nitrogen, whereas the carboxyl group has two separate oxygen atoms strung between a carbon atom and a hydrogen atom. Hence, they are not represented as O_2.

Finally, off to the west is the *R*- group, which can vary widely. It is as though the other portions of the amino acid together formed a standard suffix in the English language, such as -*tion.* To the front of that suffix can be attached all sorts of terms drawn from root words, such as *educate* or

Computer-generated model of a molecule made up of three amino acids—glycine, cysteine and alanine. Amino acids function as monomers, or individual units, that join together to form large, chainlike molecules called polymers; three amino acids bonded together are called tripeptides. *(Photo Researchers. Reproduced by permission.)*

satisfy or *revolt*—hence, *education, satisfaction,* and *revolution.* The variation in the terms attached to the front end is extremely broad, yet the tail end, *-tion,* is a single formation. Likewise the carbon, hydrogen, amino group, and carboxyl group in an amino acid are more or less constant.

A FEW ADDITIONAL POINTS. The name amino acid, in fact, comes from the amino group and the acid group, which are the most chemically reactive parts of the molecule. Each of the common amino acids has, in addition to its chemical name, a more familiar name and a three-letter abbreviation that frequently is used to identify it. In the present context, we are not concerned with these abbreviations. Amino-acid molecules, which contain an amino group and a carboxyl group, do not behave like typical molecules. Instead of melting at temperatures hotter than 392°F (200°C), they simply decompose. They are quite soluble, or capable of being dissolved, in water but are insoluble in nonpolar solvents (oil- and all oil-based products), such as benzene or ether.

RIGHT-HAND AND LEFT-HAND VERSIONS. All of the amino acids in the human body, except glycine, are either right-hand or left-hand versions of the same molecule, meaning that in some amino acids the positions of the carboxyl group and the *R*- group are switched. Interestingly, nearly all of the amino acids occurring in nature are the left-hand versions of the molecules, or the L-forms. (Therefore, the model we have described is actually the left-hand model, though the distinctions between "right" and "left"—which involve the direction in which light is polarized—are too complex to discuss here.)

Right-hand versions (D-forms) are not found in the proteins of higher organisms, but they are present in some lower forms of life, such as in the cell walls of bacteria. They also are found in some antibiotics, among them, streptomycin, actinomycin, bacitracin, and tetracycline. These antibiotics, several of which are well known to the public at large, can kill bacterial cells by interfering with the formation of proteins necessary for maintaining life and for reproducing.

AMINO ACIDS AND PROTEINS

A chemical reaction that is characteristic of amino acids involves the formation of a bond, called a peptide linkage, between the carboxyl group of one amino acid and the amino group of a second amino acid. Very long chains of amino acids can bond together in this way to form proteins, which are the basic building blocks of all living things. The specific properties of each kind of protein are largely dependent on the kind and sequence of the amino acids in it. Other aspects of the chemical behavior of protein molecules are due to interactions between the amino and the carboxyl groups or between the various *R*-groups along the long chains of amino acids in the molecule.

NUMBERS AND COMBINATIONS. Amino acids function as monomers, or individual units, that join together to form large, chainlike molecules called polymers, which may contain as few as two or as many as 3,000 amino-acid units. Groups of only two amino acids are called dipeptides, whereas three amino acids bonded together are called tripeptides. If there are more than 10 in a chain, they are termed polypeptides, and if there are 50 or more, these are known as proteins.

All the millions of different proteins in living things are formed by the bonding of only 20 amino acids to make up long polymer chains. Like the 26 letters of the alphabet that join together to form different words, depending on which letters are used and in which sequence, the 20 amino acids can join together in different combinations and series to form proteins. But whereas words usually have only about 10 or fewer letters, proteins typically are made from as few as 50 to as many as 3,000 amino acids. Because each amino acid can be used many times along the chain and because there are no restrictions on the length of the chain, the number of possible combinations for the formation of proteins is truly enormous. There are about two quadrillion different proteins that can exist if each of the 20 amino acids present in humans is used only once. Just as not all sequences of letters make sense, however, not all sequences of amino acids produce functioning proteins. Some other sequences can function and yet cause undesirable effects, as we shall see.

REAL-LIFE APPLICATIONS

DNA (deoxyribonucleic acid), a molecule in all cells that contains genetic codes for inheritance, creates encoded instructions for the synthesis of amino acids. In 1986, American medical scientist Thaddeus R. Dryja (1940–) used amino-acid sequences to identify and isolate the gene for a type of cancer known as retinoblastoma, a fact that illustrates the importance of amino acids in the body.

Amino acids are also present in hormones, chemicals that are essential to life. Among these hormones is insulin, which regulates sugar levels in the blood and without which a person would die. Another is adrenaline, which controls blood pressure and gives animals a sudden jolt of energy needed in a high-stress situation—running from a predator in the grasslands or (to a use a human example) facing a mugger in an alley or a bully on a playground. Biochemical studies of amino-acid sequences in hormones have made it

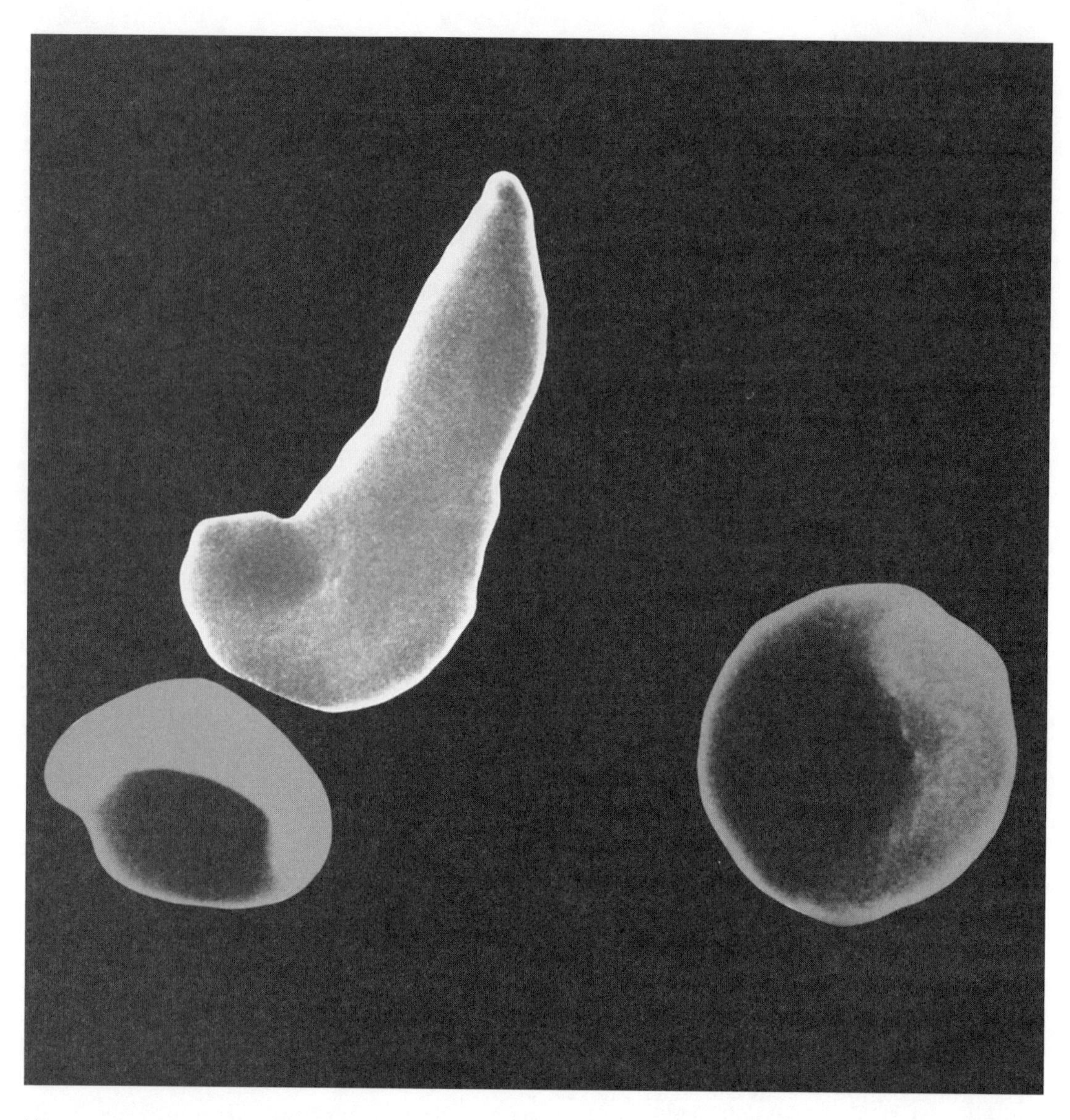

NORMAL RED BLOOD CELLS (BOTTOM) AND SICKLE CELL (TOP). SICKLE CELL ANEMIA IS A FATAL DISEASE BROUGHT ABOUT BY A SINGLE MISTAKE IN AMINO ACID SEQUENCING. WHEN RED BLOOD CELLS RELEASE OXYGEN TO THE TISSUES, THEY FAIL TO RE-OXYGENATE NORMALLY AND INSTEAD TWIST INTO THE SHAPE THAT GIVES SICKLE CELL ANEMIA ITS NAME, CAUSING OBSTRUCTION OF THE BLOOD VESSELS. *(Photograph by Dr. Gopal Murti. National Audubon Society Collection/Photo Researchers, Inc. Reproduced by permission.)*

possible for scientists to isolate and produce artificially these and other hormones, including the human growth hormone.

AMINO ACIDS AND NUTRITION

Just as proteins form when amino acids bond together in long chains, they can be broken down by a reaction called *hydrolysis,* the reverse of the formation of the peptide bond. That is exactly what happens in the process of digestion, when special digestive enzymes in the stomach enable the breaking down of the peptide linkage. (Enzymes are a type of protein—see Enzymes.) The amino acids, separated once again, are released into the small intestine, from whence they pass into the bloodstream and are carried throughout the organism. Each individual cell of the organism then can use these amino acids to assemble the new and different proteins required for its specific functions. Life thus is an ongoing cycle in which proteins are broken into individual amino-acid units, and new proteins are built up from these amino acids.

ESSENTIAL AMINO ACIDS. Out of the many thousands of possible amino acids, humans require only 20 different kinds. Two others appear in the bodies of some animal species, and approximately 100 others can be found in plants. Considering the vast numbers of amino acids and possible combinations that exist in nature, the number of amino acids essential to life is extremely small. Yet of the 20 amino acids required by humans for making protein, only 12 can be produced within the body, whereas the other eight—isoleucine, leucine, lysine, methionine, phenylalanine, threonine, tryptophan, and valine—must be obtained from the diet. (In addition, adults are capable of synthesizing arginine and histidine, but these amino acids are believed to be essential to growing children, meaning that children cannot produce them on their own.)

A complete protein is one that contains all of the essential amino acids in quantities sufficient for growth and repair of body tissue. Most proteins from animal sources, gelatin being the only exception, contain all the essential amino acids and are therefore considered complete proteins. On the other hand, many plant proteins do not contain all of the essential amino acids. For example, lysine is absent from corn, rice, and wheat, whereas corn also lacks tryptophan and rice lacks threonine. Soybeans are lacking in methionine. Vegans, or vegetarians who consume no animal proteins in their diets (i.e., no eggs, dairy products, or the like) are at risk of malnutrition, because they may fail to assimilate one or more essential amino acid.

AMINO ACIDS, HEALTH, AND DISEASE

Amino acids can be used as treatments for all sorts of medical conditions. For example, tyrosine may be employed in the treatment of Alzheimer's disease, a condition characterized by the onset of dementia, or mental deterioration, as well as for alcohol-withdrawal symptoms. Taurine is administered to control epileptic seizures, treat high blood pressure and diabetes, and support the functioning of the liver. Numerous other amino acids are used in treating a wide array of other diseases. Sometimes the disease itself involves a problem with amino-acid production or functioning. In the essay Vitamins, there is a discussion of pellagra, a disease resulting from a deficiency of the B-group vitamin known as niacin. Pellagra results from a diet heavy in corn, which, as we have noted, lacks lysine and tryptophan. Its symptoms often are described as the "three Ds": diarrhea, dermatitis (or skin inflammation), and dementia. Thanks to a greater understanding of nutrition and health, pellagra has been largely eradicated, but there still exists a condition with almost identical symptoms: Hartnup disease, a genetic disorder named for a British family in the late 1950s who suffered from it.

Hartnup disease is characterized by an inability to transport amino acids from the kidneys to the rest of the body. The symptoms at first seemed to suggest to physicians that the disease, which is present in one of about 26,000 live births, *was* pellagra. Tests showed that sufferers did not have inadequate tryptophan levels, however, as would have been the case with pellagra. On the other hand, some 14 amino acids have been found in excess within the urine of Hartnup disease sufferers, indicating that rather than properly transporting amino acids, their bodies are simply excreting them. This is a potentially very serious condition, but it can be treated with the B vitamin nicotinamide, also used to treat pellagra. Supplementation of tryptophan in the diet also has shown positive results with some patients.

SICKLE CELL ANEMIA. It is also possible for small mistakes to occur in the amino-acid sequence within the body. While these mistakes sometimes can be tolerated in nature without serious problems, at other times a single misplaced amino acid in the polymer chain can bring about an extremely serious condition of protein malfunctioning. An example of this is sickle cell anemia, a fatal disease ultimately caused by a single mistake in the amino acid sequence. In the bodies of sickle cell anemia sufferers, who are typically natives of sub-Saharan Africa or their descendants in the United States or elsewhere, glutamic acid is replaced by valine at the sixth position from the end of the protein chain in the hemoglobin molecule. (Hemoglobin is an iron-containing pigment in red blood cells that is responsible for transporting oxygen to the tissues and removing carbon dioxide from them.) This small difference makes sickle cell hemoglobin molecules extremely sensitive to oxygen deficiencies. As a result, when the red blood cells release their oxygen to the tissues, as all red blood cells do, they fail to re-oxygenate in a normal fashion and instead twist into the shape

KEY TERMS

AMINO ACIDS: Organic compounds made of carbon, hydrogen, oxygen, nitrogen, and (in some cases) sulfur bonded in characteristic formations. Strings of amino acids make up proteins.

AMINO GROUP: The chemical formation NH_2, which is part of all amino acids.

BIOCHEMISTRY: The area of the biological sciences concerned with the chemical substances and processes in organisms.

CARBOXYL GROUP: The formation -COOH, which is common to all amino acids.

COMPOUND: A substance in which atoms of more than one element are bonded chemically to one another.

DIPEPTIDE: A group of only two amino acids.

DNA: Deoxyribonucleic acid, a molecule in all cells and many viruses containing genetic codes for inheritance.

ENZYME: A protein material that speeds up chemical reactions in the bodies of plants and animals.

ESSENTIAL AMINO ACIDS: Amino acids that cannot be manufactured by the body, and which therefore must be obtained from the diet. Proteins that contain essential amino acids are known as *complete proteins.*

GENE: A unit of information about a particular heritable (capable of being inherited) trait that is passed from parent to offspring, stored in DNA molecules called *chromosomes.*

HORMONE: Molecules produced by living cells, which send signals to spots remote from their point of origin and induce specific effects on the activities of other cells.

MOLECULE: A group of atoms, usually but not always representing more than one element, joined in a structure. Compounds typically are made up of molecules.

ORGANIC: At one time chemists used the term *organic* only in reference to living things. Now the word is applied to compounds containing carbon and hydrogen.

PEPTIDE LINKAGE: A bond between the carboxyl group of one amino acid and the amino group of a second amino acid.

POLYMERS: Large, chainlike molecules composed of numerous subunits known as *monomers.*

POLYPEPTIDE: A group of between 10 and 50 amino acids.

PROTEINS: Large molecules built from long chains of 50 or more amino acids. Proteins serve the functions of promoting normal growth, repairing damaged tissue, contributing to the body's immune system, and making enzymes.

RNA: Ribonucleic acid, the molecule translated from DNA in the cell nucleus, the control center of the cell, that directs protein synthesis in the cytoplasm, or the space between cells.

SYNTHESIZE: To manufacture chemically, as in the body.

TRIPEPTIDE: A group of three amino acids.

that gives sickle cell anemia its name. This causes obstruction of the blood vessels. Before the development of a treatment with the drug hydroxyurea in the mid-1990s, the average life expectancy of a person with sickle cell anemia was about 45 years.

Amino Acids and the Distant Past

The Evolution essay discusses several types of dating, a term referring to scientific efforts directed toward finding the age of a particular item or phenomenon. Methods of dating are either relative (i.e., comparative and usually based on rock strata, or layers) or absolute. Whereas relative dating does not involve actual estimates of age in years, absolute dating does. One of the first types of absolute-dating techniques developed was amino-acid racimization, introduced in the 1960s. As noted earlier, there are "left-hand" L-forms and "right-hand" D-forms of all amino acids. Virtually all living organisms (except some microbes) incorporate only the L-forms, but once the organism dies, the L-amino acids gradually convert to the mirror-image D-amino acids.

Numerous factors influence the rate of conversion, and though amino-acid racimization was popular as a form of dating in the 1970s, there are problems with it. For instance, the process occurs at different rates for different amino acids, and the rates are further affected by such factors as moisture and temperature. Because of the uncertainties with amino-acid racimization, it has been largely replaced by other absolute-dating methods, such as the use of radioactive isotopes.

Certainly, amino acids themselves have offered important keys to understanding the planet's distant past. The discovery, in 1967 and 1968, of sedimentary rocks bearing traces of amino acids as much as three billion years old had an enormous impact on the study of Earth's biological history. Here, for the first time, was concrete evidence of life—at least, in a very simple chemical form—existing billions of years before the first true organism. The discovery of these amino-acid samples greatly influenced scientists' thinking about evolution, particularly the very early stages in which the chemical foundations of life were established.

Where to Learn More

"Amino Acids." Institute of Chemistry, Department of Biology, Chemistry, and Pharmacy, Freie Universität, Berlin (Web site). <http://www.chemie.fu-berlin.de/chemistry/bio/amino-acids_en.html>.

Goodsell, David S. *Our Molecular Nature: The Body's Motors, Machines, and Messages.* New York: Copernicus, 1996.

"Introduction to Amino Acids." Department of Crystallography, Birbeck College (Web site). <http://www.cryst.bbk.ac.uk/education/AminoAcid/overview.html>.

Michal, Gerhard. *Biochemical Pathways: An Atlas of Biochemistry and Molecular Biology.* New York: John Wiley and Sons, 1999.

Newstrom, Harvey. *Nutrients Catalog: Vitamins, Minerals, Amino Acids, Macronutrients—Beneficial Use, Helpers, Inhibitors, Food Sources, Intake Recommendations, and Symptoms of Over or Under Use.* Jefferson, NC: McFarland and Company, 1993.

Ornstein, Robert E., and Charles Swencionis. *The Healing Brain: A Scientific Reader.* New York: Guilford Press, 1990.

Reference Guide for Amino Acids (Web site). <http://www.realtime.net/anr/aminoacd.html#tryptophn>.

Silverstein, Alvin, Virginia B. Silverstein, and Robert A. Silverstein. *Proteins.* Illus. Anne Canevari Green. Brookfield, CT: Millbrook Press, 1992.

Springer Link: Amino Acids (Web site). <http://link.springer.de/link/service/journals/00726/>.

PROTEINS

CONCEPT

Most of us recognize the term protein in a nutritional context as referring to a class of foods that includes meats, dairy products, eggs, and other items. Certainly, proteins are an important part of nutrition, and obtaining complete proteins in one's diet is essential to the proper functioning of the body. But the significance of proteins extends far beyond the dining table. Vast molecules built from enormous chains of amino acids, proteins are essential building blocks for living systems—hence their name, drawn from the Greek *proteios,* or "holding first place." Proteins are integral to the formation of DNA, a molecule that contains genetic codes for inheritance, and of hormones. Most of the dry weight of the human body and the bodies of other animals is made of protein, as is a vast range of things with which we come into contact on a daily basis. In addition, a special type of protein called an enzyme has still more applications.

HOW IT WORKS

THE COMPLEXITIES OF BIOCHEMISTRY

Protein is a foundational material in the structure of most living things, and as such it is rather like concrete or steel. Just as concrete is a mixture of other ingredients and steel is an alloy of iron and carbon, proteins, too, are made of something more basic: amino acids. These are organic compounds made of carbon, hydrogen, oxygen, nitrogen, and (in some cases) sulfur bonded in characteristic formations.

Amino acids are discussed in more depth within the essay devoted to that topic, though, as noted in that essay, it is impossible to treat such a subject thoroughly without going into an extraordinarily lengthy and technical discussion. Such is the case with many topics in biochemistry, the area of the biological sciences concerned with the chemical substances and processes in organisms: the deeper within the structure of things one goes, and the smaller the items under investigation, the more complex are the properties and interactions.

THE BASICS

Amino acids react with each other to form a bond, called a *peptide linkage,* between the carboxyl group of one amino acid (symbolized as -COOH) and the amino group ($-NH_2$) of a second amino acid. In this way they can make large, chainlike molecules called *polymers,* which may contain as few as two or as many as 3,000 amino-acid units. If there are more than 10 units in a chain, the chain is called a *polypeptide,* while a chain with 50 or more amino-acid units is known as a *protein.*

All the millions of different proteins in living things are formed by the bonding of only 20 amino acids into long polymer chains. Because each amino acid can be used many times along the chain, and because there are no restrictions on the length of the chain, the number of possible combinations for the creation of proteins is truly enormous: about two quadrillion, or 2,000,000,000,000,000. Just as not all sequences of letters make sense, however, not all sequences of amino acids produce functioning proteins. In fact, the number of proteins that have significance in the functioning of nature is closer to about 100,000. This number is considerably smaller than two quadrillion—about 1/2,000,000,000th of that larger number, in fact—but it is still a very large number.

COMPONENTS OTHER THAN AMINO ACIDS. The specific properties of each kind of protein are largely dependent on the kind and sequence of the amino acids in it, yet many proteins include components other than amino acids. For example, some may have sugar molecules (sugars are discussed in the essay on Carbohydrates) chemically attached. Exactly which types of sugars are attached and where on the protein chain attachment occurs vary with the specific protein. Other proteins may have lipid, or fat, molecules chemically bonded to them. Sugar and lipid molecules always are added when synthesis of the protein's amino-acid chain is complete. Many other types of substance, including metals, also may be associated with proteins; for instance, hemoglobin, a pigment in red blood cells that is responsible for transporting oxygen to the tissues and removing carbon dioxide from them, is a protein that contains an iron atom.

SHEEP SHEARING IN NEW ZEALAND. THE ENTIRE ANIMAL WORLD IS CONSTITUTED LARGELY OF PROTEIN, AS ARE A WHOLE HOST OF ANIMAL PRODUCTS, INCLUDING LEATHER AND WOOL. *(© Adam Woolfitt/Corbis. Reproduced by permission.)*

STRUCTURES AND SYNTHESIS. Protein structures generally are described at four levels: primary, secondary, tertiary, and quaternary. Primary structure is simply the two-dimensional linear sequence of amino acids in the peptide chain. Secondary and tertiary structures both refer to the three-dimensional shape into which a protein chain folds. The distinction between the two is partly historical: secondary structures are those that were first discerned by scientists of the 1950s, using the techniques and knowledge available then, whereas an awareness of tertiary structure emerged only later. Finally, quaternary structure indicates the way in which many protein chains associate with one another. For example, hemoglobin consists of four protein chains (spirals, actually) of two slightly different types, all attached to an iron atom.

Protein synthesis is the process whereby proteins are produced, or synthesized, in living things according to "directions" given by DNA (deoxyribonucleic acid) and carried out by RNA (ribonucleic acid) and other proteins. As suggested earlier, this is an extraordinarily complex process that we do not attempt to discuss here. Following synthesis, proteins fold up into an essentially compact three-dimensional shape, which is their tertiary structure.

The steps involved in folding and the shape that finally results are determined by such chemical properties as hydrogen bonds, electrical attraction between positively and negatively charged side chains, and the interaction between polar and nonpolar molecules. Nonpolar molecules are called *hydrophobic,* or "water-fearing," because they do not mix with water but instead mix with oils and other substances in which the electric charges are more or less evenly distributed on the molecule. Polar molecules, on the other hand, are termed *hydrophilic,* or "water-loving," and mix with water and water-based substances in which the opposing electric charges occupy separate sides, or ends, of the molecule. Typically, hydrophobic amino-acid side chains tend to be on the interior of a protein, while hydrophilic ones appear on the exterior.

WANTED

GERMAN SABOTEUR

Photo taken February 19, 1936

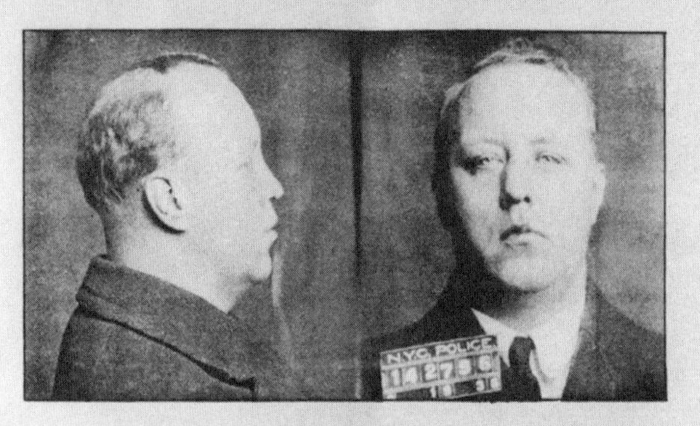

WALTER KAPPE, alias Walter Kappel

F.P.C. 16 M 28 W OOI
M 8 W III

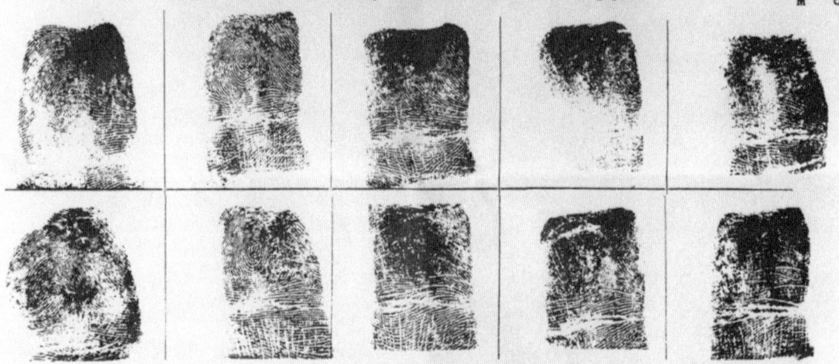

Walter Kappe is known to be connected with sabotage activities being promoted by the Nazi Government. He was born January 12, 1905 at Alfeld, Leina, Germany, and entered the United States on March 9, 1925. He filed application for United States citizenship at Kankakee, Illinois, in June, 1935. He is known to be a member of the German Literary Club, Cincinnati, Ohio, and the Teutonia Club, Chicago, Illinois. Kappe was an agent in the United States for the Ausland Organization and editor "Deutscher Weckruf und Beobachter", official organ of the German-American Bund. Kappe left the United States in 1937 and may return to the United States as an agent for Germany. This individual is described as fol-

U.S. WANTED POSTER FOR A WORLD WAR II NAZI SABOTEUR (JULY 1942). FINGERPRINTS ARE AN EXPRESSION OF OUR DNA, WHICH IS LINKED CLOSELY WITH THE OPERATION OF PROTEINS IN OUR BODIES. *(© Bettmann/Corbis. Reproduced by permission.)*

REAL-LIFE APPLICATIONS

PROTEINS ARE EVERYWHERE

Although it is very difficult to discuss the functions of proteins in simple terms, and it is similarly challenging to explain exactly *how* they function in everyday life, it is not hard at all to name quite a few areas in which these highly important compounds are applied. As we noted earlier, much of our bodies' dry weight—that is, the weight other than water, which accounts for a large percentage of the total—is protein. Our bones, for instance, are about one-fourth protein, and protein makes up a very high percentage of the material in our organs (including the skin), glands, and bodily fluids.

Humans are certainly not the only organisms composed largely of protein: the entire animal world, including the animals we eat and the microbes that enter our bodies (see Digestion and Parasites and Parasitology) likewise is constituted largely of protein. In addition, a whole host of animal products, including leather and wool, are nearly pure protein. So, too, are other, less widely used animal products, such as hormones for the treatment of certain conditions—for example, insulin, which keeps people with diabetes alive and which usually is harvested from the bodies of mammals.

Proteins allow cells to detect and react to hormones and toxins in their surroundings, and as the chief ingredient in antibodies, which help us resist infection, they play a part in protecting our bodies against foreign invaders. The lack of specific proteins in the brain may be linked to such mysterious, terrifying conditions as Alzheimer and Creutzfeldt-Jakob diseases (discussed in Disease). Found in every cell and tissue and composing the bulk of our bodies' structure, proteins are everywhere, promoting growth and repairing bone, muscles, tissues, blood, and organs.

ENZYMES. One particularly important type of protein is an enzyme, discussed in the essay on that topic. Enzymes make possible a host of bodily processes, in part by serving as catalysts, or substances that speed up a chemical reaction without actually participating in, or being consumed by, that reaction. Enzymes enable complex, life-sustaining reactions in the human body—reactions that would be too slow at ordinary body temperatures—and they manage to do so without forcing the body to undergo harmful increases in temperature. They also are involved in fermentation, a process with applications in areas ranging from baking bread to reducing the toxic content of wastewater. (For much more on these subjects, see Enzymes.)

Inside the body, enzymes and other proteins have roles in digesting foods and turning the nutrients in them—including proteins—into energy. They also move molecules around within our cells to serve an array of needs and allow healthful substances, such as oxygen, to pass through cell membranes while keeping harmful ones out. Proteins in the chemical known as chlorophyll facilitate an exceptionally important natural process, photosynthesis, discussed briefly in Carbohydrates.

PROTEINS, BLOOD, AND CRIME. The four blood types (A, B, AB, and O) are differentiated on the basis of the proteins present in each. This is only one of many key roles that proteins play where blood is concerned. If certain proteins are missing, or if the wrong proteins are present, blood will fail to clot properly, and cuts will refuse to heal. For sufferers of the condition known as hemophilia, caused by a lack of the proteins needed for clotting, a simple cut can be fatal.

Similarly, proteins play a critical role in forensic science, or the application of medical and biological knowledge to criminal investigations. Fingerprints are an expression of our DNA, which is linked closely with the operation of proteins in our bodies. The presence of DNA in bodily fluids, such as blood, semen, sweat, and saliva, makes it possible to determine the identity of the individual who perpetrated a crime or of others who were present at the scene. In addition, a chemical known as luminol assists police in the investigation of possible crime scenes. If blood has ever been shed in a particular area, such as on a carpet, no matter how carefully the perpetrators try to conceal or eradicate the stain, it can be detected. The key is luminol, which reacts to hemoglobin in the blood, making it visible to investigators. This chemical, developed during the 1980s, has been used to put many a killer behind bars.

DESIGNER PROTEINS. These are just a very few of the many applications of proteins, including a very familiar one, discussed in more depth at the conclusion of this essay: nutrition. Given the importance and complexity of proteins, it might be hard to imagine that they can be produced artificially, but, in fact, such production is taking place at the cutting edge of biochemistry today, in the field of "designer proteins."

Many such designs involve making small changes in already existing proteins: for example, by changing three amino acids in an enzyme often used to improve detergents' cleaning power, commercial biochemists have doubled the enzyme's stability in wash water. Medical applications of designer proteins seem especially promising. For instance, we might one day cure cancer by combining portions of one protein that recognizes cancer with part of another protein that attacks it. One of the challenges facing such a development, however, is the problem of designing a protein that attacks *only* cancer cells and not healthy ones.

KEY TERMS

AMINO ACIDS: Organic compounds made of carbon, hydrogen, oxygen, nitrogen and (in some cases) sulfur bonded in characteristic formations. Strings of amino acids make up proteins.

AMINO GROUP: The chemical formation -NH_2, which is part of all amino acids.

BIOCHEMISTRY: The area of the biological sciences concerned with the chemical substances and processes in organisms.

CARBOXYL GROUP: The formation -COOH, which is common to all amino acids.

DNA: Deoxyribonucleic acid, a molecule in all cells, and many viruses, containing genetic codes for inheritance.

ENZYME: A protein material that speeds up chemical reactions in the bodies of plants and animals.

ESSENTIAL AMINO ACIDS: Amino acids that cannot be manufactured by the body and therefore must be obtained from the diet. Proteins that contain essential amino acids are known as *complete proteins.*

HEMOGLOBIN: An iron-containing protein in red blood cells that is responsible for transporting oxygen to the tissues and removing carbon dioxide from them. Hemoglobin is known for its deep red color.

HERBIVORE: A plant-eating organism.

HORMONE: Molecules produced by living cells, which send signals to spots remote from their point of origin and induce specific effects on the activities of other cells.

OMNIVORE: An organism that eats both plants and other animals.

ORGANIC: At one time chemists used the term *organic* only in reference to living things. Now the word is applied to compounds containing carbon and hydrogen.

PEPTIDE LINKAGE: A bond between the carboxyl group of one amino acid and the amino group of a second amino acid.

POLYMERS: Large, chainlike molecules composed of numerous subunits known as *monomers.*

POLYPEPTIDE: A group of between 10 and 50 amino acids.

PROTEINS: Large molecules built from long chains of 50 or more amino acids. Proteins serve the functions of promoting normal growth, repairing damaged tissue, contributing to the body's immune system, and making enzymes.

RNA: Ribonucleic acid, the molecule translated from DNA in the cell nucleus, the control center of the cell, that directs protein synthesis in the cytoplasm, or the space between cells.

In the long term, scientists hope to design proteins from scratch. This is extremely difficult today and will remain so until researchers better understand the rules that govern tertiary structure. Nevertheless, scientists already have designed a few small proteins whose stability or instability has enhanced our understanding of those rules. Building on these successes, scientists hope that one day they may be able to design proteins to meet a host of medical and industrial needs.

PROTEINS IN THE DIET

Proteins are one of the basic nutrients, along with carbohydrates, lipids, vitamins, and minerals (see Nutrients and Nutrition). They can be broken down and used as a source of emergency energy if carbohydrates or fats cannot meet immediate needs. The body does not use protein from food directly: after ingestion, enzymes in the digestive system break protein into smaller peptide chains and eventually into separate amino acids. These smaller constituents then go into the bloodstream, from whence they are transported to the cells. The cells incorporate the amino acids and begin building proteins from them.

ANIMAL AND VEGETABLE PROTEINS. The protein content in plants is very small, since plants are made largely of cellulose, a type of carbohydrate (see Carbohydrates for more on this subject); this is one reason why herbivorous animals must eat enormous quantities of plants to meet their dietary requirements. Humans, on the other hand, are omnivores (unless they choose to be vegetarians) and are able to assimilate proteins in abundant quantities by eating the bodies of plant-eating animals, such as cows. In contrast to plants, animal bodies (as previously noted) are composed largely of proteins. When people think of protein in the diet, some of the foods that first come to mind are those derived from animals: either meat or such animal products as milk, cheese, butter, and eggs. A secondary group of foods that might appear on the average person's list of proteins include peas, beans, lentils, nuts, and cereal grains.

There is a reason why the "protein team" has a clearly defined "first string" and "second string." The human body is capable of manufacturing 12 of the 20 amino acids it needs, but it must obtain the other eight—known as *essential amino acids*—from the diet. Most forms of animal protein, except for gelatin (made from animal bones), contain the essential amino acids, but plant proteins do not. Thus, the nonmeat varieties of protein are incomplete, and a vegetarian who does not supplement his or her diet might be in danger of not obtaining all the necessary amino acids.

For a person who eats meat, it would be extremely difficult *not* to get enough protein. According to the U.S. Food and Drug Administration (FDA), protein should account for 10% of total calories in the diet, and since protein contains 4 calories per 0.035 oz. (1 g), that would be about 1.76 oz. (50 g) in a diet consisting of 2,000 calories a day. A pound (0.454 kg) of steak or pork supplies about twice this much, and though very few people sit down to a meal and eat a pound of meat, it is easy to see how a meat eater would consume enough protein in a day.

For a vegetarian, meeting the protein needs may be a bit more tricky, but it can be done. By combining legumes or beans and grains, it is possible to obtain a complete protein: hence, the longstanding popularity, with meat eaters as well as vegetarians, of such combinations as beans and rice or peas and cornbread. Other excellent vegetarian combos include black beans and corn, for a Latin American touch, or the eastern Asian combination of rice and tofu, protein derived from soybeans.

WHERE TO LEARN MORE

"DNA and Protein Synthesis." John Jay College of Criminal Justice, City University of New York (Web site). <http://web.jjay.cuny.edu/~acarpi/NSC/12-dna.htm>.

Inglis, Jane. *Proteins.* Minneapolis, MN: Carolrhoda Books, 1993.

Kiple, Kenneth F., and Kriemhild Coneè Ornelas. *The Cambridge World History of Food.* New York: Cambridge University Press, 2000.

"Proteins and Protein Foods." *Food Resource,* Oregon State University (Web site). <http://www.orst.edu/food-resource/protein/>.

Silverstein, Alvin, Virginia B. Silverstein, and Robert A. Silverstein. *Proteins.* Brookfield, CT: Millbrook Press, 1992.

Structural Classification of Proteins (Web site). <http://scop.mrc-lmb.cam.ac.uk/scop/>.

THINK: Teenage Health Interactive Network (Web site). <http://library.thinkquest.org/29500/nutrition/nutrition.fap.shtml>.

ENZYMES

CONCEPT

Enzymes are biological catalysts, or chemicals that speed up the rate of reaction between substances without themselves being consumed in the reaction. As such, they are vital to such bodily functions as digestion, and they make possible processes that normally could not occur except at temperatures so high they would threaten the well-being of the body. A type of protein, enzymes sometimes work in tandem with nonproteins called *coenzymes.* Among the processes in which enzymes play a vital role is fermentation, which takes place in the production of alcohol or the baking of bread and also plays a part in numerous other natural phenomena, such as the purification of wastewater.

HOW IT WORKS

AMINO ACIDS, PROTEINS, AND BIOCHEMISTRY

Amino acids are organic compounds made of carbon, hydrogen, oxygen, nitrogen, and (in some cases) sulfur bonded in characteristic formations. Strings of 50 or more amino acids are known as proteins, large molecules that serve the functions of promoting normal growth, repairing damaged tissue, contributing to the body's immune system, and making enzymes. The latter are a type of protein that functions as a catalyst, a substance that speeds up a chemical reaction without participating in it. Catalysts, of which enzymes in the bodies of plants and animals are a good example, thus are not consumed in the reaction.

CATALYSTS

In a chemical reaction, substances known as reactants interact with one another to create new substances, called products. Energy is an important component in the chemical reaction, because a certain threshold, termed the activation energy, must be crossed before a reaction can occur. To increase the rate at which a reaction takes place and to hasten the crossing of the activation energy threshold, it is necessary to do one of three things.

The first two options are to increase either the concentration of reactants or the temperature at which the reaction takes place. It is not always feasible or desirable, however, to do either of these things. Many of the processes that take place in the human body, for instance, normally would require high temperatures—temperatures, in fact, that are too high to sustain human life. Imagine what would happen if the only way we had of digesting starch was to heat it to the boiling point inside our stomachs! Fortunately, there is a third option: the introduction of a catalyst, a substance that speeds up a reaction without participating in it either as a reactant or as a product. Catalysts thus are not consumed in the reaction. Enzymes, which facilitate the necessary reactions in our bodies without raising temperatures or increasing the concentrations of substances, are a prime example of a chemical catalyst.

THE DISCOVERY OF CATALYSIS. Long before chemists recognized the existence of catalysts, ordinary people had been using the chemical process known as catalysis for numerous purposes: making soap, fermenting wine to create vinegar, or leavening bread, for instance. Early in the nineteenth century, chemists began to take note of this phenomenon. In 1812 the Russian chemist Gottlieb Kirchhoff (1764–1833) was studying the conversion of starches to sugar in the presence of strong acids when he noticed something interesting.

When a suspension of starch (that is, particles of starch suspended in water) was boiled, Kirchhoff observed, no change occurred in the starch. When he added a few drops of concentrated acid before boiling the suspension, however, he obtained a very different result. This time, the starch broke down to form glucose, a simple sugar (see Carbohydrates), whereas the acid—which clearly had facilitated the reaction—underwent no change. In 1835 the Swedish chemist Jöns Berzelius (1779–1848) provided a name to the process Kirchhoff had observed: *catalysis,* derived from the Greek words *kata* ("down") and *lyein* ("loosen"). Just two years earlier, in 1833, the French physiologist Anselme Payen (1795–1871) had isolated a material from malt that accelerated the conversion of starch to sugar, for instance, in the brewing of beer.

The renowned French chemist Louis Pasteur (1822–1895), who was right about so many things, called these catalysts *ferments* and pronounced them separate organisms. In 1897, however, the German biochemist Eduard Buchner (1860–1917) isolated the catalysts that bring about the fermentation of alcohol and determined that they were chemical substances, not organisms. By that time, the German physiologist Willy Kahne had suggested the name enzyme for these catalysts in living systems.

SUBSTRATES AND ACTIVE SITES

Each type of enzyme is geared to interact chemically with only one particular substance or type of substance, termed a substrate. The two parts fit together, according to a widely accepted theory introduced in the 1890s by the German chemist Emil Fischer (1852–1919), as a key fits into a lock. Each type of enzyme has a specific three-dimensional shape that enables it to fit with the substrate, which has a complementary shape.

The link between enzymes and substrates is so strong that enzymes often are named after the substrate involved, simply by adding *ase* to the name of the substrate. For example, lactase is the enzyme that catalyzes the digestion of lactose, or milk sugar, and urease catalyzes the chemical breakdown of urea, a substance in urine. Enzymes bind their reactants or substrates at special folds and clefts, named *active sites,* in the structure of the substrate. Because numerous interactions are required in their work of catalysis, enzymes must have many active sites, and therefore they are very large, having atomic mass figures as high as one million amu. (An atomic mass unit, or amu, is approximately equal to the mass of a proton, a positively charged particle in the nucleus of an atom.)

Suppose a substrate molecule, such as a starch, needs to be broken apart for the purposes of digestion in a living body. The energy needed to break apart the substrate is quite large, larger than is available in the body. An enzyme with the correct molecular shape arrives on the scene and attaches itself to the substrate molecule, forming a chemical bond within it. The formation of these bonds causes the breaking apart of other bonds within the substrate molecule, after which the enzyme, its work finished, moves on to another uncatalyzed substrate molecule.

COENZYMES

All enzymes belong to the protein family, but many of them are unable to participate in a catalytic reaction until they link with a nonprotein component called a coenzyme. This can be a medium-size molecule called a prosthetic group, or it can be a metal ion (an atom with a net electric charge), in which case it is known as a cofactor. Quite often, though, coenzymes are composed wholly or partly of vitamins. Although some enzymes are attached very tightly to their coenzymes, others can be parted easily; in either case, the parting almost always deactivates both partners.

The first coenzyme was discovered by the English biochemist Sir Arthur Harden (1865–1940) around the turn of the nineteenth century. Inspired by Buchner, who in 1897 had detected an active enzyme in yeast juice that he had named zymase, Harden used an extract of yeast in most of his studies. He soon discovered that even after boiling, which presumably destroyed the enzymes in yeast, such deactivated yeast could be reactivated. This finding led Harden to the realization that a yeast enzyme appar-

The conversion of cabbage to sauerkraut utilizes a particular bacterium that assists in fermentation. Here workers spread salt and pack chopped cabbage in barrels, where it will ferment for four weeks. *(© Bettmann/Corbis. Reproduced by permission.)*

ently consists of two parts: a large, molecular portion that could not survive boiling and was almost certainly a protein and a smaller portion that had survived and was probably not a protein. Harden, who later shared the 1929 Nobel Prize in chemistry for this research, termed the nonprotein a coferment, but others began calling it a coenzyme.

REAL-LIFE APPLICATIONS

The Body, Food, and Digestion

Enzymes enable the many chemical reactions that are taking place at any second inside the body of a plant or animal. One example of an enzyme is cytochrome, which aids the respiratory system by catalyzing the combination of oxygen with hydrogen within the cells. Other enzymes facilitate the conversion of food to energy and make possible a variety of other necessary biological functions. Enzymes in the human body fulfill one of three basic functions. The largest of all enzyme types, sometimes called metabolic enzymes, assist in a wide range of basic bodily processes, from breathing to thinking. Some such enzymes are devoted to maintaining the immune system, which protects us against disease, and others are involved in controlling the effects of toxins, such as tobacco smoke, converting them to forms that the body can expel more easily.

A second category of enzyme is in the diet and consists of enzymes in raw foods that aid in the process of digesting those foods. They include proteases, which implement the digestion of protein; lipases, which help in digesting lipids or fats; and amylases, which make it possible to digest carbohydrates. Such enzymes set in motion the digestive process even when food is still in the mouth. As these enzymes move with the food into the upper portion of the stomach, they continue to assist with digestion.

The third group of enzymes also is involved in digestion, but these enzymes are already in the body. The digestive glands secrete juices containing enzymes that break down nutrients chemically into smaller molecules that are more easily absorbed by the body. Amylase in the saliva begins the process of breaking down complex carbohydrates into simple sugars. While food is still in the mouth, the stomach begins producing pepsin, which, like protease, helps digest protein.

Later, when food enters the small intestine, the pancreas secretes pancreatic juice—which contains three enzymes that break down carbohydrates, fats, and proteins—into the duodenum, which is part of the small intestine. Enzymes from food wind up among the nutrients circulated to the body through plasma, a watery liquid in which red blood cells are suspended. These enzymes in the blood assist the body in everything from growth to protection against infection.

One digestive enzyme that should be in the body, but is not always present, is lactase. As we noted earlier, lactase works on lactose, the principal carbohydrate in milk, to implement its digestion. If a person lacks this enzyme, consuming dairy products may cause diarrhea, bloating, and cramping. Such a person is said to be "lactose intolerant," and if he or she is to consume dairy products at all, they must be in forms that contain lactase. For this reason, Lactaid milk is sold in the specialty dairy section of major supermarkets, while many health-food stores sell lactaid tablets.

Fermentation

Fermentation, in its broadest sense, is a process involving enzymes in which a compound rich in energy is broken down into simpler substances. It also is sometimes identified as a process in which large organic molecules (those containing hydrogen and carbon) are broken down into simpler molecules as the result of the action of microorganisms working anaerobically, or in the absence of oxygen. The most familiar type of fermentation is the conversion of sugars and starches to alcohol by enzymes in yeast. To distinguish this reaction from other kinds of fermentation, the process is sometimes termed *alcoholic* or *ethanolic* fermentation.

At some point in human prehistory, humans discovered that foods spoil, or go bad. Yet at the dawn of history—that is, in ancient Sumer and Egypt—people found that sometimes the "spoilage" (that is, fermentation) of products could have beneficial results. Hence the fermentation of fruit juices, for example, resulted in the formation of primitive forms of wine. Over the centuries that followed, people learned how to make both alcoholic beverages and bread through the controlled use of fermentation.

ALCOHOLIC BEVERAGES. In fermentation, starch is converted to simple sugars, such as sucrose and glucose, and through a complex sequence of some 12 reactions, these sugars then are converted to ethyl alcohol (the kind of alcohol that can be consumed, as opposed to methyl alcohol and other toxic forms) and carbon dioxide. Numerous enzymes are needed to carry out this sequence of reac-

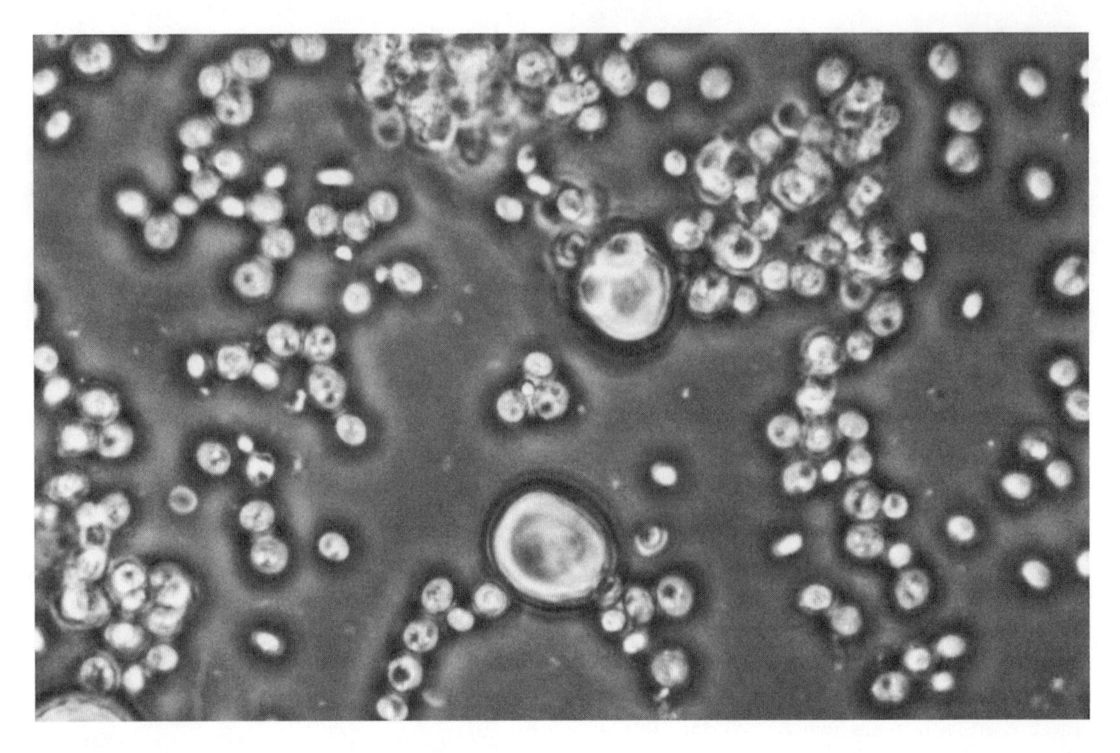

BAKER'S YEAST, SINGLE-CELL FUNGI THAT PRODUCE CARBON DIOXIDE IN THE OXIDATION OF SUGAR. CARBON DIOXIDE CAUSES BREAD TO RISE WHEN YEAST IS MIXED INTO DOUGH, A RESULT OF THE FERMENTATION OF THE SUGAR BY ENZYMES IN THE YEAST. *(© Lester V. Bergman/Corbis. Reproduced by permission.)*

tions, the most important being zymase, which is found in yeast cells. These enzymes are sensitive to environmental conditions, such that when the concentration of alcohol reaches about 14%, they are deactivated. For this reason, no fermentation product (such as wine) can have an alcoholic concentration of more than about 14%. Stronger alcoholic beverages, such as whisky, are the result of another process, distillation.

The alcoholic beverages that can be produced by fermentation vary widely, depending primarily on two factors: the plant that is fermented and the enzymes used for fermentation. Depending on the materials available to them, various peoples have used grapes, berries, corn, rice, wheat, honey, potatoes, barley, hops, cactus juice, cassava roots, and other plant materials for fermentation to produce wines, beers, and other fermented drinks. The natural product used in making the beverage usually determines the name of the synthetic product. Thus, for instance, wine made with rice—a time-honored tradition in Japan—is known as sake, while a fermented beverage made from barley, hops, or malt sugar has a name very familiar to Americans: beer. Grapes make wine, but "wine" made from honey is known as mead.

OTHER FOODS. Of course, ethyl alcohol is not the only useful product of fermentation or even of fermentation using yeast; so, too, are baked goods, such as bread. The carbon dioxide generated during fermentation is an important component of such items. When the batter for bread is mixed, a small amount of sugar and yeast is added. The bread then rises, which is more than just a figure of speech: it actually puffs up as a result of the fermentation of the sugar by enzymes in the yeast, which brings about the formation of carbon dioxide gas. The carbon dioxide gives the batter bulkiness and texture that would be lacking without the fermentation process. Another food-related application of fermentation is the production of one processed type of food from a raw, natural variety. The conversion of raw olives to the olives sold in stores, of cucumbers to pickles, and of cabbage to sauerkraut utilizes a particular bacterium that assists in a type of fermentation.

INDUSTRIAL APPLICATIONS. There is even ongoing research into the creation of edible products from the fermentation of petroleum. While this may seem a bit far-fetched, it is less difficult to comprehend powering cars with an environmentally friendly product of fer-

KEY TERMS

ACTIVATION ENERGY: A threshold that must be crossed to facilitate a chemical reaction. There are three ways to reach the activation energy: by increasing the concentration of reactants, by raising their temperature, or by introducing a catalyst, such as an enzyme.

ACTIVE SITES: Folds and clefts on the surface of an enzyme that enable attachment to its particular substrate.

AMINO ACIDS: Organic compounds made of carbon, hydrogen, oxygen, nitrogen, and (in some cases) sulfur bonded in characteristic formations. Strings of amino acids make up proteins.

BIOCHEMISTRY: The area of the biological sciences concerned with the chemical substances and processes in organisms.

CARBOHYDRATES: Naturally occurring compounds, consisting of carbon, hydrogen, and oxygen, whose primary function in the body is to supply energy. Included in the carbohydrate group are sugars, starches, cellulose, and various other substances. Most carbohydrates are produced by green plants in the process of undergoing photosynthesis.

CATALYSIS: The act or process of catalyzing, or speeding up the rate of reaction between substances.

CATALYST: A substance that speeds up a chemical reaction without participating in it. Catalysts, of which enzymes are a good example, thus are not consumed in the reaction.

COENZYME: A nonprotein component sometimes required to allow an enzyme to set in motion a catalytic reaction.

ENZYME: A protein that acts as a catalyst, a material that speeds up chemical reactions in the bodies of plants and animals without itself taking part in, or being consumed by, these reactions.

FERMENTATION: A process involving enzymes in which a compound rich in energy is broken down into simpler substances.

METABOLISM: The chemical process by which nutrients are broken down and converted into energy or are used in the construction of new tissue or other material in the body.

MOLECULE: A group of atoms, usually but not always representing more than one element, joined in a structure. Compounds typically are made up of molecules.

ORGANIC: At one time, chemists used the term *organic* only in reference to living things. Now the word is applied to compounds containing carbon and hydrogen.

PROTEINS: Large molecules built from long chains of 50 or more amino acids. Proteins serve the functions of promoting normal growth, repairing damaged tissue, contributing to the body's immune system, and making enzymes.

REACTANT: A substance that interacts with another substance in a chemical reaction, resulting in the formation of a chemical or chemicals known as the *product.*

STARCHES: Complex carbohydrates without taste or odor, which are granular or powdery in physical form.

SUBSTRATE: A reactant that typically is paired with a particular enzyme.

KEY TERMS CONTINUED

Enzymes often are named after their respective substrates by adding the suffix *ase* (e.g., the enzyme lactase is paired with the substrate lactose).

SUGARS: One of the three principal types of carbohydrate, along with starches and cellulose. Sugars can be defined as any of various water-soluble carbohydrates of varying sweetness. What we think of as "sugar" (i.e., table sugar) is actually sucrose.

VITAMINS: Organic substances that, in extremely small quantities, are essential to the nutrition of most animals and some plants. In particular, vitamins work with enzymes in regulating metabolic processes; however, they do not in themselves provide energy, and thus vitamins alone do not qualify as a form of nutrition.

mentation known as gasohol. Gasohol first started to make headlines in the 1970s, when an oil embargo and resulting increases in gas prices, combined with growing environmental concerns, raised the need for a type of fuel that would use less petroleum. A mixture of about 90% gasoline and 10% alcohol, gasohol burns more cleanly that gasoline alone and provides a promising method for using renewable resources (plant material) to extend the availability of a nonrenewable resource (petroleum). Furthermore, the alcohol needed for this product can be obtained from the fermentation of agricultural and municipal wastes.

The applications of fermentation span a wide spectrum, from medicines that go into people's bodies to the cleaning of waters containing human waste. Some antibiotics and other drugs are prepared by fermentation: for example, cortisone, used in treating arthritis, can be made by fermenting a plant steroid known as diosgenin. In the treatment of wastewater, anaerobic, or non-oxygen-dependent, bacteria are used to ferment organic material. Thus, solid wastes are converted to carbon dioxide, water, and mineral salts.

WHERE TO LEARN MORE

Asimov, Isaac. *The Chemicals of Life: Enzymes, Vitamins, Hormones.* New York: Abelard-Schulman, 1954.

"Enzymes: Classification, Structure, Mechanism." Washington State University Department of Chemistry (Web site). <http://www.chem.wsu.edu/Chem102/102-EnzStrClassMech.html>.

"Enzymes." *HordeNet: Hardy Research Group,* Department of Chemistry, The University of Akron (Web site). <http://ull.chemistry.uakron.edu/genobc/Chapter_20/>.

Fruton, Joseph S. *A Skeptical Biochemist.* Cambridge, MA: Harvard University Press, 1992.

"Introduction to Enzymes." *Worthington Biochemical Corporation* (Web site). <http://www.worthington-biochem.com/introBiochem/introEnzymes.html>.

Kornberg, Arthur. *For the Love of Enzymes: The Odyssey of a Biochemist.* Cambridge, MA: Harvard University Press, 1989.

"Milk Makes Me Sick: Exploration of the Basis of Lactose Intolerance." *Exploratorium: The Museum of Science, Art, and Human Perception* (Web site). <http://www.exploratorium.edu/snacks/milk_makes-me_sick/>.

METABOLISM

METABOLISM
DIGESTION
RESPIRATION

METABOLISM

CONCEPT

The term metabolism refers to all of the chemical reactions by which complex molecules taken into an organism are broken down to produce energy and by which energy is used to build up complex molecules. All metabolic reactions fall into one of two general categories: catabolic and anabolic reactions, or the processes of breaking down and building up, respectively. The best example of metabolism from daily life occurs in the process of taking in and digesting nutrients, but sometimes these processes become altered, either through a person's choice or through outside factors, and metabolic disorders follow. Such disorders range from anorexia and bulimia to obesity. These are all examples of an unhealthy, unnatural alteration to the ordinary course of metabolism; on the other hand, hibernation allows animals to slow down their metabolic rates dramatically as a means of conserving energy during times when food is scarce.

HOW IT WORKS

THE BODY'S FURNACE

The term metabolism, strangely enough, is related closely to devil, with which it shares the Greek root *ballein,* meaning "to throw." By adding *dia* ("through" or "across"), one arrives at devil and many related words, such as diabolical; on the other hand, the replacement of that prefix with *meta* ("after" or "beyond") yields the word metabolism. The connection between the two words has been obscured over time, but it might be helpful to picture metabolism in terms of an image that goes with that of a devil: a furnace.

Metabolism is indeed like a furnace, in that it burns energy, and that is the aspect most commonly associated with this concept. But metabolism also involves a function that a furnace does not: building new material. All metabolic reactions can be divided into either catabolic or anabolic reactions. Catabolism is the process by which large molecules are broken down into smaller ones with the release of energy, whereas anabolism is the process by which energy is used to build up complex molecules needed by the body to maintain itself and develop new tissue.

DIGESTION. One way to understand the metabolic process is to follow the path of a typical nutrient as it passes through the body. The digestive process is discussed in Digestion, while nutrients are examined in Nutrients and Nutrition as well as in Proteins, Amino Acids, Enzymes, Carbohydrates, and Vitamins. Here we touch on the process only in general terms, as it relates to metabolism.

The term digestion is not defined in the essay on that subject, because it is an everyday word whose meaning is widely known. For the present purposes, however, it is important to identify it as the process of breaking down food into simpler chemical compounds as a means of making nutrients absorbable by the body. This is a catabolic process, because the molecules of which foods are made are much too large to pass through the lining of the digestive system and directly into the bloodstream. Thanks to the digestive process, smaller molecules are formed and enter the bloodstream, from whence they are carried to individual cells throughout a person's body.

The smaller molecules into which nutrients are broken down make up the metabolic pool, which consists of simpler substances.The metabolic pool includes simple sugars, made by the breakdown of complex carbohydrates; glycerol and fatty acids, which come from the conversion of lipids, or fats; and amino acids, formed by the breakdown of proteins. Substances in the metabolic pool provide material from which new tissue is constructed—an anabolic process.

The chemical breakdown of substances in the cells is a complex and wondrous process. For instance, a cell converts a sugar molecule into carbon dioxide and water over the course of about two dozen separate chemical reactions. This is what cell biologists call a metabolic pathway: an orderly sequence of reactions, with particular enzymes (a type of protein that speeds up chemical reactions) acting at each step along the way. In this instance, each chemical reaction makes a relatively modest change in the sugar molecule—for example, the removal of a single oxygen atom or a single hydrogen atom—and each is accompanied by the release of energy, a result of the breaking of chemical bonds between atoms.

ATP AND ADP

Cells capture and store the energy released in catabolic reactions through the use of chemical compounds known as energy carriers. The most significant example of an energy carrier is adenosine triphosphate, or ATP, which is formed when a simpler compound, adenosine diphosphate (ADP), combines with a phosphate group. (A phosphate is a chemical compound that contains oxygen bonded to phosphorus, and the term *group* in chemistry refers to a combination of atoms from two or more elements that tend to bond with other elements or compounds in certain characteristic ways.)

ADP will combine with a phosphate group only if energy is added to it. In cells, that energy comes from the catabolism of compounds in the metabolic pool, including sugars, glycerol (related to fats), and fatty acids. The ATP molecule formed in this manner has taken up the energy previously stored in the sugar molecule, and thereafter, whenever a cell needs energy for some process, it can obtain it from an ATP molecule. The reverse of this process also takes place inside cells. That is, energy from an ATP molecule can be used to put simpler molecules together to make more complex molecules. For example, suppose that a cell needs to repair a rupture in its cell membrane. To do so, it will need to produce new protein molecules, which are made from hundreds or thousands of amino-acid molecules. These molecules can be obtained from the metabolic pool.

The reactions by which a compound is metabolized differ for various nutrients. Also, energy carriers other than ATP may play a part. For example, the compound known as nicotinamide adenine dinucleotide phosphate (NADPH) also has a role in the catabolism and anabolism of various substances. The general outline described here, however, applies to all metabolic reactions.

CATABOLISM AND ANABOLISM

Energy released from organic nutrients (those containing carbon and hydrogen) during catabolism is stored within ATP, in the form of the high-energy chemical bonds between the second and third molecules of phosphate. The cell uses ATP for synthesizing cell components from simple precursors, for the mechanical work of contraction and motion, and for transport of substances across its membrane. ATP's energy is released when this bond is broken, turning ATP into ADP. The cell uses the energy derived from catabolism to fuel anabolic reactions that synthesize cell components. Although anabolism and catabolism occur simultaneously in the cell, their rates are controlled independently. Cells separate these pathways because catabolism is a "downhill" process, or one in which energy is released, while anabolism is an "uphill" process requiring the input of energy.

Catabolism and anabolism share an important common sequence of reactions known collectively as the citric acid cycle, the tricarboxylic acid cycle, or the *Krebs cycle.* Named after the German-born British biochemist Sir Hans Adolf Krebs (1900–1981), the Krebs cycle is a series of chemical reactions in which tissues use carbohydrates, fats, and proteins to produce energy; it is part of a larger series of enzymatic reactions known as oxidative phosphorylation. In the latter reaction, glucose is broken down to release energy, which is stored in the form of ATP—a catabolic sequence. At the same time, other molecules produced by the Krebs cycle are used as precursor molecules for reactions that build proteins, fats, and carbohydrates—an anabolic sequence. (A precursor is a substance, cellular component, or cell from which another substance, cellular component, or cell—different in kind from the precursor—is formed.)

INTRODUCTION TO LIPIDS

As noted earlier, many practical aspects of metabolism are discussed elsewhere, particularly in the essays Digestion and Nutrients and Nutrition. Also, two types of chemical compound, proteins and carbohydrates, are so important to a variety of metabolic processes that they are examined in detail within entries of their own. In the present context, let us focus on the third major kind of nutrient, lipids or fats.

Lipids are soluble in nonpolar solvents, which is the reason why a gravy stain or other grease stain is difficult to remove from clothing without a powerful detergent or spot remover. Water molecules are polar, because the opposing electric charges tend to occupy opposite sides or ends of the molecule. In a molecule of oil, whether derived from petroleum or from animal or vegetable fat, electric charges are very small, and are distributed evenly throughout the molecule.

Whereas water molecules tend to bond relatively well, like a bunch of bar magnets attaching to one another at their opposing poles, oil and fat molecules tend not to bond. (The "bond" referred to here is the fairly weak one between molecules. Much stronger is the chemical bond *within* molecules—a bond that, when broken, brings about a release of energy, as noted earlier.) Their functions are as varied as their structures, but because they are all fat-soluble, lipids share in the ability to approach and even to enter cells. The latter have membranes that, while highly complex in structure, can be identified in simple terms as containing lipids or lipoproteins (lipids attached to proteins). The behavior of lipids and lipid-like molecules, therefore, becomes very important in understanding how a substance may or may not enter a cell. Such a substance may be toxic, as in the case of some pesticides, but if they are lipid-like, they are able to penetrate the cell's membrane. (See Food Webs for more about the biomagnification of DDT.)

In addition to lipoproteins, there are glycolipids, or lipids attached to sugars, as well as lipids attached to alcohols and some to phosphoric acids. The attachment with other compounds greatly alters the behavior of a lipid, often making them bipolar—that is, one end of the molecule is water-soluble. This is important, because it allows lipids to move out of the intestines and into the bloodstream. In the digestive process, lipids are made water-soluble either by being broken down into smaller parts or through association with another substance. The breaking down usually is done via two different processes: hydrolysis, or chemical reaction with water, and saponification. The latter, a reaction in which certain kinds of organic compounds are hydrolyzed to produce an alcohol and a salt, is used in making soap.

REAL-LIFE APPLICATIONS

PUTTING LIPIDS TO USE

Derived from living systems of plants, animals, or humans, lipids are essential to good health, not only for humans but also for other animals and even plants. Seeds, for example, contain lipids for the storage of energy. Because fat is a poor conductor of heat, lipids also can function as effective insulators, and for this reason, people living in Arctic zones seek fatty foods such as blubber. Some lipids function as chemical messengers in the body, while others serve as storage areas for chemical energy. There is a good reason why babies are born with "baby fat" and why children entering puberty often tend to become chubby: in both cases, they are building up energy reserves for the great metabolic hurdles that lie ahead, and within a few years, they will have used up those excessive fat stores.

FATS AND OILS. Fats and oils are both energy-rich compounds that are basic components of the normal diet. Both have essentially the same chemical structure—a mixture of fatty acids combined with glycerol—and are insoluble (do not dissolve) in water. While fats remain solid or at least semisolid at room temperature, however, most oils very quickly become liquid at increased temperatures. Animal fats and oils include butter, lard, tallow, and fish oil. Numerous other oils, such as cottonseed, peanut, and corn oils, are derived from plants.

Fats have two main functions: they provide some of the raw material for synthesizing (creat-

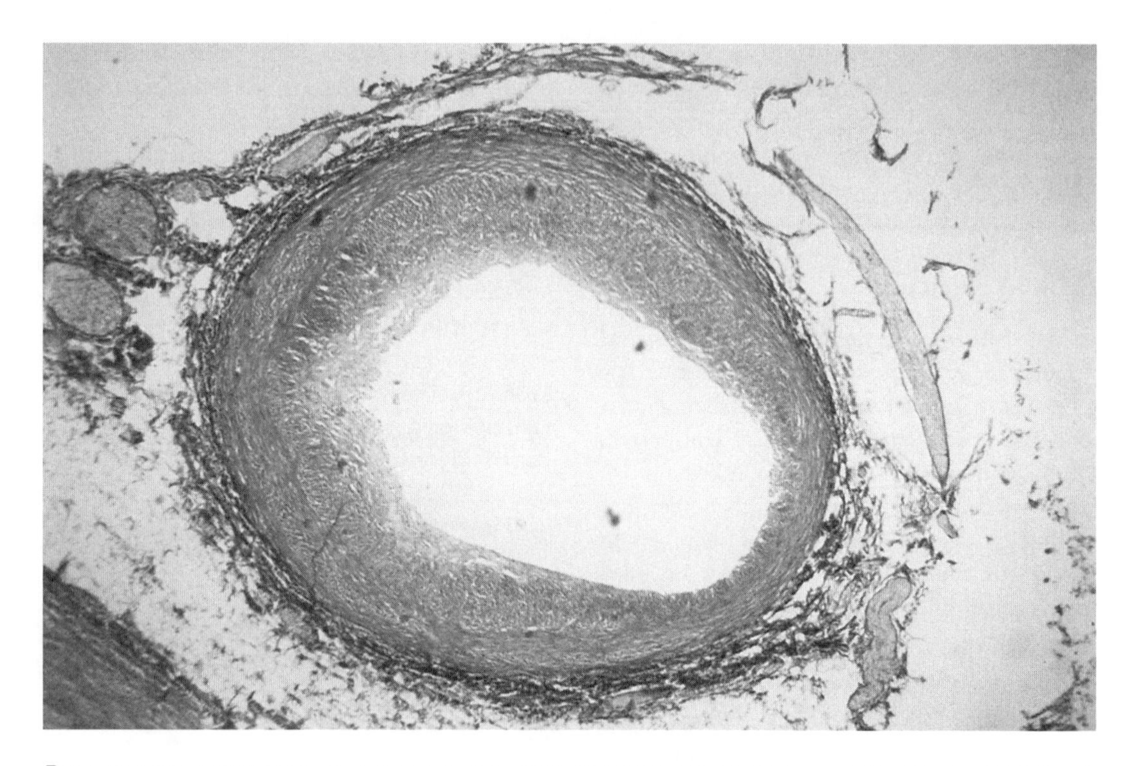

Cross-section micrograph of a blood vessel with atherosclerotic lesion, composed of cholesterol and other lipids. High levels of certain circulating fats can lead to a thickening of the artery walls and have been linked to various illnesses, including heart disease and cancer. *(© Lester V. Bergman/Corbis. Reproduced by permission.)*

ing) and repairing tissues, and they serve as a concentrated source of fuel energy. Fats, in fact, provide humans with roughly twice as much energy, per unit weight, as carbohydrates and proteins. Fats are not only an important source of day-to-day energy, but they also can be stored indefinitely as adipose (fat) tissue in case of future need. Fats also help by transporting fat-soluble vitamins, such as A and D (see Vitamins), throughout the system. They cushion and form protective pads around delicate organs, such as the heart, liver and kidneys, and the layer of fat under the skin helps insulate the body against too much heat loss. They even add to the flavor of foods that might otherwise be inedible.

NOT ALL FAT IS CREATED EQUAL. Although normal amounts of certain kinds of fat in the diet are essential to good health, unnecessarily high amounts (especially of unhealthy fats) can lead to various problems. Healthy fats include those from fatty fish, such as salmon, mackerel, or tuna, or from fat-containing vegetables, such as the avocado. In addition, many vegetable oils, particularly olive oil, can be beneficial.

Bad fats, on the other hand, are usually ones that have been tampered with through a process known as hydrogenation. This is a term describing any chemical reaction in which hydrogen atoms are added to fill in chemical bonds between carbon and other atoms, but in the case of fatty foods, hydrogenation involves the saturation of hydrocarbons, organic chemical compounds whose molecules are made up of nothing but carbon and hydrogen atoms. When they are treated with hydrogen gas, they become "saturated" with hydrogen atoms. Saturated fats, as they are called, are harder and more stable and stand up better to the heat of frying, which makes them more desirable for use in commercial products. For this reason, many foods contain hydrogenated vegetable oil; however, saturated fats have been linked to a rise in blood cholesterol levels—and to an increased risk of heart disease.

Cholesterol is a variety of lipid, and, like other lipids, some of it is essential—but only some and only of the right kind. Most cholesterol is transported through the blood in low-density lipoproteins, or LDLs, which have been nicknamed *bad cholesterol.* These lipoproteins are received by LDL receptors on the cell membranes, but if there are more LDLs than LDL receptors, the excess LDLs will be deposited in the arteries. Thus, LDLs are not really "bad" unless there are too many of them. On the other hand, "good" cholesterol (HDLs, or high-density lipoproteins) help protect against damage to the artery walls by carrying excess LDLs back to the liver.

HOW MUCH IS TOO MUCH? A certain amount of excess adipose tissue can be valuable during periods of illness, overactivity, or food shortages. Too much, however, can be unsightly and also can overwork the heart and put added stress on other parts of the body. High levels of certain circulating fats may lead to atherosclerosis, which is a thickening of the artery walls, and they have been linked to various illnesses, including cancer.

With fat, as with many things where the body is concerned, if a little is a good, this does not mean that a lot is better. In the past, nutritionists considered a diet that obtained 40% of its calories from fats a reasonable one; today, however, they recommend that no more than 30% of all calories (and preferably an even smaller percentage) come from fat. Agreement on this point, however, is far from universal. Some physicians and scientists maintain that dietary fat does not contribute as much to body fat as do carbohydrates. Carbohydrates are good for someone who needs a boost of energy that can be consumed easily by the body, such as an athlete going into competition. But for inactive people—and this includes a large portion of Americans—carbohydrates simply are stored as fat.

Experts do not even agree on the answer to a question much simpler than "How much is too much fat in the diet?"—the question "How much is too much fat on the body?" Some doctors classify a person as obese whose weight is at least 20% more than the recommended weight for his or her height, but others say that standard height-and-weight charts are misleading. After all, muscle weighs more than fat, and it is conceivable that a very muscular athlete with very little body fat might qualify as "overweight" compared with the recommended weight for his or her height.

BODY FAT, THE SEXES, AND NATURE. Because of the complexity of the issue, many experts contend that the proportion of fat to muscle, measured by the skinfold "pinch" test, is a better measure of obesity. (Being obese is not the same as being overweight: the muscular athlete described in the last paragraph is *overweight* but not *obese*, a term that implies an excess of body fat.) In healthy adults, fat typically should account for about 18–25% of the body weight in females and 15–20% in males.

The reason for the difference between men and women is that fat naturally accumulates in a woman's buttocks and thighs, because nature "assumes" that she will bear children, in which case such excess fat will be useful. This is why women over the age of about 25 often complain that when they and their husbands or boyfriends embark on a fitness program together, the men usually see results faster. The reason is that there is no genetic or evolutionary benefit to be gained from a man having fat around his waist, which is where men usually gain. If anything—since our genetic codes and makeup have changed little since prehistory—the well-being and propagation of the human species are best served by a lean, muscular male capable of killing animals to feed and protect his family. All of this means, of course, that men should not gloat if they see better results from a regular workout program; instead, they should just recognize that nature is at work in their wives' or girlfriends' bodies as in their own.

METABOLIC DISORDERS

Enzymes, as we noted earlier, are critical participants in metabolic reactions. They are like relay runners in a race, in this case a race along the metabolic pathways whereby nutrients are turned into energy or new bodily material. Therefore, if an enzyme is missing or does not function as it should, it can create a serious metabolic disorder. An example is phenylketonuria (PKU), caused by the lack of an enzyme known as phenylalanine hydroxylase. This enzyme is responsible for converting the amino acid phenylalanine to a second amino acid, tyrosine; when this does not happen, phenylalanine builds up in the body. It is converted to a compound called phenylpyruvate, which impairs normal brain development, resulting in severe mental retardation.

Other examples of metabolic disorders include alkaptonuria, thalassemia, porphyria, Tay-Sachs disease, Hurler syndrome, Gaucher disease, galactosemia, Cushing syndrome, dia-

DIANA, THE LATE PRINCESS OF WALES, SUFFERED FROM BULIMIA, A DISORDER THAT PROMPTS A PERSON TO GO ON EATING BINGES AND THEN "PURGE," EITHER BY VOMITING OR BY TAKING LARGE AMOUNTS OF LAXATIVES. *(© Corbis. Reproduced by permission.)*

betes mellitus, hyperthyroidism, and hypothyroidism. Most of these conditions affect a small population; however, diabetes mellitus (discussed in Noninfectious Diseases) is one of the leading killers in America. At present, no cures for metabolic disorders exist. The best approach is to diagnose such conditions as early as possible and then arrange a person's diet to deal as effectively as possible with that disorder.

EATING DISORDERS

Eating disorders are a different matter, because they are psychological rather than physiological conditions. No one is sure what causes eating disorders, but researchers think that family dynamics, biochemical abnormalities, and modern American society's preoccupation with thinness all may contribute. Eating disorders are virtually unknown in parts of the world where food is scarce, but in wealthy lands, such as the United States, problems of overeating, self-induced starvation, or forced purging have gained considerable attention.

Anorexia nervosa, bulimia, and obesity are the most well known types of eating disorder. The word *anorexia* comes from the Greek for "lack of appetite," but the problem for people with anorexia is not that they are not hungry. On the contrary, they are starving, but unlike poor people in the Third World, they are not starving as the result of a shortage of food but because they are denying themselves nutrition. They do this because they fear gaining weight, even when they are so severely underweight that they look like skeletons.

The name of a related condition, *bulimia,* literally means "hungry as an ox." People with this problem go on eating binges, often gorging on junk food. Then they force their bodies to get rid of the food, either by vomiting or by taking large amounts of laxatives. A third type of eating disorder, obesity, also is characterized by uncontrollable overeating, but in this case the person does not force the body to eject the food that has been consumed. That, at least, makes obesity more healthy than bulimia, but there is nothing healthy about accumulating vast amounts of body fat, as severely obese people do.

ANOREXIA AND BULIMIA. Young people are more likely than older people to suffer anorexia or bulimia, conditions that typically become apparent at about the age of 20 years. Although both men and women can experience the problem, in fact, only about 5% of people with these eating disorders are male. And though anorexia and bulimia are closely related—particularly inasmuch as they are psychological in origin but can exact a heavy biological toll—there are several important differences.

People who have anorexia or bulemia often come from families with overprotective parents who have unrealistically high expectations of their children. Frequently, high expectations go hand in hand with a wealthy background, and certainly anorexia and bulimia are not conditions that typically affect the poor. Anorexia and bulimia often seem to develop after some stressful experience, such as moving to a new town, changing schools, or going through puberty. Low self-esteem, fear of losing control, and fear of growing up are common characteristics of people with these conditions. Their need for approval manifests in a quest to meet or exceed our culture's idealized concept of extreme thinness. This quest is a part of our popular culture, promoted by waiflike models whose sunken eyes stare out of fashion magazines.

Like anorexia, bulimia results in starvation, but there are behavioral, physical, and psychological differences between the two. Bulimia is both less and more dangerous: on the one hand, people who have it tend to be of normal weight or are overweight, and unlike those with anorexia, they are aware of the fact that they have a problem. On the other hand, because the effects of their behavior are not so readily apparent, it is easier for a person with bulimia to persist in the pattern of bingeing and purging for much longer.

Approximately one in five persons with bulimia has a problem with drug or alcohol use, and they pursue their binges in a way not unlike that of a guilty addict or alcoholic hiding the spent needles or empty bottles from family members. They may go from restaurant to restaurant to avoid being seen eating too much in any one place, or they may pretend to be shopping for a large dinner party when, in fact, they intend to eat all the food themselves. Because of the expense of consuming so much food, some resort to shoplifting.

During a binge, people suffering from bulimia favor high-carbohydrate foods, such as doughnuts, candy, ice cream, soft drinks, cookies, cereal, cake, popcorn, and bread, and they consume many times the number of calories they would normally consume in one day. No matter what their normal eating habits, they tend to eat quickly and messily during a binge, stuffing the food into their mouths and gulping it down, sometimes without even tasting it. Some say they get a feeling of euphoria during binges, similar to the "runner's high" that some people get from exercise. Then, when they have gorged themselves, they force the food back out, either by causing themselves to vomit or by taking large quantities of laxatives.

Regular self-induced vomiting can cause all sorts of physical problems, such as damage to the stomach and esophagus, chronic heartburn, burst blood vessels in the eyes, throat irritation, and erosion of tooth enamel from the acid in vomit. Excessive use of laxatives can induce muscle cramps, stomach pains, digestive problems, dehydration, and even poisoning, while bulimia, in general, brings about vitamin deficiencies and imbalances of critical body fluids, which, in turn, can lead to seizures and kidney failure.

The self-imposed starvation of people with anorexia likewise takes a heavy toll on the body. The skin becomes dry and flaky, muscles begin to waste away, bones stop growing and may become brittle, and the heart weakens. Seeking to protect itself in the absence of proper insulation from fat, the body sprouts downy hair on the face, back, and arms in response to lower body temperature. In women, menstruation stops, and permanent infertility may result. Muscle cramps, dizziness, fatigue, and even brain damage as well as kidney and heart failure are possible. An estimated 10% to 20% of people with anorexia die either as a direct result of starvation or by suicide.

To save people with anorexia, force-feeding may be necessary. Some 70% of anorexia patients who are treated for about six months return to normal body weight, but about 15-20% can be expected to relapse. Bulimia is not as likely as anorexia to reach life-threatening stages, so hospitalization typically is not necessary. Treatment generally calls for psychotherapy and sometimes the administration of antidepressant drugs. Unlike people with anorexia, those with bulimia usually admit they have a problem and want help overcoming it.

OBESITY. Unlike anorexia or bulimia, obesity is more of a problem among people from lower-income backgrounds. This probably relates to a lack of education concerning nutrition, combined with the fact that healthier food is more expensive; by contrast, unhealthy items, such as white sugar, corn meal, and fatty cuts of pork and other meats can fill or overfill a person's stomach inexpensively. In addition, though men and women both tend to gain weight as they age, women are almost twice as likely as men to be obese.

Some cases of obesity relate to metabolic problems, while others stem from compulsive eating, which is psychologically motivated. Some studies suggest that obese people are much more likely than others to eat in response to stress, loneliness, or depression. And just as emotional pain can lead to obesity, obesity can lead to psychological scars. From childhood on, obese people are taunted and shunned, and throughout life they may face discrimination in school and on the job.

Physically, obesity is a killer, especially for those who are morbidly obese—that is, people whose obesity endangers their health. Obesity is

BEARS, WHICH WE THINK OF AS THE CLASSIC HIBERNATING ANIMAL, ARE ACTUALLY JUST DEEP SLEEPERS. A HIBERNATING ANIMAL SHOWS A DRASTIC REDUCTION IN METABOLISM AND THEN AWAKES RELATIVELY SLOWLY, WHEREAS A SLEEPING ANIMAL DECREASES ITS METABOLISM ONLY SLIGHTLY AND CAN WAKE UP ALMOST INSTANTLY IF DISTURBED. *(© Dan Guravich/Corbis. Reproduced by permission.)*

a risk factor for diabetes, high blood pressure, arteriosclerosis, angina pectoralis (chest pains due to inadequate blood flow to the heart), varicose veins, cirrhosis of the liver, and kidney disease. Obese people are about 1.5 times more likely to have heart attacks than are other people, and the overall death rate among people ages 20-64 is 50% higher for the obese than for people of ordinary weight.

HIBERNATION

Having looked at several unnatural ways in which people alter their metabolisms, let us close with an example of a very natural way that animals sometimes temporarily change theirs. This is hibernation, a state of inactivity in which an animal's heart rate, body temperature, and breathing rate are decreased as a way to conserve energy through the cold months of winter. A similar state, known as estivation, is adopted by some desert animals during the dry months of summer.

Hibernation is a technique that animals have developed, as a result of natural selection over the generations (see Evolution), to adapt to harsh environmental conditions. When food is scarce, a nonhibernating animal would be like a business operating at a loss—that is, using more energy maintaining its body temperature and searching for food than it would receive from consuming the food. Hibernating animals use 70-100 times less energy than when they are active, allowing them to survive until food is once again plentiful.

CONTRAST WITH SLEEP. Many animals sleep more often when food is scarce, but only a few truly hibernate. Bears, which many people think of as the classic hibernating animal, are actually just deep sleepers. By contrast, true hibernation occurs only in small mammals, such as bats and woodchucks and a few birds, among them nighthawks. Some insects also practice a form of hibernation. Hibernation differs from sleep, in that a hibernating animal shows a drastic reduction in metabolism and then awakes relatively slowly, whereas a sleeping animal decreases its metabolism only slightly and can wake up almost instantly if disturbed. Also, hibernating animals do not show periods of rapid eye movement (REM), the stage of sleep associated with dreaming in humans.

THE PROCESS OF HIBERNATION. Animals prepare for hibernation in the

KEY TERMS

ADIPOSE: Of or relating to animal fat.

AMINO ACIDS: Organic compounds made of carbon, hydrogen, oxygen, nitrogen, and (in some cases) sulfur bonded in characteristic formations. Strings of amino acids make up proteins.

ANABOLISM: The metabolic process by which energy is used to build up complex molecules that the body needs to maintain itself and develop new material.

ATOM: The smallest particle of an element, consisting of protons, neutrons, and electrons. An atom can exist either alone or in combination with other atoms in a molecule.

ATP: Adenosine triphosphate, an energy carrier formed when a simpler compound, adenosine diphosphate (ADP), combines with a phosphate group.

BLOOD SUGAR: The glucose in the blood.

CARBOHYDRATES: Naturally occurring compounds, consisting of carbon, hydrogen, and oxygen, whose primary function in the body is to supply energy. Included in the carbohydrate group are sugars, starches, cellulose, and various other substances. Most carbohydrates are produced by green plants in the process of undergoing photosynthesis.

CATABOLISM: The metabolic process by which large molecules are broken down into smaller ones with the release of energy. Compare with *anabolism.*

COMPOUND: A substance in which atoms of more than one element are bonded chemically to one another.

DIGESTION: The process of breaking food down into simpler chemical compounds as a means of making the nutrients absorbable by the body or organism.

ENZYME: A protein material that speeds up chemical reactions in the bodies of plants and animals without itself taking part in, or being consumed by, these reactions.

GLUCOSE: A monosaccharide (sugar) that occurs widely in nature and which is the form in which animals usually receive carbohydrates. Also known as *dextrose, grape sugar,* and *corn sugar.* See also *blood sugar.*

HYDROCARBON: Any organic chemical compound whose molecules are made up of nothing but carbon and hydrogen atoms.

LIPIDS: Fats and oils, which dissolve in oily or fatty substances but not in water-based liquids. In the body, lipids supply energy in slow-release doses, protect organs from shock and damage, and provide insulation for the body, for instance from toxins.

METABOLIC PATHWAY: An orderly sequence of reactions, with particular enzymes acting at each step along the way. Metabolic pathways may be either linear or circular, and sometimes they are linked, meaning that the product of one pathway becomes a reactant in another.

METABOLIC POOL: A group of relatively simple substances (e.g., amino acids) formed by the breakdown of relatively complex nutrients.

METABOLISM: The chemical process by which nutrients are broken down and converted into energy or are used in the

KEY TERMS CONTINUED

construction of new tissue or other material in the body. All metabolic reactions are either catabolic or anabolic.

MOLECULE: A group of atoms, usually but not always representing more than one element, joined in a structure. Compounds typically are made up of molecules.

NUTRIENT: Materials essential to the survival of organisms. They include proteins, carbohydrates, lipids (fats), vitamins, and minerals.

NUTRITION: The series of processes by which an organism takes in nutrients and makes use of them for its survival, growth, and development. The term *nutrition* also can refer to the study of nutrients, their consumption, and their use in the organism's body.

ORGANIC: At one time chemists used the term *organic* only in reference to living things. Now the word is applied to compounds containing carbon and hydrogen.

PHOSPHATE GROUP: A group (that is, a combination of atoms from two or more elements that tend to bond with other elements or compounds in certain characteristic ways) that includes a phosphate, or a chemical compound that contains oxygen bonded to phosphorus.

PRODUCT: A substance or substances formed from the interaction of reactants in a chemical reaction.

PROTEINS: Large molecules built from long chains of 50 or more amino acids. Proteins serve the functions of promoting normal growth, repairing damaged tissue, contributing to the body's immune system, and making enzymes.

REACTANT: A substance that interacts with another substance in a chemical reaction, resulting in the formation of a chemical or chemicals known as the *product(s).*

SUGARS: One of the three principal types of carbohydrate, along with starches and cellulose. Sugars can be defined as any of various water-soluble carbohydrates of varying sweetness. What we think of as "sugar" (i.e., table sugar) is actually sucrose; "blood sugar," on the other hand, is glucose.

TISSUE: A group of cells, along with the substances that join them, that forms part of the structural materials in plants or animals.

fall by storing food; usually this storage is internal, in the form of fat reserves. A woodchuck in early summer may have only about 5% body fat, but as fall approaches, changes in the animal's brain chemistry cause it to feel hungry and to eat constantly. As a result, the woodchuck's body fat increases to about 15% of its total weight. In other animals, such as the dormouse, fat may constitute as much as 50% of the animal's weight by the time hibernation begins. A short period of fasting follows the feeding frenzy, to ensure that the digestive tract is emptied completely before hibernation begins.

Going into hibernation is a gradual process. Over a period of days, an animal's heart rate and breathing rate drop slowly, eventually reaching rates of just a few beats or breaths per minute. Their body temperatures also drop from levels of about 100°F (38°C) to about 60°F (15°C). The lowered body temperature makes fewer demands on metabolism and food stores. Electric activity in the brain ceases almost completely during hibernation, although some areas—those that respond to external stimuli, such as light, temperature, and noise—remain active. Thus, the hibernating animal can be aroused under extreme conditions.

Periodically—perhaps every two weeks or so—the hibernating animal awakes and takes a few deep breaths to refresh its air supply. If the weather is particularly mild, some animals may venture from their lairs. An increase in heart rate signals that the time for arousal, or ending hibernation, is near. Blood vessels dilate, particularly around the heart, lungs, and brain, and this leads to an increased breathing rate. Eventually, the increase in circulation and metabolic activity spreads throughout the body, and the animal resumes a normal waking state.

WHERE TO LEARN MORE

Bouchard, Claude. *Physical Activity and Obesity.* Champaign, IL: Human Kinetics, 2000.

"KEGG Metabolic Pathways." *KEGG: Kyoto Encyclopedia of Genes and Genomes—GenomeNet,* Bioinformatics Center, Institute for Chemical Research, Kyoto University (Web site). <http://www.genome.ad.jp/kegg/metabolism.html>.

Medline Plus: Food, Nutrition, and Metabolism Topics. Medline, National Library of Medicine, National Institutes of Health (Web site). <http://www.nlm.nih.gov/medlineplus/foodnutritionandmetabolism.html>.

Metabolic Pathways of Biochemistry. George Washington University (Web site). <http://www.gwu.edu/~mpb/>.

Metabolism (Web site). <http://www.ultranet.com/~jkimball/BiologyPages/M/Metabolism.html>.

Michal, Gerhard. *Biochemical Pathways: An Atlas of Biochemistry and Molecular Biology.* New York: Wiley, 1999.

Pasternak, Charles A. *The Molecules Within Us: Our Body in Health and Disease.* New York: Plenum, 1998.

Pathophysiology of the Digestive System (Web site). <http://arbl.cvmbs.colostate.edu/hbooks/pathphys/digestion/>.

Spallholz, Julian E. *Nutrition, Chemistry, and Biology.* Englewood Cliffs, NJ: Prentice-Hall, 1989.

Wolinsky, Ira. *Nutrition in Exercise and Sport.* 3d ed. Boca Raton, FL: CRC Press, 1998.

DIGESTION

CONCEPT

Digestion is the process whereby the foods we eat pass through our bodies and are directed toward the purposes of either providing the body with energy or building new cellular material, such as fat or muscle. The parts of food that the body cannot use, along with other wastes from the body, are eliminated in the form of excrement. Aspects of digestion, particularly the production of waste and intestinal gas, are not exactly topics for polite conversation, yet without these and other digestive processes, life for humans and other organisms would be impossible. The functioning of digestion itself is like that of a well-organized, cohesive sports team or even of a symphony orchestra: there are many parts and players, each with an indispensable role.

HOW IT WORKS

NUTRIENTS

For digestion to occur, of course, it is necessary first to have something to digest—namely, nutrients. What follows is a cursory overview of nutrients and nutrition, subjects covered in much more depth within the essay of that name. Nutrients include proteins, carbohydrates, fats, minerals, and vitamins. In addition to these nutrients, animal life requires other materials, not usually considered nutrients, which include water, oxygen, and something that greatly aids the process of food digestion and elimination of wastes: fiber.

PROTEINS AND CARBOHYDRATES. Proteins are large molecules built from long chains of amino acids, which are organic compounds made of carbon, hydrogen, oxygen, nitrogen, and (in some cases) sulfur-bonded in characteristic formations. Proteins serve the functions of promoting normal growth, repairing damaged tissue, contributing to the body's immune system, and making enzymes. (An enzyme is a protein material that speeds up chemical reactions in the bodies of plants and animals.) Good examples of dietary proteins include eggs, milk, cheese, and other dairy products. Incomplete proteins, or ones lacking essential amino acids—those amino acids that are not produced by the human body—include peas, beans, lentils, nuts, and cereal grains.

Carbohydrates are compounds that consist of carbon, hydrogen, and oxygen. Their primary function in the body is to supply energy. When a person ingests more carbohydrates than his or her body needs at the moment, the body converts the excess into a compound known as glycogen. It then stores the glycogen in the liver and muscle tissues, where it remains, a potential source of energy for the body to use in the future, though if it is not used soon, it may be stored as fat. The carbohydrate group comprises sugars, starches, cellulose (a type of fiber), and various other chemically related substances.

LIPIDS, VITAMINS, AND MINERALS. Lipids include all fats and oils and are distinguished by the fact that they are soluble (i.e., capable of being dissolved) in oily or fatty substances but not in water. In the body, lipids supply energy much as carbohydrates do, only much more slowly. Lipids also protect the organs from shock and damage and provide the body with insulation from cold, toxins, and other threats. Processed, saturated fats (fats that have been enhanced artificially to make them more firm) are extremely unhealthy, and consumption of some types of animal fat (e.g., pork fat) is also inadvisable. On the other hand, vegetable fats, such as those in avocados and olive oil, as well as the animal fats in such fish as tuna, mackerel, and salmon can be highly beneficial.

Vitamins are organic substances that, in extremely small quantities, are essential to the nutrition of most animals and some plants. In particular, they work with enzymes in regulating metabolic processes—that is, the chemical processes by which nutrients are broken down and converted into energy or used in the construction of new tissue or other material in the body. Vitamins do not in themselves provide energy, however, and thus they do not qualify as a form of nutrition. Much the same is true of minerals, except that these are inorganic substances, meaning that they do not contain chemical compounds made of carbon and hydrogen.

THE DIGESTIVE SYSTEM

To supply the body with the materials it needs for energy and the building of new tissue, nutrients have to pass through the digestive system. The latter is composed of organs (an organ being a group of tissues and cells, organized into a particular structure, that performs a specific function within an organism) and other structures through which nutrients move. The nutrients pass first through the mouth and then through the esophagus, stomach, small intestine, and large intestine, or colon. Collectively, these structures are known as the alimentary canal.

Nutrients advance through the alimentary canal to the stomach and small intestine, and waste materials continue from the small intestine to the colon (large intestine) and anus. Along the way, several glands play a role. A gland is a cell or group of cells that filters material from the blood, processes that material, and secretes it either for use again in the body or to be eliminated as waste. Among the glands that play a part in the digestive process are the salivary glands, liver, gallbladder, and pancreas. (The last three are examples of glands that are also organs.) The glands with a role in digestion secrete digestive juices containing enzymes that break down nutrients chemically into smaller molecules that are absorbed more easily by the body. There are also hormones involved in digestion-there are, for example, glandular cells in the lining of the stomach that make the hormone gastrin.

FROM THE MOUTH TO THE STOMACH. The first stage of digestion is ingestion, in which food is taken into the mouth and then broken down into smaller pieces by the chewing action of the teeth. To facilitate movement of the food through the mouth and along the tongue, it is necessary for saliva to be present. Usually, the sensations of sight, taste, and smell associated with food set in motion a series of neural responses that induce the formation of saliva by the salivary glands in the mouth. Amylase, an enzyme in the saliva, begins the process of breaking complex carbohydrates into simple sugars. (The terms *simple* and *complex* in this context refer to chemical structures.)

By the time it is ready to be swallowed, food is in the form of a soft mass known as a bolus. The action of swallowing pulls the food down through the pharynx, or throat, and into the esophagus, a tube that extends from the bottom of the throat to the top of the stomach. (Note that for the most part, we are using human anatomy as a guide, but many aspects of the digestive process described here also apply to other higher animals, particularly mammals.) The esophagus does not take part in digestion but rather performs the function of moving the bolus into the stomach.

A wavelike muscular motion termed peristalsis, which consists of alternating contractions and relaxations of the smooth muscles lining the esophagus, moves the bolus through this passage. At the place where the esophagus meets the stomach, a powerful muscle called the *esophageal sphincter* acts as a valve to keep food and stomach acids from flowing back into the esophagus and mouth. (Although the most well-known sphincter muscle in the body is the one surrounding the anus, sometimes known simply as "*the* sphincter," in fact, *sphincter* is a general term for a muscle that surrounds, and is able to control the size of, a bodily opening.)

FROM THE STOMACH TO THE SMALL INTESTINE. Chemical digestion begins in the stomach, a large, hollow, pouchlike muscular organ. While food is still in the mouth, the stomach begins its production of gastric juice, which contains hydrochloric acid and pepsin, an enzyme that digests protein. Gastric juice is the material that breaks down the food.

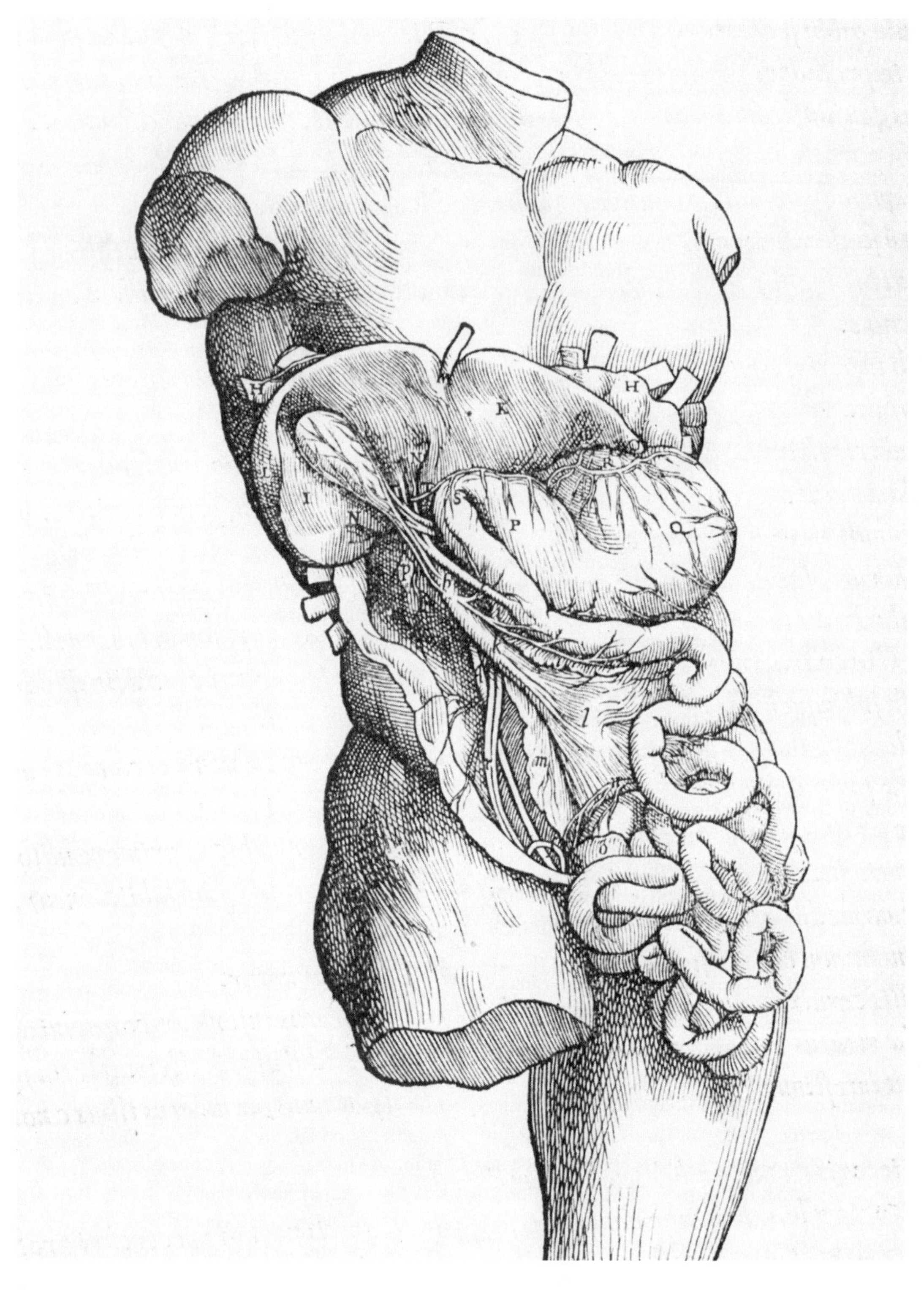

A SIXTEENTH-CENTURY ANATOMICAL DIAGRAM OF THE INTERNAL ORGANS, SHOWING THE STOMACH, LIVER, INTESTINE, AND GALLBLADDER. *(© Corbis. Reproduced by permission.)*

Once nerves in the cheeks and tongue are stimulated by the food, they send messages to the brain, which, in turn, alerts nerves in the stomach wall, stimulating the secretion of gastric juice before the bolus itself arrives in the stomach. Once the bolus touches the stomach lining, it triggers a second release of gastric juice, along with mucus that helps protect the stomach lining from the action of the hydrochloric acid. Three layers of powerful stomach muscles churn food into a thick liquid called *chyme,* which is pumped gradually through the pyloric sphincter, which connects the stomach small intestine.

THE SMALL INTESTINE. The names of the small and large intestines can be confusing, rather like those of Upper and Lower Egypt in ancient history. In both cases, the adjectives seem to refer to one thing but actually refer to something else entirely. Thus, it so happens that Upper Egypt was south of Lower Egypt (because it was "upper" in elevation, not latitude), while the small intestine is, in fact, much longer than the large intestine. The reason is that *small* refers to its diameter rather than its length: though it is about 23 ft. (7 m) long, the small intestine is only 1 in. (2.5 cm) in diameter, while the large intestine, only 5 ft. (1.5 m) in length, is 3 in. (7.6 cm) across.

The small intestine, which connects the stomach and large intestine, is in three sections: the duodenum, jejunum, and ileum. About 1 ft. (0.3 m) long, the duodenum breaks down chyme from the stomach with the aid of the pancreas and gallbladder. The pancreas, a large gland located below the stomach, secretes pancreatic juice, which contains three enzymes that break down carbohydrates, fats, and proteins, into the duodenum through the pancreatic duct. The gallbladder empties bile, a yellowish or greenish fluid from the liver, into the duodenum when chyme enters that portion of the intestine. Although bile does not contain enzymes, it does have bile salts that help dissolve fats.

Digested carbohydrates, fats, proteins, and most of the vitamins, minerals, and iron in food are absorbed in the jejunum, which is about 4 ft. (1.2 m) long. Aiding this absorption are up to five million tiny finger-like projections called *villi,* which greatly increase the surface area of the small intestine, thus accelerating the rate at which nutrients are absorbed into the bloodstream. The remainder of the small intestine is taken up by the ileum, which is smaller in diameter and has thinner walls than the jejunum. It is the final site for absorption of some vitamins and other nutrients, which enter the circulatory system in plasma, a watery liquid in which red blood cells also are suspended.

As it moves through the circulatory system, plasma takes with it amino acids, enzymes, glycerol (a form of alcohol found in fats), and fatty acids, which it directs to the body's tissues for energy and growth. Plasma also contains waste products from the breakdown of proteins, including creatinine, uric acid, and ammonium salts. These constituents are moved to the kidneys, where they are filtered from the blood and excreted in the urine. But, of course, urine is not the only waste product excreted by the body; there is also the solid waste, processed through the large intestine, or colon.

THE LARGE INTESTINE AND BEYOND. Like the small intestine, the large intestine is in segments. It rises up on the right side of the body (the ascending colon), crosses over to the other side underneath the stomach (the transverse colon), descends on the left side, (the descending colon), and forms an S shape (the sigmoid colon) before reaching the rectum and anus. In addition to its function of pumping solid waste, the large intestine removes water from the waste products—water that, when purified, will be returned to the bloodstream. In addition, millions of bacteria in the large intestine help produce certain B vitamins and vitamin K, which are absorbed into the bloodstream along with the water.

After leaving the sigmoid colon, waste passes through the muscular rectum and then the anus, the last point along the alimentary canal. In all, the movement of food through the entire length of the alimentary tract takes from 15 to 30 hours, with the majority of that time being taken up by activity in the colon. Food generally spends about three to five hours in the stomach, another four to five hours in the small intestine, and between five and 25 hours in the large intestine.

The transit time, or the amount of time it takes for food to move through the system, is a function of diet: for a vegetarian who eats a great deal of fiber, it will be on the short end, while for a meat eater who has just consumed a dinner of prime rib, it will take close to the maximum time. People who eat diets heavy in red meat or junk foods are also likely to experience a buildup, over time, of partially digested material on the linings of their intestines. Obviously, this is not a healthy situation, and to turn it around, a person may have to change his or her diet and perhaps even undergo some sort of colon-cleansing program. There is an easy way to test transit time in one's system: simply eat a large serving of corn or red beets, and measure how long it takes for these to fully work their way through the digestive system.

REAL-LIFE APPLICATIONS

DIGESTIVE DISORDERS

It is hard to watch more than a few minutes of commercial television without seeing advertisements for fast foods and other varieties of junk food or stomach-relief medicine or both. There is a connection, of course: a society glutted on greasy drive-through burgers and thick-crust pizzas needs something to cure the upset stomachs that result.

Indigestion is a general condition that, as its name suggests, involves an inability to digest food properly. Heartburn, sometimes called acid indigestion, is a specific type of indigestion that occurs when the stomach produces too much hydrochloric acid. The latter is essential to digestion, but if a person eats a giant Polish sausage, a spicy-hot bowl of jambalaya, or some other hard-to-digest food (as opposed to a healthy meal of baked fish with brown rice and spinach, for instance), the stomach may produce too much of the acid.

Heartburn is so named because it causes a sharp pain behind the breastbone, which might feel like a heart attack. It also may produce acid reflux, in which the stomach acid backs up into the esophagus. If you have ever experienced what might be called a *leap of vomit,* in which a burp is associated with the rise of burning, foul-tasting bile through the esophagus, then you have first-hand knowledge of acid reflux and heartburn.

Digestive tract diseases, such as dyspepsia, sometimes can cause chronic indigestion, but more often than not, people experience indigestion as a result of eating too quickly or too much, consuming high-fat foods, or eating in a stressful situation. (This is why you might feel sick to your stomach when eating lunch at school on the day of a difficult test, a fight, or a romantic trauma, such as a breakup or asking for a first date.) Smoking, excessive drinking, fatigue, and the consumption of medications that irritate the stomach lining also can contribute to indigestion. In addition, it is a good idea not to eat too soon before going to bed, since this can produce heartburn.

COMMERCIAL ANTACIDS. Most nonprescription stomach-relief medicine is in the form of an antacid, which, as the term suggests, is a substance that works against acids in the stomach. Chemically, the opposite of an acid is a base, or an alkaline substance, the classic example being sodium bicarbonate or sodium hydrogen carbonate ($NaHCO_3$)—that is, baking soda. Baking soda alone can perform the function of an antacid, but the taste is rather unpleasant, and for this reason most antacid products combine it with other chemicals to enhance the flavor.

One famous commercial stomach remedy actually *uses* acid. This is Alka-Seltzer, but the presence of citric acid has more to do with marketing than with the chemistry of the stomach. The citric acid, often used as a sweetener, imparts a more pleasant flavor than the bitter taste of alkaline antacids, and, moreover, when Alka-Seltzer tablets are placed in water, the acid reacts chemically with the sodium bicarbonate to create the product's trademark fizz.

Ultimately, all antacids (Alka-Seltzer included) work because the bases in the product react with the acids in the stomach. This is a chemical process called neutralization, in which the acid and base cancel out each other, producing water and a salt in the process. (Table salt, or sodium chloride, is just one of many salts, all of which are formed by the chemical bonding of a metal with a nonmetal—in the case of table salt, sodium and chlorine, respectively.) Thanks to this process, acid in the stomach of a heartburn sufferer is neutralized.

ULCERS. There are some digestive disorders that cannot be cured by Alka-Seltzer or its many competitors, such as Rolaids, Maalox, Mylanta, Tums, Milk of Magnesia, or Pepto-Bismol. Instead of just occasional indigestion or heartburn, a person may be afflicted with a sore in one part of the digestive tract, which may be either a stomach ulcer or a duodenal ulcer. Stomach ulcers, which form in the lining of the stomach, are called peptic ulcers because they form with the help of stomach acid and pepsin. Duodenal ulcers, which are more common, tend to be smaller than stomach ulcers and heal more quickly. Any ulcer, whether a small sore or a deep cavity, leaves a scar in the alimentary canal.

Until the early 1990s, physicians generally maintained that personal behavior and conditions, such as stress and poor diet, were the principal factors behind ulcer. Medical researchers eventually came to believe, however, that the culprit was a certain bacterium, which can live

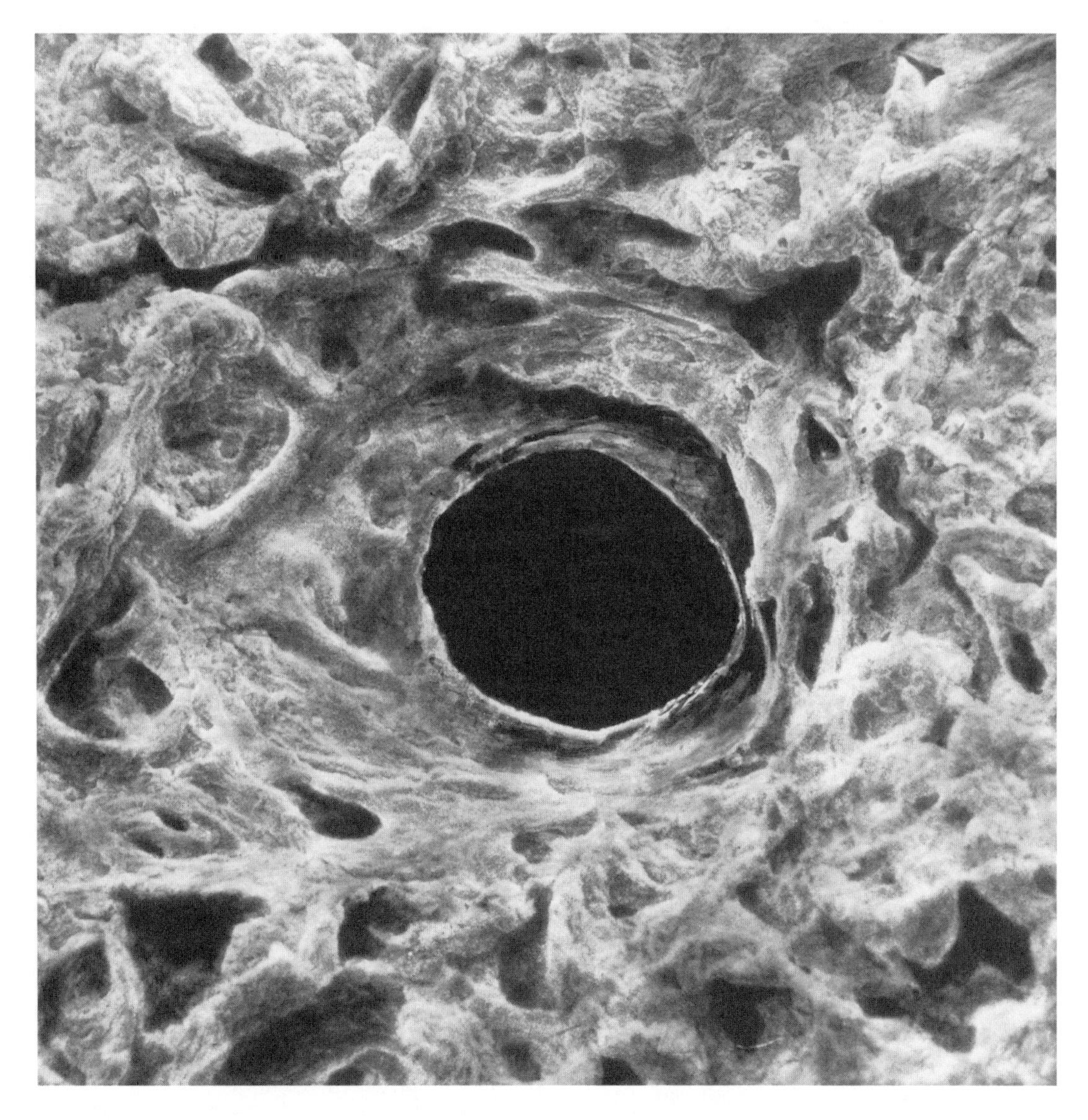

SCANNING ELECTRON MICROGRAPH OF A STOMACH ULCER. INSTEAD OF JUST OCCASIONAL INDIGESTION, A PERSON MAY BE AFFLICTED WITH AN ULCER, OR A SORE IN ONE PART OF THE DIGESTIVE TRACT. STOMACH ULCERS ARE CALLED PEPTIC ULCERS BECAUSE THEY FORM WITH THE HELP OF STOMACH ACID AND PEPSIN. *(Photo Researchers. Reproduced by permission.)*

undetected in the mucous lining of the stomach. This bacterium irritates and weakens the lining, making it more susceptible to damage by stomach acids. As many as 80% of all stomach ulcers may be caused by such a bacterial infection. With this newfound knowledge, ulcer patients today are more likely to be treated with antibiotics and antacids rather than special diets or expensive medicines.

WAYS TO IMPROVE DIGESTION

There are numerous ways to improve digestion, by changing either the way one eats or the things one eats. In the first category, it is important to eat only when you are really hungry and to eat slowly and chew food thoroughly. Drinking liquids with a meal is probably not a good practice; it is better to wait until you are finished, so as not to interfere with the action of digestive fluids. Certainly, smoking and excessive drinking have a negative impact on digestion, whereas regular exercise has a positive influence.

In the category of diet, it is a good idea to minimize one's intake of red meat, such as steak. Although most people find red meat tasty, and it can be a good supplier of dietary iron in limited proportions, the digestion of red meat requires the production of much more stomach acid, and thus it places a great burden on the digestive system. It is also wise to eliminate as many processed foods as possible, including sweets and junk foods, and to eat as many natural foods as one can manage.

In general, one can hardly go wrong with raw vegetables, which are just about the best thing a person can eat—not only because of their digestive properties but also because many of them are packed so full of vitamins and minerals. (Note that vegetables are best when raw and fresh, since cooking removes many of the nutrients. Canned vegetables usually are both nutrient-poor and full of sodium or even synthetic chemicals. Frozen vegetables are much better than canned ones, but they still do not compare nutritionally with fresh vegetables.)

A good diet includes a great deal of fiber, indigestible material that simply passes through the system, assisting in the peristaltic action of the alimentary canal and in the process of eliminating waste. Cellulose, found in most raw fruits and vegetables, is an example of fiber, also called bulk or roughage. Yogurt may be a beneficial food, because it includes "good" bacteria (a topic we discuss near the conclusion of this essay) that assist the digestive process. Some foods, such as raw bean sprouts, papaya, figs, and pineapple, contain enzymes that appear to assist the body in digesting them.

In addition, one of the greatest "foods" for aiding digestion is not a food at all, but water, of which most people drink far too little. Some experts claim that a person should drink eight 8-oz. (0.24 l) glasses a day, but others maintain that a person should drink half as many ounces of water as his or her weight in pounds. In other words, a person who weighed 100 lb. (45.36 kg) would drink 50 oz. (1.48 l) of water a day, whereas a person who weighed 150 lb. (68.04 kg) would drink 75 oz. (2.22 l). A good rule of thumb for metric users, instead of the 2:1 pounds-to-ounces ratio, would be 30:1 kilograms to liters. Note also that tap water may contain chemicals or other impurities, and therefore consuming it in large quantities is not advisable. A much better alternative is bottled or filtered water.

HUMAN WASTE

Many years ago, both a serious book and a comedic movie had the title *Everything You Always Wanted to Know about Sex (But Were Afraid to Ask).* Just as the title of David R. Reuben's 1969 book inspired Woody Allen's 1972 movie, which was very loosely based on it, the "Everything you always wanted to know ... " motif inspired a whole array of imitators. Often such titles play on the very fact that hardly anyone wants to know, and certainly no one is afraid to ask, about the topic in question. A good example is an on-line article by central Asia authority Mark Dickens entitled "Everything You Always Wanted to Know about Tocharian But Were Afraid To Ask" (http://www.oxuscom.com/eyawtkat.htm). The joke, of course, is that most people have never heard of Tocharian, a central Asian language. Nonetheless, one subject ranks with sex as something everyone wonders about but most are afraid to ask.

Even the name of that topic creates problems, since people have so many euphemisms for it: "number two," for instance, or BM (short for bowel movement). There are baby- and child-oriented terms for this process and product, the bodily control of which can be a major problem for a very young human being, and, of course, there is at least one grown-up term for it that will not be mentioned here. People even have nicknames for animal dung, such as pies, patties, or chips. For the sake of convenience, let us call the process defecation, and the product human waste (or excrement or feces) and admit that everyone has wondered how something as pleasant as food can, after passing through the alimentary canal, turn into something as unpleasant as the final product.

The average person excretes some 7 lb. (3.2 kg) of feces per day, an amount equal to a little more than 1 ton (0.91 tonnes) per year. This waste is made up primarily of indigestible materials as well as water, salts, mucus, cellular debris from the intestines, bacteria, and cellulose and other types of fiber. Like the human body itself, these waste products are mostly water: about 75%, compared with 25% solid matter. Much of what goes into producing excrement has nothing to do with what enters the digestive system, so even if a person were starving he or she would continue to excrete feces.

COLORATION. What about the color and the smell? The color of feces comes from

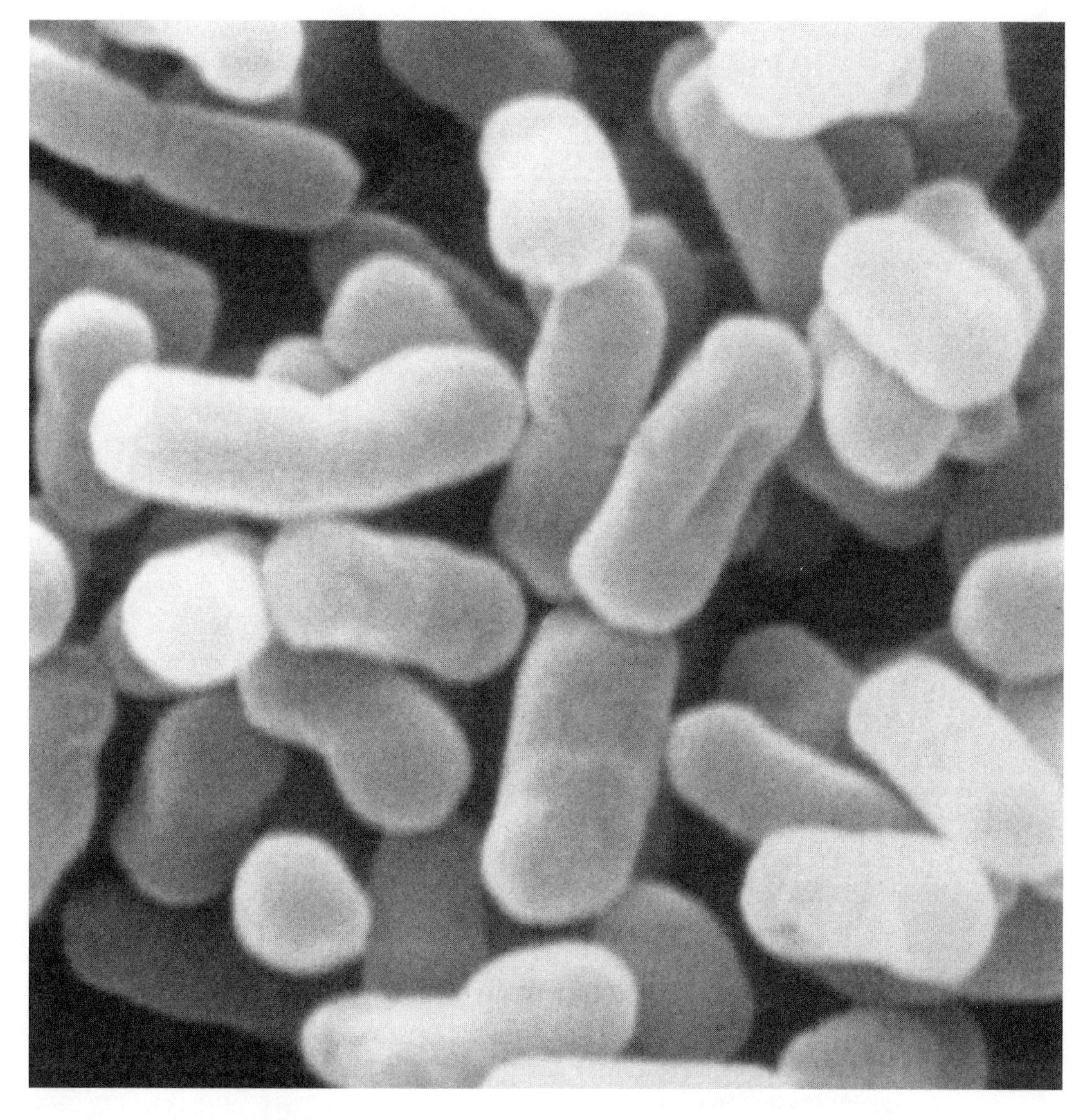

ESCHERICHIA COLI IS A TYPE OF BACTERIA THAT LIVES IN THE INTESTINAL TRACT, AIDING THE DIGESTIVE PROCESS BY SUPPRESSING THE GROWTH OF HARMFUL BACTERIA AND SYNTHESIZING VITAMINS. *(© 1997 Custom Medical Stock Photo. Reproduced by permission.)*

bilirubin, a reddish-yellow pigment found in blood and bile, which passes through the liver and enters the small intestine via the gallbladder. Later, as it passes through the large intestine, it is degraded by the action of bacteria, a process that turns it brown and gives feces its characteristic color.

Not surprisingly, disorders involving the red blood cells, liver, or gallbladder can change the color of human waste. A person with gallstones or hepatitis (a disease characterized by inflammation of the liver) is likely to excrete grayish-brown feces, while anemia (a condition that involves a lack of red blood cells) may be associated with a yellowish stool. A person experiencing bleeding in the gastrointestinal tract (the stomach and intestines) may produce waste the color of black tar. In addition, foods with distinctive colors and textures also can affect the appearance of stools.

FRIENDLY BACTERIA. Before addressing the smell of feces, which is the result of action by bacteria in the colon, it is worth saying a few words about those single-cell organisms themselves. This is especially important in light of the fact that bacteria have a bad reputation that is not entirely deserved. Without question, there are harmful microbes in the world, but a world completely free of these organisms

KEY TERMS

ALIMENTARY CANAL: The entire length of tube that extends from the mouth to the anus, including the esophagus, stomach, and small and large intestines. Nutrients pass through the alimentary canal to the stomach and small intestine, and waste materials from these nutrients (and from other sites in the body) pass from the small intestine to the colon (large intestine) and anus.

AMINO ACIDS: Organic compounds made of carbon, hydrogen, oxygen, nitrogen, and (in some cases) sulfur bonded in characteristic formations. Strings of amino acids make up proteins.

BILE: A yellowish or greenish digestive fluid excreted by the liver.

BOLUS: A term for a chewed mass of food making its way through the initial portions of the alimentary canal.

CARBOHYDRATES: Naturally occurring compounds, consisting of carbon, hydrogen, and oxygen, whose primary function in the body is to supply energy. Included in the carbohydrate group are sugars, starches, cellulose, and various other substances.

CELLULOSE: A polysaccharide that is the principal material in the cell walls of plants. Cellulose also is found in such natural fibers as cotton and is used as a raw material in manufacturing such products as paper.

COLON: The large intestine, through which waste materials pass on their way to excretion through the anus.

COMPOUND: A substance in which atoms of more than one element are bonded chemically to one another.

ENZYME: A protein material that speeds up chemical reactions in the bodies of plants and animals.

FIBER: Indigestible material in food that simply passes through the digestive system, assisting in the peristaltic action of the alimentary canal and in the processing of waste. Examples of fiber, also called *bulk* or *roughage,* include cellulose.

GASTROINTESTINAL TRACT: The stomach and intestines.

GLAND: A cell or group of cells that filters material from the blood, processes that material, and secretes it either for use again in the body or to be eliminated as waste.

GLUCOSE: A type of sugar that occurs widely in nature. Glucose is the form in which animals usually receive carbohydrates.

GLYCOGEN: A white polysaccharide that is the most common form in which carbohydrates are stored in animal tissues, particularly muscle and liver tissues.

GUT: A term that refers to all or part of the alimentary canal. Although the word is considered a bit crude in everyday life, physicians and biological scientists concerned with this part of the anatomy use it regularly.

would be one in which humans and other animals would be unable to live. In fact, we have a mutually beneficial relationship, a type of symbiosis (see Symbiosis) with the microorganisms in our alimentary canals, particularly in the colon.

Bacteria live in the guts—a term that refers to all or part of the alimentary canal—of most ani-

KEY TERMS CONTINUED

HEMOGLOBIN: An iron-containing pigment in red blood cells that is responsible for transporting oxygen to the tissues and removing carbon dioxide from them.

LIPIDS: Fats and oils, which dissolve in oily or fatty substances but not in water-based liquids. In the body, lipids supply energy in slow-release doses, protect organs from shock and damage, and provide insulation for the body, for instance, from toxins.

METABOLISM: The chemical process by which nutrients are broken down and converted into energy or used in the construction of new tissue or other material in the body.

MINERALS: Inorganic substances that, in a nutritional context, serve a function similar to that of vitamins. Minerals may include chemical elements, particularly metallic ones, such as calcium or iron, as well as some compounds.

ORGAN: A group of tissues and cells, organized into a particular structure, that performs a specific function within an organism.

ORGANIC: At one time, chemists used the term *organic* only in reference to living things. Now the word is applied to compounds containing carbon and hydrogen.

PERISTALSIS: A series of involuntary muscle contractions that force bolus, and later waste, through the alimentary canal.

POLYSACCHARIDE: A complex sugar, in which the molecules are composed of many glucose subunits arranged in a chain. Polysaccharides can be broken down chemically to produce simple sugars, or monosaccharides.

PROTEINS: Large molecules built from long chains of amino acids. Proteins serve the functions of promoting normal growth, repairing damaged tissue, contributing to the body's immune system, and making enzymes.

SPHINCTER: A general term for a muscle that surrounds and is able to control the size of a bodily opening.

SYMBIOSIS: A biological relationship in which (usually) two species live in close proximity to each other and interact regularly in such a way as to benefit one or both of the organisms.

TISSUE: A group of cells, along with the substances that join them, that forms part of the structural materials in plants or animals.

VITAMINS: Organic substances that, in extremely small quantities, are essential to the nutrition of most animals and some plants. In particular, vitamins work with enzymes in regulating metabolic processes; they do not in themselves provide energy, however, and thus vitamins alone do not qualify as a form of nutrition.

mals, where they assist in such difficult digestive activities as the processing of chewed grasses. The latter is heavy in cellulose, and to digest it, cows, sheep, deer, and other grass eaters (known as ruminants) have stomachs with several compartments. The first of these compartments is called the rumen, and it serves as home to millions of bacteria, which assist in breaking down the heavy fibers.

Humans' bacterial symbiotic partners (actually, this is a type of symbiosis known as mutualism, in which both creatures benefit) include bacteria of the species *Escherichia coli,* or *E. coli.* The name is no doubt familiar to most readers from its appearance in the news in connection with horror stories involving *E. coli* poisoning in food or local water supplies. Certainly, *E. coli* can be extremely harmful when it is outside the human gut, but inside the gut it is humans' friend.

E. coli is a coprophile (literally, "excrement lover"), meaning that it depends on feces for survival. Fecal matter itself can contain all manner of harmful substances associated with the decomposition of foods or with the body's efforts to rid itself of toxins (including pathogens, or disease-carrying parasites—see Parasites and Parasitology), so anything associated with feces is dirty and potentially dangerous. It is for this reason that *E. coli* can cause serious illness or death if it gets into other parts of the body.

As long as it stays where it belongs, however, *E. coli* not only aids in the digestive process but also provides the body with vitamin K, essential for proper blood clotting, as well as vitamin B_{12}, thiamine, and riboflavin. Every person carries millions and millions of these helpful fellow travelers; even though a single bacterium weighs almost nothing in human terms, the combined weight of all the helpful, "good" bacteria in our guts is a staggering 7 lb. (3.2 kg).

INTESTINAL GAS. As those bacteria do their work, they generate vast quantities of gases, which are by-products of the chemical processes that play a part in breaking down the foods passing through the gut. Among these gaseous products are hydrogen sulfide, a foul-smelling substance that can be toxic in large quantities. Unlike carbon monoxide, which has no odor, few people are in danger of dying from inhalation of hydrogen sulfide—even though it is abundant in nature—because the smell is enough to dissuade anyone from inhaling it for long periods of time. (Incidentally, gas companies include traces of hydrogen sulfide with natural gas. By itself, natural gas is odorless, but when a leak occurs, a homeowner will smell hydrogen sulfide and alert the gas company.)

Hydrogen sulfide is just one of many unpleasant-smelling chemical products that result from bacterial action on solids in the gut. Others include indole, skatole, ammonia, and mercaptans, though the most distinctive-smelling of all are indole and skatole, which come primarily from the digestion of an amino acid known as tryptophan. In addition, the particular foods a person eats, as well as the specific bacterial residents (some harmful) of his or her gut, can affect the odor of intestinal gas.

When gases pass outside the rectum, the result is flatulence (of course, there are other, less polite words for it), which is the subject of much schoolboy humor. Even inside the body, intestinal gas can make noise and cause embarrassment, in the form of borborygmus—intestinal rumbling caused by moving gas. As for the flammability of intestinal gas, it probably results from the high proportion of hydrogen, an extremely flammable gas.

WHERE TO LEARN MORE

Avraham, Regina. *The Digestive System.* New York: Chelsea House Publishers, 1989.

Ballard, Carol. *The Stomach and Digestive System.* Austin, TX: Raintree Steck-Vaughn, 1997.

Digestive System Diseases. Karolinska Institutet (Web site). <http://www.mic.ki.se/Diseases/c6.html>.

The Human Body's Digestive System Theme Page. Community Learning Network (Web site). <http://www.cln.org/themes/digestive.html>.

Medline Plus: Digestive System Topics. National Library of Medicine, National Institutes of Health (Web site). <http://www.nlm.nih.gov/medlineplus/digestivesystem.html>.

Morrison, Ben. *The Digestive System.* New York: Rosen Publishing Group, 2001.

National Institute of Diabetes and Digestive and Kidney Diseases of the National Institutes of Health (Web site). <http://www.niddk.nih.gov/index.htm

Parker, Steve, and Ian Thompson. *Digestion.* Brookfield, CT: Copper Beech Books, 1997.

Pathophysiology of the Digestive System. Colorado State University (Web site). <http://arbl.cvmbs.colostate.edu/hbooks/pathphys/digestion/>.

Richardson, Joy. *What Happens When You Eat?* Illus. Colin Maclean and Moira Maclean. Milwaukee, WI: Gareth Stevens Publishing, 1986.

RESPIRATION

CONCEPT

Respiration is much more than just breathing; in fact, the term refers to two separate processes, only one of which is the intake and outflow of breath. At least cellular respiration, the process by which organisms convert food into chemical energy, requires oxygen; on the other hand, some forms of respiration are anaerobic, meaning that they require no oxygen. Such is the case, for instance, with some bacteria, such as those that convert ethyl alcohol to vinegar. Likewise, an anaerobic process can take place in human muscle tissue, producing lactic acid—something so painful that it feels as though vinegar itself were being poured on an open sore.

HOW IT WORKS

Forms of Respiration

Respiration can be defined as the process by which an organism takes in oxygen and releases carbon dioxide, one in which the circulating medium of the organism (e.g., the blood) comes into contact with air or dissolved gases. Either way, this means more or less the same thing as breathing. In some cases, this meaning of the term is extended to the transfer of oxygen from the lungs to the bloodstream and, eventually, into cells or the release of carbon dioxide from cells into the bloodstream and thence to the lungs, from whence it is expelled to the environment. Sometimes a distinction is made between external respiration, or an exchange of gases with the external environment, and internal respiration, an exchange of gases between the body's cells and the blood, in which the blood itself "bathes" the cells with oxygen and receives carbon dioxide to transfer to the environment.

This is just one meaning—albeit a more familiar one—of the word respiration. Respiration also can mean cellular respiration, a series of chemical reactions within cells whereby food is "burned" in the presence of oxygen and converted into carbon dioxide and water. This type of respiration is the reverse of photosynthesis, the process by which plants convert dioxide and water, with the aid of solar energy, into complex organic compounds known as carbohydrates. (For more about carbohydrates and photosynthesis, see Carbohydrates.)

How Gases Move Through the Body

Later in this essay, we discuss some of the ways in which various life-forms breathe, but suffice it to say for the moment—hardly a surprising revelation!—that the human lungs and respiratory system are among the more complex mechanisms for breathing in the animal world. In humans and other animals with relatively complex breathing mechanisms (i.e., lungs or gills), oxygen passes through the breathing apparatus, is absorbed by the bloodstream, and then is converted into an unstable chemical compound (i.e., one that is broken down easily) and carried to cells. When the compound reaches a cell, it is broken down and releases its oxygen, which passes into the cell.

On the "return trip"—that is, the reverse process, which we experience as exhalation—cells release carbon dioxide into the bloodstream, where it is used to form another unstable chemical compound. This compound is carried by the bloodstream back to the gills or lungs, and, at the end of the journey, it breaks down and releases the carbon dioxide to the surrounding environment. Clearly, the one process is a mirror image of the other, with the principal difference being the fact that oxygen is the key chemical component in the intake process, while carbon dioxide plays the same role in the process of outflow.

HEMOGLOBIN AND OTHER COMPOUNDS. In humans the compound used to transport oxygen is known by the name hemoglobin. Hemoglobin is an iron-containing protein in red blood cells that is responsible for transporting oxygen to the tissues and removing carbon dioxide from them. In the lungs, hemoglobin, known for its deep red color, reacts with oxygen to form oxyhemoglobin. Oxyhemoglobin travels through the bloodstream to cells, where it breaks down to form hemoglobin and oxygen, and the oxygen then passes into cells. On the return trip, hemoglobin combines with carbon dioxide to form carbaminohemoglobin, an unstable compound that, once again, breaks down—only this time it is carbon dioxide that it releases, in this case to the surrounding environment rather than to the cells.

In other species, compounds other than hemoglobin perform a similar function. For example, some types of annelids, or segmented worms, carry a green blood protein called chlorocruorin that functions in the same way as hemoglobin does in humans. And whereas hemoglobin is a molecule with an iron atom at the center, the blood of lobsters and other large crustaceans contains hemocyanin, in which copper occupies the central position. Whatever the substance, the compound it forms with oxygen and carbon dioxide must be unstable, so that it can break down easily to release oxygen to the cells or carbon dioxide to the environment.

CELLULAR RESPIRATION

Both forms of respiration involve oxygen, but cellular respiration also involves a type of nutrient—materials that supply energy, or the materials for forming new tissue. Among the key nutrients are carbohydrates, naturally occurring compounds that consist of carbon, hydrogen, and oxygen. Included in the carbohydrate group are sugars, starches, cellulose, and various other substances.

Glucose is a simple sugar produced in cells by the breakdown of more complex carbohydrates, including starch, cellulose, and such complex sugars as sucrose (cane or beet sugar) and fructose (fruit sugar). In cellular respiration, an organism oxidizes glucose (i.e., combines it with oxygen) so as to form the energy-rich compound known as adenosine triphosphate (ATP). ATP, critical to metabolism (the breakdown of nutrients to provide energy or form new material), is the compound used by cells to carry out most of their ordinary functions. Among those functions are the production of new cell parts and chemicals, the movement of compounds through cells and the body as a whole, and growth.

In cellular respiration, six molecules of glucose ($C_6H_{12}O_6$) react with six molecules of oxygen (O_2) to form six molecules of carbon dioxide (CO_2), six molecules of water (H_2O), and 36 molecules of ATP. This can be represented by the following chemical equation:

$$6C_6H_{12}O_6 + 6\ O_2 \rightarrow\ 6\ CO_2 + 6\ H_2O + 36\ ATP$$

The process is much more complicated than this equation makes it appear: some two dozen separate chemical reactions are involved in the overall conversion of glucose to carbon dioxide, water, and ATP.

THE MECHANICS OF BREATHING

All animals have some mechanism for removing oxygen from the air and transmitting it into the bloodstream, and this same mechanism typically is used to expel carbon dioxide from the bloodstream into the surrounding environment. Types of animal respiration, in order of complexity, include direct diffusion, diffusion into blood, tracheal respiration, respiration with gills, and finally, respiration through lungs. Microbes, fungi, and plants all obtain the oxygen they use for cellular respiration directly from the environment, meaning that there are no intermediate organs or bodily chemicals, such as lungs or blood. More complex organisms, such as sponges, jellyfish, and terrestrial (land) flatworms, all of which have blood, also breathe through direct diffusion. The latter term describes an exchange of oxygen and carbon dioxide directly between an organism, or its bloodstream, and the surrounding environment.

More complex is the method of diffusion into blood whereby oxygen passes through a moist layer of cells on the body surface and then

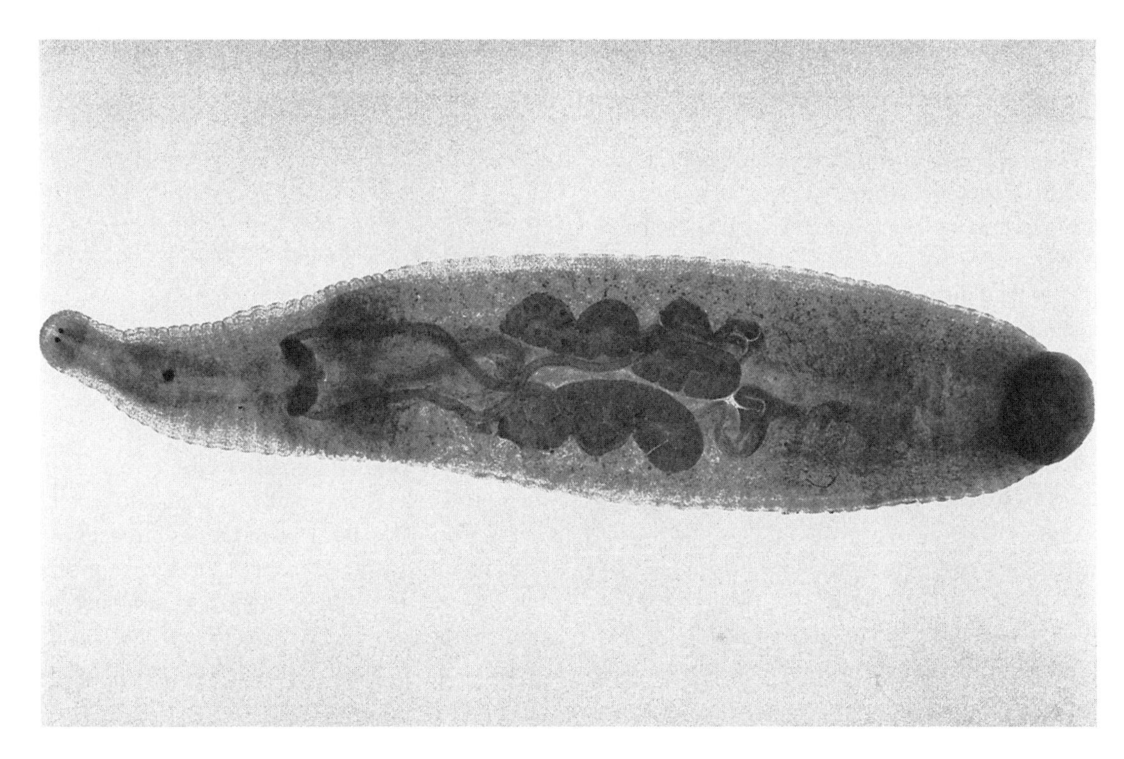

ONE FORM OF RESPIRATION IS DIFFUSION INTO BLOOD, WHEREBY OXYGEN PASSES THROUGH A MOIST LAYER OF CELLS ON THE BODY SURFACE, THEN THROUGH CAPILLARY WALLS AND INTO THE BLOODSTREAM, WHERE IT MOVES ON TO TISSUES AND CELLS. AMONG THE ORGANISMS THAT RELY ON DIFFUSION INTO BLOOD ARE ANNELIDS, A GROUP THAT INCLUDES LEECHES. *(© Lester V. Bergman/Corbis. Reproduced by permission.)*

through capillary walls (capillaries are small blood vessels that form a network throughout the body) and into the bloodstream. Once oxygen is in the blood, it moves throughout the body to different tissues and cells. Among the organisms that rely on diffusion into blood are annelids, a group that includes earthworms, various marine worms, and leeches.

In tracheal respiration air moves through openings in the body surface called spiracles. It then passes into special breathing tubes called tracheae that extend into the body. The tracheae divide into many small branches that are in contact with muscles and organs. In small insects, air simply moves into the tracheae, while in large insects, body movements assist tracheal air movement. Insects and terrestrial arthropods (land-based organisms with external skeletons) use this method of respiration.

Much more complicated than tracheae, gills are specialized tissues with many infoldings. Each gill is covered by a thin layer of cells and filled with blood capillaries. These capillaries take up oxygen dissolved in water and expel carbon dioxide dissolved in blood. Fish and other aquatic animals use gills, as did the early ancestors of humans and other higher animals. A remnant of this chapter from humans' evolutionary history can be seen in the way that an embryo breathes in its mother's womb, not by drawing in oxygen through its lungs but through gill-like mechanisms that disappear as the embryo develops.

LUNGS. Lungs are composed of many small chambers or air sacs surrounded by blood capillaries. Thus, they work with the circulatory system, which transports oxygen from inhaled air to all tissues of the body and also transports carbon dioxide from body cells to the lungs to be exhaled. After air enters the lungs, oxygen moves into the bloodstream through the walls of these capillaries. It then passes from the lung capillaries to the different muscles and organs of the body.

Although they are common to amphibians, reptiles, birds, and mammals, lungs differ enormously throughout the animal kingdom. Frogs, for instance, have balloon-like lungs that do not have a very large surface area. By contrast, if the entire surface of an adult male human's lungs were spread flat, it would cover about 750 sq. ft. (70 m^2), approximately the size of a handball court. The reason is that humans have about 300 million gas-filled alveoli, tiny protrusions inside the lungs that greatly expand the surface area for gas exchange.

Birds have specialized lungs that use a mechanism called crosscurrent exchange, which allows air to flow in one direction only, making for more efficient oxygen exchange. They have some eight thin-walled air sacs attached to their lungs, and when they inhale, air passes through a tube called the bronchus and enters posterior air sacs—that is, sacs located toward the rear. At the same time, air in the lungs moves forward to anterior air sacs, or ones located near the bird's front. When the bird exhales, air from the rear air sacs moves to the outside environment, while air from the front moves into the lungs. This efficient system moves air forward through the lungs when the bird inhales and exhales and makes it possible for birds to fly at high altitudes, where the air has a low oxygen content.

Humans and other mammals have lungs in which air moves in and out through the same pathway. This is true even of dolphins and whales, though they differ from humans in that they do not take in nutrition through the same opening. In fact, terrestrial mammals, such as the human, horse, or dog, are some of the only creatures that possess two large respiratory openings: one purely for breathing and smelling and the other for the intake of nutrients as well as air (i.e., oxygen in and carbon dioxide out).

REAL-LIFE APPLICATIONS

Anaerobic Respiration

Activity that involves oxygen is called aerobic; hence the term aerobic exercise, which refers to running, calisthenics, biking, or any other form of activity that increases the heart rate and breathing. Activity that does not involve oxygen intake is called anaerobic. Weightlifting, for instance, will increase the heart rate and rate of breathing if it is done intensely, but that is not its purpose and it does not depend on the intake and outflow of breath. For that reason, it is called an anaerobic exercise—though, obviously, a person has to keep breathing while doing it.

In fact, a person cannot consciously stop breathing for a prolonged period, and for this reason, people cannot kill themselves simply by holding their breath. A buildup of carbon dioxide and hydrogen ions (electrically charged atoms) in the bloodstream stimulates the breathing centers to become active, no matter what we try to do. On the other hand, if a person were underwater, the lungs would draw in water instead of air, and though water contains air, the drowning person would suffocate.

ANAEROBIC BACTERIA. Some creatures, however, do not need to breathe air but instead survive by anaerobic respiration. This is true primarily of some forms of bacteria, and indeed scientists believe that the first organisms to appear on Earth's surface were anaerobic. Those organisms arose when Earth's atmosphere contained very little oxygen, and as the composition of the atmosphere began to incorporate more oxygen over the course of many millions of years, new organisms evolved that were adapted to that condition.

The essay on paleontology discusses Earth's early history, including the existence of anaerobic life before the formation of oxygen in the atmosphere. The appearance of oxygen is a result of plant life, which produces it as a by-product of the conversion of carbon dioxide that takes place in photosynthesis. Plants, therefore, are technically anaerobic life-forms, though that term usually refers to types of bacteria that neither inhale nor exhale oxygen. Anaerobic bacteria still exist on Earth and serve humans in many ways. Some play a part in the production of foods, as in the process of fermentation. Other anaerobic bacteria have a role in the treatment of sewage. Living in an environment that would kill most creatures—and not just because of the lack of oxygen—they consume waste materials, breaking them down chemically into simpler compounds.

HUMANS AND ANAEROBIC RESPIRATION. Even in creatures, such as humans, that depend on aerobic respiration, anaerobic respiration can take place. Most cells are able to switch from aerobic to anaerobic respiration when necessary, but they generally are not able to continue producing energy by this process for very long. For example, a person who exercises vigorously may be burning up glucose faster than oxygen is being pumped to the cells, meaning that cellular respiration cannot take place quickly enough to supply all the energy the body needs. In that case, cells switch over to

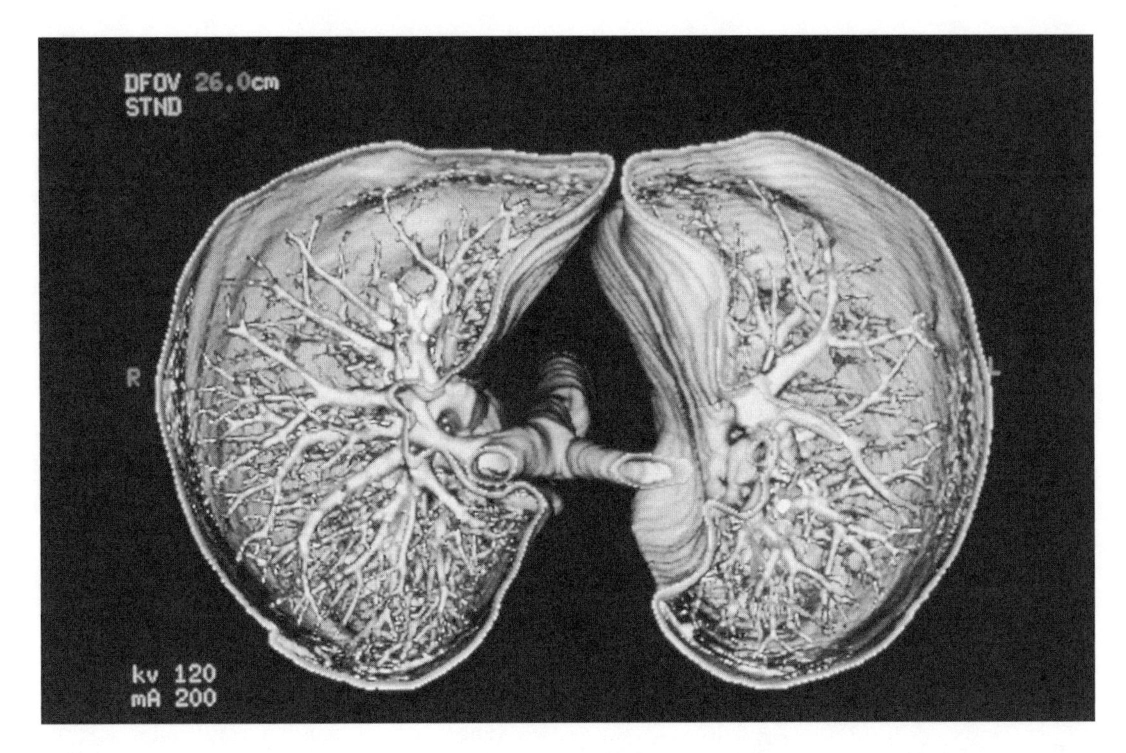

COMPUTERIZED TOMOGRAPHIC VIEW OF THE LUNGS, SHOWING THE TRACHEA (CENTER) SPLITTING TO FORM THE TWO MAIN BRONCHI, WHICH LEAD TO THE LUNGS. INSIDE THE LUNGS, MANY BRANCHING BRONCHI TERMINATE IN ALVEOLI, AIR SACS WHERE GAS EXCHANGE TAKES PLACE. *(© BSIP/Gems Europe/Photo Researchers. Reproduced by permission.)*

anaerobic respiration, which results in the production of lactic acid, or $C_3H_6O_3$. One advantage of anaerobic respiration is that it can take place very quickly and in short bursts, as opposed to aerobic respiration, which is designed for slower and steadier use of muscles. The disadvantage is that anaerobic respiration produces lactic acid, which, when it builds up in muscles that are overworked, causes soreness and may even lead to cramps.

LACTIC ACID IN THE BODY. Eventually, the buildup of lactic acid is carried away in the bloodstream, and the lactic acid is converted to carbon dioxide and water vapor, both of which are exhaled. But if lactic acid levels in the bloodstream rise faster than the body can neutralize them, a state known as lactic acidosis may ensue. Lactic acidosis rarely happens in healthy people and, more often than not, is a result of the body's inability to obtain sufficient oxygen, as occurs in heart attacks or carbon monoxide or cyanide poisoning or in the context of diseases such as diabetes.

The ability of the body to metabolize lactic acid is diminished significantly by alcohol, which impairs the liver's ability to carry out normal metabolic reactions. For this reason, alcoholics often have sore muscles from lactic acid buildup, even though they may not exercise. Lactic acid also can lead to a buildup of uric acid crystals in the joints, in turn causing gout, a very painful disease.

LACTIC ACID IN FOOD AND INDUSTRY. Lactic acid is certainly not without its uses, and it is found throughout nature. When lactose, or milk sugar, is fermented by the action of certain bacteria, it causes milk to sour. The same process is used in the manufacture of yogurt, but the reaction is controlled carefully to ensure the production of a consumable product. Lactic acid also is applied by the dairy industry in making cheese. Molasses contains lactic acid, a product of the digestion of sugars by various species of bacteria, and lactic acid also is used in making pickles and sauerkraut, foods for which a sour taste is desired.

A compound made from lactic acid is used as a food preservative, but the applications of lactic acid extend far beyond food production. Lactic acid is important as a starting material for making drugs in the pharmaceutical industry. Additionally, it is involved in the manufacturing of lacquers and inks; is used as a humectant, or moisturizer, in some cosmetics; is applied as a mordant, or a chemical that helps fabrics accept dyes, to textiles; and is employed in tanning leather.

RESPIRATORY DISORDERS

In almost any bodily system, there are bound to be disorders, or at least the chance that disorders may occur. This is particularly the case with something as complex as the respiratory system, because the more complex the system, the more things that can go wrong. Among the respiratory disorders that affect humans is a whole range of ailments from the common cold to emphysema, and from the flu to cystic fibrosis.

THE COMMON COLD. Colds are among the most common conditions that affect the respiratory system, though what we call the common cold is actually an invasion by one of some 200 different types of virus. Thus, it is really not one ailment but 200, though these are virtually identical, but the large number of viral causative agents has made curing the cold an insurmountable task.

When you get a cold, viruses establish themselves on the mucus membrane that coats the respiratory passages that bring air to your lungs. If your immune system is unsuccessful in warding off this viral infection, the nasal passages become inflamed, swollen, and congested, making it difficult to breathe.

Coughing is a reflex action whereby the body attempts to expel infected mucus or phlegm. It is essential to removing infected secretions from the body, but of course it plays no role in actually bringing a cold to an end. Nor do antibiotics, which are effective against bacteria but not viruses (see Infection). Only when the body builds up its own defense to the cold—assuming the sufferer has a normally functioning immune system—is the infection driven away.

INFLUENZA AND ALLERGIES. Influenza, a group of viral infections that can include swine flu, Asian flu, Hong Kong flu, and Victoria flu, is often far more serious than the common cold. A disease of the lungs, it is highly contagious, and can bring about fever, chills, weakness, and aches. In addition, influenza can be fatal: a flu epidemic in the aftermath of World War I, spread to far corners of the globe by returning soldiers, killed an estimated 20 million people.

Respiratory ailments often take the form of allergies such as hay fever, symptoms of which include sneezing, runny nose, swollen nasal tissue, headaches, blocked sinuses, fever, and watery, irritated eyes. Hay fever is usually aggravated by the presence of pollen or ragweed in the air, as is common in the springtime. Other allergy-related respiratory conditions may be aggravated by dust in the air, and particularly by the feces of dust mites that live on dust particles.

BRONCHIAL AILMENTS. Allergic reactions can be treated by antihistamines (see The Immune System for more about allergies), but simple treatments are not available for such complex respiratory disorders as asthma, chronic bronchitis, and emphysema. All three are characterized by an involuntary constriction in the walls of the bronchial tubes (the two divisions of the trachea or windpipe that lead to the right and left lungs), which causes the tubes to close in such a way that it becomes difficult to breathe.

Emphysema can be brought on by cigarette smoking, and indeed some heavy smokers die from that ailment rather than from lung cancer. On the other hand, a person can contract a bronchial illness without engaging in smoking or any other activity for which the sufferer could ultimately be blamed. Indeed, small children may have asthma. One treatment for such disorders is the use of a bronchodilator, a medicine used to relax the muscles of the bronchial tubes. This may be administered as a mist through an inhaler, or given orally like other medicine.

TUBERCULOSIS AND PNEUMONIA. More severe is tuberculosis, an infectious disease of the lungs caused by bacteria. Tuberculosis attacks the lungs, leading to a chronic infection with such symptoms as fatigue, loss of weight, night fevers and chills, and persistent coughing that brings up blood. Without treatment, it is likely to be fatal. Indeed, it was a significant cause of death until the introduction of antibiotics in the 1940s, and it has remained a problem in underdeveloped nations. Additionally, thanks to mutation in the bacteria themselves, strains of the disease are emerging that are highly resistant to antibiotics.

Another life-threatening respiratory disease is pneumonia, an infection or inflammation of the lungs caused by bacteria, viruses, mycoplasma (microorganisms that show similarities to

KEY TERMS

AEROBIC: Oxygen-breathing.

ANAEROBIC: Non-oxygen-breathing.

ATP: Adenosine triphosphate, an energy carrier formed when a simpler compound, adenosine diphosphate (ADP), combines with a phosphate group.

CAPILLARY: A very small blood vessel. Capillaries form networks throughout the body.

CARBOHYDRATES: Naturally occurring compounds, consisting of carbon, hydrogen, and oxygen, whose primary function in the body is to supply energy. Included in the carbohydrate group are sugars, starches, cellulose, and various other substances. Most carbohydrates are produced by green plants in the process of undergoing photosynthesis.

CELLULAR RESPIRATION: A process that, when it takes place in the presence of oxygen, involves the intake of organic substances, which are broken down into carbon dioxide and water, with the release of considerable energy.

CIRCULATORY SYSTEM: The parts of the body that work together to move blood and lymph. They include the heart, blood vessels, blood, and the lymphatic glands, such as the lymph nodes.

COMPOUND: A substance in which atoms of more than one element are bonded chemically to one another.

FERMENTATION: A process, involving enzymes, in which a compound rich in energy is broken down into simpler substances.

GLUCOSE: A monosaccharide (sugar) that occurs widely in nature and which is the form in which animals usually receive carbohydrates. Also known as *dextrose, grape sugar,* and *corn sugar.*

HEMOGLOBIN: An iron-containing protein in human red blood cells that is responsible for transporting oxygen to the tissues and removing carbon dioxide from them. Hemoglobin is known for its deep red color.

LYMPH: The portion of the blood that includes white blood cells and plasma but not red blood cells.

LYMPH NODES: Masses of tissue, at certain places in the body, that act as filters for blood.

METABOLISM: The chemical process by which nutrients are broken down and converted into energy or are used in the construction of new tissue or other material in the body. All metabolic reactions are either catabolic or anabolic.

MONOSACCHARIDE: The simplest type of carbohydrate. Monosaccharides, which cannot be broken down chemically into simpler carbohydrates, also are known as *simple sugars.*

NUTRIENT: Materials that supply energy or the materials to form new tissue for organisms. They include proteins, carbohydrates, lipids (fats), vitamins, and minerals.

PHOSPHATE GROUP: A group (that is, a combination of atoms from two or more elements that tend to bond with other elements or compounds in certain characteristic ways) involving a phosphate, or a chemical compound that contains oxygen bonded to phosphorus.

KEY TERMS CONTINUED

PHOTOSYNTHESIS: The biological conversion of light energy (that is, electromagnetic energy) from the Sun to chemical energy in plants. In this process carbon dioxide and water are converted to carbohydrates and oxygen.

RESPIRATION: A term that can refer either to cellular respiration (see definition) or, more commonly, to the process by which an organism takes in oxygen and releases carbon dioxide. Sometimes a distinction is made between external respiration, or an exchange of gases with the external environment, and internal respiration, an exchange of gases between the body's cells and the blood.

SIMPLE SUGAR: A monosaccharide, or simple carbohydrate.

TISSUE: A group of cells, along with the substances that join them, that forms part of the structural materials in plants or animals.

both viruses and bacteria), and fungi, as well as such inorganic agents as inhaled dust or gases. Symptoms include pleurisy (chest pain), high fever, chills, severe coughing that brings up small amounts of mucus, sweating, blood in the sputum (saliva and mucus expelled from the lungs), and labored breathing.

In 1936, pneumonia was the principal cause of death in the United States. Since then, it has been controlled by antibiotics, but as with tuberculosis, resistant strains of bacteria have developed, and therefore the number of cases has increased. Today, pneumonia and influenza combined are among the most significant causes of death in the United States (see Diseases).

LUNG CANCER AND CYSTIC FIBROSIS. Respiratory ailments may also take the form of lung cancer, which may or may not be a result of smoking. Cigarette smoking and air pollution are considered to among the most significant causes of lung cancer, yet people have been known to die of the disease without being smokers or having been exposed to significant pollution.

One particularly serious variety of respiratory illness is cystic fibrosis, a genetic disorder that causes a thick mucus to build up in the respiratory system and in the pancreas, a digestive organ. (For more about genetic disorders, see Heredity; for more on role of the pancreas, see Digestion.) In the United States, the disease affects about one in every 3,900 babies born annually. No cure for cystic fibrosis exists, and the disease is invariably fatal, with only about 50% of sufferers surviving into their thirties.

Lung complications are the leading cause of death from cystic fibrosis, and most symptoms of the disease are related to the sticky mucus that clogs the lungs and pancreas. People with cystic fibrosis have trouble breathing, and are highly susceptible to bacterial infections of the lungs. Coughing, while it may be irritating and painful if you have a cold, is necessary for the expulsion of infected mucus, but mucus in the lungs of a cystic fibrosis is too thick to be moved. This makes it easy for bacteria to inhabit the lungs and cause infection.

WHERE TO LEARN MORE

Bryan, Jenny. *Breathing: The Respiratory System.* New York: Dillon Press, 1993.

Cellular Metabolism and Fermentation. Estrella Mountain Community College (Web site). <http://gened.emc.maricopa.edu/bio/bio181/BIOBK/BioBookGlyc.html>.

Kimball, Jim. "The Human Respiratory System." *Kimball's Biology Pages* (Web site). <http://www.ultranet.com/~jkimball/BiologyPages/P/Pulmonary.html>.

Levesque, Mireille, Letitia Fralick, and Joni McDowell. "Respiration in Water: An Overview of Gills." University of New Brunswick (Web site). <http://www.unb.ca/courses/biol4775/SPAGES/SPAGE13.HTM>.

Llamas, Andreu. *Respiration and Circulation.* Milwaukee: Gareth Stevens, 1998.

Paustian, Timothy. *Anaerobic Respiration.* Department of Bacteriology, University of Wisconsin–Madison (Web site). <http://www.bact.wisc.edu/microtextbook/Metabolism/RespAnaer.html>.

Roca, Núria, and Marta Serrano. *The Respiratory System, the Breath of Life.* Illus. Antonio Tenllado. New York: Chelsea House Publishers, 1995.

Silverstein, Alvin, and Virginia B. Silverstein. *The Respiratory System.* New York: Twenty-First Century Books, 1994.

SCIENCE OF EVERYDAY THINGS

REAL-LIFE BIOLOGY

NUTRITION

FOOD WEBS

NUTRIENTS AND NUTRITION

VITAMINS

FOOD WEBS

CONCEPT

The idea of a food chain is common in everyday life, so much so that it has become a metaphor applied in many situations. A high achiever in business or other endeavors is said to be "at the top of the food chain," and images of big fish eating little fish abound in cartoons. Yet in the study of the biological sciences, the concept of food chains is part of the much larger idea of a food web. Whereas a food chain is a linear series of organisms dependent on each other for food, a food web is an interconnected set of food chains in the same ecosystem. Food webs make possible the transfer of energy from plants through herbivores to carnivores and omnivores, and ultimately to the detritivores and decomposers that enrich the soil with organic waste. Just as a food web can transfer materials essential to the life of organisms, it is also a devastatingly efficient conduit for the transfer of poisons.

HOW IT WORKS

FOOD WEBS IN CONTEXT

An ecosystem is a community of independent organisms along with the inorganic components (chiefly soil, water, air, and rocks) that make up their environment. A biome is a large ecosystem, characterized by its dominant life-forms—for example, the Amazonian rain forest.

That portion of an ecosystem composed only of living things, as opposed to the formerly living or never living components, is known as a biological community. This community includes creatures from all five kingdoms of living organisms (including bacteria, algae, and fungi), whereas the term biota typically refers only to plant and animal life within a biological community or ecosystem. (For more on these subjects, see Ecosystems and Ecology and Biological Communities.)

TROPHIC LEVELS

The organisms in a biological community are linked in their need to obtain energy from food, which derives from the Sun through plant life. (There are, however, some communities, in areas such as deep-ocean rifts, that are not dependent on sunlight at all.) The Sun's energy is electromagnetic and travels in the form of radiation, which Earth receives as light and heat. Plants, known as primary producers, convert this electromagnetic energy into chemical energy through the process of photosynthesis.

The plants are eaten by herbivores (plant-eating animals), known also as primary consumers, examples of which include squirrels, rabbits, mice, deer, cows, horses, sheep, and seed-eating birds. These creatures, in turn, are eaten by secondary consumers, which are either carnivores, which are creatures that eat only meat, or omnivores—creatures, such as humans, that eat meat and plants.

There may even be tertiary, or third-level, consumers. These are animals that eat secondary consumers; examples are mountain lions and hawks, both of which eat such second-order consumers as snakes and owls. Human societies that eat dogs or cats, as well as those that engage in cannibalism, also behave as tertiary consumers. (See Biological Communities for a biological explanation of what otherwise is considered an abhorrent and immoral practice—not to mention a dangerous one, due to the risk of such diseases as kuru, a type of spongiform encephalopathy.) In any case, the further along in the chain of trophic levels or stages of the food web, the fewer consumers there are.

ENERGY AND TROPHIC LEVELS. It is fairly obvious that when a creature is "higher on the food chain" (to use the common expression), it has fewer natural predators. The reason for this is that at each successive trophic level, there are simply fewer organisms; this, in turn, is due to the fact that the energy available to each level is progressively smaller, and the organisms themselves progressively larger. *This,* in turn, stems from one of the most intriguing, maddening concepts in the entire universe: the second law of thermodynamics, which we discuss shortly.

Because of the diminishing number of organisms at each trophic level, the food web often is depicted as a pyramid, a concept we explore further later in this essay. The number of organisms begins to increase again at the next trophic level beyond secondary or tertiary consumers, that of decomposers. Large omnivores and carnivores may not be prey for other creatures in life, but everything dies eventually, and anything that has ever lived is food for detritivores, or organisms that feed on waste matter.

DETRITIVORES AND DECOMPOSERS. Detritivores, which range in size and complexity from maggots to vultures, may not be the most appealing creatures on Earth, but without them life itself would suffer. By consuming the remains of formerly living things, they break organic material down into inorganic substances. In other words, their internal systems chemically process compounds containing the element carbon in characteristic structures. They then release that carbon into the atmosphere and soil in such a way that what remains is inorganic material that enriches the soil for the growth of new plant life.

But detritivores are not the last stop on the food web. The final trophic level, before the cycle comes back around to plants, contains the largest number of organisms in the entire food web—perhaps billions and billions, even in a space smaller than a coffee cup. These are decomposers, or organisms that, as with detritivores, obtain their energy from the chemical breakdown of dead organisms. The decomposers, however, break down the nutrients in decayed organic matter to a far greater extent than do detritivores.

Typically, decomposers are microorganisms, including bacteria and fungi, and they process materials in such a way that complex compounds undergo the chemical reaction of decomposition. Through decomposition, compounds are broken down into simpler forms, or even into their constituent elements, which provide the environment with nutrients necessary to the growth of more plant life.

CATEGORIZING THE TROPHIC LEVELS. The organisms in the food web can be viewed in three groups: producers (plants), consumers (primary- and secondary-consuming animals, whether herbivores, carnivores, or omnivores), and decomposers (that is, both detritivores and true decomposers). Producers and consumers are part of a larger structure known as the grazing food web, in which food is "on its way up the food chain," as it were. Decomposers and detritivores make up the decomposer food web, which brings food back "down" to the soil.

Producers also are called *autotrophs,* from Greek roots meaning "self-feeders," because they are not dependent on other organisms as a source of energy. Beyond the level of the primary producers, all consumers are known as *heterotrophs,* or "other-feeders." These creatures feed on other organisms to obtain their energy and are classified according to the types of food they eat—herbivores, carnivores, omnivores, as we already have discussed. Detritivores and decomposers also are considered heterotrophs.

ORGANISMS AND ENERGY

Rather than depending on other organisms for energy, autotrophs obtain energy from the Sun and carbon dioxide from the atmosphere. From these components, they build the large organic molecules that they need to survive. Green plants do this through the process of photosynthesis, a chemical reaction that can be represented as follows:

solar energy + carbon dioxide (CO_2) + water (H_2O) $\rightarrow$ glucose (sugar: $C_6H_{12}O_6$) + oxygen (O_2)

Actually, in order to produce what chemists call a *balanced equation,* it would be necessary to show this equation as a reaction between the energy and six molecules each of carbon dioxide and water, which would produce a glucose molecule and six oxygen molecules. In any case, what we have described here is an amazing thing and one of the great wonders of nature. Sunlight aids plants in converting carbon dioxide, which they receive from the respiration of animals, along with water (which also may come from animal respiration, though this is not necessarily the case), into a sugar molecule for the plant's sustenance. Furthermore, oxygen, essential to the life of virtually all animals, also is produced—yet from the standpoint of the plant, it is simply a waste by-product!

THE SECOND LAW OF THERMODYNAMICS. The productivity of plants, which is measured in terms of biomass (the combined mass of all organisms at a particular trophic level in a food web), determines the amount of "fixed," or usable, energy available to other trophic levels on the food web. The amount of energy available always will be less for each successive trophic level, through the point where consumers end and decomposers begin—that is, through the level of the secondary or perhaps tertiary consumer.

If there is any scientific equivalent of the curse in the Garden of Eden (the punishment for the sins of Adam and Eve, according to Judeo-Christian belief), it is the second law of thermodynamics. Just as the expulsion from Paradise in the biblical story ensured that life would be much more difficult for humans than it would have been in Eden, so the second law thwarts all ambitions toward transcending the limits of physical reality.

The first law of thermodynamics states that it is impossible to obtain more energy from a system than is put into it. Thus, for instance, a car will go only as far as is allowed by the amount of energy that is pumped into its tank. The first law, discovered in the mid–nineteenth century, effectively ruled out any hopes of a perpetual-motion machine, but the second law, derived a few decades later, delivered even worse news.

Though it can be stated in a number of ways, the second law essentially means that it is impossible to extract as much energy from a system as one puts into it. Thus, in the case of an automobile, most of the energy contained in the gas does not go toward moving the car; rather, it is dissipated in the form of heat and sound, as a natural by-product of operating the engine. Even without running an air conditioner or other energy-consuming device, only about 30% of the energy from the gas goes to turning the wheels.

THE ECOLOGICAL PYRAMID. What this means for the food web is that there is bound to be a loss of energy in the transfer from one trophic level to another. Organisms never manage to retrieve 100% of the energy from the materials they eat; in fact, the figure is more like 10%. A rabbit that eats a carrot gets only about 10% of the energy in it, and an owl that eats the rabbit gets only about 10% of the energy from the rabbit, or 1% of the energy in the carrot. Because of these diminishing returns, there are always fewer organisms at each successive trophic level on the grazing food web. This fact is expressed in a model known as the *ecological pyramid*, or *energy pyramid*, which shows that as the amount of total energy decreases with each trophic level, so does the biomass. As a result, it may take 1,000 carrots to support 100 rabbits, 10 owls, and one hawk.

The picture changes as the shift is made from the grazing web to the decomposer web. Detritivores and decomposers are extraordinarily efficient feeders, reworking detritus over and over and extracting more fixed energy as they do. Eventually, they break the waste down into simple inorganic chemicals, which, as we have noted, then may be reused by the primary producers. The number of organisms in the decomposing food web dwarfs that of all others combined, though decomposers themselves are very small, and their combined population takes up very little physical space.

REAL-LIFE APPLICATIONS

KEYSTONE SPECIES

The keystone in an archway is a wedge-shaped stone at the top of the arch. It's position is extraordinarily important: if the keystone is removed, the arch will collapse. Thus, the keystone has become an often-used metaphor in other circumstances, as, for instance, in the nickname of Pennsylvania, the "Keystone State." In the realm of ecology, the term *keystone species* refers to those organisms that, like a keystone in

THE TERM KEYSTONE SPECIES REFERS TO AN ORGANISM THAT PLAYS A CRITICAL ROLE IN ITS ENVIRONMENT, ONE THAT MAY BECOME APPARENT ONLY ONCE IT IS REMOVED FROM AN ECOSYSTEM. ON THE WEST COAST OF NORTH AMERICA, FOR INSTANCE, REMOVAL OF A CERTAIN SPECIES OF STARFISH CAUSED A RAPID GROWTH IN THE NUMBERS AND BIOMASS OF THE MUSSEL UPON WHICH THE STARFISH FED. *(© Stuart Westmorland/Corbis. Reproduced by permission.)*

architecture or engineering, play a critical role in their environments.

Within a food web, for instance, a keystone species can have a powerful influence, one that is far out of proportion to its relative biomass or productivity. Typically, a keystone species is a top predator (that is, a large secondary or even tertiary consumer), though occasionally an herbivore can occupy the keystone position. Often, the role of the keystone species becomes apparent only once it is removed, either experimentally or by natural forces, from an ecosystem.

STARFISH AND MUSSELS. In temperate ecosystems on the west coast of North America, for instance, removal of a certain species of starfish (*Pisaster ochraceous*) was found to cause a rapid growth in the numbers and biomass of the mussel *Mytilus californianus.* The latter then forced out other species and proceeded to dominate the biological community. As it turned out, the starfish acted as a keystone predator by consuming these mussels.

Specifically, the starfish prevented the mussel from gaining dominance that it otherwise would have gained, owing to its competitive superiority in relation to other species within this particular coastal ecosystem. Yet the starfish could not eliminate the mussel, because it was incapable of feeding on larger individuals of that species. The result was that the community enjoyed a much greater degree of diversity and complexity than it would have if the mussel had been allowed to dominate.

SEA OTTERS AND KELP FORESTS. Another keystone species in a geographic area close to that of the starfish we have just described is the sea otter, native to western North America. Its principal food source is the sea urchin, an herbivore that, in turn, survives by consuming kelp, a large form of algae. By controlling the numbers and densities of sea urchins, sea otters allowed kelp to retain a relatively large biomass within the community, thus facilitating the growth of "kelp forests."

When humans began hunting sea otters for their fur during the late eighteenth and into the nineteenth centuries, however, the ecological effect soon was felt in the form of declining kelp forests. Fortunately, hunting did not render the species extinct, and since the 1930s, sea otters have been colonizing many of their former habitats. This colonization has resulted in a corresponding increase in the density of surrounding kelp forests.

INDICATOR SPECIES

A concept similar to that of the keystone species is the idea of an indicator species: plants or animals that, by their presence, abundance, or chemical composition, demonstrate a distinctive aspect of the character or quality of the environment. For instance, in an ecosystem affected by pollution, examination of indicator plant species may reveal the pollution patterns. By their presence, indicator species also may serve to show the quality or integrity of an ecosystem. Such is the case, for instance, with the spotted owl, or *Strix occidentalis,* and other species that depend on old-growth forests. (See Succession and Climax for more on this subject.) Because the needs of these species are so particular, their presence or absence can illustrate the health or lack thereof of the biome in question.

Other indicator plants also can be used to determine the presence of valuable mineral deposits in the soil, because those minerals make their way into the tissues of the plants themselves. Nickel concentrations as great as 10% have been found in the tissues of Russian plants from the mustard family, and a mintlike species called *Becium homblei* has proved useful for locating copper deposits in parts of Africa. Since the plant can tolerate more than 7% copper in soil (a great amount and many times the percentage of copper in the human body, for instance), it can and does live near enormous copper deposits.

INDICATING TOXINS. Some plants can serve as indicators of serpentine minerals, varieties of compounds that can be toxic in large concentrations. In California, for instance, where serpentine soils are not uncommon, there plant species unique to specific ecosystems high in serpentine mineral content. Elsewhere, there are types of lichens that are sensitive to toxic gases, such as sulfur dioxide, and thus these lichens can be monitored as a way of keeping tabs on air pollution.

In semiarid regions where soils contain large quantities of the element selenium, plants can accumulate such large concentrations of the element that they become poisonous to primary consumers (for example, rabbits) who eat them. The result may be temporary or even permanent blindness. In such situations, legumes of the genus *Astragalus,* which can accumulate as much as 15,000 ppm (parts per million)—a comparatively enormous concentration—serve as indicators. Their heavy selenium concentration gives them a noticeably unpleasant smell.

Aquatic invertebrates and fish often have been surveyed for what they can show as to the quality of water and the health of aquatic ecosys-

BEACHGOERS ARE SPRAYED WITH DDT IN THE FIRST PUBLIC TEST OF A NEW MACHINE FOR DISTRIBUTING THE INSECTICIDE IN 1945. ONCE PESTICIDES SUCH AS DDT HAVE BEEN SPRAYED IN A REGION, RAIN CAN WASH THEM INTO CREEKS, LAKES, AND OTHER BODIES OF WATER, WHERE THEY ARE ABSORBED BY CREATURES THAT EITHER DRINK OR SWIM IN THE WATER AND THUS ENTER THE FOOD WEB. *(© Bettmann/Corbis. Reproduced by permission.)*

tems. The presence of a creature known by the disgusting name sewage worm, or tubificid worm (*Tubificidae*), indicates the degradation of water quality by the introduction of sewage or other outside organic matter that consumes oxygen. Tubificid worms are virtually anaerobic, or non-oxygen-breathing, unlike the animals that would occupy a similar niche in a healthy aquatic environment.

BIOACCUMULATION AND BIOMAGNIFICATION

The subject of indicator species leads us, naturally enough, to the topic of pollution in the food web, which can be seen on a small scale in the phenomenon of bioaccumulation and which manifests on a much broader scale as biomagnification. One of the key concepts in ecological studies is the idea that a disturbance in one area can lead to serious consequences elsewhere. The interconnectedness of components in the environment thus makes it impossible for any event or phenomenon to be truly isolated.

Nowhere is this principle better illustrated that in the processes of bioaccumulation and biomagnification. The first of these is the buildup of toxins, and particularly chemical pollutants, in the tissues of individual organisms. Part of the danger in these toxins is the fact that the organism cannot easily process them either by metabolizing them (i.e., incorporating them into the metabolic system, as one does food or water) or by excreting them through urine or other substances produced by the body.

The only way for the organism to release toxins, in fact, is by passing them on to other members of the food web. Because organisms at each successive trophic level must consume more biomass to meet their energy requirements, they experience an increase in contamination, a phenomenon known as biomagnification. The following example illustrates how toxins enter the food web and gradually make their way down the line, growing in proportion as they do.

Particles of pollutant may stick to algae, for instance, which are so small that the toxin does little damage at this level of the food web. But even a small herbivore, such as a zooplankton, takes in larger quantities of the pollutant when it consumes the algae, and so begins the cycle of biomagnification. By the time the toxin has passed from a zooplankton to a small fish, the amount of pollutant in a single organism might be 100 times what it was at the level of the algae. The reason, again, is that the fish can consume 10 zooplankton that each has consumed 10 algae.

These particular numbers, of course, are used simply for the sake of convenience, as were those cited earlier in relation to the ecological pyramid. Note, incidentally, the similarity of the relationships between this "pyramid of poison" and the ecological pyramid, whereby energy, which is beneficial, passes between trophic levels. The higher the trophic level, the smaller the amounts of energy that the organism extracts from its food—but the higher the amount of toxic content. By the time the toxins have passed on to a few more levels in the food web, they might be appearing in concentrations as great as 10,000 times their original amount.

DDT BIOMAGNIFICATION. Among the most prominent examples of chemical pollutants that are bioaccumulated are such pesticides as DDT (dichlorodiphenyltrichloroethane). DDT is an insecticide of the hydrocarbon family, a large group of chemical compounds of which the many varieties of petroleum are examples. Because it is based in hydrocarbons, DDT is hydrophobic ("water-fearing") and instead mixes with oils—including the fat of organisms.

In the twenty or so years leading up to 1972, Americans used vast amounts of DDT for the purpose of controlling mosquitoes and other pests. DDT appeared to be a remarkably successful killer, and in fact it turned out to be a little too successful. As it found its way into water sources, DDT entered the bodies of fish, and then those of predatory birds such as osprey, peregrine falcons, brown pelicans, and even the bald eagle.

The detrimental effect of DDT on America's national symbol, a bird protected by law since 1940, aptly illustrates the ravages exacted by this powerful insecticide. DDT levels in birds became so high that the birds' eggshells were abnormally thin, and adult birds sitting on the nest would accidentally break the shells of unhatched chicks. As a result, baby birds died, and populations of these species dwindled. Environmentalists in the late 1960s and early 1970s raised public awareness of this phenomenon, and this led to the banning of DDT spraying in 1972. The period since that time has seen dramatic increases in bird populations.

THE HUMAN FACTOR. Because the species of bird affected were not ones that people normally consume for food, DDT biomagnification did not have a wide-ranging effect on human populations. However, tests showed that some DDT had made its way into the fat deposits of some members of the population. In any case, bioaccumulation and biomagnification have threatened humans, for instance in the late 1940s and early 1950s, when cows fed on grass that had been exposed to nuclear radiation, and this radioactive material found its way into milk.

Traces of the radioactive isotope strontium-90, a by-product of nuclear weapons testing in the atmosphere from the late 1940s onward, fell to earth in a fine powder that coated the grass. Later the cows ate the grass, and strontium-90 wound up in the milk they produced. Because of its similarities to calcium, the isotope became incorporated into the teeth and gums of children who drank the milk, posing health concerns that helped bring an end to atmospheric testing in the early 1960s.

Humans themselves can serve as repositories for contaminants, a particular concern for a mother nursing her baby. Assuming the mother's own system has been contaminated by toxins, it is likely that her milk contains traces of the harmful chemical, which will be passed on to her child. Obviously, this is a very serious matter. Nursing mothers, babies, and their loved ones are not the only people affected by bioaccumulation and the storing of toxins in fatty tissues. In fact, some physicians and nutritionists maintain that one of the reasons for the buildup of fat on a person's body (though certainly not the only reason) may be as a response to toxins, the idea being that the body produces fat cells as a means of keeping the toxins away from the bloodstream.

Another case of large-scale bioaccumulation occurred during the 1970s and 1980s, when fish such as tuna were found to contain bioaccumulated levels of mercury. In the face of such concerns, governments have intervened in several ways, including the banning of DDT spraying, as mentioned earlier. The U.S. federal government and the governments of some states have issued warnings against the consumption of certain types of fish, owing to bioaccumulated levels of toxic pollutants. Bioaccumulation is particularly serious in the case of species that live a long time,

KEY TERMS

AUTOTROPHS: Primary producers. *Autotroph* means "self-feeder," and these organisms are distinguished by the fact that they do not depend on other organisms as a source of energy. Instead, they obtain energy from the Sun and carbon dioxide from the atmosphere, and from these constituents they build the large organic molecules that they need to survive.

BIOACCUMULATION: The buildup of toxic chemical pollutants in the fatty tissues of organisms.

BIOGEOCHEMICAL CYCLES: The changes that particular elements undergo as they pass back and forth through the various earth systems (e.g., the biosphere) and particularly between living and nonliving matter. The elements involved in biogeochemical cycles are hydrogen, oxygen, carbon, nitrogen, phosphorus, and sulfur.

BIOLOGICAL COMMUNITY: The living components of an ecosystem.

BIOMAGNIFICATION: The increase in bioaccumulated contamination at higher levels of the food web. Biomagnification results from the fact that larger organisms consume larger quantities of food—and, hence, in the case of polluted materials, more toxins.

BIOMASS: The combined mass of all organisms at a particular trophic level in a food web.

BIOME: A large ecosystem, characterized by its dominant life-forms.

BIOSPHERE: A combination of all living things on Earth—plants, mammals, birds, amphibians, reptiles, aquatic life, insects, viruses, single-cell organisms, and so on—as well as all formerly living things that have not yet decomposed.

BIOTA: A combination of all flora and fauna (plant and animal life, respectively) in a region.

CARNIVORE: A meat-eating organism, or an organism that eats *only* meat (as distinguished from an omnivore).

DECOMPOSER FOOD WEB: That portion of the food web occupied by detritivores and decomposers. (Compare with *grazing food web.*)

DECOMPOSERS: Organisms that obtain their energy from the chemical breakdown of dead organisms as well as from animal and plant waste products. The principal forms of decomposer are bacteria and fungi.

DECOMPOSITION REACTION: A chemical reaction in which a compound is broken down into simpler compounds, or into its constituent elements. In the biosphere, this often is achieved through the help of detritivores and decomposers.

DETRITIVORES: Organisms that feed on waste matter, breaking organic material down into inorganic substances that then can become available to the biosphere in the form of nutrients for plants. Their function is similar to that of decomposers; however, unlike decomposers—which tend to be bacteria or fungi—detritivores are relatively complex organisms, such as earthworms or maggots.

ECOLOGY: The study of the relationships between organisms and their environments.

KEY TERMS CONTINUED

ECOSYSTEM: A community of interdependent organisms along with the inorganic components of their environment.

ENERGY TRANSFER: The flow of energy between organisms in a food web.

FIRST LAW OF THERMODYNAMICS: A law of physics stating that the amount of energy in a system remains constant, and therefore it is impossible to perform work that results in an energy output greater than the energy input.

FOOD CHAIN: A series of singular organisms in which each plant or animal depends on the organism that precedes it. Food chains rarely exist in nature; therefore, scientists prefer the term *food web.*

FOOD WEB: A term describing the interaction of plants, herbivores, carnivores, omnivores, decomposers, and detritivores in an ecosystem. Each of these organisms consumes nutrients and passes them along to other organisms (or, in the case of the decomposer food web, to the soil and environment). The food web may be thought of as a bundle or network of food chains, but since the latter rarely exist separately, scientists prefer the concept of a food web to that of a food chain.

GRAZING FOOD WEB: That portion of the food web occupied by autotrophs, herbivores, carnivores, and omnivores. (Compare with *decomposer food web.*)

HERBIVORE: A plant-eating organism.

HETEROTROPHS: Secondary consumers, or "other-feeders." These creatures feed on other organisms to obtain their energy and are classified according to the types of food they eat. Thus, they are known as herbivores, carnivores, and so on.

INDICATOR SPECIES: A plant or animal that, by its presence, abundance, or chemical composition, demonstrates a particular aspect of the character or quality of the environment.

KEYSTONE SPECIES: A species that plays a crucial role in the functioning of its ecosystem or that has a disproportionate influence on the structure of its ecosystem.

NICHE: A term referring to the role that a particular organism plays within its biological community.

OMNIVORE: An organism that eats both plants and other animals.

ORGANIC: At one time chemists used the term *organic* only in reference to living things. Now the word is applied to most compounds containing carbon, with the exception of carbonates (which are minerals) and oxides, such as carbon dioxide.

PHOTOSYNTHESIS: The biological conversion of light energy (that is, electromagnetic energy) from the Sun to chemical energy in plants.

PRIMARY CONSUMERS: Animals that eat green plants. (Compare with *secondary consumers.*)

PRIMARY PRODUCERS: Green plants that depend on photosynthesis for their nourishment.

SECONDARY CONSUMERS: Animals that eat other animals.

SECOND LAW OF THERMODYNAMICS: A law of physics, which can be stat-

KEY TERMS CONTINUED

ed in several ways, all of which mean the same thing. According to one version of the second law, it is impossible to transfer energy with perfect efficiency, because some energy always will be lost in the transfer. This is the same as saying that it is impossible to extract from a system the same amount of energy that was put into it.

TROPHIC LEVELS: Various stages within a food web. For instance, plants are on one trophic level, herbivores on another, and so on.

because a longer life allows for much longer periods of bioaccumulation. For this reason, some governments warn against consuming fish over a certain age or size: the older and larger the creature, the more contaminated it is likely to be.

WHERE TO LEARN MORE

A to Z of Food Chains and Webs (Web site). <http://www.education.leeds.ac.uk/~edu/technology/epb97/forest/azfoodcw.htm>.

Bioaccumulation and Biomagnification (Web site). <http://www.marietta.edu/~biol/102/2bioma95.html>.

Busch, Phyllis S. *Dining on a Sunbeam: Food Chains and Food Webs.* Photos Les Line. New York: Four Winds Press, 1973.

Extension Toxicology Network (EXTOXNET): Toxicology Information Briefs (Web site). <http://ace.orst.edu/info/extoxnet/tibs/bioaccum.htm>.

Food Chains and Food Webs: An Introduction (Web site). <http://www.si.edu/sites/educate/troprain/foodchai.htm>.

Food Webs: Build Your Own (Web site). <http://www.gould.edu.au/foodwebs/kids_web.htm>.

Fox, Nicols. *Spoiled: The Dangerous Truth About a Food Chain Gone Haywire.* New York: Basic Books, 1997.

Introduction to Biogeography and Ecology: Trophic Pyramids and Food Webs. Fundamentals of Physical Geography (Web site). <www.geog.ouc.bc.ca/physgeog/contents/9o.html>.

Pimm, Stuart L. *Food Webs.* New York: Chapman and Hall, 1982.

Wallace, Holly. *Food Chains and Webs.* Chicago: Heinemann Library, 2001.

NUTRIENTS AND NUTRITION

CONCEPT

In the modern world people are accustomed to hearing a great deal about nutrients and nutrition. Words such as protein, carbohydrate, vitamins, minerals, and fats are a regular part of daily life, yet few people who talk about these nutrients really know what they are. In fact, these are the basic building blocks of nutrition, whereby animal life is sustained. Whereas plants can get their energy directly from the Sun and the atmosphere, animals (including humans) depend on other organisms to provide them with nutrition. These other organisms include plants, which generate carbohydrates as a result of photosynthesis, as well as other animals that eat plants and thereby build proteins and fats. Plants also may contain proteins and fats, and both plants and animals contain vitamins and minerals. These nutrients, consumed in the proper forms and proportions, sustain life and prevent the miseries of malnutrition—a condition that can involve either undernourishment or overnourishment.

HOW IT WORKS

Nutrients and Nutrition

In order to live, animals must consume nutrients, of which there are five major classes: carbohydrates, proteins, lipids or fats, vitamins, and minerals. In addition to these constituents, of course, animal life requires other materials for its sustenance-water, oxygen, and fiber, which aids in the digestive processing of foods-but these components usually are not regarded as nutrients.

Nutrition itself is the series of processes by which an organism takes in nutrients and makes use of them for its survival, growth, and development. The term nutrition also can refer to the study of nutrients, their consumption, and their processing in the bodies of organisms. Here the general term organism has been used, but for the most part the present essay is concerned with animal nutrition, or at least the nutrition of primary consumers (animals that eat plants) and secondary consumers (animals that eat other animals).

AUTOTROPHS AND THEIR NUTRIENTS. By contrast, plants and a few other types of organism are autotrophs, or primary producers in the food web. Autotroph means "self-feeder," and these organisms are distinguished by the fact that they do not depend on other organisms as a source of energy. Instead, plants obtain energy from the Sun and carbon dioxide from the atmosphere, and from these materials they build the large organic molecules that they need to survive.

Though plants are the most obvious example of an autotroph, they are not the only ones. In the deep oceans, far from any plant life, primary consumers depend on phytoplankton, which are microscopic organisms that encompass a range of bacteria and algae. Nonplant autotrophs may use means different from those employed by plants in generating their own food. For example, there are certain nonplant autotrophic organisms that live in the deep oceans near hydrothermal vents, which are cracks in the ocean floor caused by volcanic activity. These organisms, unlike most autotrophs, do not need sunlight to survive. Instead, they build their own nutrients in a sunless world, using sulfur compounds found near the vents.

CHEMICAL ELEMENTS AND NUTRITION

An element is a chemical substance made of only one kind of atom, whereas in a compound, atoms of more than one element are chemically bonded to one another. Unlike compounds, elements cannot be broken chemically into other substances. There are approximately 90 elements that occur in nature, and many of these elements—but not nearly all—are important to nutrition.

ELEMENTS IN THE HUMAN BODY AND BIOGEOCHEMICAL CYCLES. Even when we rule out obviously harmful elements, such as lead or uranium, there are still numerous chemical elements that play a part in the nutrition of living things. This can be illustrated by a glance at the abundance of various chemical elements in the human body, which include oxygen, carbon, and hydrogen. Oxygen alone accounts for a whopping 65% of the human body's mass, and carbon (18%), hydrogen (10%), and oxygen together make up 93% of the mass in the human body.

A great deal of oxygen and hydrogen, of course, is found in that most useful of all chemical compounds, water. In this vein, it should be noted that all the elements that take part in biogeochemical cycles, which are essential to the functioning of Earth, appear in relatively large proportions within the human body. These elements are hydrogen, oxygen, carbon, nitrogen, phosphorus, and sulfur. (For more about biogeochemical cycles and the elements involved in them, including their proportion within the human body's mass, see The Biosphere.)

Carbon is present in all living things, and its presence in certain forms is key to distinguishing organic from inorganic substances. Contrary to popular belief, organic substances are not just living things, their parts, and their products. Something that has never been living still can be considered organic, provided that it contains compounds that include carbon. (The only exceptions would be carbonates and carbon oxides, two groups of carbon-based compounds that are excluded from the ranks of organic substances.) As we shall see, carbon, along with oxygen and hydrogen, plays a key role in nutrition.

Most of the remaining 7% of the body's mass is composed of ten other elements. Among these elements are the other three involved in biogeochemical cycles, whose names are italicized: *nitrogen* (3%), calcium (1.4%), *phosphorus* (1.0%), magnesium (0.50%), potassium (0.34%), *sulfur* (0.26%), sodium (0.14%), chlorine (0.14%), iron (0.004%), and zinc (0.003%). Note that many of these elements are found in vitamin and mineral supplements that people might take on a daily basis to augment the essential nutrients in their bodies. There are exceptions, however, such as sodium, of which most people already ingest too much in the form of salt.

TRACE ELEMENTS. Generally speaking, it is safe to assume that any element that appears naturally in the human body is healthful as a nutritional component. This rule of thumb goes only so far, however: chlorine, for instance, is poisonous in large quantities, whereas in the very small proportions found in the human body, it can be essential to health and well-being. It is certainly possible to ingest some elements in unhealthy quantities, a fact that is particularly true of trace elements.

Copper is an example of a trace element, so named because only traces of them are present in the human body. In tiny quantities, copper is beneficial to human health, but if that small amount is exceeded, the effects can run the gamut from sneezing to diarrhea. In the proper proportions, however, trace elements are essential: without enough iodine, for instance, goiter, a large swelling of the thyroid gland in the neck area, can develop. Chromium helps the body metabolize sugars, which is why people concerned with losing weight or toning their bodies through exercise may take a chromium supplement. Even arsenic, which is lethal in large quantities, is a trace element in the human body, and medicines for treating such illnesses as "sleeping sickness" contain tiny amounts of arsenic. Other trace elements include cobalt, fluorine, manganese, molybdenum, nickel, selenium, silicon, and vanadium.

REAL-LIFE APPLICATIONS

NUTRIENTS

If you glance at the side of a cereal box, or virtually any other food product manufactured in the United States, chances are that you will see a table

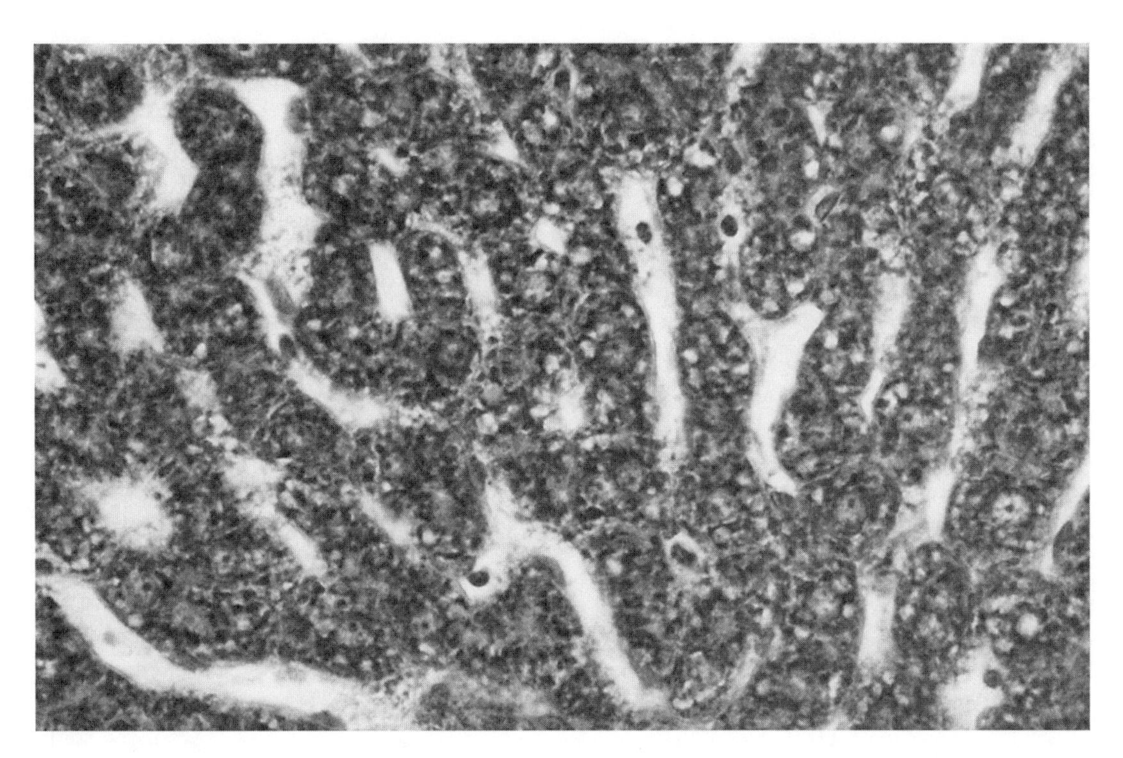

WHEN A PERSON INGESTS MORE CARBOHYDRATES THAN THE BODY NEEDS, THE BODY CONVERTS THE EXCESS INTO A COMPOUND KNOWN AS GLYCOGEN. IT THEN STORES THE GLYCOGEN IN THE LIVER (SHOWN HERE) AND MUSCLE TISSUES, WHERE IT REMAINS, A POTENTIAL SOURCE OF ENERGY FOR THE FUTURE. *(© Lester V. Bergman/Corbis. Reproduced by permission.)*

listing proportions of nutrients per serving. This is the "Nutrition Facts" label, which replaced the old "Nutrition Information Per Serving" label in the early 1990s. Both forms of information label were administered by the United States Food and Drug Administration (FDA), with the newer labeling format being far more extensive in terms of the information it provides.

Usually these tables show the amount of nutrients both in terms of mass (usually rendered in metric components, such as grams or milligrams) and as a proportion of recommended daily value according to the FDA. These listings must include information about some components such as calories, fat, sodium, sugars, certain vitamins, and so on. In addition to these mandatory listings, labels may contain information that the manufacturer chooses to provide concerning other food components, such as potassium or insoluble fiber.

Today, even fast-food restaurants such as McDonald's, which is probably not the first name that comes to mind when one thinks of healthy eating, provide extensive nutritional information to customers. Additionally, makers of fast food or supermarket "junk food" such as potato chips have introduced offerings that represent a nod to nutritional concerns. These include products that are fat- or sodium-free, or are otherwise geared toward greater health consciousness.

Clearly, diet is a significant concern to Americans, the most well-fed group of people that has ever existed, and terms from the Nutrition Facts label—proteins, carbohydrates, fats, minerals, and vitamins—are household words, known to almost everyone but understood by only a few. In much of the remainder of this essay, we explore these concepts, discussing what they mean in very basic scientific terms, as well as in terms of their significance in the diets of humans and other animals.

PROTEINS. Proteins are large molecules built from long chains of amino acids, which are organic compounds made of carbon, hydrogen, oxygen, nitrogen, and (in some cases) sulfur bonded in characteristic formations. Proteins serve the functions of promoting normal growth, repairing damaged tissue, contributing to the body's immune system, and making enzymes. (An enzyme is a protein material that speeds up chemical reactions in the bodies of plants and animals.)

Proteins in the human body contain about 20 different amino acids, of which the body is able to manufacture 12 from the foods we eat. The other eight, which the body requires for protein production but is unable to manufacture on its own, are known as the *essential amino acids.* When a protein contains all of the essential amino acids, it is known as a complete protein. Among the best forms of complete protein are fish, red meat, and poultry as well as eggs, milk, cheese, and other dairy products. Fittingly, a protein that lacks at least one of the essential amino acids is known as an incomplete protein. Examples include peas, beans, lentils, nuts, and cereal grains. These can, however, be combined in such a way as to make a complete protein, beans and rice being a good example.

CARBOHYDRATES. Carbohydrates are natural compounds that consist of carbon, hydrogen, and oxygen and whose primary function in the body is to supply energy. When a person ingests more carbohydrates than his or her body needs at the moment, the body converts the excess into a compound known as glycogen. It then stores the glycogen in the liver and muscle tissues, where it remains, a potential source of energy for the body to use in the future.

Sugars, starches, cellulose, and various other chemically related substances are part of the carbohydrate group. Most carbohydrates are produced by green plants in the process of undergoing photosynthesis. Nutritionally, the carbohydrates include sugar in its various forms as well as another class of food that people do not always think of as carbohydrates: fruits. Additionally, such starchy foods as potatoes, rice, and wheat products (bread, pasta, and so on) rank as important carbohydrates, while cereal grains and corn are examples of foods that contain both starchy carbohydrates and proteins.

LIPIDS. All fats and oils are lipids; these substances are distinguished by the fact that they are soluble only in compounds made of nonpolar molecules. Water is an example of a polar molecule, because the oxygen and hydrogen atoms tend to occupy opposite "ends" of the molecule, with one end exerting a negative electric charge and the other end a positive one. Therefore, water molecules tend to stick closely together. On the other hand, oil molecules, which consist of carbon and hydrogen, are nonpolar, because the atoms of the two elements do not tend to drift to opposite ends of the molecule. As a result, oil has the slippery texture for which it is known.

With their affinity for nonpolar molecules, lipids are soluble, or capable of dissolving, in oily or fatty substances but not in water. In the body, lipids, like carbohydrates, supply energy, only in different ways and on a different timetable. When burned, a gram of lipid actually produces about three times as much energy as a gram of carbohydrate, but this energy release takes place much more slowly. Among the other functions performed by lipids in the body are protection of organs from shock and damage and the provision of insulation for the body, for instance, from toxins. (It is for this reason that toxins often are stored in fat cells. See Food Webs for more about DDT bioaccumulation in the fat cells of animals.)

VITAMINS AND MINERALS. Vitamins are organic substances that, in extremely small quantities, are essential to the nutrition of most animals and some plants. In particular, they work with enzymes in regulating metabolic processes, but they do not in themselves provide energy; thus, vitamins alone do not qualify as a form of nutrition. Much the same is true of minerals, except that they are inorganic substances. And whereas vitamins are usually chemically complex (the formula for vitamin A, for instance, is $C_{20}H_{29}OH$), minerals may be as simple as a single element—for instance, iron or calcium.

Though the body can produce some vitamins, in general, vitamins and minerals are substances that the body is incapable of making for itself. Therefore, for optimal health, it is necessary to include them in the diet on a regular, if not daily, basis. They also have in common the fact that the body needs them only in very small quantities, for which reason they are sometimes known as micronutrients.

Vitamin A, for example, is a substance necessary to the functioning of the eye's retina in adjusting to light, and thus proper vitamin A levels are essential for night vision. Without vitamin A, a person can be afflicted with a condition known as night blindness, as well as with dryness of the skin. Vitamin A is also essential to bone growth. This vitamin occurs naturally in such foods as green and yellow vegetables, eggs, fruits, and liver and particularly in fish liver oils, such as cod liver oil.

Calcium, a mineral, helps build strong bones and teeth. It also has a role in the normal functioning of nerve and muscle activity. Ninety-nine percent of the body's calcium is stored in the skeleton and teeth, while the remainder circulates in the bloodstream, where it helps make possible muscle contractions. Bones are 70% calcium by weight, which gives them their strength and rigidity. Calcium, which is even more prevalent than iron, is the most abundant metallic element in the human body. Good dietary sources of calcium include milk and milk products, eggs, such leafy green vegetables as spinach, and sardines.

THE U.S. DEPARTMENT OF AGRICULTURE FOOD PYRAMID

The U.S. Department of Agriculture (USDA) has developed a diagram, called the *food pyramid,* to illustrate the components needed in a healthy diet. The bottom and widest level of the pyramid contains the cereal foods, such as breads, pastas, and rice. Primarily carbohydrates, these foods are a major source of energy, and therefore the USDA recommends 6-11 servings of 1-2 oz (30–60 g) from this food group. As to the exact number of servings, this is a function of such variables as age, gender, weight, and degree of regular physical activity.

The second level of the food pyramid, which is smaller than the first, consists of fruits and vegetables. These foods, which are also primarily carbohydrates, are especially important in supplying vitamins and minerals. A secondary function is the delivery of indigestible fiber, which improves the functioning and health of the large intestine, or colon. From this group, the USDA recommends 5-9 servings a day.

At the third level of the pyramid are proteins, including meats, eggs, beans, nuts, and milk products. According to the USDA, the percentage of these foods in one's diet should be much smaller than the percentage of carbohydrates. Smaller still is the quantity of servings at the top level, which contains the lipids. The small space allotted to this food emphasizes the fact that fats and oils should be consumed in small quantities for optimum health.

FAT AND THE AMERICAN DIET

What we have just described is the orthodox view of nutrition in the United States as the nation entered the twenty-first century. By that time, however, physicians, nutritionists, dieticians, and other specialists had begun to question the emphasis on carbohydrates in the USDA food pyramid and other mainstream diets. For a young person, whose body is still growing, the food pyramid is a good dietary plan. But for a person past the early twenties, particularly those who are overweight or suffering from a condition such as diabetes, other approaches may be needed.

The average American adult is considerably overweight for his or her height and age group, a fact for which a number of practices can be blamed. Among these practices are inactivity. Things have changed a great deal since our ancestors spent their whole lives in a flurry of physical activity, hunting animals for food and remaining ever on the move. Our bodies themselves—a product of natural selection that took place over countless generations (see Evolution)—know nothing about this change. They are still programmed to perform as they did 10,000 years ago, storing fat for use in lean times.

Thus, inactivity breeds obesity, a condition that cannot be addressed successfully by diets aimed simply at reducing consumption. In such a situation, the body simply holds on to its fat more fiercely, and this is one reason why a starvation diet is less than useless as a means of bringing about healthy weight loss. Starving oneself also reduces lean muscle mass, which further slows the metabolism and makes it still harder to burn fat. In fact, one of the best ways to lose weight is by combining resistance exercise (i.e., weight lifting) with proper eating.

Then, of course, there are the things Americans eat: junk food and fried foods, for instance. Eating junk food, pumped full of chemicals and white sugar, is like dumping garbage into a gas tank. As for fried foods, an American seldom realizes how much is in his or her diet until making a visit overseas. In Germany, for instance, virtually nothing is fried, and though people eat hearty meals with plenty of sausage, potatoes, bread, and beer, obesity is far less of a problem in Germany than in America. (Furthermore, the American traveler is likely to have far better bowel movements on the high-fiber German diet than on the greasy, fatty, salty American diet.)

THE CASE FOR PROTEINS

As healthy as a diet based on the food pyramid is, an overweight adult who stuck religiously to it

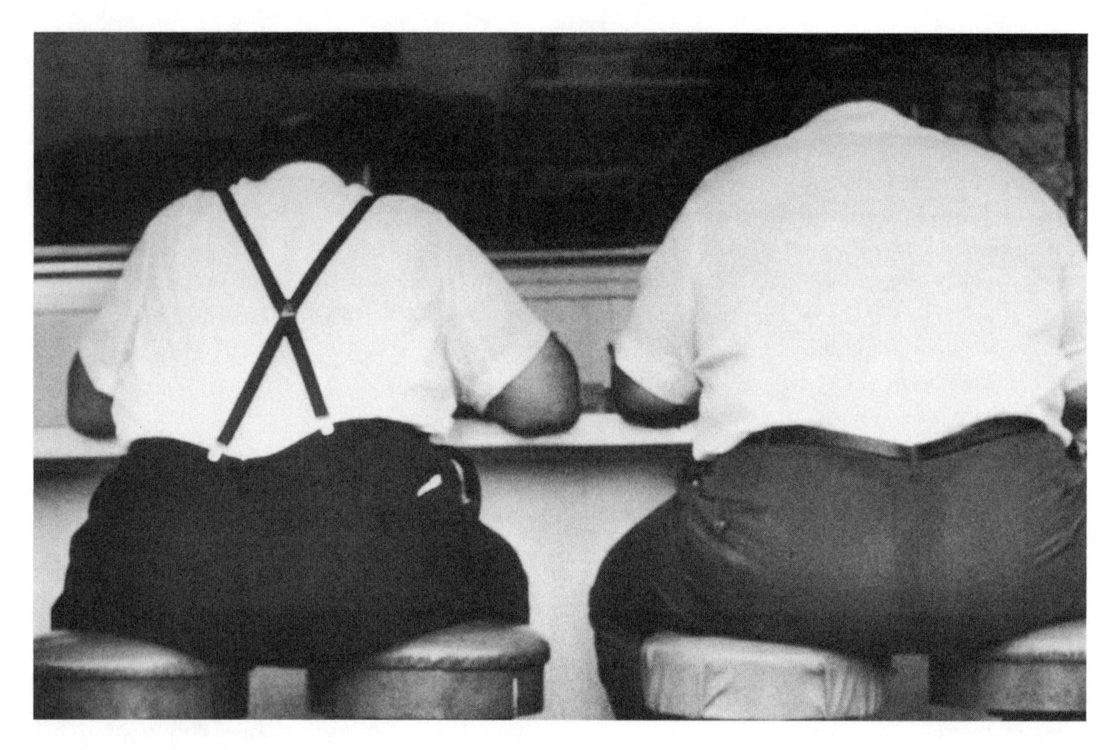

THE AVERAGE AMERICAN ADULT IS CONSIDERABLY OVERWEIGHT, A RESULT OF INACTIVITY COUPLED WITH POOR DIET. THINGS HAVE CHANGED A GREAT DEAL SINCE OUR ANCESTORS SPENT THEIR LIVES IN A FLURRY OF PHYSICAL ACTIVITY, HUNTING ANIMALS FOR FOOD AND EVER ON THE MOVE, BUT THE HUMAN BODY IS STILL PROGRAMMED TO PERFORM AS IT DID 10,000 YEARS AGO, STORING FAT FOR USE IN LEAN TIMES. *(© Bettmann/Corbis. Reproduced by permission.)*

most likely would remain overweight, even if he or she combined it with a regular program of aerobic exercise (e.g., jogging). Thus, by the late twentieth century, numerous experts across a wide spectrum of health professions began to challenge the old orthodoxy that held up carbohydrates as the central component of a healthy diet. Instead, a growing body of opinion favored diets that rely more on proteins and even healthy fats, such as those in broiled or grilled salmon or olive oil.

Whereas carbohydrate consumption can help an athlete gain a burst of energy, for most people carbohydrates are simply the raw materials for fat, which the body will store when it discovers that it does not need the carbohydrates themselves as an immediate source of energy. Furthermore, the brain has a mechanism for signaling the body that it has consumed enough protein, whereas there is no such mechanism where carbohydrates are concerned. To test this, try eating a meal of just protein: chances are that you will feel you have had enough fairly quickly. On the other hand, try eating a meal of just starches; you will find that you can eat and eat and eat, piling on calories as you do.

MALNUTRITION

As we noted earlier, corn has both protein and carbohydrate components, but this does not mean that a diet heavy in corn and corn products is a healthy one. Such a diet was not uncommon among poor people in the American South during the late nineteenth and the early twentieth centuries and among even poorer people in Mexico and other parts of Latin America to the present day. With Southern foods, such as grits, hominy, and cornbread, and Latin American foods, such as corn tortillas and polenta, it is quite possible to eat cheaply from a diet that relies primarily on corn. Yet someone who does so is at risk of serious health problems, because corn is lacking in two essential amino acids, lysine and tryptophan.

This is just one example of malnutrition, a condition that develops when the body does not obtain the right amount of the vitamins, minerals, and other nutrients it needs to maintain healthy tissues and organ function. Malnutrition occurs in people who are either undernourished or overnourished. Undernourishment is a consequence of consuming too few essential nutrients or using or excreting them more rapidly than

KEY TERMS

AMINO ACIDS: Organic compounds composed of carbon, hydrogen, oxygen, nitrogen, and (in some cases) sulfur bonded in characteristic formations. Strings of amino acids make up proteins.

AUTOTROPHS: Primary producers. *Autotroph* means "self-feeder," and these organisms are distinguished by the fact that they do not depend on other organisms as a source of energy. Instead, they need only sunlight, water, and a few simple chemical compounds, from which they build the large organic molecules that they need to survive.

BIOGEOCHEMICAL CYCLES: The changes that particular elements undergo as they pass back and forth through the various earth systems (e.g., the biosphere) and particularly between living and nonliving matter. The elements involved in biogeochemical cycles are hydrogen, oxygen, carbon, nitrogen, phosphorus, and sulfur.

CARBOHYDRATES: Naturally occurring compounds, consisting of carbon, hydrogen, and oxygen, whose primary function in the body is to supply energy. Included in the carbohydrate group are sugars, starches, cellulose, and various other substances. Most carbohydrates are produced by green plants in the process of undergoing photosynthesis.

COMPLETE PROTEIN: A protein that includes all eight essential amino acids.

COMPOUND: A substance in which atoms of more than one element are bonded chemically to one another.

ELEMENT: A substance made up of only one kind of atom. Unlike compounds, elements cannot be broken down chemically into simpler substances.

ENZYME: A protein material that speeds up chemical reactions in the bodies of plants and animals.

ESSENTIAL AMINO ACIDS: Amino acids that cannot be manufactured by the body and which therefore must be obtained from the diet. Proteins that contain essential amino acids are known as *complete proteins.*

FOOD WEB: A term describing the interaction of plants, herbivores, carnivores, omnivores, decomposers, and detritivores in an ecosystem. Each of these organisms consumes nutrients and passes them along to other organisms (or, in the case of the decomposer food web, to the soil and environment). The food web may be thought of as a bundle or network of food chains, but since the latter rarely exist separately, scientists prefer the concept of a food web to that of a food chain.

LIPIDS: Fats and oils. With their affinity for nonpolar molecules, lipids dissolve in oily or fatty substances but not in water-based liquids. In the body, lipids supply energy in slow-release doses, protect organs from shock and damage, and provide insulation for the body, for instance, from toxins.

MALNUTRITION: Any one of several conditions that develop when the body does not obtain the right amount of vitamins, minerals, and other nutrients it needs to maintain healthy tissues and organ function. Malnutrition may result from undernourishment *or* overnourishment.

KEY TERMS CONTINUED

MINERALS: Inorganic substances that, in a nutritional context, serve a function similar to that of vitamins. Minerals may include chemical elements—particularly metallic ones, such as calcium or iron—as well as some compounds.

NUTRIENT: Materials essential to the survival of organisms. They include proteins, carbohydrates, lipids (fats), vitamins, and minerals.

NUTRITION: The series of processes by which an organism takes in nutrients and makes use of them for its survival, growth, and development. The term *nutrition* also can refer to the study of nutrients, their consumption, and their use in the organism's body.

ORGANIC: At one time chemists used the term *organic* only in reference to living things. Now the word is applied to most compounds containing carbon, with the exception of carbonates (which are minerals), and oxides, such as carbon dioxide.

PHOTOSYNTHESIS: The biological conversion of light energy (that is, electromagnetic energy) from the Sun to chemical energy in plants. In this process carbon dioxide and water are converted to sugars.

PRIMARY CONSUMERS: Animals that eat green plants. Compare with *secondary consumers.*

PRIMARY PRODUCERS: Green plants that depend on photosynthesis for their nourishment.

PROTEINS: Large molecules built from long chains of amino acids. Proteins serve the functions of promoting normal growth, repairing damaged tissue, contributing to the body's immune system, and making enzymes.

SECONDARY CONSUMERS: Animals that eat other animals.

TRACE ELEMENT: A chemical element that appears within an organism or other natural system in very small quantities, or traces. In the human body, for instance, such trace elements as copper or iodine are essential to health, though they comprise far less than 1% of the body's mass.

VITAMINS: Organic substances that, in extremely small quantities, are essential to the nutrition of most animals and some plants. In particular, vitamins work with enzymes in regulating metabolic processes; however, they do not in themselves provide energy, and thus vitamins alone do not qualify as a form of nutrition.

they can be replaced. This brings about the all-too-familiar scenarios most of us associate with malnutrition: scenes of starving children in the third world, their bellies distended from kwashiorkor (which we discuss next) and thus abnormally large, far out of proportion to their bony arms and legs.

Such scenes of horror, and the virtually unimaginable misery of which they provide a hint, are almost beyond the comprehension of the average American. In America even the poorest of people are reasonably well fed, certainly compared with the poor of Africa, Asia, or Latin America. But that does not mean that malnutrition is not a problem in the affluent Western world. In the United States undernourishment may not be nearly as much of a problem, but dietary imbalances and excesses are, and they have become associated with many of the leading causes of death and disability. Overnutrition results from eating too much, eating too many of the wrong things (including junk foods, especially those containing white sugar), not exercising enough, or taking too many vitamins or other dietary replacements. As with many things, where vitamins are concerned, it is not the case that if a little is good, a lot is better. Vitamins can only be absorbed by the body if ingested on a full stomach, and an excess of fat-soluble vitamins such as A or D could actually be toxic. (See Vitamins for more on this subject.)

KWASHIORKOR. For any nutrient that the body requires, there is a corresponding disease, ailment, or condition (or many of them) that will develop if a person's body is deprived of that nutrient. For instance, people whose diets lack protein may become susceptible to a condition known as *kwashiorkor,* characterized by apathy, exhaustion, fatigue, wasting of muscle tissue, and edema (swelling, a result of water collecting in the body.)

Kwashiorkor, in fact, is the condition that causes swelling in the stomachs of children suffering from malnutrition. When people eat a diet consisting mainly of starchy vegetables, as is common in parts of Africa, Latin America, and southern Asia, they consume an appropriate amount of calories, but they do not get certain essential amino acids that are important for growth. This is particularly serious for children. Nursing babies usually do not suffer from kwashiorkor, because they receive adequate protein from their mother's milk. A child who is weaned, however, is at much greater risk.

In much of the third world, where human populations are high relative to the number of animals raised for meat, meat itself is a luxury. A child of a poor family is unlikely to have meat in his or her diet, and without the mother's milk the child is left virtually without sources of protein. This situation may have informed the choice of the word *kwashiorkor* to describe this condition. Derived from the Ga language of coastal Ghana in Africa, the word is based on *kwàsìokó,* a Ga term for "the influence a child is said to be under when a second child comes." In regions where birth control is unavailable or simply is not used, it is easy to imagine how an older child would be forced to give up the mother's milk to make room for a second child.

The results are devastating for the child deprived of protein sources. The most shocking outward manifestation of the condition is, of course, the swollen, bloated abdomen, brought about by a decrease in the amounts of the protein albumin (also found in egg whites) in the bloodstream. Skin discoloration may occur, along with severe diarrhea, an enlarged liver, and atrophy, or withering, of muscles and glands. Kwashiorkor can bring about retarded mental and physical development as well, but at least there is a treatment: adding proteins to the diet. For this reason, aid organizations supply powdered milk, which, if administered to these undernourished children in time, can save them from further damage to their bodies and minds.

MALNUTRITION IN THE UNITED STATES. Malnutrition exists even in the United States and not only in the form of overnourishment. Particularly among the nation's poor, undernourishment is serious, though it is not necessarily because food is unavailable to poor families. Rather, the right food may be unavailable, because the health foods described at the conclusion of this essay tend to be expensive. Furthermore, lack of education and information ensures that poor families maintain destructive dietary practices that leave their bodies deprived of essential nutrients. Low-cost, high-bulk items, such as sugar, white flour, and corn meal, may provide immediate satisfaction to the stomach, but they leave the body undernourished in the long run.

Nevertheless, the most prevalent form of malnutrition in America is still overnourishment. Implied in the term *overnourishment,* of course, is the idea that the person so afflicted is not eating the right kinds of things, because it is hard to be overnourished with a proper diet. Junk foods, such as one finds in the candy aisle of a convenience store or in a fast-food restaurant, contain carbohydrates and some proteins (not to mention *plenty* of fats); usually, these nutrients are presented in such a way as to maximize taste and minimize nutritional value. The result is "empty calories" that simply go to the waistline, buttocks, and abdomen of the consumer.

America is awash in food, but very little of it is the kind of food that sustains the health of the body or prolongs life. People have been conditioned by advertising to believe that white bread is the only kind of bread (even though the process of refining white flour leaches out most of the nutrients); that it is possible to have "zero fat" cookies sweetened with white sugar (which simply turns to fat as soon as the body digests it); and that a diet of burgers, fries, and soft drinks is "normal." The result is an overnourished, overweight populace whose life spans are being cut short.

There is another way, however, as a visit to a well-stocked health-food grocery store (or even the organic/health food section of a mainstream grocery store) will illustrate. After trying genuine whole wheat or sprouted wheat instead of white bread, fructose or cane-juice sweeteners in place of refined sugar, free-range chicken and grass-fed beef rather than meats pumped full of chemicals, many Americans might be surprised to learn just how good food can taste—and how good they can feel.

WHERE TO LEARN MORE

Anderson, Jean, and Barbara B. Deskins. *The Nutrition Bible: A Comprehensive, No-Nonsense Guide to Foods, Nutrients, Additives, Preservatives, Pollutants, and Everything Else We Eat and Drink.* New York: William Morrow, 1995.

Food and Nutrition Information Center, National Agricultural Library, U.S. Department of Agriculture (Web site). <http://www.nal.usda.gov/fnic/>.

Hands, Elizabeth S. *Nutrients in Food.* Philadelphia: Lippincott, Williams and Wilkins, 2000.

Kiple, Kenneth F., and Kriemhild Coneè Ornelas. *The Cambridge World History of Food.* New York: Cambridge University Press, 2000.

Nutrition.gov (Web site). <http://www.nutrition.gov/>.

Nutrition and Healthy Eating Advice. About.com (Web site). <http://nutrition.about.com/>.

Patent, Dorothy Hinshaw. *Nutrition: What's in the Food We Eat.* Illus. William Muñoz. New York: Holiday House, 1992.

Renders, Eileen. *Food Additives, Nutrients, and Supplements A–Z: A Shopper's Guide.* Santa Fe, NM: Clear Light Publishers, 1999.

Schwarzbein, Diana, and Nancy Deville. *The Schwarzbein Principle: The Truth About Losing Weight, Being Healthy, and Feeling Younger.* Deerfield Beach, FL: Health Communications, 1999.

U.S. FDA Center for Food Safety and Applied Nutrition (Web site). <http://vm.cfsan.fda.gov/list.html>.

VITAMINS

CONCEPT

Most of us have been told to take our vitamins, but few people know why, and despite all the talk about them in modern culture, vitamins remain something of a mystery. Vitamins are organic substances, essential for maintaining life functions and preventing disease among humans and animals and even some plants. They are found in very small quantities in food; certain health specialists recommend taking vitamin supplements to augment the supplies in food, while others insist that a well-balanced diet provides all the vitamins that an ordinary person needs. Some vitamins, such as vitamin C and the B complex, are water-soluble, which means that they are excreted easily and must be ingested every day. Others, such as vitamins A, D, E, and K, are fat-soluble and therefore are retained in the body's fatty tissues. With such vitamins, there may be a danger of taking too much, but in the case of most vitamins, the greatest harm comes from not receiving enough. Vitamin deficiencies can be the cause of rickets, pellagra, and other diseases that have plagued the poor in the Western world and the third world in the past and in the present.

HOW IT WORKS

AN INTRODUCTION TO VITAMINS

Once they were called *vitamines,* but for reasons that we address later, the "e" was dropped, and they became known as *vitamins.* There is also a reason for the strange alphabet of vitamins (A, B, C, D, E, K), which, like the change in spelling, came out of the early days of scientific research into the subject during the first third of the twentieth century. Though people did not know about vitamins per se until that time, folk wisdom certainly had taken account of the fact that certain foods are essential to the health and well-being of humans and animals.

Vitamins may be defined as organic substances, found in food, that are essential in very small quantities for the health of most animals and some plants. Organic substances, discussed in The Biosphere, are compounds (substances in which atoms of more than one element are chemically bonded to one another) containing hydrogen and carbon. Primarily, vitamins work with enzymes (protein materials that speed up chemical reactions in the bodies of plants and animals) in regulating metabolic processes—that is, processes that convert food to energy. They do not in themselves provide energy, however, and thus vitamins alone do not qualify as a form of nutrition.

Organisms require vitamins only in very small amounts: the total amount of vitamin mass a person needs in one day, for instance, is only about 0.0011 lb. (0.5 g). Yet vitamins are absolutely essential to the maintenance of health and for disease prevention, and most animals are not capable of synthesizing or manufacturing vitamins on their own. Nonetheless, most animals can produce vitamin C, though there are exceptions—humans included.

Animals depend on plants for their nutrition, either directly or indirectly (i.e., either by consuming the plant or by consuming an animal that has consumed the plant). Plants, on the other hand, are autotrophs, meaning that they can meet their nutritional needs with only sunlight, water, and a few chemical compounds. Among the nutrients plants produce are vitamins, which they pass on to animals that consume them directly or indirectly. (See Food Webs for more about autotrophs and the relationship of animal consumers to them.)

CLASSIFYING VITAMINS

Numerous vitamin groups are necessary for the nutritional needs of humans, and though only minute amounts of each are required to achieve their purpose, without them life could not be maintained. Some vitamins, including A, D, E, and K, are fat-soluble, meaning that they are found in fattier foods and in body fat. Thus, they can be stored in the body; for this reason, it is not necessary to include them in the diet every day. In fact, it could be dangerous to do so, since it is possible that they would build up to toxic levels in the tissues. Other vitamins, the most notable of which are vitamin C and the many vitamins in what is known as the "B complex," are water-soluble. They are found in the watery parts of food and body tissue, and because they are excreted regularly in the urine, they cannot be stored by the body. Instead, they must be consumed on a daily basis. This difference in solubility is extremely important to the way the vitamins function within an organism and in the ways and amounts in which they are consumed.

THE NAMES OF VITAMINS. Vitamins originally were classified in terms of their solubility in water or in fat, and these distinctions remain important for the reasons outlined above. Today, vitamins are known primarily by letters of the alphabet, a fact that harks back to a naming system developed as more and more vitamins were discovered in the early years of the twentieth century.

As scientists detected the existence of more vitamins, or what they thought were vitamins, they assigned to them successive letters of the alphabet: A, B, C, and so on. Eventually, however, they discovered that some substances originally thought to be vitamins were not vitamins, and they removed them from the roster. For example, what used to be called vitamin F is simply an essential fatty acid, a necessary component of the diet of a mammal but not the same thing as a vitamin. In other cases, what were once believed to be individual vitamins later were subsumed into the B complex. Among these substances are riboflavin, formerly termed vitamin G, and biotin, once called vitamin H. The result is that today the only alphabetical vitamin names are A, the B complex, C, D, E, and K.

REAL-LIFE APPLICATIONS

FAT-SOLUBLE VITAMINS

We are accustomed in modern life to being told that fat is bad for us, but to quote a much-cited line from the American composer George Gershwin's opera *Porgy and Bess,* "It ain't necessarily so." Fat is not inherently bad for people; in fact, a certain amount in the diet is essential. The problem in America today is the type of fat that people consume. There is a big difference between the healthy, natural, unsaturated fats one might find, say, in fresh salmon, and the highly processed and saturated fats in a bag of potato chips. (The term *saturated* means that every gap in which a hydrogen atom could fit in a string of carbon and hydrogen atoms has been filled. This helps make fats firm, for use in such products as shortening.)

Such fat is extremely harmful, because the body is not able to process it; even so, a certain amount of natural fat in the diet can be highly beneficial. This is true in large part because fat can serve as a medium for the fat-soluble vitamins A, D, E, and K, which are deposited in the body's fat cells. But as we noted earlier, it is important not to overdose on fat-soluble vitamins, because then what is inherently healthy can become extremely unhealthy.

VITAMIN A. In 1596 the Dutch explorer Willem Barents (1550–1597) and his shipwrecked crew spent a grueling winter on the island of Novaya Zemlya in the Arctic Ocean north of Russia. They had sailed from Holland in search of the Northeast Passage, which, like the more famous Northwest Passage above Canada, offered the prospect of a short, relatively direct sea route from Europe to Asia and the Americas. The problem was that the ice made sailing the northern seas virtually impossible. It would be almost three centuries before a crew managed to negotiate the Northeast Passage, by which time the European powers had long since given up all hopes of using it as a viable sailing route. (The same was true of the Northwest Passage, which was not traversed until 1906.)

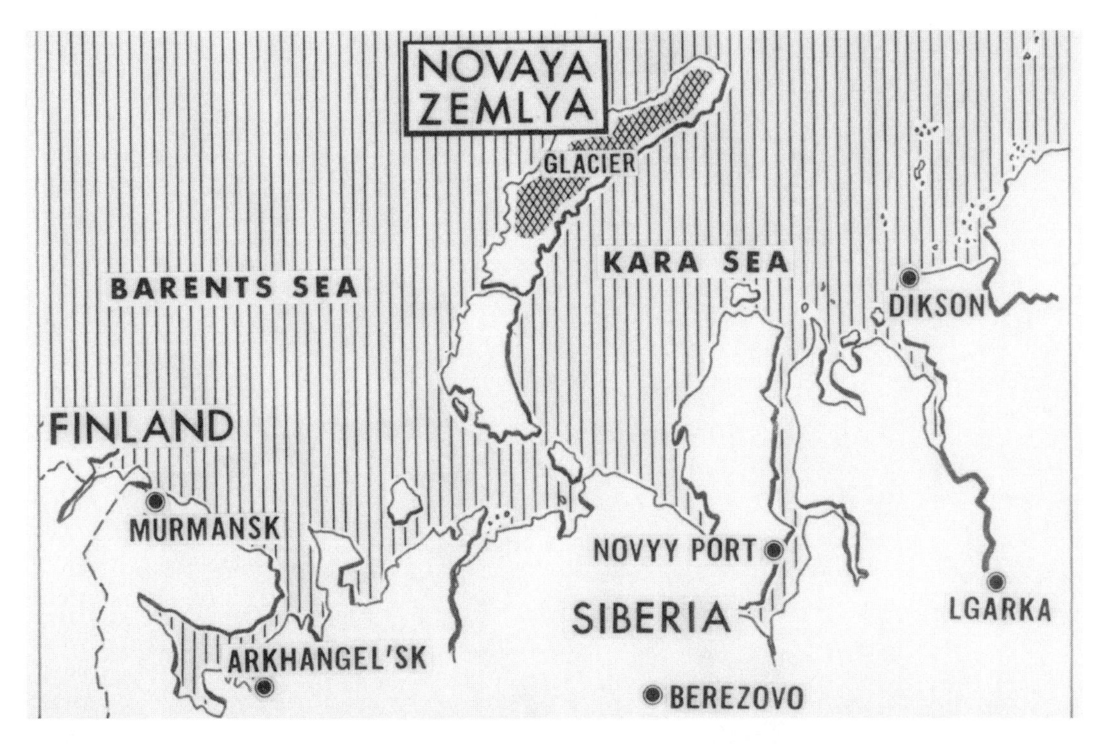

THE DUTCH EXPLORER WILLEM BARENTS AND HIS SHIPWRECKED CREW EXPERIENCED VITAMIN A POISONING WHEN THEY ATE POLAR BEARS TO STAY ALIVE ON THE ISLAND OF NOVAYA ZEMLYA IN THE ARCTIC OCEAN. POLAR BEAR LIVER CONTAINS ABOUT 450 TIMES THE RECOMMENDED DAILY DOSE OF VITAMIN A; SYMPTOMS OF POISONING INCLUDE PAINFUL JOINTS, BONE THICKENING, PEELING OF THE SKIN, AND LIVER DISEASE. *(© Bettmann/Corbis. Reproduced by permission.)*

Barents and his men knew none of that, nor would they have cared in that miserable winter of 1596–1597. All they cared about was survival, the chances for which seemed slim—and not just because of the almost inhuman cold or the fact that their ship had been cracked to pieces by the ice. Men were dying of scurvy, a vitamin-deficiency disease we discuss later in the context of vitamin C, as well as from the cold. Yet there were a few blessings, mainly in the form of available wood for fuel and animals for food. The men killed polar bears and ate their meat, and no doubt they were thankful just to stay alive. They could not have guessed, however, that they were actually killing themselves with an overdose of vitamin A.

Just 1 lb. (0.454 kg) of polar bear liver contains about 450 times the recommended daily dose of vitamin A, and the men in Barents's expedition were absorbing far more of the vitamin than they should have. In time, they began to experience the effects of vitamin A poisoning: painful joints, bone thickening, peeling of the skin over the entire body, and chronic liver disease. When spring came, the men managed to make it off the island, but many of them—Barents included—never lived to see Holland again, in part because the side effects of vitamin A toxicity had weakened them.

So why take vitamin A at all? Because it is necessary for proper growth of bones and teeth, for the maintenance and functioning of skin and mucous membranes, and for the ability to see in dim light. There is some evidence that it can help prevent cataracts (a clouding of the lens in the eye) and cardiovascular disease, a condition of the heart and circulatory system. Furthermore, when taken at the onset of a cold, vitamin A can ward off the illness and fight its symptoms.

One of the first signs of vitamin A deficiency is "night blindness," in which the rods of the eye (necessary for night vision) fail to function normally. Extreme cases of vitamin A deficiency can lead to total blindness. Other symptoms include dry and scaly skin, problems with the mucous linings of the digestive tract and urinary system, and abnormal growth of teeth and bones. The bodies of healthy adults who have an adequate diet can store several years' supply of this vitamin, but young children, who have not had time to build up such a large reserve, suffer from

DURING THE GREAT DEPRESSION, A CHILD IN THE SOUTHERN UNITED STATES SHOWS SIGNS OF RICKETS, STEMMING FROM VITAMIN D DEFICIENCY. UNDER THE INFLUENCE OF THIS DEBILITATING AND DISFIGURING DISEASE, THE LEGS BECOME BOWED BY THE WEIGHT OF THE BODY, AND THE WRISTS AND ANKLES THICKEN. *(© Marion Post Wolcott/Corbis. Reproduced by permission.)*

deprivation much more quickly if they do not consume enough.

Vitamin A is present in meats (mainly liver), fish oil, egg yolks, butter, and cheese. Although plants do not have vitamin A, dark green leafy vegetables and yellow fruits and vegetables (e.g., carrots, sweet potatoes, cantaloupe, corn, and peaches) contain a substance called beta-carotene, which is converted to vitamin A in the intestine and then absorbed by the body. It is nearly impossible to ingest beta-carotene in toxic amounts, unlike vitamin A from animal sources, since the body will not convert excess amounts to toxic levels of vitamin A.

VITAMIN D. Vitamin D is actually two different substances, D_2 and D_3. (There was no D_1, since the substance designated thus at one time turned out to be a mixture of several compounds, including calciferol, or D_2) Both forms of vitamin D are activated, or made effective, by sunlight, and for this reason vitamin D often is called the *sunshine vitamin.* It is hard to suffer a vitamin D deficiency if one gets enough sunshine in combination with consuming such foods as eggs (specifically, the yolk), such fatty fish as salmon, and enriched milk. (Milk does not naturally contain vitamin D, but the vitamin is sometimes included as an additive.)

Vitamin D lets the body utilize calcium and phosphorus in bone and tooth formation, and a deficiency causes a bone disease called rickets. Under the influence of this physically debilitating and disfiguring disease, legs become bowed by the weight of the body, and the wrists and ankles thicken. The teeth are badly affected and, for a young child, take much longer to mature. Infants and children are most likely to suffer the effects of rickets, but since all milk and infant formulas have vitamin D added to them, the condition is seen rarely in the industrialized world today. In the brutal early days of the Industrial Revolution, however (i.e., in England *ca.* 1760–1830), crowded slum conditions in areas where there was little or no sunlight made possible many cases of rickets.

Whereas rickets primarily affects children, adults may suffer from a disease called osteomalacia, caused by a deficiency of vitamin D, calcium, and phosphorous. Sometimes seen in the Middle East and other parts of Asia, osteomalacia brings with it rheumatic pain and causes the bones to become soft and deformed. As with rickets, the treatment for osteomalacia is a combination of calcium, phosphorous, and vitamin D. On the other hand, as with all fat-soluble vitamins, a person may take in excessive amounts of vitamin D, which has its own ill effects: nausea, diarrhea, weight loss, and pain in the bones and joints. Damage to the kidneys and blood vessels also can occur as calcium deposits build up in these tissues.

VITAMIN E. Composed of at least seven similar chemicals called the tocopherols, vitamin E is found in green leafy vegetables, wheat germ and other plant oils, egg yolks, and meat. The main function of this vitamin is to act as an antioxidant, to counteract the harmful effects oxygen can have on tissues. It may seem strange to speak of oxygen causing harm, since it is essential to life, but oxidation is an extremely powerful chemical reaction that, under various conditions, can manifest as rotting or putrefaction, rusting, or even combustion and explosion. When an apple turns brown a few minutes after you have cut it open, it is the result of oxidation.

Oxidation also may be linked to the effects of aging in humans as well as to other conditions, such as cancer, hardening of the arteries, and rheumatoid arthritis. It appears that oxygen molecules, which draw electrons to them, extract these electrons from the membranes in human cells. Over time, this can cause a gradual breakdown in the body's immune system. Antioxidants, such as vitamin E or beta carotene, therefore may be important in preserving human health and well-being.

Vitamin E is particularly important for counteracting oxidation in fats. When they are oxidized, fats form a highly reactive substance called peroxide, which is often very damaging to cells. Vitamin E is more reactive (i.e., more likely to form or break chemical bonds) than the fatty acid molecule, and, therefore, the vitamin reacts instead of the fat. Because cell membranes are composed partly of fat molecules, vitamin E is vitally important in maintaining the nervous, circulatory, and reproductive systems and in protecting the kidneys, lungs, and liver.

Because vitamin E is so common in foods, it is very difficult to suffer from a deficiency of this vitamin unless a person avoids consuming fats altogether—another example of why a no-fat diet is not a healthy one. The effects of vitamin E deficiency, all of which are apparently linked to the loss of its antioxidant protection, include cramping in the legs, fibrocystic breast disease (a condition that involves the formation of lumps and cysts in the breasts), and even muscular dystrophy. The seriousness of the latter two diseases only serves to highlight the importance of vitamin E to the body.

VITAMIN K. Like vitamin D, vitamin K is composed of two groups of compounds, vitamins K_1 and K_2. There is also a substance called K_3, but this vitamin is actually menadione, a synthetic compound from which the other forms of K are derived. You can find vitamin K in many plants, especially green leafy ones such as spinach, and in liver. Vitamin K is also made by the bacteria that live in the intestine—the "good" bacteria that help make possible the processing of food through the body.

Vitamin K appears to be critical to blood clotting, thanks to its role in assisting the formation of a chemical called prothrombin in the liver.

THE AMERICAN CHEMIST LINUS PAULING (SHOWN TOSSING AN ORANGE), WINNER OF THE NOBEL PRIZES IN CHEMISTRY AND IN PEACE, HELPED POPULARIZE VITAMIN C, ALSO KNOWN AS ASCORBIC ACID. *(© Roger Ressmeyer/ Corbis. Reproduced by permission.)*

Deficiencies of this vitamin rarely occur as the result of an incomplete diet; instead, it is usually a consequence of liver damage and the blood's inability to process the vitamin. The deficiency manifests in unusual bleeding or large bruises under the skin or in the muscles. Adults in the West seldom experience vitamin K deficiencies, but newborn infants have been known to suffer from brain hemorrhage owing to a lack of this vitamin.

WATER-SOLUBLE VITAMINS

B VITAMINS. The two water-soluble vitamins, as we shall see, have played a major part in medical history. Actually, there are more than two water-soluble vitamins, because vitamin B is really a complex of about a dozen vitamins—hence, the name *B complex.* Among them are vitamin B_1 (thiamine), vitamin B_2 (riboflavin), vitamin B_6 (pyridoxine), and vitamin B_{12} (cobalamin). A few others—for example, niacin (vitamin B_7) and pantothenic acid (vitamin B_3), are known better by names other than their "B names," while biotin and folate, or folic acid, are not known by "B names" at all.

Vitamin B_1, present in whole grains, nuts, legumes (e.g., peas), pork, and liver, helps the body release energy from carbohydrates. More than 4,000 years ago, the Chinese described a disease we know today as beriberi, which affects the nervous and gastrointestinal systems and causes nausea, fatigue, and mental confusion. The cause of beriberi is a deficiency of thiamine, or B_1, found in the husks or bran of rice and grains. White rice, which most people find more pleasing to the palate than brown rice, is the result of a milling and polishing process in which the husks—and along with them, this important nutrient—are removed. Manufacturers today produce "enriched" rice, flour, and other grain products by adding thiamine back in, but until scientists discovered the importance of thiamin in grain husks, many people, especially in the Far East, suffered the effects of beriberi. (Early research on beriberi will be discussed later.)

Vitamin B_2 helps the body release energy from fats, proteins, and carbohydrates. It can be obtained from whole grains, organ meats (e.g., liver), and green leafy vegetables. Lack of this vitamin causes severe skin problems. Vitamin B_6 is important in the building of body tissue as well as in protein metabolism and the synthesis of hemoglobin (an iron-containing pigment in red blood cells that is responsible for transporting oxygen to the tissues and removing carbon dioxide). A deficiency can cause depression, nausea, and vomiting. Vitamin B_{12} is necessary for the proper functioning of the nervous system and in the formation of red blood cells. It can be obtained from meat, fish, and dairy products. Anemia (a lack of red blood cells, which produces a lethargic condition), nervousness, fatigue, and even brain degeneration, can result from vitamin B_{12} deficiency.

Niacin is also highly important to human health, as we explain later in the context of the disease pellagra. Pantothenic acid helps release energy from fats and carbohydrates and is found in large quantities in egg yolks, liver, eggs, nuts, and whole grains. Deficiency of this vitamin causes anemia. Biotin, widely available from grains, legumes, and liver, plays a part in the release of energy from carbohydrates and in the formation of fatty acids. A lack of biotin causes dermatitis, or skin inflammation.

VITAMIN C. The American chemist and peace activist Linus Pauling (1901–1994), winner of the Nobel Prize in chemistry (1954) and peace (1962), helped popularize vitamin C, also known as *ascorbic acid.* It was Pauling who originated the idea, now widespread in society, that massive doses of vitamin C can ward off the common cold. Pauling went further, by maintaining that vitamin C offers protection against some forms of cancer. While scientific studies have been unable to prove this theory, they do suggest that the vitamin can at least reduce the severity of the symptoms associated with colds.

Most animals can synthesize this vitamin in the liver, where glucose (a type of sugar that occurs widely in nature) is converted to ascorbic acid. This is not the case with at least four types of animal: monkeys, guinea pigs, Indian fruit bats, and humans, all of which must obtain vitamin C from their diets. Citrus fruits, berries, and some vegetables (e.g., tomatoes and peppers) are good sources of vitamin C. It is a fragile vitamin, one that is oxidized or destroyed easily. Food storage or food processing can render it ineffective; so, too, can soaking vitamin C–containing fruits and vegetables in water for long periods.

VITAMIN DEFICIENCIES AND HISTORY

Most of the early history in the study of vitamins centered around what are now known as the *water-soluble vitamins.* Although vitamins as such were not discovered until early in the twentieth century, it was common knowledge long before that time that substances in certain foods were necessary for good health. An important turning point came in the mid–eighteenth century, with the work of the Scottish physician James Lind (1716–1794) on a vitamin deficiency condition that jeopardized England's vast merchant and military navies.

At a time when England had emerged as the world's leading sea power, even Her Majesty's sailing crews were at the mercy of a condition known as scurvy. Common among crews who had been at sea too long, scurvy could result in swollen joints, bleeding gums, loose teeth, and an inability to recover from wounds. Scientists today recognize scurvy as resulting from a deficiency of vitamin C, available in such citrus fruits as oranges. At the time, however, the concept of vitamins was unknown, and sailors at sea continued to live on a diet that consisted primarily of salted meats and hard biscuits—items that could be stored easily without spoilage in an era before refrigeration.

In 1746, Lind, a ship's doctor, observed that 80 of 350 seamen aboard his ship came down with scurvy during a 10-week cruise. Conducting a controlled experiment, he took 12 of the sailors in whom scurvy had developed and divided them into six groups. He gave each pair different substances, such as nutmeg, cider, seawater, and vinegar; the final pair was given lemons or oranges. The two men given the oranges and lemons both completely recovered in about a week. Not only was this a milestone in the history of vitamin research, but it also was the first example of a clinical trial, or the testing of a medication by careful and well-documented experimentation in which other variables or factors are kept unchanged.

It would be another half-century before the British navy adopted Lind's techniques. Another Scottish physician, Sir Gilbert Blane (1749–1834), had long fought for the adoption of Lind's methods, and finally, in 1796, he persuaded the navy to give each sailor a daily ration of lemons. At that time, the term lime was common for both lemons and limes, and, as a result, British sailors became known as *limeys.* Eventually, the treatment spread to the population as a whole, but outbreaks of scurvy continued until after World War I, when doctors isolated vitamin C as the controlling factor in scurvy prevention.

BERIBERI IN THE DUTCH EAST INDIES. In 1897 the Dutch government sent the physician Christiaan Eijkman (1858–1930) as part of a government commission to the Dutch East Indies (now Indonesia), which had been long afflicted by a condition known as beriberi. There are various forms of this disease, including infant beriberi, which can kill a breast-feeding baby after the fifth month, as well as various juvenile and adult forms. In the childhood and adult versions of the disease, there is a preliminary condition of fatigue, loss of appetite, and a numb, tingling feeling in the legs. This condition can lead to either wet beriberi, characterized by the accumulation of fluid throughout the body and a rapid heart rate that can bring about sudden death, or dry beriberi,

KEY TERMS

AMINO ACIDS: Organic compounds made of carbon, hydrogen, oxygen, nitrogen, and (in some cases) sulfur bonded in characteristic formations. Strings of amino acids make up proteins.

ANEMIA: A condition marked by a lack of red blood cells or hemoglobin or a shortage in total blood volume, any one of which can produce a lethargic condition.

ANTIOXIDANT: An enzyme, or some other organic substance, that is capable of counteracting the negative impact of oxygen (which draws electrons to it) on living tissue.

CARBOHYDRATES: Naturally occurring compounds, consisting of carbon, hydrogen, and oxygen, whose primary function in the body is to supply energy. Included in the carbohydrate group are sugars, starches, cellulose, and various other substances.

COMPOUND: A substance in which atoms of more than one element are bonded chemically to one another.

ENZYME: A protein material that speeds up chemical reactions in the bodies of plants and animals.

GLUCOSE: A type of sugar that occurs widely in nature. Glucose is the form in which animals usually receive carbohydrates.

HEMOGLOBIN: An iron-containing pigment in red blood cells that is responsible for transporting oxygen to the tissues and removing carbon dioxide.

METABOLISM: The chemical process by which nutrients are broken down and converted into energy or used in the construction of new tissue or other material in the body.

MINERALS: Inorganic substances that, in a nutritional context, serve a function similar to that of vitamins. Minerals may include chemical elements, particularly metallic ones, such as calcium or iron, as well as some compounds.

ORGANIC: At one time chemists used the term *organic* only in reference to living things. Now the word is applied to compounds containing carbon and hydrogen.

PROTEINS: Large molecules built from long chains of amino acids. Proteins serve the functions of promoting normal growth, repairing damaged tissue, contributing to the body's immune system, and making enzymes.

TISSUE: A group of cells, along with the substances that join them, that form part of the structural materials in plants or animals.

VITAMINS: Organic substances that, in extremely small quantities, are essential to the nutrition of most animals and some plants. In particular, vitamins work with enzymes in regulating metabolic processes; however, they do not in themselves provide energy, and thus vitamins alone do not qualify as a form of nutrition.

which is marked by a loss of sensation and weakness in the legs. The patient first needs to walk with the aid of a stick and then becomes bedridden and easy prey to infectious diseases.

Experimenting with birds, Eijkman noticed that some of the fowl experienced paralysis and polyneuritis (a disorder affecting several nerves at once), as in the dry form of beriberi. The director of the hospital had told Eijkman that he could not feed the birds with table scraps from the dining hall, where the diet was heavy in polished rice. The doctor thus was forced to feed his birds whole rice, and something amazing happened: the birds began to regain their ability to move and experienced no recurrence of paralysis.

Eijkman's colleagues rejected his claim that the birds had contracted some form of beriberi, though, in fact, he was correct. He was incorrect, however, in his supposition that the polished rice contained a toxin that was missing from the whole, unpolished rice. After Eijkman and the rest of the medical commission left the East Indies, another Dutch physician, Gerrit Grijns (1865–1944), stayed on to study the disease. He discovered that when the chickens were taken off the rice diet completely and fed meat instead, they did not show signs of the characteristic paralysis; if the meat was overcooked, however, the condition reappeared. In 1901, Grijns showed that beriberi could be cured by putting the rice polishings back into the rice. As it turned out, the husks and the meat contained vitamin B_1, also present in wheat germ, whole grain and enriched bread, legumes, peanuts, and nuts.

PELLAGRA IN THE AMERICAS. A vitamin-deficiency disease often associated with poverty, pellagra produces symptoms known as the "three Ds": diarrhea, dermatitis, and dementia, or mental deterioration. It was first identified in 1762 by the Spanish physician Gaspar Casal (*ca.* 1680–1759), who wrote about the "mal de la rosa"—so called because of the reddened dermatitis that appeared around the back of the victim's neck—that afflicted sufferers in one particular region of Spain.

Casal was far ahead of his time in maintaining that inadequate nutrition caused pellagra, but for many centuries the belief persisted that the disease resulted from infection. The breakthrough came with the work of the American physician Joseph Goldberger (1874–1929), a member of the U.S. Public Health Service who studied the high numbers of pellagra cases among poor blacks and whites in the southern United States. Goldberger established that pellagra stems from insufficient niacin, which is required to release energy from glucose.

Niacin is present in whole grains, meat, fish, and dairy products. One of the foods from which niacin is not easily available is corn, and it so happened that corn products constituted a major part of the diet in areas suffering from high rates of pellagra. The poor of Spain and Latin America subsisted on a diet heavy in corn products, such as tortillas and polenta, while their counterparts in the southern United States survived on cornbread, grits, hominy, and other variants of corn. Although such foods made it possible to fill the stomach cheaply, this diet was killing people in large numbers because it was not delivering the essential B vitamin niacin.

MODERN KNOWLEDGE OF VITAMINS. As it turned out, corn, in fact, does contain niacin, but to release the niacin from the large, fibrous molecules in corn, it is necessary to treat the corn with an alkaline solution, such as limewater. This is just one example of the vast knowledge that has accumulated since the time when the Polish-American biochemist Casimir Funk (1884–1967) coined the term *vitamine* in 1912. The first half of the word came from the Latin *vita,* or "life," and the second half reflected Funk's belief that all these substances belonged to a group of chemicals known as *amines.* Scientists later dropped the "e" when they discovered that not all vitamins contain an amine group.

In the director Stephen Spielberg's acclaimed 1987 film *Empire of the Sun,* one of the characters asks another, "Do you believe in vitamins?" The question reflects the relative newness of vitamins as an idea at the time when the movie was set, during World War II. After the war, interest in commercially produced vitamin supplements exploded, particularly among America's middle classes. Pauling's promotion of vitamin C, coming in the 1950s and 1960s, found a willing audience.

Although vitamin supplements remain a big business today, opinions vary as to the importance of enhancing one's diet with them. Many nutritionists insist that eating a well-balanced diet, consisting of the major food substances, is an effective and economical way to obtain nutrients for health. On the other hand, advocates of health foods and alternative medicine (medical practices that are not recognized officially by groups of university-trained and state-licensed medical doctors) insist that the recommended daily allowances established by the U.S. Food and Drug Administration do not provide sufficient vitamins for an average person.

The American Dietetic Association (ADA) recommends that nutrient needs come from a variety of foods taken from different dietary sources rather than from self-prescribed vitamin supplements. The organization makes allowances, however, for supplement usage by people who need extra doses of key vitamins and minerals. Examples include the use of iron supplements by women experiencing heavy menstrual bleeding as well as supplements of iron, folic acid, and calcium by pregnant women.

WHERE TO LEARN MORE

Apple, Rima D. *Vitamania: Vitamins in American Culture.* New Brunswick, NJ: Rutgers University Press, 1996.

Brody, Jane E., and Denise Grady. *The New York Times Guide to Alternative Health: A Consumer Reference.* New York: Times Books/Henry Holt, 2001.

Duyff, Roberta Larson. *The American Dietetic Association's Complete Food and Nutrition Guide.* Minneapolis, MN: Chronimed Publishing, 1998.

Kiple, Kenneth F., and Kriemhild Coneè Ornelas. *The Cambridge World History of Food.* New York: Cambridge University Press, 2000.

Nardo, Don. *Vitamins and Minerals.* New York: Chelsea House Publishers, 1994.

Reference Guide for Vitamins (Web site). <http://www.realtime.net/anr/vitamins.html>.

Snyder, Carl H. *The Extraordinary Chemistry of Ordinary Things.* New York: John Wiley and Sons, 1998.

Vitamins and Coenzymes. Indiana State University (Web site). <http://www.indstate.edu/thcme/mwking/vitamins.html>.

The Vitamin Collection. Molecular Expressions: Exploring the World of Optics and Microscopy, Florida State University (Web site). <http://micro.magnet.fsu.edu/vitamins/>.

Vitamin-Deficiency Diseases. Medic Planet (Web site). <http://www.medic-planet.com/MP_article/internal_reference/Vitamin-deficiency_diseases>.

Vitamins and Minerals Topic Page. Food and Nutrition Information Center National Agricultural Library, U.S. Department of Agriculture (Web site). <http://www.nal.usda.gov/fnic/etext/000068.html>.

SCIENCE OF EVERYDAY THINGS

REAL-LIFE BIOLOGY

GENETICS

GENETICS

HEREDITY

GENETIC ENGINEERING

MUTATION

GENETICS

CONCEPT

Genetics is the area of biological study concerned with heredity and with the variations between organisms that result from it. It demands an understanding of numerous terms, such as DNA (deoxyribonucleic acid), a molecule in all cells that contains blueprints for genetic inheritance; genes, units of information about particular heritable traits, which are made from DNA; and chromosomes, DNA-containing bodies, located in the cells of most living things, that hold most of the organism's genes. The vocabulary of genetics goes far beyond these three terms, as we shall see, but these are the core concepts. Among the areas in which genetics is applied is forensic science, or the application of science to matters of law—specifically, through "DNA fingerprinting," whereby samples of skin, blood, semen, and other materials can be used to prove or disprove a suspect's innocence. Another fascinating application of genetics is the Human Genome Project, an effort whose goals include the location and identification of every gene in the human body.

HOW IT WORKS

GENETICS AND HEREDITY

Genetics and heredity, the subject of another essay in this book, are closely related ideas. Whereas heredity is the transmission of genetic characteristics from ancestor to descendant through the genes, genetics is concerned with hereditary traits passed down from one generation to the next. It is very hard, if not impossible, to separate the two concepts completely, yet the entire body of knowledge encompassed by these topics is so large and so complex that it is best to separate them as much as possible. For this reason, the Heredity essay is concerned with such issues as how traits are passed on and why they appear in a particular generation but not another. That essay addresses the topics of alleles, dominant and recessive genes, and so on. It also briefly discusses the history of studies in areas that encompass genetics, heredity, and the mechanics thereof. In general, the Heredity essay is concerned with the larger patterns of inheritance over the generations, while the present one examines inheritance at a level smaller than the microscopic—that is, from the molecular or biochemical level.

SOMATIC AND GERM CELLS

Heredity begins with the cell, the smallest basic unit of all life. The information for heredity is carried within the cell nucleus, which is the control center not only in physical terms (it is usually located near the middle of the cell) but also because it contains the chromosomes. Within these threadlike structures is the genetic information organized in DNA molecules.

There are two basic types of cell in a multicellular organism: somatic, or body, cells, and germ, or reproductive, cells. The somatic cells are the primary components of most organisms, making up everything except some of the the cells in reproductive organs. The somatic cells of humans have 23 pairs of chromosomes, or 46 chromosomes overall, and are thus known as diploid cells. As the cells grow, they reproduce themselves by a process called *mitosis,* whereby a diploid cell splits to produce new diploids, each of which is a replica of the original. Thus cells grow and are replaced, making possible the formation of specific tissues and organs, such as muscles and nerves. Without mitosis, an organism's cells would not regenerate, resulting not only in cell death but possibly even the death of the entire organism. Mitosis is also the means of reproduction for organisms that reproduce asexually (see Reproduction).

A germ cell, by contrast, undergoes a process of cell division known as *meiosis,* whereby it becomes a haploid cell—a cell with half the basic number of chromosomes, which for a human would be 23 unpaired chromosomes. The sperm cells in a male and the egg cells in a female are both haploid germ cells: each contains only 23 chromosomes, and each is prepared to form a new diploid by fusion with another haploid. Sperm cells and egg cells are known as gametes, mature male or female germ cells that possess a haploid set of chromosomes and are prepared to form a new diploid by undergoing fusion with a haploid gamete of the opposite sex.

When egg and sperm fuse, they form a zygote, in which the diploid chromosome number is restored, with the zygote possessing the same chromosomes as *both* the sperm and the egg. This cell carries all the genetic information needed to grow into an embryo and eventually a full-grown human, with the specific traits and attributes passed on by the parents. Not all offspring of the same parents are the same, of course, and this is because the sperm cells and egg cells vary in their genetic codes—that is, in their DNA blueprints.

THE DNA BLUEPRINT

To understand genes and their biological function in heredity, it is necessary to understand the chemical makeup and structure of DNA. The complete DNA molecule often is referred to as the *blueprint for life,* because it carries all the instructions, in the form of genes, for the growth and functioning of organisms. This fundamental molecule is similar in appearance to a spiral staircase, which also is called a double helix. The sides of the DNA ladder are made up of alternate sugar and phosphate molecules, like links in a chain. The rungs, or steps, of DNA are made from a combination of four different chemical bases. Two of these, adenine and guanine, are known as purines, and the other two, cytosine and thymine, are pyrimidines. The four letters designating these bases—A, G, C, and T—are the alphabet of the genetic code, and each rung of the DNA molecule is made up of a combination of two of these letters.

DNA SEQUENCES. In this genetic code A always combines with T and C with G, to form what is called a base pair. Specific sequences of these base pairs make up the genes. Although a four-letter alphabet may seem rather small for constructing the extensive vocabulary that defines the myriad life-forms on Earth, in practice, the sequences of these base pairs make for almost limitless combinations. For any sequence, there are four possibilities as to the first two letters (AT, TA, CG, or GC) and four more possibilities for the second two letters. Thus, just for a four-letter sequence, there are 16 possibilities, and for each pair of letters added to the sequence, the total is multiplied by four. Given the long strings of base pairs that form DNA sequences, the numbers can be extremely large.

The more complex an organism, from bacteria to humans, the more rungs, or genetic sequences, appear on the ladder. The entire genetic makeup of a human, for example, may contain three billion base pairs, with the average gene unit being 2,000-200,000 base pairs long. Each one of these combinations has a different meaning, providing the code not only for the type of organism but also for specific traits, such as brown hair and blue eyes, dimples, detached earlobes, and so on and on. Except for identical twins, no two humans have exactly the same genetic information.

DNA REPLICATION, PROTEIN SYNTHESIS, AND RNA

Genetic information is duplicated during the process of DNA replication, which is initiated by proteins in the cells. To produce identical genetic information during cell mitosis, the DNA hydrogen bonds between the two strands arebroken, splitting the DNA in half lengthwise. This process begins a few hours before the initiation of cell mitosis, and once it is completed, each half of the DNA ladder is capable of forming a new DNA molecule with an identical genetic code. It can do this because of specific chemical catalysts (a substance that enables a chemical reaction without taking part in it) that help synthesize the complementary strand.

Catalysts formed from proteins are known as enzymes, and the functioning of specific cells and organisms is conducted by enzymes synthesized by the cells. Cells contain hundreds of different proteins, complex molecules that make up more than half of all solid body tissues and control most biological processes within and among these tissues. A cell functions in accordance with the particular protein—one of thousands of different types—it contains. It is the genetic base-pair sequence in DNA that determines, or "codes for," the specific arrangement of amino acids to build particular proteins.

Since the sites of protein production lie outside the cell nucleus, coded messages pass from the DNA in the nucleus to the cytoplasm, the material inside the cell that is external to the nucleus. This transfer of messages is achieved by RNA, or ribonucleic acid, and specifically by messenger RNA, or mRNA. Other types of RNA molecules are involved in linking the amino-acids together in a sequence form to shape the protein. (For more about amino acids, proteins, and enzymes, see the respective essays devoted to each subject.)

MUTATION

Once a protein has been created for a specific function, it cannot be changed. This is why the theory of acquired characteristics (the idea that changes in an organism's overall anatomy, as opposed to changes in its DNA, can be passed on to offspring) is a fallacy. People may have genes that make it easier for them to acquire certain traits, such as larger muscles or the ability to play the piano through exercise or practice, respectively, but the traits themselves, if they are acquired during the life of the individual and are not encoded in the DNA, are not heritable.

There is only one way in which changes that take place during the life of an organism can be passed on to its offspring, and that is if those changes are encoded in the organism's DNA. This is known as *mutation.* Suppose lung cancer develops in a man as a result of smoking; unless a tendency to cancer is already a part of his genetic makeup, he cannot genetically pass the disease on to his unborn children. But if the tobacco has acted as a mutagen, a substance that brings about mutation, it is possible that his DNA can be altered in such a way as to pass on either the tendency toward lung cancer or some other characteristic.

Because DNA is extremely stable chemically, it rarely mutates, or experiences an alteration in its physical structure, during replication. But because there are so many strands of DNA in the world, and so much material in the strands, mutation is bound to happen eventually—and, to an extent, at least, this is a good thing. Mutation is the engine that drives evolution, and a certain amount of genetic variation is necessary if species are to adapt by natural selection to a changing environment. If it were not for mutation, neither humans nor the many millions of other species that exist would ever have appeared.

Mutation often occurs when chromosome segments from two parents physically exchange places with each other during the process of meiosis. This is known as *genetic recombination.* Genes also can change by mutations in the DNA molecule, which take place when a mutagen alters the chemical or physical makeup of DNA. The mutations that result are of two types, corresponding to the two basic varieties of cell: somatic mutations, which occur solely within the affected individual, and germinal mutations, which happen in the DNA of germ cells, producing altered genes that may be passed on to the next generation.

The odd thing about mutations is that while most of them are harmful, the few that are beneficial are, as we have noted, the driving force behind the evolution of life-forms that successfully adapt to their environments. Thus, while most germinal mutations bring about congenital disorders (birth defects) ranging from physical abnormalities to deficiencies in body or mind to diseases, every once in a while a germinal mutation results in an improvement, such as a change in body coloring that acts as camouflage. If the trait improves an individual organism's chances for survival within a particular environment, it may become a permanent trait of the species, because the offspring with this gene have a greater chance of survival and thus will pass on the trait to succeeding generations. (For more about mutation, see the essay by that title. See also Evolution for a discussion of the role played by mutation and natural selection in the evolution of species.)

REAL-LIFE APPLICATIONS

THE GENETICS REVOLUTION

In the modern world genetics plays a part in more dramatic breakthroughs than any other

A SCIENTIST STUDIES THE ARRANGEMENT OF CHROMOSOME PAIRS, THE THREADLIKE DNA-CONTAINING BODIES IN CELLS THAT CARRY GENETIC INFORMATION. THE HUMAN GENOME PROJECT IS WORKING TO COMPLETE A MAP OUTLINING THE LOCATION AND FUNCTION OF THE GENES IN HUMAN CELLS. *(© BSIP/V & L /Photo Researchers. Reproduced by permission.)*

field of biological study. These breakthroughs have an impact in a wide variety of areas, from curing diseases to growing better vegetables to catching criminals. The field of genetics is in the midst of a revolution, and at the center of this exciting (and, to some minds, terrifying) phenomenon is the realm of genetic engineering: the alteration of genetic material by direct intervention in genetic processes. In agriculture, for instance, genes are transplanted from one organism to another to produce what are known as *transgenic* animals or plants. This approach has been used to reduce the amount of fat in cattle raised for meat or to increase proteins in the milk produced by dairy cattle. Fruits and vegetables also have been genetically engineered so that they do not bruise easily or have a longer shelf life.

Not all of the work in genetics is genetic engineering per se; in the realm of law, for instance, the most important application of genetics is genetic fingerprinting. A genetic fingerprint is a sample of a person's DNA that is detailed enough to distinguish it from the DNA of all others. The genetic fingerprint can be used to identify whether a man is the father of a particular child (i.e., to determine paternity), and it can be applied in the solving of crimes. If biological samples can be obtained from a crime scene—for example, skin under the fingernails of a murder victim, presumably the result of fighting against the assailant in the last few moments of life—it is possible to determine with a high degree of accuracy whether that sample came from a particular suspect. The use of DNA in forensic science is discussed near the conclusion of this essay.

THE REVOLUTION IN MEDICINE. Some of the biggest strides in genetic engineering and related fields are taking place, not surprisingly, in the realm of medicine. Genetic engineering in the area of health is aimed at understanding the causes of disease and developing treatments for them: for example, recombinant DNA (a DNA sequence from one species that is combined with the DNA of another species) is being used to develop antibiotics, hormones, and other disease-preventing agents. Vaccines also have been genetically re-engineered to trigger an immune response that will protect against specific diseases. One approach is to remove genetic material from a diseased organism, thus making the material weaker and initiating an immune response without causing the disease. (See Immunity and Immunology for more about how vaccines work.)

Gene therapy is another outgrowth of genetics. The idea behind gene therapy is to introduce specific genes into the body either to correct a genetic defect or to enhance the body's capabilities to fight off disease and repair itself. Since many inherited or genetic diseases are caused by the lack of an enzyme or protein, scientists hope one day to treat the unborn child by inserting genes to provide the missing enzyme. (For more about inherited disorders, see the essays Disease, Noninfectious Diseases, and Mutation.)

THE HUMAN GENOME PROJECT. One of the most exciting developments in genetics is the initiation of the Human Genome Project, designed to provide a complete genetic map outlining the location and function of the 40,000 or so genes that are found in human cells. (A genome is all of the genetic material in the chromosomes of a particular organism.) With the completion of this map, genetic researchers will have easy access to specific genes, to study how the human body works and to develop therapies for diseases. Gene maps for other species of animals also are being developed.

The project had its origins in the 1990s, with the efforts of the United States Department of Energy (DOE) and the National Institutes of Health (NIH). The NIH connection is probably clear enough, but the DOE's involvement at first might seem strange. What, exactly, does genetics have to do with electricity, petroleum, and other concerns of the DOE? The answer is that the DOE grew out of agencies, among them the Atomic Energy Commission (AEC), established soon after the explosion of the two atomic bombs over Japan in 1945. Even at that early date, educated nonscientists understood that the radioactive fallout produced from nuclear weaponry can act as a mutagen; therefore, Congress instructed the AEC to undertake a broad study of genetics and mutation and the possible consequences of exposure to radiation and the chemical by-products of energy production.

Eventually, scientists in the AEC and, later, the DOE recognized that the best way to undertake such a study was to analyze the entire scope of the human genome. The project formally commenced on October 1, 1990, and is scheduled for completion in the middle of the first decade of the twenty-first century. Upon completion, the Human Genome Project will provide a vast store of knowledge and no doubt will lead to the curing of many diseases.

Still, there are many who question the Human Genome Project in particular, and genetic engineering in general, on ethical grounds, fearing that it could give scientists or govern-

THE EXPLOSION OF AN ATOMIC BOMB OVER NAGASAKI, JAPAN, DURING WORLD WAR II (SEPTEMBER 1, 1945). THE RADIOACTIVE FALLOUT PRODUCED FROM NUCLEAR WEAPONRY CAN ACT AS A MUTAGEN, ALTERING THE CHEMICAL OR PHYSICAL MAKEUP OF DNA. *(© Bettmann/Corbis. Reproduced by permission.)*

ments too much power, unleash a Nazi-style eugenics (selective breeding) program, or result in horrible errors, such as the creation of deadly new diseases. In fact, it is impossible to search "genetic engineering" on the World Wide Web without coming across the Web sites of literally dozens and dozens of agencies, activist groups, and individuals opposed to genetic engineering and the mapping of the human genome. For more about the Human Genome Project, genetic engineering, and their opponents, see Genetic Engineering.

GENETICS IN FORENSIC SCIENCE

Forensic science, as we noted earlier, is the application of science to matters of law. It is based on the idea that a criminal always leaves behind some kind of material evidence that, through careful analysis, can be used to determine the identity of the perpetrator—and to exonerate someone falsely accused. Among those forms of material evidence of interest to forensic scientists working in the field of genetics are blood, semen, hair, saliva, and skin, all of which contain DNA that can be analyzed. In addition, there are areas of forensic science that rely on biological study, though not in the area of genetics: blood typing as well as the analysis of fingerprints or bite marks, both of which have patterns that are as unique to a single individual as DNA is.

One of the first detectives to use science, including biology and medicine, in solving crimes was a fictional character: Sherlock Holmes, whose creator, the British writer Sir Arthur Conan Doyle (1859–1930), happened to be a physician as well. The first full-fledged (and real) police practitioner of forensic science was the French police official Alphonse Bertillon (1853–1914), who developed an identification system that consisted of a photograph and 11 body measurements, including dimensions of the head, arms, legs, feet, hands, and so on, for each individual. Bertillon claimed that the likelihood of two people having the same measurements for all 11 traits was less than one in 250 million. In 1894 fingerprints, which were easier to use and more unique than body measurements, were added to the Bertillon system.

O. J. SIMPSON REACTS TO THE JURY'S VERDICT. DESPITE DNA EVIDENCE THAT LINKED BLOOD AT THE CRIME SCENE WITH BLOOD IN HIS CAR, THE JURY FOUND HIM NOT GUILTY OF DOUBLE MURDER. *(AP/WIde World Photos. Reproduced by permission.)*

Fingerprints, unlike DNA, are unique to the individual; indeed, identical twins have the same DNA but different fingerprints. Mark Twain (1835–1910) could not have known this in 1894, when he published *The Tragedy of Pudd'nhead Wilson, and the Comedy of Those Extraordinary Twins.* Nonetheless, the story involves a murder committed by one man and blamed on his twin, who eventually is exonerated on the basis of fingerprint evidence—still a new concept at the time. In some situations, however, fingerprint evidence may be unavailable, and though law-enforcement agencies have developed extraordinary techniques for analyzing nearly invisible (i.e., latent) prints, sometimes this is still not enough.

THE SIMPSON CASE AND THE CONTROVERSY OVER DNA EVIDENCE. For example, in the infamous murder of Nicole Brown Simpson and Ron Goldman on June 12, 1994, fingerprint evidence would have been ineffective in the case against the suspect, the former football star and actor O. J. Simpson. Since Nicole Simpson was his ex-wife, the appearance of his prints at the scene of her murder in her Los Angeles home could be explained away easily, even though she had taken out a restraining order against her former husband (who she had accused of spousal abuse) some time before the murder. Rather than fingerprints, the prosecution in his murder trial used DNA evidence connecting blood at the crime scene with blood found in Simpson's vehicle. (Some of this blood was apparently his own, since he had mysterious cuts on his hands that he could not explain to police officers.)

A jury found Simpson not guilty on October 3, 1995, and jurors later claimed that the prosecution had failed to make a strong case using DNA evidence. Furthermore, they cited police contamination of the DNA evidence, which had been established in their minds by Simpson's defense team, as a cause for reasonable doubt concerning Simpson's guilt. In fact, assuming that the defense was fully justified in this claim, that would have meant only that the DNA samples would have been *less* (not more) likely to convict Simpson.

KEY TERMS

ACQUIRED CHARACTERISTICS: Sometimes known as *acquired characters* or *Lamarckism,* after one of its leading proponents, the French natural philosopher Jean Baptiste de Lamarck (1744–1829), the theory of acquired characteristics is a fallacy that should not be confused with mutation. Acquired characteristics theory maintains that changes that occur in an organism's overall anatomy (as opposed to changes in its DNA) can be passed on to offspring.

AMINO ACIDS: Organic compounds made of carbon, hydrogen, oxygen, nitrogen, and (in some cases) sulfur bonded in characteristic formations. Strings of amino acids make up proteins.

BASE PAIR: A pair of chemicals that form the "rungs" on a DNA molecule, which has the shape of a spiral staircase. A base pair always consists of a type of chemical called a purine on one side and a chemical termed a pyrimidine on the other. This means that DNA base pairs always consist of adenine linked with thymine and guanine with cytosine.

BIOCHEMISTRY: The area of the biological sciences concerned with the chemical substances and processes in organisms.

BODY CELL: See *somatic cell.*

CHROMOSOME: A DNA-containing body, located in the cells of most living things, that holds most of the organism's genes.

CONGENITAL DISORDER: An abnormality of structure or function or a disease that is present at birth. Congenital disorders also are called *birth defects.*

CYTOPLASM: The material inside a cell that is external to the nucleus.

DIPLOID: A term for a cell that has the basic number of doubled chromosome cells. In humans, somatic cells, which are diploid cells, have 23 pairs of chromosomes, for a total of 46 chromosomes.

DNA: Deoxyribonucleic acid, a molecule in all cells, and many viruses, that contains genetic codes for inheritance.

DOMINANT: In genetics, a term for a trait that can manifest in the offspring when inherited from only one parent. Its opposite is *recessive.*

ENZYME: A protein material that speeds up chemical reactions in the bodies of plants and animals without itself taking part in or being consumed by those reactions.

FORENSIC SCIENCE: The application of science to matters of law and legal or police procedure.

GAMETE: A mature male or female germ cell that possesses a haploid set of chromosomes and is prepared to form a new diploid by undergoing fusion with a haploid gamete of the opposite sex.

GENE: A unit of information about a particular heritable trait. Usually stored on chromosomes, genes contain specifications for the structure of a particular polypeptide or protein.

GENETIC ENGINEERING: The alteration of genetic material by direct intervention in genetic processes.

GENETIC FINGERPRINT: A sample of a person's DNA that is detailed enough

KEY TERMS CONTINUED

to distinguish it from all other people's DNA.

GENETIC RECOMBINATION: A process whereby chromosome segments from two parents physically exchange places with each other during the process of meiosis. This is one of the ways that mutation occurs.

GENETICS: The area of biological study concerned with heredity, with hereditary traits passed down from one generation to the next through the genes, and with the variations between organisms that result from heredity.

GENOME: All of the genetic material in the chromosomes of a particular organism.

GERM CELL: One of two basic types of cells in a multicellular organism. In contrast to somatic, or body, cells, germ cells are involved in reproduction.

GERMINAL MUTATION: A mutation that occurs in the germ cells, meaning that the mutation can be passed on to the organism's offspring.

HAPLOID: A term for a cell that has half the number of chromosome cells that appear in a diploid, or somatic, cell. In humans, germ cells, which are haploid cells, have 23 unpaired chromosomes, as opposed to the 23 paired chromosomes (46 overall) that appear in a somatic cell.

HEREDITY: The transmission of genetic characteristics from ancestor to descendant through the genes.

HERITABLE: Capable of being inherited.

MEIOSIS: The process of cell division that produces haploid genetic material. Compare with *mitosis*.

MITOSIS: A process of cell division that produces diploid cells. Compare with *meiosis*.

MRNA: Messenger ribonucleic acid, a molecule of RNA that carries the genetic information for producing proteins.

MUTAGEN: A chemical or physical factor that increases the rate of mutation.

MUTATION: Alteration in the physical structure of an organism's DNA, resulting in a genetic change that can be inherited.

NATURAL SELECTION: The process whereby some organisms thrive and others perish, depending on their degree of adaptation to a particular environment.

NUCLEIC ACIDS: Acids, including DNA and RNA, that are made up of nucleotide chains.

NUCLEOTIDE: A compound formed from one of several types of sugar joined with a base of purine or pyrimidine (see *base pair*) and a phosphate group. Nucleotides are the basis for nucleic acids.

NUCLEUS: The control center of a cell, where DNA is stored.

POLYPEPTIDE: A group of between 10 and 50 amino acids.

PROTEINS: Large molecules built from long chains of 50 or more amino acids. Proteins serve the functions of promoting normal growth, repairing damaged tissue, contributing to the body's immune system, and making enzymes.

RECESSIVE: In genetics, a term for a trait that can manifest in the offspring only if it is inherited from both parents. Its opposite is *dominant*.

KEY TERMS CONTINUED

REPRODUCTIVE CELL: See *germ cell.*

RNA: Ribonucleic acid, a molecule translated from DNA in the cell nucleus that directs protein synthesis in the cytoplasm. See also *mRNA.*

SOMATIC CELL: One of two basic types of cells in a multicellular organism. In contrast to germ cells, somatic cells (also known as *body cells*) do not play a part in reproduction; rather, they make up the tissues, organs, and other parts of the organism.

SOMATIC MUTATION: A mutation that occurs in cells other than the reproductive, or sex, cells. These mutations, as contrasted with germinal mutations, cannot be transmitted to the next generation.

SYNTHESIZE: To manufacture chemically, as in the body.

TRANSLOCATION: A mutation in which chromosomes exchange parts.

ZYGOTE: A diploid cell formed by the fusion of two gametes.

At the same time, a number of legitimate concerns regarding the use of DNA evidence were raised by experts for the defense in the Simpson trial. Samples can become contaminated and thus difficult to read; small samples are difficult for analysts to work with effectively; and results are often open to interpretation. Furthermore, the outcome of the Simpson case illustrates the fact that findings based on DNA evidence are not readily understood by non-specialists, and may not make the best basis for a case-particularly in one so fraught with controversy. The prosecution based its case almost entirely on extremely technical material, explained in excruciating detail by experts who had devoted their lives to studying areas that are far beyond the understanding of the average person. Attempting to wow the jurors with science, the prosecution instead seemed to create the impression that DNA evidence was some sort of hocus-pocus invented to frame an innocent man. Simpson went free, though the jury in a 1996 civil trial (which took a much simpler approach, eschewing complicated DNA testimony) found him guilty.

DNA EVIDENCE SUCCESS STORIES. Because of the Simpson case, the use of DNA evidence gained something of a bad name. Nonetheless, it has been successful in less high profile cases, beginning in 1986, when English police tracked down a rapist and murderer by collecting blood samples from some 2,000 men. One of them, named Colin Pitchfork, paid another man to provide a sample in his place. This attracted the attention of the police, who tested his DNA and found their man.

Since that time, DNA evidence has been used in more than 24,000 cases and has aided in the conviction of about 700 suspects. The DNA in such cases is not always obtained from a human subject. In the investigation of the May 1992 murder of Denise Johnson in Arizona, a homicide detective found two seed pods from a paloverde tree in the bed of a pickup truck owned by the suspect, Mark Bogan. The accused man admitted having known the victim but denied ever having been near the site where her body was found. It so happened that there was a paloverde tree at the site, and testing showed that the DNA in the pods on his truck bed matched that of the tree itself. Bogan became the first suspect ever convicted by a plant.

On the other hand, in some cases, DNA evidence has cleared a suspect falsely accused. Such was the case with Kerry Kotler, convicted in 1981 for rape, robbery, and burglary and sentenced to 25–50 years in jail. In 1988, Kotler began petitioning for DNA analysis, which subsequently showed that his DNA did not match that of the rapist, who had left a semen sample in the victim's underwear. Kotler was released in December 1992 and in March 1996 was awarded $1.5 million in damages for his wrongful imprisonment. The story does not end there, however.

Kotler's case turned out to be one of the more bizarre in the annals of forensic DNA testing. Perhaps he did not commit the first rape, but a month after he received the damage award, he was on his way back to prison for the August 1995 rape of another victim. This time prosecutors showed that Kotler's semen matched samples taken from his victim's clothing—and to prove their case, they used DNA testing.

WHERE TO LEARN MORE

Department of Energy Human Genome Program (Web site). <http://www.ornl.gov/hgmis/>.

The DNA Files/National Public Radio (Web site). <http://www.dnafiles.org/>.

Fridell, Ron. *DNA Fingerprinting: The Ultimate Identity.* New York: Franklin Watts, 2001.

Genetics Education Center, University of Kansas Medical Center (Web site). <http://www.kumc.edu/gec/>.

Henig, Robin Marantz. *The Monk in the Garden: The Lost and Found Genius of Gregor Mendel, the Father of Genetics.* Boston: Houghton Mifflin, 2000.

Lerner, K. Lee, and Brenda Wilmoth Lee. *World of Genetics.* Detroit: Gale Group, 2002.

National Human Genome Research Institute (Web site). <http://www.nhgri.nih.gov>.

Schwartz, Jeffrey H. *Sudden Origins: Fossils, Genes, and the Emergence of Species.* New York: John Wiley and Sons, 1999.

Tudge, Colin. *The Impact of the Gene: From Mendel's Peas to Designer Babies.* New York: Hill and Wang, 2001.

Virtual Library on Genetics, Oak Ridge National Laboratory (Web site). <http://www.ornl.gov/TechResources/Human_Genome/genetics.html>.

HEREDITY

CONCEPT

Heredity is the transmission of genetic characteristics from ancestor to descendant through the genes. As a subject, it is tied closely to genetics, the area of biological study concerned with hereditary traits. The study of heritable traits helps scientists discern which are dominant and therefore are likely to be passed on from one parent to the next generation. On the other hand, a recessive trait will be passed on only if both parents possess it. Among the possible heritable traits are genetic disorders, but study in this area is ongoing, and may yield many surprises.

HOW IT WORKS

HEREDITY AND GENETICS

As discussed at the beginning of the essay on genetics, the subjects of genetics and heredity are inseparable from each other, but there are so many details that it is extremely difficult to wrap one's mind around the entire concept. It is advisable, then, to break up the overall topic into more digestible bits. One way to do this is to study the biochemical foundations of genetics as a subject in itself, as is done in Genetics, and then to investigate the impact of genetic characteristics on inheritance in a separate context, as we do here.

Also included in the present essay is a brief history of genetic study, which reveals something about the way in which these many highly complex ideas fit together. Many brilliant minds have contributed to the modern understanding of genetics and heredity; unfortunately, within the present context, space permits the opportunity to discuss only a few key figures. The first—a man whose importance in the study of genetics is comparable to that of Charles Darwin (1809–1882) in the realm of evolutionary studies—was the Austrian monk and botanist Gregor Mendel (1822–1884).

GENES. For thousands of years, people have had a general understanding of genetic inheritance—that certain traits can be, and sometimes are, passed along from one generation to the next—but this knowledge was primarily anecdotal and derived from casual observation rather than from scientific study. The first major scientific breakthrough in this area came in 1866, when Mendel published the results of a study on the hybridization of plants in which he crossed pea plants of the same species that differed in only one trait.

Mendel bred these plants over the course of several successive generations and observed the characteristics of each individual. He found that certain traits appeared in regular patterns, and from these observations he deduced that the plants inherited specific biological units from each parent. These units, which he called *factors,* today are known as *genes,* or units of information about a particular heritable trait. From his findings, Mendel formed a distinction between genotype and phenotype that is still applied by scientists studying genetics. Genotype may be defined as the sum of all genetic input to a particular individual or group, while phenotype is the actual observable properties of that organism. We return to the subjects of genotype and phenotype later in this essay.

MUTATION AND DNA. Although Mendel's theories were revolutionary, the scientific establishment of his time treated these new ideas with disinterest, and Mendel died in obscurity. Then, in 1900, the Dutch botanist Hugo De Vries (1848–1935) discovered Mendel's writings, became convinced that his predecessor had made an important discovery, and proceeded to take Mendel's theories much further. Unlike the Austrian monk, De Vries believed that genetic changes occur in big jumps rather than arising from gradual or transitional steps. In 1901 he gave a name to these big jumps: mutations. Today a mutation is defined as an alteration of a gene, which contains something neither De Vries nor Mendel understood: deoxyribonucleic acid, or DNA.

Actually, DNA, a molecule that contains genetic codes for inheritance, had been discovered just four years after Mendel presented his theory of factors. In 1869 the Swiss biochemist Johann Friedrich Miescher (1844–1895) isolated a substance from the remnants of cells in pus. The substance, which contained both nitrogen and phosphorus, separated into a protein and an acid molecule and came to be known as nucleic acid. A year later he discovered DNA itself in the nucleic acid, but more than 70 years would pass before a scientist discerned its purpose.

THE DISCOVERY OF CHROMOSOMES. In the meantime, another major step in the history of genetics was taken just two years after De Vries outlined his mutation theory. In 1903 the American surgeon and geneticist Walter S. Sutton (1877–1916) discovered chromosomes, threadlike structures that split and then pair off as a cell divides in sexual reproduction. Today we know that chromosomes contain DNA and hold most of the genes in an organism, but that knowledge still lay in the future at the time of Sutton's discovery.

In 1910 the American geneticist Thomas Hunt Morgan (1866–1945) confirmed the relationship between chromosomes and heredity through experiments with fruit flies. He also discovered a unique pair of chromosomes called the sex chromosomes, which determine the sex of offspring. From his observation that a sex-specific chromosome was always present in flies that had white eyes, Morgan deduced that specific genes reside on chromosomes. A later discovery showed that chromosomes could mutate, or change structurally, resulting in a change of characteristics that could be passed on to the next generation.

DNA MAKES ITS APPEARANCE. All this time, scientists knew about the existence of DNA without guessing its function. Then, in the 1940s, a research team consisting of the Canadian-born American bacteriologist Oswald Avery (1877–1955), the American bacteriologist Maclyn McCarty (1911–), and the Canadian-born American microbiologist Colin Munro MacLeod (1909–1972) discovered the blueprint function of DNA. By taking DNA from one type of bacteria and inserting it into another, they found that the second form of bacteria took on certain traits of the first.

The final proof that DNA was the specific molecule that carries genetic information came in 1952, when the American microbiologists Alfred Hershey (1908–1997) and Martha Chase (1927–) showed that transferring DNA from a virus to an animal organ resulted in an infection, just as if an entire virus had been inserted. But perhaps the most famous DNA discovery occurred a year later, when the American biochemist James D. Watson (1928–) and the English biochemist Francis Crick (1916–) solved the mystery of the exact structure of DNA. Their goal was to develop a DNA model that would explain the blueprint, or language, by which the molecule provides necessary instructions at critical moments in the course of cell division and growth. To this end, Watson and Crick focused on the relationships between the known chemical groups that compose DNA. This led them to propose a double helix, or spiral staircase, model, which linked the chemical bases in definite pairs. Using this twisted-ladder model, they were able to explain how the DNA molecule could duplicate itself, since each side of the ladder contains a compound that fits with a compound on the opposite side. If separated, each would serve as the template for the formation of its mirror image.

Autosomes and Sex Chromosomes

Genetic information is organized into chromosomes in the nucleus, or control center, of the cell. Human cells have 46 chromosomes each, except for germ, or reproductive, cells (i.e., sperm cells in males and egg cells in females), which each have 23 chromosomes. Each person receives 23 chromosomes from the mother's egg and 23 chromosomes from the father's sperm. Of these 23 chromosomes, 22 are called autosomes, or

IF TWO GROUPS OF THE SAME SPECIES ARE SEPARATED FOR A LONG TIME, GENETIC DRIFT MAY LEAD TO THE FORMATION OF DISTINCT SPECIES, AS WHEN THE COLORADO RIVER CUT OPEN THE GRAND CANYON AND ISOLATED THE KAIBAB SQUIRREL OF THE NORTH RIM FROM THE ABERT SQUIRREL (SHOWN HERE) OF THE SOUTH. *(© W. Perry Conway/Corbis. Reproduced by permission.)*

non-sex chromosomes, meaning that they do not determine gender. The remaining chromosome, the sex chromosome, is either an X or a Y. Females have two Xs (XX), and males have one of each (XY), meaning that females can pass only an X to their offspring, whereas males can pass either an X or a Y. (This, in turn, means that the sperm of the father determines the gender of the offspring.)

ALLELES

The 44 autosomes have parallel coded information on each of the two sets of 22 autosomes, and this coding is organized into genes, which provide instructions for the synthesis (manufacture) of specific proteins. Each gene has a set locus, or position, on a particular chromosome, and for each locus, there are two slightly different forms of a gene. These differing forms, known as alleles, each represent slightly different codes for the same trait. One allele, for instance, might say "attached earlobe," meaning that the bottom of the lobe is fully attached to the side of the head and cannot be flapped. Another allele, however, might say "unattached earlobe," indicating a lobe that is not fully attached and therefore can be flapped.

DOMINANT AND RECESSIVE ALLELES. Each person has two alleles of the same gene—the genotype for a single locus. These can be written as uppercase or lowercase letters of the alphabet, with capital letters defining dominant traits and lowercase letters indicating recessive traits. A dominant trait is one that can manifest in the offspring when inherited from only one parent, whereas a recessive trait must be inherited from both parents in order to manifest. For instance, brown eyes are dominant and thus would be represented in shorthand with a capital B, whereas blue eyes, which are recessive, would be represented with a lowercase b. Genotypes are either homozygous (having two identical alleles, such as BB or bb) or heterozygous (having different alleles, such as Bb). The phenotype, however—that is, the actual eye color—must be one or the other, because both sets of genes cannot be expressed together.

Unless there is some highly unusual mutation, a child will not have one brown eye and one blue eye; instead, the dominant trait will overpower the recessive one and determine the eye color of the child. If an individual's genotype is BB or Bb, that person definitely will have brown eyes; the only way for the individual to have blue eyes is if the genotype is bb—meaning that both parents have blue eyes. Oddly, two parents with brown eyes could produce a child with blue eyes. How is that possible? Suppose both the mother and the father had the heterozygous alleles Bb—a dominant brown and a recessive blue. There is then a 25% chance that the child could inherit both parents' recessive genes, for a bb genotype—and a blue-eyed phenotype.

LEARNING FROM HEREDITARY LAW. What we have just described is called *genetic dominance,* or the ability of a single allele to control phenotype. This principle of classical Mendelian genetics does not explain everything. For example, where height is concerned, there is not necessarily a dominant or recessive trait; rather, offspring typically have a height between that of the parents, because height also is determined by such factors as diet. (Also, more than one pair of genes is involved.) Hereditary law does, however, help us predict everything from hair and eye color to genetic disorders. As with the blue-eyed child of brown-eyed parents, it is possible that neither parent will show signs of a genetic disorder and yet pass on a double-recessive combination to their children. Again, however, other factors—including genetic ones—may come into play. For example, Down syndrome (discussed in Mutation) is caused by abnormalities in the number of chromosomes, with the offspring possessing 47 chromosomes instead of the normal 46.

REAL-LIFE APPLICATIONS

POPULATION GENETICS

Studies in heredity and genetics can be applied not only to an individual or family but also to a whole population. By studying the gene pool (the sum of all the genes shared by a population) for a given group, scientists working in the field of population genetics seek to explain and understand specific characteristics of that group. Among the phenomena of interest to population geneticists is genetic drift, a natural mechanism for genetic change in which specific traits coded in alleles change by chance over time, especially in small populations, as when organisms are isolated on an island. If two groups of the same species are separated for a long time, genetic drift may lead even to the formation of distinct species from what once was a single life-form. When the Colorado River cut open the Grand Canyon, it separated groups of squirrels that lived in the high-altitude pine forest. Over time, populations ceased to interbreed, and today the Kaibab squirrel of the north rim and the Abert squirrel of the south are different species, no more capable of interbreeding than humans and apes.

Where humans are concerned, population genetics can aid, for instance, in the study of genetic disorders. As discussed in Mutation, certain groups are susceptible to particular conditions: thus, cystic fibrosis is most common among people of northern European descent, sickle cell anemia among those of African and Mediterranean ancestry, and Tay-Sachs disease among Ashkenazim, or Jews whose ancestors lived in eastern Europe. Studies in population genetics also can supply information about prehistoric events. As a result of studying the DNA in fossil records, for example, some scientists have reached the conclusion that the migration

THERE ARE SEVERAL THOUSAND GENETIC DISORDERS, CLASSIFIED INTO AUTOSOMAL DOMINANT, AUTOSOMAL RECESSIVE, SEX-LINKED, AND MULTIFACTORIAL TYPES. DWARFISM, FOR EXAMPLE, CAN BE CAUSED BY ACHONDROPLASIA, TRANSMITTED BY A GENE INHERITED FROM ONE PARENT (DOMINANT), OR BY A GROWTH HORMONE DEFICIENCY, TRANSMITTED BY GENES INHERITED FROM BOTH PARENTS (RECESSIVE). *(© Photo Researchers. Reproduced by permission.)*

of peoples from Siberia to North America in about 11,000 B.C. took place in two distinct waves.

GENETIC DISORDERS

There are several thousand genetic disorders, which can be classified into one of several groups: autosomal dominant disorders, which are transmitted by genes inherited from only one parent; autosomal recessive disorders, which are transmitted by genes inherited from both parents; sex-linked disorders, or ones associated with the X (female) and Y (male) chromosome; and multifactorial genetic disorders. If one parent has an autosomal dominant disorder, the offspring have a 50% chance of inheriting that disease. Approximately 2,000 autosomal dominant disorders have been identified, among them Huntington disease, achondroplasia (a type of

KEY TERMS

ALLELE: For any locus, one of two (or more) slightly different forms of a gene. These differing forms mean that alleles code for different versions of the same trait.

AUTOSOMES: The 22 non-sex chromosomes.

CHROMOSOME: A DNA-containing body, located in the cells of most living things, that holds most of the organism's genes.

DNA: Deoxyribonucleic acid, a molecule in all cells, and many viruses, that contains genetic codes for inheritance.

DOMINANT: In genetics, a term for a trait that can manifest in the offspring when inherited from only one parent. Its opposite is *recessive.*

GENE: A unit of information about a particular heritable trait. Usually stored on chromosomes, genes contain specifications for the structure of a particular polypeptide or protein.

GENE POOL: The sum of all the genes shared by a population, such as that of a species.

GENETIC DISORDER: A condition, such as a hereditary disease, that can be traced to an individual's genetic makeup.

GENETIC DOMINANCE: The ability of a single allele to control phenotype.

GENOTYPE: The sum of all genetic input to a particular individual or group.

GERM CELL: One of two basic types of cells in a multicellular organism. In contrast to somatic, or body, cells, germ cells are involved in reproduction.

HEREDITY: The transmission of genetic characteristics from ancestor to descendant through the genes.

HETEROZYGOUS: Having two different alleles—for example, Bb.

HOMOZYGOUS: Having two identical alleles, such as BB or bb.

LOCUS: The position of a particular gene on a specific chromosome.

MUTATION: Alteration in the physical structure of an organism's DNA, resulting in a genetic change that can be inherited.

NUCLEUS: The control center of a cell, where DNA is stored.

PHENOTYPE: The actual observable properties of an organism, as opposed to its genotype.

RECESSIVE: In genetics, a term for a trait that can manifest in the offspring only if it is inherited from both parents. Its opposite is *dominant.*

SEX CHROMOSOMES: Chromosomes that determine gender. Human females have two X chromosomes (XX), and males have an X and a Y (XY).

SYNTHESIZE: To manufacture chemically, as in the body.

dwarfism), Marfan syndrome (extra-long limbs), polydactyly (extra toes or fingers), some forms of glaucoma (a vision disorder), and hypercholesterolemia (high levels of cholesterol in the blood).

The first two are discussed in Mutation. Marfan syndrome, or arachnodactyly ("spider arms"), is historically significant because it is believed that Abraham Lincoln suffered from that condition. Some scientists even maintain that his case of Marfan, a disease sometimes accompanied by eye and heart problems, was so severe that he probably would have died six months or a year after the time of his actual death by assassination at age 56 in April 1865.

RECESSIVE GENE DISORDERS. Just as a person has a 25% chance of inheriting two recessive alleles, so two parents who each have a recessive gene for a genetic disorder stand a 25% chance of conceiving a child with that disorder. Among the approximately 1,000 known recessive genetic disorders are cystic fibrosis, sickle cell anemia, Tay-Sachs disease, galactosemia, phenylketonuria, adenosine deaminase deficiency, growth hormone deficiency, Werner syndrome (juvenile muscular dystrophy), albinism (lack of skin pigment), and autism. Several of these conditions are discussed briefly elsewhere, and albinism is treated at length in Mutation. Note that all of the disorders mentioned earlier, in the context of population genetics, are recessive gene disorders. Phenylketonuria (see Metabolism) and galactosemia are examples of metabolic recessive gene disorders, in which a person's body is unable to carry out essential chemical reactions. For example, people with galactosemia lack an enzyme needed to metabolize galactose, a simple sugar that is found in lactose, or milk sugar. If they are given milk and other foods containing galactose early in life, they eventually will suffer mental retardation.

SEX-LINKED GENETIC DISORDERS. Dominant sex-linked genetic disorders affect females, are usually fatal, and—fortunately—are rather rare. An example is Albright hereditary osteodystrophy, which brings with it seizures, mental retardation, and stunted growth. On the other hand, several recessive sex-linked genetic disorders are well known, though at least one of them, color blindness, is relatively harmless. Among the more dangerous varieties of these disorders, which are passed on to sons through their mothers, the best known is hemophilia, discussed in Noninfectious Diseases. Many recessive sex-linked genetic disorders affect the immune, muscular, and nervous systems and are typically fatal. An example is severe combined immune deficiency syndrome (SCID), which is characterized by a very poor ability to combat infection. The only known cure for SCID is bone marrow transplantation from a close relative. Short of a cure, patients may be forced to live enclosed in a large plastic bubble that protects them from germs in the air. From this sad fact derives the title of an early John Travolta movie, *The Boy in the Plastic Bubble* (1976), based on the true story of the SCID victim Tod Lubitch. (The ending, in which Travolta, as Tod, leaves his bubble and literally rides off into the sunset with his beautiful neighbor Gina, is more Hollywood fiction than fact. Lubitch actually died in his early teens, shortly after receiving a bone marrow transplant.)

MULTIFACTORIAL GENETIC DISORDERS. Scientists often find it difficult to determine the relative roles of heredity and environment in certain medical disorders, and one way to answer this question is with statistical and twin studies. Identical and fraternal twins who have been raised in different and identical homes are evaluated for multifactorial genetic disorders. Multifactorial genetic disorders include medical conditions associated with diet and metabolism, among them obesity, diabetes, alcoholism, rickets, and high blood pressure. Other such multifactorial conditions are a tendency toward certain infectious diseases, such as measles, scarlet fever, and tuberculosis; schizophrenia and some other psychological illnesses; clubfoot and cleft lip; and various forms of cancer. The tendency of a particular person to be susceptible to any one of these disorders is a function of that person's genetic makeup, as well as environmental factors.

BREEDING WITHIN THE FAMILY

If there is one thing that most people know about heredity and breeding, it is that a person should never marry or conceive offspring with close relatives. Aside from moral restrictions, there is the fear of the genetic defects that would result from close interbreeding. How close is too close? Certainly, first cousins are off-limits as potential mates, though second or third cousins (people who share the same great-grandparents and the same great-great-grandparents, respectively) are probably far enough apart. Hence, the phrase "kissin' cousins," meaning a relative who is a distant enough to be considered a potential partner.

What kind of defects? Hemophilia, mentioned earlier, is popularly associated with royalty because several members of European ruling houses around the turn of the nineteenth century had it. Common wisdom maintains that the tendency toward the disease resulted from the fact that royalty were apt to marry close relatives. In fact, hemophilia has nothing to do with royalty per se and certainly bears no relation to marriages between close relatives. Research findings gathered over the course of more than three decades, beginning in 1965, indicate that many views about first cousins marrying may be more a matter of tradition than of scientific fact. According to information published in the *Journal of Genetic Counseling* and reported in the *New York Times* in April 2002, first cousins who have children together face only a slightly higher risk than parents who are completely unrelated. For example, within the population as a whole, the risk that a child will be born with a serious defect, such as cystic fibrosis, is 3-4%, while first cousins who conceive a child typically add another 1.7-2.8 percentage points of risk. Although this represents nearly double the risk, it is still a very small factor.

Researchers were quick to point out that mating should not take place between persons more closely related than first cousins. According to Denise Grady in the *New York Times,* "The report made a point of saying that the term 'incest' should not be applied to cousins, but only to sexual relations between siblings or between parents and children." First cousins, on the other hand, are a quite different matter, a fact borne out by the long history of people who married their first cousins. One example was Charles Darwin, who fathered many healthy children with his cousin, Emma Wedgwood.

WHERE TO LEARN MORE

Ackerman, Jennifer. *Chance in the House of Fate: A Natural History of Heredity.* Boston: Houghton Mifflin, 2001.

Center for the Study of Multiple Birth (Web site). <http://www.multiplebirth.com/>.

Clark, William R., and Michael Grunstein. *Are We Hardwired?: The Role of Genes in Human Behavior.* New York: Oxford University Press, 2000.

The Gene School (Web site). <http://library.thinkquest.org/19037/heredity.html>.

Genetic Disorders (Web site). <http://dir.yahoo.com/Health/Diseases_and_Conditions/Genetic_Disorders/>.

Grady, Denise. "Few Risks Seen to the Children of First Cousins." New York Times, April 4, 2002.

Hawley, R. Scott, and Catherine A. Mori. *The Human Genome: A User's Guide.* San Diego: Academic Press, 1999.

Heredity and Genetics. The Biology Project at the University of Arizona (Web site). <http://student.biology.arizona.edu/sciconn/heredity/worksheet_heredity.html>.

Reproduction and Heredity (Web site). <http://www.usoe.k12.ut.us/curr/science/sciber00/7th/genetics/sciber/intro.htm>.

Ridley, Matt. *Genome: The Autobiography of a Species in 23 Chapters.* New York: HarperCollins, 1999.

Wynbrandt, James, and Mark D. Ludman. *The Encyclopedia of Genetic Disorders and Birth Defects.* New York: Facts on File, 2000.

GENETIC ENGINEERING

CONCEPT

Genetic engineering is the alteration of genetic material by direct intervention in genetic processes with the purpose of producing new substances or improving functions of existing organisms. It is a very young, exciting, and controversial branch of the biological sciences. On the one hand, it offers the possibility of cures for diseases and countless material improvements to daily life. Hopes for the benefits of genetic engineering are symbolized by the Human Genome Project, a vast international effort to categorize all the genes in the human species. On the other hand, genetic engineering frightens many with its potential for misuse, either in Nazi-style schemes for population control or through simple bungling that might produce a biological holocaust caused by a man-made virus. Symbolic of the alarming possibilities is the furor inspired by a single concept on the cutting edge of genetic engineering: cloning.

HOW IT WORKS

DNA

Any discussion of genetics makes reference to DNA (deoxyribonucleic acid), a molecule that contains genetic codes for inheritance. DNA resides in chromosomes, threadlike structures found in the nucleus, or control center, of every cell in every living thing. Chromosomes themselves are made up of genes, which carry codes for the production of proteins. The latter, of which there are many thousands of different varieties, make up the majority of the human body's dry weight.

Although it is central to the latest advances in modern genetic research, DNA was discovered more than 130 years ago. In 1869 the Swiss biochemist Johann Friedrich Miescher (1844–1895) isolated a substance, containing both nitrogen and phosphorus, that separated into a protein and an acid molecule. He called it *nucleic acid,* and in this material he discovered DNA. Some 74 years would pass, however, before scientists recognized the function of the nucleic acid Miescher had discovered. Then, in 1944, a research team led by the Canadian-born American bacteriologist Oswald Avery (1877–1955) found that by taking DNA from one type of bacterium and inserting it into another, the second bacterium took on certain traits of the first. This experiment, along with other experiments and research, proved that DNA serves as a blueprint for the characteristics and functions of organisms.

THE DOUBLE HELIX. Nine years later, in 1953, the American biochemist James D. Watson (1928–) and the English biochemist Francis Crick (1916–) solved the mystery of DNA's structure and explained the means by which it provides necessary instructions at critical moments in the course of cell division and growth. They proposed a double helix, or spiral staircase, model, which linked the chemical bases of DNA in definite pairs. Using this twisted-ladder model, they were able to explain how the DNA molecule could duplicate itself, since each side of the ladder is identical to the other; if separated, each would serve as the template for the formation of its mirror image.

The sides of the DNA ladder are composed of alternating sugar and phosphate molecules, like links in a chain, and consist of four different chemical bases: adenine, guanine, cytosine, and

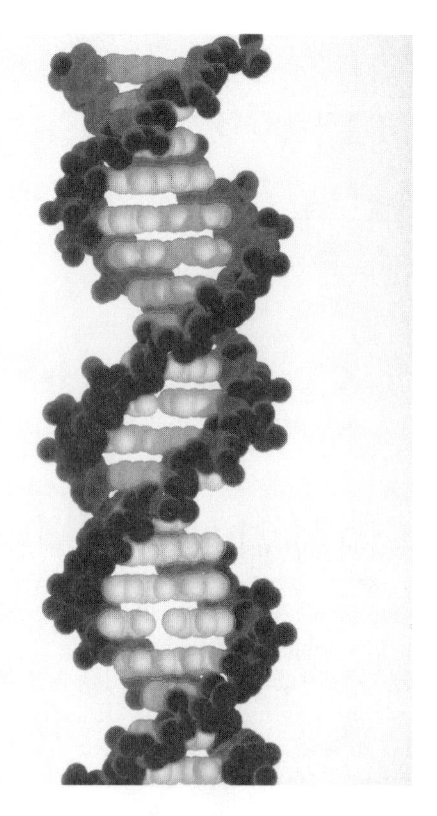

COMPUTER MODEL OF DNA, SHOWING ITS DOUBLE HELIX, OR SPIRAL STAIRCASE, FORM, WHICH LINKS THE CHEMICAL BASES OF DNA IN PAIRS. EACH SIDE OF THE LADDER IS IDENTICAL TO THE OTHER; IF SEPARATED, EACH WOULD SERVE AS THE TEMPLATE FOR THE FORMATION OF ITS MIRROR IMAGE. *(© Kenneth Eward/Grafz/Science Source/Photo Researchers. Reproduced by permission.)*

thymine. The four letters designating these bases—A, G, C, and T—are the alphabet of the genetic code, and each rung of the DNA molecule is made up of a combination of two of these letters. Owing to specific chemical affinities, A always combines with T and C with G, to form what is called a base pair. Specific sequences of these base pairs, which are bonded to each other by atoms of hydrogen, constitute the genes.

ENDLESS COMBINATIONS. A four-letter alphabet may seem rather small for constructing the extensive vocabulary that defines the myriad life-forms on Earth. If one stops to consider the exponential operations involved, however, it is easy to understand how large the range of possibilities can become. For any sequence, there are four possibilities for the first two letters (AT, TA, CG, or GC) and four more possibilities for the second two letters. Thus, just for a four-letter sequence, there are 16 possibilities, and for each pair of letters added to the sequence, the total is multiplied by four.

To see where this might lead, imagine that you started with a penny and tried to quadruple your funds every day. The first day there would not be a dramatic increase, since you would have to earn only $0.04, and even by day 4 you would need only $2.56 to meet your goal. But as the quadrupling process continued, day by day the sums of money would get bigger ($655.36 on day 8) and bigger ($16,772.16 on day 12) and bigger ($687,194,767.36 on day 18). Given the fact that the human body contains an almost unfathomable number of genes, each of which may be between 2,000 and 200,000 base pairs long, one can begin to imagine just how large the number of possibilities would become.

Each one of these combinations has a different meaning, providing the code for all manner of specific traits, such as brown hair and blue eyes, dimples, unattached earlobes, and so on. Except for identical twins, no two humans have exactly the same genetic information. What follows are just a few facts about the human genome—that is, all of the genetic material in the chromosomes of the human organism:

Some Facts About the Human Genome

- The human body contains about 100 trillion cells.
- Each cell has a DNA code consisting of some 1.5 billion base pairs.
- The DNA in each cell, if stretched to its full length, would be 6 ft. (1.8 m) long—yet it fits into a space about 0.0004 in. (0.0001 cm) across, smaller than the head of a pin.
- If all of the DNA in the human body were stretched end to end, it would reach to the Sun and back more than 600 times.
- If a person attempted to recite the entire human genome, with all its base pairs, at the rate of one letter per second, 24 hours a day, it would take a century.
- Every second scientists working on the Human Genome Project are decoding some 12,000 letters of DNA.
- Our DNA is 98% identical to that of chimpanzees.
- Only 0.2% of all human DNA differs between individuals; in other words, people are 99.8% the same, and all the vast differences between people are a product of just 1/500th of the total DNA.
- Despite all that scientists know about DNA, a staggering 97% of all human DNA has no known function.

PRINCIPLES OF GENETIC ENGINEERING

Just as DNA is at the core of studies in genetics, recombinant DNA (rDNA)—that is, DNA that has been genetically altered through a process known as *gene splicing*—is the focal point of genetic engineering. In gene splicing, a DNA strand is cut in half lengthwise and joined with a strand from another organism or perhaps even another species. Use of gene splicing makes possible two other highly significant techniques. Gene transfer, or incorporation of new DNA into an organism's cells, usually is carried out with the help of a microorganism that serves as a vector, or carrier. Gene therapy is the introduction of normal or genetically altered genes to cells, generally to replace defective genes involved in genetic disorders.

DNA also can be cut into shorter fragments through the use of restriction enzymes. (An enzyme is a type of protein that speeds up chemical reactions.) The ends of these fragments have an affinity for complementary ends on other DNA fragments and will seek those out in the target DNA. By looking at the size of the fragment created by a restriction enzyme, investigators can determine whether the gene has the proper genetic code. This technique has been used to analyze genetic structures in fetal cells and to diagnose certain blood disorders, such as sickle cell anemia.

GENE TRANSFER. Suppose that a particular base-pair sequence carries the instruction "make insulin"; if a way could be found to insert that base sequence into the DNA of bacteria, for example, those bacteria would be capable of manufacturing insulin. This, in turn, would greatly improve the lives of people with type 1 diabetes, who depend on insulin shots to aid their bodies in processing blood sugar. (See Noninfectious Diseases for more about diabetes.)

Although the concept of gene transfer is relatively simple, its execution presents considerable technical obstacles. The first person to surmount these obstacles was the American biochemist Paul Berg (1926–), often referred to as the "father of genetic engineering." In 1973 Berg developed a method for joining the DNA from two different organisms, a monkey virus known as *SV40* and a virus called *lambda phage.* Although the accomplishment was clearly a breakthrough, Berg's method was difficult. Then, later that year, the American biochemists Stanley Cohen (1922–) at Stanford University, and Herbert Boyer (1936–) at the University of California at San Francisco discovered an enzyme that greatly increased the efficiency of the Berg procedure. The gene-transfer technique developed by Berg, Boyer, and Cohen formed the basis for much of the ensuing progress in genetic engineering.

REAL-LIFE APPLICATIONS

BIG BUSINESS IN DNA

Ever since the breakthrough discoveries of Watson, Crick, and others in the 1950s made genetic engineering a possibility, the new field has promised increasingly bigger payoffs. These payoffs take the form of improvements to human life and profits to those who facilitate those improvements. The possible applications of genetic engineering are virtually limitless—as are the profits to be made from genetic engineering as a business. As early as the 1970s, entrepreneurs (independent businesspeople) recognized the commercial potential of genetically engineered products, which promised to revolutionize life, technology, and commerce as computers also were doing. Thus was born one of the great buzzwords of the late twentieth century: biotechnology, or the use of genetic engineering for commercial purposes.

Several early biotechnology firms were founded by scientists involved in fundamental research: Boyer, for example, teamed up with the venture capitalist Robert Swanson in 1976 to form Genentech (Genetic Engineering Technology). Other pioneering companies, including Cetus, Biogen, and Genex, likewise were founded through the collaboration of scientists and businesspeople. Today biotechnology promises a revolution in numerous areas, such as agriculture. Recombinant DNA techniques enable scientists to produce plants that are resistant to freezing temperatures, that will take longer to ripen, that will develop their own resistance to pests, and so on. By 1988 scientists had tested more than two dozen kinds of plants engineered to have special properties such as these. Yet no field of biotechnology and genetic engineering is as significant as the applications to health and the cures for diseases.

MEDICINES AND CURES. The use of rDNA allows scientists to produce many products that were previously available only in limited quantities: for example, insulin, which we referred to earlier. Until the 1980s the only source of insulin for people with diabetes came from animals slaughtered for meat and other purposes. The supply was never high enough to meet demand, and this drove up prices. Then, in 1982, the U.S. Food and Drug Administration (FDA) approved the sale of insulin produced by genetically altered organisms—the first such product to become available. Since 1982 several additional products, such as human growth hormone, have been made with rDNA techniques.

One of the most exciting potential applications of genetic engineering is the treatment of genetic disorders, which are discussed in Heredity, through the use of gene therapy. Among the more than 3,000 such disorders, quite a few of which are quite serious or even fatal, many are the result of relatively minor errors in DNA sequencing. Genetic engineering offers the potential to provide individuals with correct copies of a gene, which could make possible a cure for that condition. In the 1980s scientists began clinical trials of a procedure known as *human gene therapy* to replace defective genes. The technique, still very much in the developmental stage, offers the hope of cures for diseases that medicine has long been powerless to combat.

In 2001 scientists at the Weizmann Institute in Israel brought together two of the most exciting fields of research, biotechnology and computers, to produce the DNA-processing nanocomputer. It is an actual computer, but it is so small that a trillion of them would fit in a test tube. It consists of DNA and DNA-processing enzymes, both dissolved in liquid; thus its input, output, and software are all in the form of DNA molecules. The purpose of the nanocomputer is to analyze DNA, detecting abnormalities in the human body and creating remedies for them.

The Human Genome Project

At the center of genetic studies, with vast potential applications to genetic engineering, is the Human Genome Project (HGP), an international effort to analyze and map the DNA of humans and several other organisms. As discussed in the essay Genetics, the HGP began with efforts by the Atomic Energy Commission, a predecessor to the U.S. Department of Energy, to study the genetic effects of radioactive nuclear fallout. In 1990 the Department of Energy in cooperation with the National Institutes of Health (NIH), launched the project. At about the same time, the governments of the United Kingdom, Japan, Russia, France, and Italy initiated their own, similar undertakings, which are coordinated with American efforts.

The purpose of the project is to locate each human gene and determine its specific structure and function. Such knowledge will provide the framework for studies in health, disease, biology, and medicine during the twenty-first century and no doubt will make possible the cures for countless diseases. Although great strides have been made in gene therapy in a relatively short time, its potential usefulness has been limited by lack of scientific data concerning the multitude of functions that genes control in the human body. For gene therapy to advance to its full potential, scientists must discover the biological role of each of these genes and locate each base pair of which they are comprised.

A PROGRESS REPORT. Scientists participating in the project have identified an average of one new gene a day, and this rate of discovery has increased. At the time of its establishment in 1990 (under the leadership of James D. Watson, who served as director until 1992), HGP was expected to reach completion by 2005. In 2002, however, the project's leadership predicted completion by some time in the following year. Along the way, they had discovered that the human genome, originally believed to include 100,000 to as many as 150,000 genes, actually consists of about 30,000 to 40,000 genes.

Both HGP and a private firm, Celera Genomics (founded 1998), had undertaken the study of the human genome, and in June 2000 the entities jointly reported that they had finished the initial sequencing of the three billion-odd base pairs in the human genome. By that point, researchers also had completed thorough DNA sequences for many other organisms. The basis for the latter undertaking is that humans share many genes with other life-forms. With the completion of initial sequencing, scientists working on the HGP undertook the effort of determining the exact sequence of the base pairs that make them up all human genes. Long before completion, the project had yielded some information. Some of the genes identified through the HGP include one that guides reproduction of the human immunodeficiency virus (HIV), which causes the acquired immunodeficiency syndrome, better known as AIDS (see Infectious Diseases). Researchers also have located a gene that predisposes people to obesity as well as genes associated with such inherited disorders as Huntington disease, Lou Gehrig disease (also called amyotrophic lateral sclerosis, or ALS), and some colon and breast cancers.

A WORLD OF CONTROVERSY

The HGP has numerous implications—and not all of them, in the minds of some critics, are positive. The NIH inadvertently created cause for concern when, in 1991, it attempted to patent certain forms of DNA. While patenting is touted as a necessary financial incentive for research initiatives, critics maintain that it restricts access to the information generated and to the use of those discoveries. That places a great deal of control in the hands of the private firm that funded the research and may limit the spread of real benefits that result from discovery.

Some scientists and politicians have raised the concern that the ability to produce detailed genetic information on people could give too much power to the people who possess that knowledge. Most states do not have laws protecting citizens against the misuse of genetic information, for instance, by employers and insurers. In the absence of effective legal remedies, genetic testing may be used to bar people from employment or insurance coverage. Insurers may even make mandatory testing a requirement for coverage. Existing laws may not be adequate to protect people's privacy: whereas the individual may be protected from having to provide potentially damaging genetic information, such information still can be obtained by testing the individual's relatives.

NIGHTMARISH IMAGES. The idea of genetic information being used to control a person's destiny calls to mind all sorts of nightmarish images, such as those raised by the movie *Gattaca* (1997). Set in a dystopian, or anti-utopian, future, the film depicts a character who employs elaborate means to conceal his true identity in order to hold on to a job that otherwise would be forbidden to him because of his DNA. In fact, one does not have to go to fiction

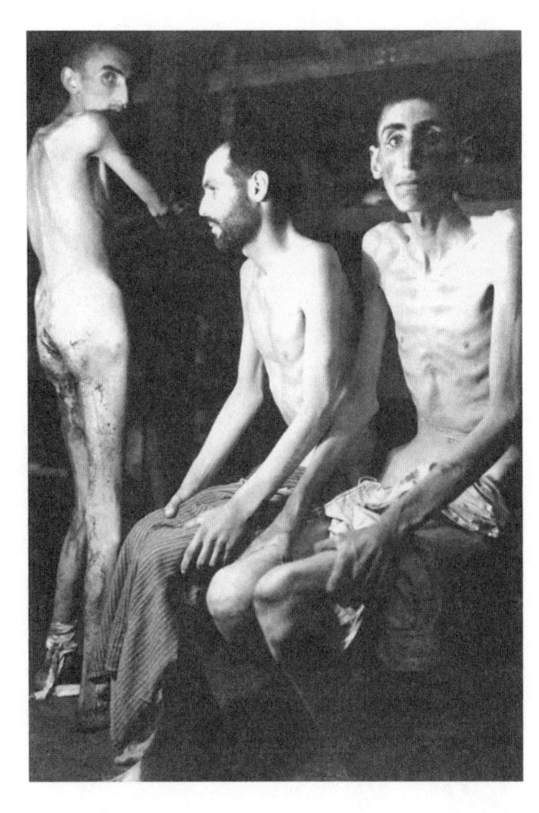

STARVING PRISONERS OF A NAZI CONCENTRATION CAMP, ONE FRIGHTENING EXAMPLE OF THE ATTEMPT TO USE GENETICS AS A FORM OF SOCIAL CONTROL. NAZI GERMANY PRACTICED MASS MURDER OF JEWS AND OTHERS WHO WERE DEEMED TO HAVE "UNDESIRABLE" TRAITS. *(© Corbis. Reproduced by permission.)*

to find examples of societies and movements that have used genetics as a form of social control.

The most frightening example of this was Nazi Germany, which practiced mass murder not only of Jews and other ethnic and social groups but also of people who suffered from mental retardation or other "undesirable" traits. The purpose of DNA was discovered only a year before the collapse of the Nazi empire in 1945, and one can only imagine what Adolf Hitler and his minions would have done with this knowledge if they had had access to it. (DNA is one of several scientific and technological concepts that came to fruition at the end of World War II but which Hitler, fortunately, was unable to use to his advantage. Others include rocketry, nuclear weaponry, radar, computers, and television.)

Nazism was actually an especially repugnant version of a movement known as *eugenics,* which had its origins in the late nineteenth and early twentieth centuries. Based on the idea that populations could be improved by encouraging people with "positive" traits to reproduce while discour-

ON FEBRUARY 24, 1997, THE FIRST MAMMAL CLONED FROM AN ADULT CELL, A LAMB NAMED DOLLY (SHOWN HERE AS AN ADULT), WAS BORN IN EDINBURGH, SCOTLAND. *(Photograph by Jeff Mitchell. Archive Photos. Reproduced by permission.)*

aging reproduction among those with less desirable traits, eugenics was at one time a mainstream movement whose adherents included the distinguished U.S. Supreme Court justice Oliver Wendell Holmes, Jr. (1841–1935).

OTHER CONCERNS. The specter of eugenics raises the threat that a single human, or a group of humans, could "play God" with the lives of others. Another dramatic fear associated with genetic engineering is the threat that a genetically re-engineered virus could turn out to be extremely virulent, or deadly, and spread. There are other, more mundane questions of ethics: for instance, is it appropriate for scientists to establish private, for-profit corporations to benefit from discoveries they made while working for public-sponsored research institutions? No wonder, then, that the budget for the HGP in the United States includes a small allocation (3% of its total) toward study of the ethical, legal, and social implications (ELSI) of the project. The ELSI Working Group is charged with studying the issues of fairness, privacy, delivery of health care, and education. Meanwhile, there is a vast body of opposition to genetic engineering, biotechnology, and the HGP. And no aspect of the larger subject is more upsetting to certain individuals, as well as special interest groups, as that of cloning.

CLONING

A clone is a cell, group of cells, or organism that contains genetic information identical to that of the parent cell or organism. It is a form of asexual reproduction (see Reproduction), and as such it is not as new as it seems; what *is* new, however, is humans' ability to manipulate cloning at the genetic level. The first clones produced by humans as long as 2,000 years ago were plants developed from grafts and stem cuttings. By cloning—a process that calls into play complex laboratory techniques and the use of DNA replication—people usually mean a relatively recent scientific advance. Among these techniques is the ability to isolate and copy (that is, to clone) individual genes that direct an organism's development.

THE PROMISE OF CLONING. The cloning of specific genes can provide large numbers of copies of that gene for use in genetic and taxonomic research as well as in the practical areas of medicine and farming. In the latter field, the goal is to clone plants with specific traits that make them superior to naturally occurring organisms. For example, in 1985 scientists conducted field tests using clones of plants whose genes had been altered in the laboratory to generate resistance to insects, viruses, and bacteria. New strains of plants resulting from cloning could produce crops that can grow in poor soil or even underwater and fruits and vegetables with improved nutritional qualities and longer shelf lives. A cloning technique known as *twinning* could induce livestock to give birth to twins or even triplets, and on the environmental front cloning might help save endangered species from extinction.

In the realm of medicine and health, cloning has been used to make vaccines and hormones. It has become possible, by combining two different kinds of cells (such as mouse and human cancer cells), to produce large quantities of specific antibodies, via the immune system, to fight off disease. When injected into the bloodstream, these cloned antibodies seek out and attack disease-causing cells anywhere in the body. By attaching a tracer element to the cloned antibodies, scientists can locate hidden cancers, and by attaching specific cancer-fighting drugs, the treatment dose can be transported directly to the cancer cells.

EXPERIMENTS IN CLONING. The modern era of laboratory cloning began in 1958 when the British plant physiologist F. C. Steward (1904–1993) cloned carrot plants from mature single cells placed in a nutrient culture containing hormones. The first cloning of animal cells took place in 1964, when the British molecular biologist John B. Gurdon (1933–1989) took nuclei from intestinal cells of toad tadpoles and injected them into unfertilized eggs. The cell nuclei in the eggs had been destroyed with ultraviolet light, but when the eggs were incubated, Gurdon found that 1–2% of the eggs developed into fertile, adult toads.

The first successful cloning of mammals occurred nearly 20 years later, when scientists in Switzerland and the United States successfully cloned mice using a method similar to Gurdon's approach. Their method required one extra step, however: after taking the nuclei from the embryos of one type of mouse, they transferred them into the embryos of another type of mouse. The latter served as a surrogate, or replacement, mother. The cloning of cattle livestock was tried first in 1988, when embryos from prize cows were transplanted to unfertilized cow eggs whose own nuclei had been removed. An even greater breakthrough transpired on February 24, 1997, with the birth of a lamb named Dolly in Edinburgh, Scotland. Dolly was no ordinary sheep: she was the first mammal born from the cloning of an adult cell. Thus, she had been produced by asexual reproduction in the form of genetically engineered cloning rather than by anything resembling a normal process. Nonetheless, she proved her own ability to reproduce the old-fashioned way when, on April 23, 1998, she gave birth to a daughter named Bonnie.

ARE HUMANS NEXT? Though Dolly's and Bonnie's births excited hopes, they also inspired fears. If large mammals could be cloned, could humans? As early as 1993 an attempt had been made at cloning human embryos as part of studies on *in vitro* (out of the body) fertilization. The purpose was to develop fertilized eggs in test tubes and then to implant them into the wombs of women having difficulty becoming pregnant. These fertilized eggs, however, did not develop to a stage that was suitable for transplantation into a human uterus. Then, on October 13, 2001, scientists at Advanced Cell Technology in Worcester, Massachusetts, successfully cloned a human embryo. They had not created human life, as it might sound; what they had developed instead was a source for nerve and other tissues that could be harvested for use in medicine and research. Still, the news—overshadowed though it was in America, where people were still reeling from the September 11 terrorist attacks—was earth-shattering. Human cells had been reproduced, and once again it appeared that the production of human clones might be possible.

It is easy to understand how people might respond with alarm to such frightening news with alarm. Such fears have a great deal more to do with Hollywood than they do with science. In fact, the accomplishment of the Massachusetts firm, while impressive from a scientific standpoint, was fairly modest compared with the Frankenstein-like image presented by anti–genetic engineering scaremongers. "Cloned an embryo" actually sounds a great deal more dramatic than what the Massachusetts scientists achieved, with just one embryo reaching the size of six cells before the cells stopped dividing. This is hardly the beginnings of a clone army.

At any rate, the cloning practiced at the Massachusetts firm was therapeutic cloning, involving the production of genetic material for the treatment of specific conditions. It is a far cry from reproductive cloning, which entails implanting a cloned embryo in a uterus—and even that is still a long way from the clichéd image of clones produced in a test tube without any parents other than the biological material used to create them.

Such ideas are related much more closely to those highlighted in Aldous Huxley's 1932 novel *Brave New World* than they are to scientific realities. And even if humans wanted to develop such technology, it would be many, many years in the future. As for "creating life," to do so is probably not even possible; if it is, such an achievement is about as far off as travel to another solar system. This is not to say that all fears of cloning and genetic engineering are unwarranted; on the contrary, it is good to have a healthy level of skepticism. But it is also good to be an equal-opportunity skeptic and therefore to question ideas in the

KEY TERMS

AMINO ACIDS: Organic compounds made of carbon, hydrogen, oxygen, nitrogen, and (in some cases) sulfur bonded in characteristic formations. Strings of amino acids make up proteins.

BASE PAIR: A pair of chemicals that form the "rungs" on a DNA molecule, which has the shape of a spiral staircase.

BIOTECHNOLOGY: A name for the industry built around the application of genetic engineering techniques.

CHROMOSOME: A DNA-containing body, located in the cells of most living things, that holds most of the organism's genes.

CLONE: A cell, group of cells, or organism that contains genetic information identical to that of its parent cell or organism.

CLONING: A specialized genetic process whereby clones are produced. Cloning is a form of asexual reproduction.

DNA: Deoxyribonucleic acid, a molecule in all cells, and many viruses, that contains genetic codes for inheritance.

ENZYME: A protein material that speeds up chemical reactions in the bodies of plants and animals without itself taking part in or being consumed by those reactions.

GENE: A unit of information about a particular heritable trait. Usually stored on chromosomes, genes contain specifications for the structure of a particular polypeptide or protein.

GENE SPLICING: A process whereby recombinant DNA is formed by cutting a DNA strand in half lengthwise and joining it with a strand from another organism or perhaps even another species.

GENE THERAPY: The introduction of normal or genetically altered genes to cells, typically to replace defective genes involved in genetic disorders.

GENE TRANSFER: Incorporation of new DNA into an organism's cells, usually with the help of a microorganism that serves as a vector.

GENETIC DISORDER: A condition, such as a hereditary disease, that can be traced to an individual's genetic makeup.

GENETIC ENGINEERING: The alteration of genetic material by direct intervention in genetic processes with the purpose of producing new substances or improving functions of existing organisms.

NUCLEUS: The control center of a cell, where DNA is stored.

PROTEINS: Large molecules built from long chains of 50 or more amino acids. Proteins serve the functions of promoting normal growth, repairing damaged tissue, contributing to the body's immune system, and making enzymes.

RDNA: Recombinant DNA, or DNA that has been genetically altered through gene splicing.

RESTRICTION ENZYMES: Enzymes that break DNA into fragments at particular sites.

VECTOR: In the context of genetics, a vector is a microorganism or virus that is used to transfer DNA from one organism to another.

popular culture—including opposition to genetic engineering.

WHERE TO LEARN MORE

Barash, David P. *Revolutionary Biology: The New, Gene-Centered View of Life.* New Brunswick, NJ: Transaction Publishers, 2001.

Chadwick, Ruth F. *The Concise Encyclopedia of the Ethics of New Technologies.* San Diego: Academic Press, 2001.

"Cloning." *New Scientist* (Web site). <http://www.newscientist.com/hottopics/cloning/>.

Department of Energy Human Genome Program (Web site). <http://www.ornl.gov/hgmis/>.

Genetic Engineering and Cloning: Improving Nature or Uncorking the Genie? (Web site). <http://library.thinkquest.org/19697/>.

Genetics Education Center, University of Kansas Medical Center (Web site). <http://www.kumc.edu/gec/>.

Hyde, Margaret O., and John F. Setaro. *Medicine's Brave New World: Bioengineering and the New Genetics.* Brookfield, CT: Twenty-First Century Books, 2001.

Judson, Karen. *Genetic Engineering: Debating the Benefits and Concerns.* Berkeley Heights, NJ: Enslow Publishers, 2001.

National Human Genome Research Institute (Web site). <http://www.nhgri.nih.gov>.

Twenty Facts About the Human Genome. Wellcome Trust (Web site). <http://www.wellcome.ac.uk/en/genome/thgfac.htm>.

Wade, Nicholas. *Life Script: How the Human Genome Discoveries Will Transform Medicine and Enhance Your Health.* New York: Simon & Schuster, 2001.

MUTATION

CONCEPT

A word familiar to all fans of science fiction, *mutation* refers to any sudden change in DNA—deoxyribonucleic acid, the genetic blueprint for an organism—that creates a change in an organism's appearance, behavior, or health. Unlike in the sci-fi movies, however, scientists typically use the word mutant as an adjective rather than as a noun, as, for example, in the phrase "a mutant strain." Mutation is a phenomenon significant to many aspects of life on Earth and is one of the principal means by which evolutionary change takes place. It is also the cause of numerous conditions, ranging from albinism to cystic fibrosis to dwarfism. Mutation indicates a response to an outside factor, and the nature of that factor can vary greatly, from environmental influences to drugsto high-energy radiation.

HOW IT WORKS

DNA, CHROMOSOMES, AND MUTATIONS

Deoxyribonucleic acid, or DNA, is a molecule in the cells of all life-forms that contains genetic codes for inheritance. DNA, discussed elsewhere in this book, is as complex in structure as it is critically important in shaping the characteristics of the organism to which it belongs, and therefore it is not surprising that a subtle alteration in DNA can produce significant results. Alterations to DNA are called mutations, and they can result in the formation of new characteristics that are heritable, or capable of being inherited.

Every cell in the body of every living organism contains DNA in threadlike structures called chromosomes. Stretches of DNA that hold coded instructions for the manufacture of specific proteins are known as genes, of which the human race has approximately 40,000 varieties. If the DNA of a particular gene is altered, that gene may become defective, and the protein for which it codes also may be missing or defective. Just one missing or abnormal protein can have an enormous effect on the entire body: albinism, for instance, is the result of one missing protein.

Mutations also can be errors in all or part of a chromosome. Humans normally have 23 pairs of chromosomes, and an extra chromosome can have a tremendous negative impact. For example, there should be two of chromosome 21, as with all other chromosomes, but if there are three, the result is Down syndrome. People with Down syndrome have a unique physical appearance and are developmentally disabled. Nor is an extra chromosome the only chromosomal abnormality that causes problems: if chromosomes 9 and 22 exchange materials, a phenomenon known as *translocation,* the result can be a certain type of leukemia. Down syndrome also results from translocation.

Germinal mutations are those that occur in the egg or sperm cells and therefore can be passed on to the organism's offspring. Somatic mutations are those that happen in cells other than the sex cells, and they cannot be transmitted to the next generation. This is an important distinction to keep in mind in terms of both the causes and the effects of mutation. If only the somatic cells of the organism are affected, the mutation will not appear in the next generation; on the other hand, if a germinal mutation is involved, what was once an abnormality may become so common in certain populations that it emerges as the norm.

THE ROLE OF MUTATION IN EVOLUTION

Most of the forms of mutation we discuss in this essay appear suddenly (i.e., in a single generation) and affect just a few generations. Yet even such seemingly "normal" characteristics as our ten fingers and ten toes or our two eyes or our relatively hairless skin (compared with that of apes) are ultimately the product of mutations that took shape over the many hundreds of millions of years during which animal life has been evolving. Evolution, in fact, is driven by mutation, along with natural selection (see Evolution).

Over the eons, advantageous mutations, examples of which we look at later, have allowed life to develop and diversify from primitive cells into the multitude of species—including *Homo sapiens*—that exist on Earth today. If DNA replicated perfectly every time, without errors, the only life-forms existing now would be those that existed about three billion years ago: single-cell organisms. Mutations, therefore, are critical to the development of diverse life-forms, a phenomenon known as *speciation* (see Speciation). Mutations that allow an organism to survive and reproduce better than other members of its species are always beneficial, though a mutation that may be beneficial in some circumstances can be harmful in others. Mutations become especially important when an organism's environment is changing—something that has happened often over the course of evolutionary history. And though we cannot watch evolution taking place, we can see how mutations are used among domesticated plants and animals, as discussed later.

REAL-LIFE APPLICATIONS

ETHNICITY AND MUTATION

Every single human trait—blue eyes, red hair, cystic fibrosis, a second toe longer than the big toe, and so on—is the result of some genetic mutation somewhere back down the line. Traits that are shared by all people must have arisen long ago, while other traits occur only in certain populations of people. Traits may be as innocuous as eye color or hair texture or as grave as a shared tendency toward a particular disease. Cystic fibrosis, for instance, is most common in people of northern European descent, while sickle cell anemia (see Amino Acids) occurs frequently in those of African and Mediterranean ancestry. A fatal disorder known as Tay-Sachs is found primarily in Jewish people whose ancestors came from Eastern Europe. In many cases, the particular mutation, while harmful in one regard, proved to be a useful one for that population. We know, for example, that while two copies of the mutant sickle cell anemia gene cause illness, one copy confers resistance to malaria—a very useful trait to people living in the tropics, where malaria is common.

THE PIMA "FAT-STORAGE MUTATION." Researchers have noted a high incidence of obesity among the Pima, a Native American tribe whose ancestral homeland is along the Gila and Salt rivers in Arizona. The Pima tend to eat a diet that is no more fatty than that of the average American—which, of course, means that it is plenty fatty, complete with chips, bologna, ice cream, and all the other high-calorie, low-nutrient foods that most Americans consume. But whereas the average American is overweight, the average Pima is more dramatically so. This suggests that long ago, when the ancestors of the Pima had to face repeated periods of famine in the dry lands of the American Southwest, survival favored the individual or individuals who had a mutation for fat storage. It so happens that today, there is more than enough food at the local supermarket, but by now the Pima as a group has the fat-storage gene. Therefore, many members of the tribe have to undergo strict dietary and exercise regimens so as not to become grossly overweight and susceptible to heart disease and other ailments.

FAVORABLE MUTATIONS

As with other mutations relating to ethnic groups, scientists have hypothesized that some advantage must be conferred upon people with single copies of the cystic fibrosis gene or the Tay-Sachs disease gene. Though many mutations are harmful, others prove to be beneficial to a species by helping it adapt to a particular environmental influence. Useful mutations, in fact, are the driving force behind evolution.

The processes of evolution are usually much too slow for people to discern, but it is possible to

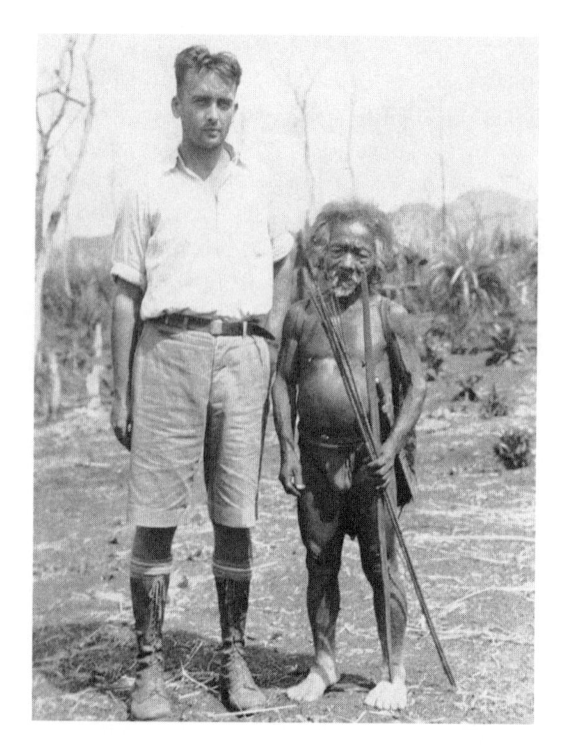

PYGMIES, A GROUP OF PEOPLE IN SOUTHERN AFRICA, APPEAR TO BE MIDGETS THROUGH A GERMINAL MUTATION, BUT IN MOST POPULATIONS THE MUTATION IS SOMATIC, OCCURRING ONLY OCCASIONALLY IN FAMILIES WHOSE OTHER MEMBERS ARE OF ORDINARY SIZE. *(© Bettmann/Corbis. Reproduced by permission.)*

observe the effects of selective breeding when applied to domesticated animals and plants. The artificial selection of pigeons by breeders, in fact, provided the English naturalist Charles Darwin (1809–1882) with a model for his theory of natural selection, discussed in Evolution. Likewise, animal and plant breeders use mutations to produce new or improved strains of crops and livestock. Careful breeding in this manner has spawned the many different breeds of dogs, cats, and horses—each with their characteristic coloring, size, temperament, and so on—that we know today. It also has resulted in crops that are resistant to drought or insects or which have a high yield per acre. Likewise, goldfish, yellow roses, and Concord grapes are all descendants of ancestors with specific mutations.

DISEASES AND MUTATION

The majority of mutations, however, are less than favorable, and this is illustrated by the relationship between mutation and certain hereditary diseases. An example is Huntington disease, a condition that strikes people in their forties or fifties and slowly disables their nervous systems. It produces shaking and a range of other symptoms, including depression, irritability, and apathy, and is usually fatal. The gene associated with Huntington's is dominant.

The horrible degenerative brain condition known as Creutzfeldt-Jakob disease, discussed in Diseases, is usually caused by another mutation. (Though it can be caused by infection, most cases of the disease are the result of heredity.) As with some of the other conditions we have mentioned, this one seems to affect particular groups more than others. Whereas the worldwide incidence of this rare condition is about one in one million, among Libyan Jews the rate is higher. The disease is a type of spongiform encephalopathy, so named because it produces characteristic spongelike patterns on the surface of the brain. Spongiform encephalopathies are caused by the appearance of a prion, a deviant form of protein whose production typically is caused by a mutation.

Most hereditary diseases are, by definition, linked with a mutation. Such is the case with hemophilia, for instance (see Noninfectious Diseases), and with cystic fibrosis, a lethal disorder that clogs the lungs with mucus and typically kills the patient before the age of 30 years. Cystic fibrosis, like Huntington, occurs when a person inherits two copies of a mutated gene. In 1989 researchers found the source of cystic fibrosis on chromosome 7, where an infinitesimal change in the DNA sequence leads to the production of an aberrant protein.

CONGENITAL DISORDERS

In the past, all manner of superstitions arose to explain why a child was born, for instance, with a cleft palate, a situation in which the two sides of the roof of the mouth fail to meet, causing a speech disorder that may be mild or severe. Once known as a *harelip,* the cleft palate was said to have formed as a result of the mother's being frightened by a hare while she was carrying the child. In fact, it is just one example of a congenital disorder, an abnormality of structure or function or a disease that is present at birth. Congenital disorders, which also are called birth defects, may be the result of several different factors, mutation being one of the most significant. Among the many examples of congenital disorder are the hereditary diseases we have already mentioned, as well as dwarfism, Down syn-

AN EXTRA CHROMOSOME 21 CAUSES DOWN SYNDROME. PEOPLE WITH THIS CONGENITAL DISORDER HAVE UNIQUE PHYSICAL FEATURES, SUCH AS A WIDE, FLAT FACE AND SLANTED EYES, AND ARE DEVELOPMENTALLY DISABLED. *(© Laura Dwight/Corbis. Reproduced by permission.)*

drome, albinism, and numerous other conditions.

DWARVES AND MIDGETS. The term *dwarf* has many associations from fairy tales—an example of the combined fascination and revulsion with which people with congenital disorders have long been treated—but it also is used to describe persons of abnormally short stature. A dwarf is distinguished from a midget in a number of ways, all of which indicate that the features of a midget are less removed from the norm. Midgets, while small, have bodies with proportions in the ordinary range. Likewise, the intelligence and sexual development of an adult midget are similar to those of other adults, and a midget or midget couple typically produces children of ordinary size. Pygmies, a group of people in southern Africa, appear to be midgets through a germinal mutation, but in many populations the mutation is somatic, occurring only occasionally in families whose other members are of ordinary size.

Dwarfs, by contrast, have several different disorders. One variety of dwarfism, known in the past as *cretinism,* is characterized by a small, abnormally proportioned body and an impaired mind. On the other hand, several forms of hereditary dwarfism carry with them no ill effect on the mental capacity. For example, people with the type of dwarfism known as *achondroplasia* have short limbs and unusually large heads, but the life span and intelligence of someone with this condition are quite normal. In the case of diastrophic dwarfism, the brain is fine, but the skeleton is deformed, and the risk of death from respiratory failure is high in infancy. Persons with diastrophic dwarfism who survive early childhood, however, are likely to enjoy a normal life span.

DOWN SYNDROME. Like people with many other congenital disorders, those with Down syndrome used to be called by a name that now is considered crude and insensitive: *mongoloid.* The term, when used with a capital *M,* refers to people of east Asian descent and is analogous to other broad racial groupings: Caucasoid, Negroid, and Australoid. In the case of people with Down syndrome, *mongoloid* referred to the unusual facial features that mark someone with that condition.

A person with Down syndrome (caused by an extra chromosome in the 21st chromosomal pair) is likely to have a wide, flat face and eyes that are slanted, sometimes with what is known

AN ALBINO NORTHERN FUR SEAL. A CONDITION THAT RESULTS FROM AN INHERITED DEFECT IN MELANIN METABOLISM (MELANIN IS RESPONSIBLE FOR THE COLORING OF SKIN), ALBINISM IS MARKED BY AN ABSENCE OF PIGMENT FROM THE HAIR, SKIN, AND EYES. *(© John Francis/Corbis. Reproduced by permission.)*

as an *inner epicanthal fold*—all facial characteristics common among people who are racially Mongoloid. Numerous other facial features identify a person with Down syndrome as someone who suffers from a specific congenital disorder, including a short neck, ears that are set low, a small nose, large tongue and lips, and a chin that slopes. People with Down syndrome are apt to have poor muscle tone and possess abnormal ridge patterns on their palms and fingers and the soles of their feet. Heart and kidney problems are common with Down syndrome as well, but one feature is most common of all: mental retardation. The condition occurs in about one of 1,000 live births among women under age 40 but about one in 40 live births to older women. Overall, the incidence is about one in 800 live births. As noted earlier, the cause of Down syndrome is translocation, but the reason translocation occurs is not known.

ALBINISM. Compared with dwarfism or Down syndrome, albinism is not nearly as severe in terms of its effect on a person's functioning. A condition that results from an inherited defect in melanin metabolism (melanin is responsible for the coloring of skin), albinism is marked by an absence of pigment from the hair, skin, and eyes. The hair of an albino tends to be whitish blond, the skin an extremely pale white, and the eyes pinkish. Albinism occurs among other animals: hence the white rats, rabbits, and mice almost everyone has seen. Domestic white chickens, geese, and horses are partial albinos that retain pigment in their eyes, legs, and feet. As was once true of people with other congenital disorders, human albinos once inspired fear and awe. Sometimes they were killed at birth, and in the mid–nineteenth century, albinos were exhibited in carnival sideshows. In these cruel spectacles, sometimes whole families were put on display, touted as a unique race of "night people" who lived underground and came out only when the light was dim enough not to hurt their eyes.

On the other hand, some ethnic groups experience enough albino births that another one causes no excitement. For example, among the San Blas Indians of Panama, one in approximately 130 births is an albino, compared with one in 17,000 for humans as a whole. Albinism comes about when melanocytes (melanin-producing cells) fail to produce melanin. In tyrosinase-negative albinism, the most common form, the enzyme tyrosinase (a catalyst in the conversion of tyrosine to melanin) is missing from the

KEY TERMS

AMINO ACIDS: Organic compounds made of carbon, hydrogen, oxygen, nitrogen, and (in some cases) sulfur bonded in characteristic formations. Strings of amino acids make up proteins.

CHROMOSOME: DNA-containing bodies, located in the cells of most living things, that hold most of the organism's genes.

CONGENITAL DISORDER: An abnormality of structure or function or a disease that is present at birth. Congenital disorders also are called *birth defects.*

DNA: Deoxyribonucleic acid, a molecule in all cells, and many viruses, containing genetic codes for inheritance.

GENE: A unit of information about a particular heritable trait. Usually stored on chromosomes, genes contain specifications for the structure of a particular polypeptide or protein.

GERMINAL MUTATION: A mutation that occurs in the egg or sperm cells, which therefore can be passed on to the organism's offspring.

HERITABLE: Capable of being inherited.

MUTAGEN: A chemical or physical factor that increases the rate of mutation.

MUTATION: Alteration in the physical structure of an organism's DNA, resulting in a genetic change that can be inherited.

NATURAL SELECTION: The process whereby some organisms thrive and others perish, depending on their degree of adaptation to a particular environment.

ORGANIC: At one time chemists used the term *organic* only in reference to living things. Now the word is applied to compounds containing carbon and hydrogen.

POLYPEPTIDE: A group of between 10 and 50 amino acids.

PROTEINS: Large molecules built from long chains of 50 or more amino acids. Proteins serve the functions of promoting normal growth, repairing damaged tissue, contributing to the body's immune system, and making enzymes.

RNA: Ribonucleic acid, the molecule translated from DNA in the cell nucleus, the control center of the cell, that directs protein synthesis in the cytoplasm, or the space between cells.

SOMATIC MUTATION: A mutation that occurs in cells other than the reproductive, or sex, cells. These mutations, as contrasted with germinal mutations, cannot be transmitted to the next generation.

SPECIATION: The divergence of evolutionary lineages and creation of new species.

TRANSLOCATION: A mutation in which chromosomes exchange parts.

melanocytes. When the enzyme is missing, no melanin is produced. In tyrosinase-positive albinism, a defect in the body's tyrosine transport system impairs melanin production. One in every 34,000 persons in the United States has tyrosinase-negative albinism. It is equally common among blacks and whites, while more blacks than whites are affected by tyrosinase-positive albinism. Native Americans have a particularly high incidence of both forms of albinism.

MUTAGENS AND OTHER CAUSES

As might be expected, cells that divide many, many times in a lifetime are more at risk of errors and mutations than cells that divide less frequently. In a human female, egg cells are fully formed at birth, and they never divide thereafter. By contrast, sperm cells are being produced constantly, and the older a man is, the more frequently his sperm-producing cells have divided. By age 20 they will have divided 200 times and by age 45 about 770 times. This has led scientists to hypothesize that when a baby is born with a congenital disorder caused by an error in cell division, the father is the parent more likely to have contributed the gene with the mutation.

This is just one example of why mutation occurs. Many mutations are caused by mutagens—chemical or physical factors that increase the rate of mutation. Some mutagens occur naturally, and some are synthetic. Cosmic rays from space, for instance, are natural, but they are mutagenic. Some naturally occurring viruses are considered mutagenic, since they can insert themselves into host DNA. Hydrogen and atomic bombs are man-made, and they emit harmful radiation, which is a mutagen. Recreational drugs, tobacco, and alcohol also can be mutagens in the bodies of pregnant women. The first mutagens to be identified were carcinogens, or cancer-causing substances. Carcinogens in chimney soot were linked with the chimney sweep's cancer of late eighteenth-century England, discussed in Noninfectious Diseases. In fact, cancer itself is a kind of mutation, involving uncontrolled cell growth. Other environmental factors that are known to bring about mutations include exposure to pesticides, asbestos, and some food additives, many of which have been banned.

WHERE TO LEARN MORE

"Are Mutations Harmful?" Talk. Origins (Web site). <http://www.talkorigins.org/faqs/mutations.html>.

Human Gene Mutation Database, Institute of Medical Genetics, University of Wales College of Medicine (Web site). <http://archive.uwcm.ac.uk/uwcm/mg/hgmd0.html>.

Kimball, Jim. *Mutations. Kimball's Biology Pages* (Web site). <http://www.ultranet.com/~jkimball/BiologyPages/M/Mutations.html>.

"Mutations." Brooklyn College, City University of New York (Web site). <http://www.brooklyn.cuny.edu/bc/ahp/BioInfo/SD.Mut.HP.html>.

Patterson, Colin. *Evolution.* Ithaca: Comstock Publishing Associates, 1999.

Reilly, Philip. *Abraham Lincoln's DNA and Other Adventures in Genetics.* Cold Spring Harbor, NY: Cold Spring Harbor Laboratory Press, 2000.

Twyman, Richard M. *Advanced Molecular Biology: A Concise Reference.* Oxford, UK: Bios Scientific Publishers, 1998.

Weinberg, Robert A. *One Renegade Cell: How Cancer Begins.* New York: Basic Books, 1998.

REPRODUCTION AND BIRTH

REPRODUCTION
SEXUAL REPRODUCTION
PREGNANCY AND BIRTH

REPRODUCTION

CONCEPT

The term reproduction encompasses the entire variety of means by which plants and animals produce offspring. Reproductive processes fall into two broad groupings: sexual and asexual, the latter being the means by which bacteria and algae reproduce. Many plants reproduce sexually by means of pollination, and some plants alternate between sexual and asexual forms of reproduction. Other creatures, such as bees and ants, reproduce through a form of reproduction called parthenogenesis, which is neither fully sexual nor asexual.

HOW IT WORKS

Asexual Reproduction

Asexual reproduction involves only one organism, as opposed to two in sexual reproduction. It occurs when a single cell divides to form two daughter cells that are genetically identical to the parent cell. This process is known as fission, and it may take the form either of binary fission, in which two new cells are produced, or multiple fission, which results in the creation of many new cells. Since there is no fusion of two different cells, the daughter cells produced by asexual reproduction are genetically identical to the parent cell. Asexual reproduction usually takes place by mitosis, a process during which the chromosomes in a cell's nucleus are duplicated before cell division. (Mitosis, chromosomes, and many other topics referred to in this essay are discussed in considerably more detail in Genetics.)

Whereas sexual reproduction is extremely complex—and human sexual reproduction is much more so, freighted as it is with degrees of meaning that go far beyond mere biology—asexual reproduction is a fairly simple, cellular process. Of course, nothing in nature is really simple, and, in fact, the dividing and replication of DNA (deoxyribonucleic acid, the genetic blueprint material found in each cell) is a complicated subject; however, that subject, too, is discussed in the essay Genetics. DNA is located at the cell nucleus, which is the cell's control center, and the nucleus is the first part of the cell to divide in asexual reproduction. After the nucleus splits, the cytoplasm, or the cellular material external to the nucleus, then divides. The result is the formation of two new daughter cells whose nuclei have the same number and kind of chromosomes as the parent.

The adaptive advantage of asexual reproduction is that organisms can reproduce quickly and by doing so colonize favorable environments rapidly. (See Evolution for more about the importance of adaptation and environment in shaping species.) For example, some bacteria can double their numbers every 20 minutes. In addition to bacteria, which are discussed in more detail in Infection, other life-forms that reproduce asexually include protozoa (varieties of which are examined in Parasites and Parasitology), blue-green algae, yeast, dandelions, and flatworms.

Sexual Reproduction

Sexual reproduction involves the union of two organisms rather than the splitting of one. Like asexual reproduction, it is a process that takes place at the cellular level. In sexual reproduction it is not binary fission that occurs, but the fusion of two cells. Nor are the two cells identical;

rather, the cells—known as gametes—can be identified as either male or female according to the makeup of their chromosomes. The male gamete is called a sperm cell, and the female gamete is termed an egg cell. In sexual reproduction, the sperm cell fuses, or bonds, with the egg cell to produce a cell that is genetically different from either of the parent cells. This process of fusion is known as fertilization, and the fertilized egg is called a zygote. Gametes are produced in the male testes and female ovaries by a splitting process called meiosis. (Meiosis and other terms mentioned briefly in these paragraphs are discussed in much more detail in Genetics.)

Meiosis produces haploid cells, or ones that have half the number of chromosomes as are in a normal cell for that species. When the haploid sperm and egg cells fuse at fertilization, however, the chromosomes from both combine, so that the normal number of chromosomes appears in the zygote. The shuffling of the parents' genetic material that happens during meiosis allows for new gene combinations in offspring that account for variations between offspring (which is why you don't look just like your siblings) and which, over time, can improve a species' chances of survival.

REAL-LIFE APPLICATIONS

EXAMPLES OF ASEXUAL REPRODUCTION

As we noted earlier, bacteria, blue-green algae, most protozoa, yeast, and flatworms all reproduce asexually, as do mosses and starfish. (The last actually reproduce both sexually and asexually by means of alternation of generations, discussed later.) The products of asexual reproduction are known as clones—an example of the fact, discussed in Genetic Engineering, that cloning and the concept of clones are not as new as one might imagine. (See that essay for much more about artificial cloning.) A starfish can regenerate and eventually produce a whole new organism from a single severed appendage, while flatworms divide in two and regenerate to form two new flatworms. This formation of a separate organism is obviously much more complex than the simple splitting of single bacteria cells, but it is still a form of asexual reproduction.

VEGETATIVE PROPAGATION. Strawberries reproduce by forming growths called runners, which grow horizontally and generate new stalks. At some point, the runner decomposes, leaving a new plant that is a clone of the original. This is an example of vegetative propagation, a term for a number of processes by which crop plants are produced asexually. Vegetative propagation is used for such crops as potatoes, bananas, raspberries, pineapples, and some flowering plants. Its advantage to farmers is that the crops will be more uniform than those grown from seed. Furthermore, some plants are difficult to cultivate from seed, and the vegetative propagation of those plants makes it possible to grow crops that otherwise would not be available for commercial marketing.

In reproducing potatoes through vegetative propagation, farmers plant the so-called eyes to produce duplicates of the parent. With banana plants, the farmer separates the suckers that grow from the root of the plant and plants them. The farmer raising raspberry bushes bends the branches and covers them with soil, whereupon a process not unlike that of the runner growth of mosses takes place: the branches eventually grow into a separate plant, with their own root system, and ultimately can be detached from the parent plant.

BETWEEN ASEXUAL AND SEXUAL

The example of vegetative propagation suggests that there is not a sharp dividing line between sexual and asexual reproduction—that is, that many organisms can reproduce either way. This is true even of humans, who, in theory, could be cloned, though the technology to do so—let alone resolution of the ethical issues of the procedure—lies in the far distant future. (See Genetic Engineering for more on this subject.) Even humans, however, can use external fertilization, which is sexual reproduction without sexual intercourse (see Sexual Reproduction).

Plants go through a process known as alternation of generations, in which they alternate as sexual and asexual reproducers, or gametophytes and sporophytes, respectively. In the asexual stage, the sporophyte produces diploid reproductive cells called spores, which develop into gametophytes. These gametophytes produce haploid gametes, which then unite sexually to form a diploid zygote that grows into a sporophyte. In

A PINEAPPLE TOP PLACED ON SOIL CAN ROOT AND DEVELOP INTO A PLANT. NUMEROUS CROP PLANTS, AMONG THEM, POTATOES, BANANAS, RASPBERRIES, AND PINEAPPLES, ARE PRODUCED ASEXUALLY, A PROCESS CALLED VEGETATIVE PROPAGATION. *(© Corbis. Reproduced by permission.)*

plain English, this means that the asexual "grandparent" generates a sexually reproducing "child," which in turn produces a "grandchild" that is asexual, like its grandparent.

At one phase in the alternation of generation for mosses, for instance, male and female moss plants grow from spores. Male moss plants produce sperm cells, which, when the moss receives rainfall, are able to propagate because they have a medium (water) in which to move. They fertilize the female plants, producing zygotes. The zygote grows on top of the female moss plant, which helps to store moisture and thus provides a hospitable environment in which the zygote can develop. The zygote eventually produces haploid spores, which it releases into the air. These tiny spores, carried by the wind, float away from their point of origin until they come to rest, and soon the cycle begins once again.

PARTHENOGENESIS. There are also organisms, including bees, ants, wasps, and other insects, that reproduce in a way that is neither fully sexual nor asexual. This is parthenogenesis, a type of reproduction in which a gamete develops without fertilization. In other words, a sex cell is reproduced without actual intercourse between male and female. The gamete is almost always female—a fact indicated in the name itself, which comes from *parthenos,* Greek for "maiden."

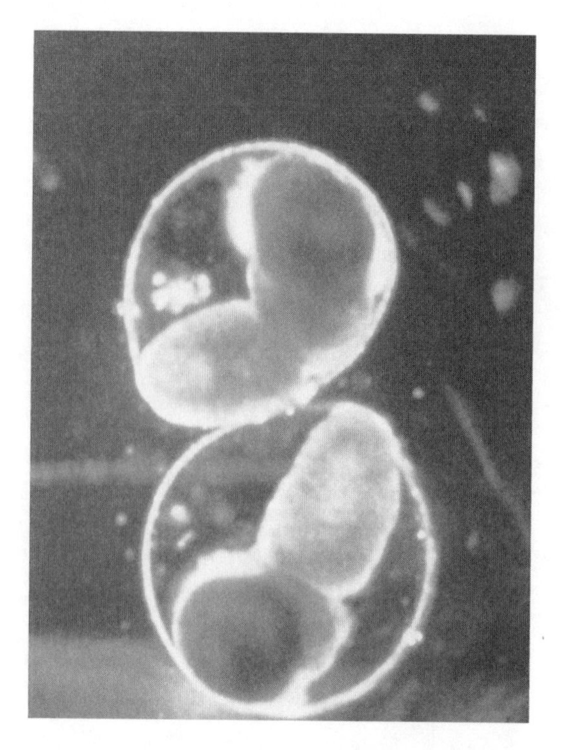

THE ACORN WORM UNDERGOES ASEXUAL REPRODUCTION BY BUDDING; SMALL PIECES FRAGMENT FROM THE TRUNK, AND EACH GROWS INTO A NEW WORM. *(© Lester V. Bergman/Corbis. Reproduced by permission.)*

The Parthenon in Athens, like the city itself, is named after the goddess Athena (also called Minerva), who was known by the nickname Parthenos. She is said to have been born fully formed, having sprung from the head of her father, Zeus, dressed in armor and ready for battle. Thus, her own birth was a form of parthenogenesis, a word whose second half (a name well known from the Bible) means "beginning."

POLLEN AND POLLINATION

Pollen is a fine, powdery substance consisting of microscopic grains containing the male gametophyte of certain plants that reproduce sexually. These plants include angiosperms, a type of plant that produces flowers during sexual reproduction, and gymnosperms, which reproduce sexually through the use of seeds that are exposed and not hidden in an ovary, as with an angiosperm. Pollen is designed for long-distance dispersal from the parent plant, so that fertilization can occur. Pollination is the transfer of pollen from the male reproductive organs to the female reproductive organs of a plant, and it precedes fertilization. In other words, pollination is the equivalent of sexual intercourse for seed-bearing plants. Actually, cross-pollination, or the transfer of pollen from one plant to another, would perhaps be analogous to sexual intercourse in animals. Pollination occurs in seed-bearing plants, as opposed to the more primitive spore-producing plants, such as ferns and mosses. Gymnosperms, such as pines, firs, and spruces, produce male and female cones, whereas angiosperms produce flowers containing a male organ called the stamen and a female organ called the pistil. Both types of plants rely on insects and other creatures to aid in the pollen transfer.

DARWIN'S MOTH. The German physician and botanist Rudolf Jakob Camerarius (1665–1721) was the first scientist to demonstrate that plants reproduce sexually, and he pioneered the study of pollination. One of the scientists influenced by his work was the English naturalist Charles Darwin (1809–1882), who discussed the subject in *The Various Contrivances by which Orchids Are Fertilized by Insects* (1862). Darwin wrote this book partly to support the ideas on evolution presented in his much more well known book *Origin of Species* (1859). In *Various Contrivances,* he suggested that orchids and their insect pollinators evolved by interacting with one another over many generations.

As an example, he discussed *Angraecum sesquipedale,* an orchid native to Madagascar. Darwin had not seen the plant in its native habitat, however; he had looked only at its dried leaves. The white flower of this orchid has a foot-long (30 cm) tubular spur with a small drop of nectar at its base, and from observing this, he hypothesized that the orchid had been pollinated by an insect with a foot-long tongue. This hypothesis, he wrote, "has been ridiculed by some entomologists," or scientists who study insects. After all, no such creature had been found in Madagascar. But then, around the turn of the nineteenth century—some two decades after Darwin's death—it was found. A Madagascan moth was discovered that had a foot-long tongue that uncoils to sip the nectar of *A. sesquipedale* as it cross-pollinated the flowers.

PLANTS AND THEIR POLLINATORS. Angiosperms and gymnosperms are discussed in Ecosystems and Ecology, where each is compared in terms of its degree of adaptation to its environment. Angiosperms seem to be the hands-down winner: by enlisting the aid of

KEY TERMS

ALTERNATION OF GENERATIONS: A process whereby plant generations alternate as sexual and asexual reproducers—gametophytes and sporophytes, respectively.

ASEXUAL REPRODUCTION: One of the two major varieties of reproduction (along with sexual reproduction), In contrast to sexual reproduction, which involves two organisms, asexual reproduction involves only one. Asexual reproduction occurs when a single cell divides through mitosis to form two daughter cells, which are genetically identical to the parent cell.

BINARY FISSION: The process in asexual reproduction whereby a single cell divides to form two daughter cells that are genetically identical to the parent cell.

CLONE: A cell, group of cells, or organism that contains genetic information identical to that of its parent cell or organism.

CLONING: A specialized genetic process whereby clones are produced. Cloning is a form of asexual reproduction.

CHROMOSOME: A DNA-containing body, located in the cells of most living things, that holds most of the organism's genes.

CROSS-POLLINATION: The transfer of pollen from one plant to another.

CYTOPLASM: The material inside a cell that is external to the nucleus.

DIPLOID: A term for a cell that has the basic number of doubled chromosomes.

DNA: Deoxyribonucleic acid, a molecule in all cells, and many viruses, that contains genetic codes for inheritance.

EGG CELL: A female gamete.

ENZYME: A protein material that speeds up chemical reactions in the bodies of plants and animals without itself taking part in or being consumed by those reactions.

FERTILIZATION: The process of cellular fusion that takes place in sexual reproduction. The nucleus of a male reproductive cell, or gamete, fuses with the nucleus of a female gamete to produce a zygote.

GAMETE: A reproductive cell—that is, a mature male or female germ cell that possesses a haploid set of chromosomes and is prepared to form a new diploid by undergoing fusion with a haploid gamete of the opposite sex. Sperm and egg cells are, respectively, male and female gametes.

GAMETOPHYTE: In alternation of generations, a gametophyte is a plant that reproduces sexually.

GERM CELL: One of two basic types of cells in a multicellular organism. In contrast to somatic or body cells, germ cells play a part in reproduction.

HAPLOID: A term for a cell that has half the number of chromosomes that appear in a diploid or somatic cell.

MEIOSIS: The process of cell division that produces haploid genetic material. Compare with *mitosis.*

MITOSIS: A process of cell division that produces diploid cells, as in asexual reproduction. Compare with *meiosis.*

NUCLEUS: The control center of a cell, where DNA is stored.

OVARY: Female reproductive organ that contains the eggs.

KEY TERMS CONTINUED

OVULE: Female haploid gametophyte of seed plants, which develops into a seed upon fertilization by a pollen grain.

PARTHENOGENESIS: A type of reproduction that involves the development of a gamete without fertilization. In other words, a sex cell (usually female) is reproduced without actual intercourse between male and female.

POLLEN: Male haploid gametophyte of seed plants (including angiosperms and gymnosperms), which unites with the ovule to form a seed. Pollen is a fine, powdery substance consisting of microscopic grains.

POLLINATION: The transfer of pollen from the male reproductive organs to the female reproductive organs of a plant. Pollination precedes fertilization. See also *cross-pollination.*

REGENERATION: A biological process among some lower animals whereby a severed body part is restored by the growth of a new one.

SEXUAL REPRODUCTION: One of the two major varieties of reproduction (along with asexual reproduction). In contrast to asexual reproduction, which involves a single organism, sexual reproduction involves two. Sexual reproduction occurs when male and female gametes undergo fusion, a process known as fertilization, and produce cells that are genetically different from those of either parent.

SOMATIC CELL: One of two basic types of cells in a multicellular organism. In contrast to germ cells, somatic cells (also known as body cells) are not involved in reproduction; rather, they make up the tissues, organs, and other parts of the organism.

SPERM CELL: A male gamete.

SPOROPHYTE: In alternation of generations, a sporophyte is a plant that reproduces asexually.

ZYGOTE: A diploid cell formed by the fusion of two gametes.

insects and other pollinators, they manage to pollinate much more efficiently than gymnosperms, which have to produce vast quantities of pollen for each grain that reaches its target. Typically, pollination benefits the animal pollinator by supplying it with sweet nectar and, of course, benefits the plant by providing direct transfer of pollen from one plant to the pistil of another plant. For this reason, specific plant and animal species have developed a relationship of mutualism, a form of symbiosis in which each participant reaps benefits (see Symbiosis). In many cases, plant and pollinator have evolved together, and it is possible to determine which animal pollinates a certain flower species simply by studying the morphologic features (shapes), color, and odor of the flower.

For example, some flowers are pure red, or nearly pure red, and have very little odor. In most such situations, the pollinator is a bird species, since birds have excellent vision in the red region of the spectrum but a rather undeveloped sense of smell. It so happens that Europe, which has no pure red native flowers, also has no bird-pollinated native flower species. Not all bird-pollinated flowers are red, but they are all characterized by striking, and sometimes contrasting, colors that readily catch the eye. Examples of plants pollinated by birds include the cardinal flower, the red columbine, the hibiscus, the eucalyptus, and varieties of orchid, cactus, and pineapple.

Some flowering plants have a very strong odor but are very dark, or at least drab, in color.

These flowers and plants—examples include the saguaro cactus, century plant, or cup-and-saucer vine—are often pollinated by bats, which have very poor vision, are typically active during the night, and have a very well developed sense of smell. The flowers of many plant species are marked with special pigments called flavonoids, which absorb ultraviolet light and appear to direct the pollinator toward the pollen and nectar. These pigments are invisible to humans and most animals, but bees' eyes have special ultraviolet photoreceptors that enable the bees to detect patterns and so pollinate these flowers.

WHERE TO LEARN MORE

"Asexual Reproduction Lab." Lester B. Pearson College of the Pacific (Web site). <http://www.pearson-college.uwc.ca/pearson/biology/asex/asex.htm>.

Canine Reproduction (Web site). <http://www.labbies.com/canine_reproduction_table_of_con.htm>.

CRES: The Center for Reproduction of Endangered Species/San Diego Zoo (Web site). <http://www.sandiegozoo.com/conservation/cres_home.html>.

Elia, Irene. *The Female Animal.* New York: Henry Holt, 1988.

"Flowering Plant Reproduction." Estrella Mountain Community College (Web site). <http://gened.emc.maricopa.edu/bio/bio181/BIOBK/BioBookflowers.html>.

Kevles, Bettyann. *Females of the Species: Sex and Survival in the Animal Kingdom.* Cambridge, MA: Harvard University Press, 1986.

Kimball, Jim. "Asexual Reproduction." *Kimball's Biology Pages* (Web site). <http://www.ultranet.com/~jkimball/BiologyPages/A/AsexualReproduction.html>.

Maxwell, Kenneth E. *The Sex Imperative: An Evolutionary Tale of Sexual Survival.* New York: Plenum, 1994.

The Pollination Home Page (Web site). <http://pollinator.com/>.

Reproduction (Web site). <http://www.factmonster.com/ce6/sci/A0841565.html>.

Topoff, Howard R. *The Natural History Reader in Animal Behavior.* New York: Columbia University Press, 1987.

Walters, Mark Jerome. *The Dance of Life: Courtship in the Animal Kingdom.* New York: Arbor House, 1988.

SEXUAL REPRODUCTION

CONCEPT

Sexual reproduction is one of the two major ways, along with asexual reproduction, that plants and animals create offspring and thus propagate the species. Critical to sexual reproduction is the process of fertilization, whereby the male and female sex cells fuse, or bond. Fertilization may be of two types, either internal or external, and though humans normally fertilize by the first of those means, they may use the second, in the form of *in vitro* fertilization. Humans, of course, have by far the most complicated reproductive process, inasmuch as it is surrounded by a vast societal, interpersonal, and moral framework that is not a factor in animal reproduction. Part of that framework is the activity that precedes sexual reproduction: attraction, courtship, and so forth. Although no other animal's courtship rituals rival those of humans for sophistication, some of them are quite impressive in their complexity.

HOW IT WORKS

The Reproductive System

The contrast between sexual and asexual reproduction is examined in Reproduction, an essay that also provides examples of plant reproduction through pollination. The present essay is concerned primarily with human sexual reproduction and secondarily with animal sexual reproduction. Some technical aspects of reproduction at the cellular level require consultation of processes explained in Genetics; here we confine our technical discussion to reproduction at the level of organs, fluids, and other bodily components. Reproduction is facilitated by the reproductive system, a group of organized structures that can be subdivided into male and female reproductive systems. During puberty, which typically occurs between the ages of 10 and 14 years, the reproductive systems of both sexes mature. This phase is marked in part by the release of eggs (female sex cells) in the female ovary and the formation of sperm (male sex cells) in the male testes. Reproduction can take place only when a sperm unites with an egg, a process called *fertilization.*

THE MALE REPRODUCTIVE SYSTEM. The testes are the pair of male reproductive glands located in the scrotum, a skin-covered sac that hangs from the groin. Each testis produces sperm cells, while the testes as a whole secrete testosterone. Testosterone is a hormone—a type of molecule that sends signals to spots remote from its point of origin to induce specific effects on the activities of other cells. Testosterone is associated with masculinity, though females secrete it in much smaller quantities as well. In males, testosterone secretion is critical to the development of secondary sexual characteristics—those unique traits that mark a person as a male or female, though they do not occur in the sexual organs themselves. A deepened voice is an example of a male secondary sex characteristic evident at puberty.

Sperm cells produced in the testes move to the epididymis, a coiled tube at the base of the penis where they are stored and matured. During ejaculation, or the ejection of sperm from the penis during orgasm, sperm travel from the epididymis through a long tube called the *vas deferens* to the urethra. This single tube, which extends from the bladder to the tip of the penis, is also the means by which urine passes out of the body. Liquid secretions from various glands combine with sperm (itself a gooey substance that is barely liquid) to form the semen, or seminal fluid. Ejaculated semen may contain as many as 400 million sperm.

THE FEMALE REPRODUCTIVE SYSTEM

The female system is much more complicated than the male version and has a role in all stages of reproduction. Whereas the male system primarily delivers semen to the vagina, the female system plays a critical part from fertilization until long after the birth of offspring. It produces ova, or eggs, receives sperm from the penis, houses and provides nutrients to the developing zygote (fertilized egg) and later the embryo and fetus, gives birth to offspring, and feeds those offspring after birth.

The visible part of the female reproductive system, which, of course, is not even half of the entire picture, includes the opening of the vagina and the external genital organs, or vulva. The vagina, a muscular tube extending from the uterus to the outside of the body, is the receptacle for sperm ejaculated during sexual intercourse and also forms part of the birth canal that will be used later, when the offspring comes to term. The external genital organs, known collectively as the *vulva,* include the labia, folds of skin on both sides of the openings to the vagina and urethra; the clitoris, a small, sensitive organ that is comparable to the male penis inasmuch as it swells when stimulated; and the mons pubis, a mound of fatty tissue above the clitoris.

THE OVARIES AND MENSTRUATION. Eggs are produced in the ovaries, oval-shaped organs in the groin that also generate sex hormones. At birth, a female's ovaries contain hundreds of thousands of undeveloped eggs, each surrounded by a group of cells to form a follicle, or sac; however, only about 360-480 follicles reach full maturity. During puberty the action of hormones causes several follicles to develop each month. Normally, just one follicle fully matures, rupturing and releasing an ovum through the ovary wall in a process called *ovulation.* The mature egg enters one of the paired fallopian tubes, where it may be fertilized by a sperm and move on to the uterus to develop into a fetus. The lining of the uterus, called the *endometrium,* prepares for pregnancy each month by thickening, but if fertilization does not take place, the endometrium is shed during menstruation.

FERTILIZATION

During sexual intercourse, a man releases approximately 300 million sperm into a woman's vagina, but only one of the sperm can fertilize the ovum. The successful sperm cell must enter the uterus, swim up the fallopian tube (a trumpet-shaped passageway between the ovary and the uterus) to meet the ovum, and then pass through a thick coating, known as the *zona pellucida,* that surrounds the egg. The head of the sperm cell contains enzymes (a type of protein that speeds up chemical reactions—see Enzymes) that break through the zona pellucida and allow the sperm to penetrate the egg. Once the head of the sperm is inside the egg, the tail falls off, and the outside of the egg thickens to prevent another sperm from entering. Many variables affect whether fertilization occurs after intercourse among humans. One factor is a woman's ovulatory, or menstrual, cycle. Human eggs can be fertilized for only a few days after ovulation, which typically occurs only once every 28 days. (To learn about what happens after fertilization, see Pregnancy and Birth.)

REAL-LIFE APPLICATIONS

EXTERNAL FERTILIZATION

Most land animals use some form of internal fertilization similar to that which we have described for humans. External fertilization, on the other hand, is more common among aquatic animals, who simply dump their sperm and eggs into the water and let currents mix the two male and female cells together. The sea urchin is a typical example: a male sea urchin releases several billion sperm into the water, and these sperm then swim toward eggs released in the same area. Fertilization occurs within seconds when sperm come into contact and fuse with eggs. As noted in Reproduction, external fertilization is essentially sexual reproduction without sexual intercourse. For humans the process of reproduction by external means may lack the intimacy of internal reproduction, but since 1978 a form of external fertilization has offered the opportunity of con-

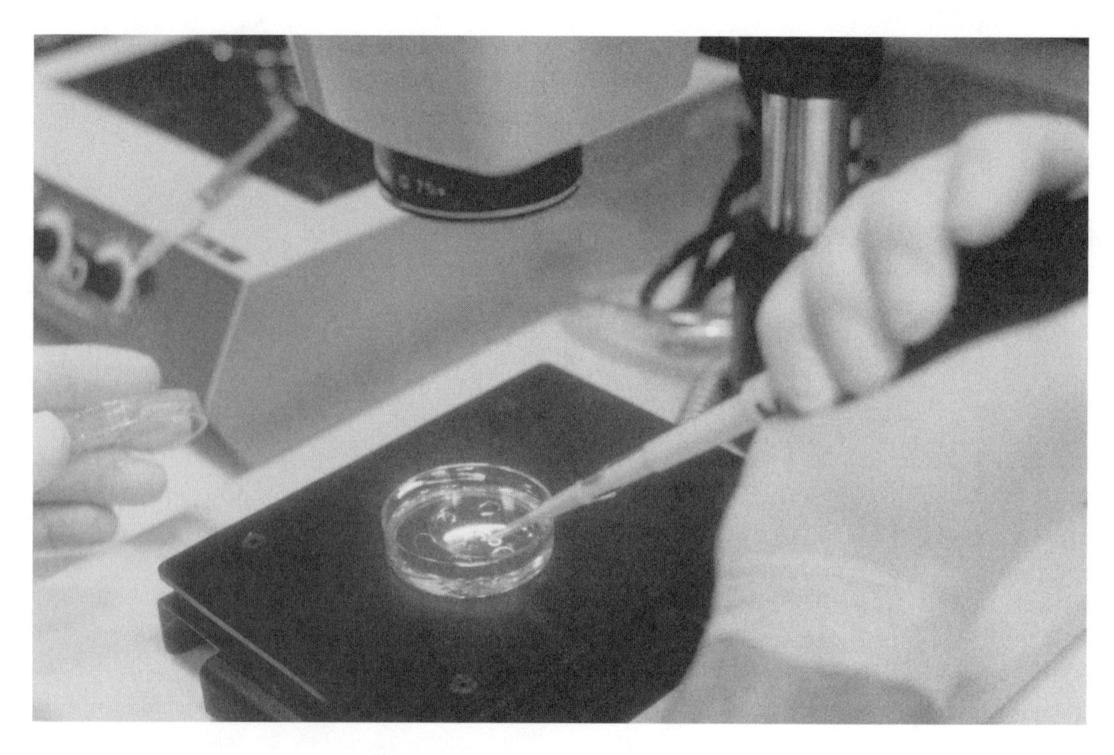

DURING IN VITRO FERTILIZATION, EGGS ARE REMOVED SURGICALLY FROM A FEMALE'S REPRODUCTIVE TRACT AND FERTILIZED IN A TEST TUBE OR PETRI DISH WITH SPERM (SHOWN HERE). AFTER THE FERTILIZED EGGS HAVE DIVIDED TWICE AND AN EMBRYO IS STARTING TO FORM, THEY ARE REINTRODUCED TO THE FEMALE'S BODY. *(© Owen Franken/Corbis. Reproduced by permission.)*

ceiving children to couples who otherwise might have remained childless.

IN VITRO FERTILIZATION. This form of external fertilization is known as *in vitro,* or "in glass"—that is, in a glass test tube or petri dish. The term is contrasted with *in vivo,* meaning "in a living organism," or *in utero,* "in the uterus." During *in vitro* fertilization, eggs are removed surgically from a female's reproductive tract. They then can be fertilized in a test tube or petri dish by sperm taken from the woman's partner or another male. After the fertilized eggs have divided twice, indicating that the operation has "taken" and an embryo is starting to form, they are reintroduced to the female's body. If all goes well, the embryo and fetus eventually result in a normal birth.

In vitro fertilization has been performed successfully on a variety of domestic animals since the 1950s but on humans only since the late 1970s. Two English physicians, Patrick Steptoe (1913–1988) and Robert G. Edwards (1925–), developed a method for stimulating ovulation with hormone treatment and then retrieving the nearly mature eggs and placing them in a petri dish to mature. In their breakthrough 1978 operation, they added male sperm to the egg in the petri dish, where fertilization took place, and then implanted the eight-celled embryo in the mother's body. The result of this extraordinary operation was a healthy baby named Louise Brown.

A WORLD WITHOUT SEX? The birth of Louise, whose twentieth birthday was celebrated in Britain with great fanfare, excited fears and controversy similar to those surrounding cloning (see Genetic Engineering). As with cloning, the fear was that the technology of *in vitro* fertilization would lead to the depersonalized manufacturing of human beings associated with some nightmarish future society, but this has not come to pass for several reasons. One is that *in vitro* fertilization is successful only about 15% of the time. Another, much more significant reason is that there are few women capable of producing a child through internal fertilization who would want to conceive without the intimacy of sexual intercourse.

There are exceptions, of course, in real life (two women in a same-sex relationship who want a child, for example) as well as in fiction. In the latter category, the character of Jenny Fields in the American novelist John Irving's *World According to Garp*—published, ironically, in the same year as the birth of Louise Brown—would undoubtedly have conceived a child by *in vitro* means if those means had been available to her. As it was, Jenny, who conceived the book's title character during World War II, manages to do so by having intercourse purely for the purpose of producing offspring. She does this by choosing a wounded, brain-damaged airman who will never remember having had sex with her and who dies shortly thereafter.

COURTSHIP

There is something far greater than mere biochemistry involved in sexual reproduction, which is why humans enjoy the intimacy of producing a baby. This is also why courtship and mating rituals are a significant part of human life, providing a means by which a male and female join as partners. Such is true even in modern America and the West in general, where the aftermath of the 1960s sexual revolution has left behind a world largely stripped of its former mystery. For better or worse, sex before marriage is a common part of life today in a way that it was not before the 1960s, with the advent of birth control and "free love," and today there is little stigma attached to the conception of a child outside wedlock. That much has changed—and changed dramatically—but humans still practice courtship.

Males still have to prove to females that they are suitable mates, usually by displaying their physical prowess or some other attribute associated with masculinity. Females still do most of the choosing, and despite all the changes in views toward male and female roles, the ideas of basic male and female differences seem to be hard-wired into the minds of most people. This is particularly so of people who are heterosexual and especially those who either have children or intend to have them. Such attitudes are not surprising, since humans, while being something more than animals, are still animals as well.

ANIMAL COURTSHIP. Among animals, courtship is a complex set of behaviors that leads to mating. Courtship behavior communicates to each potential mate that the other is not a threat and serves to reveal to each that the species, gender, and physical condition of the other are suitable for mating. During courtship,

AMONG ANIMALS, COURTSHIP IS A COMPLEX SET OF BEHAVIORS THAT LEADS TO MATING. HERE MALE FRIGATE BIRDS DISPLAY THEIR INFLATED SCARLET POUCHES TO ATTRACT FEMALES. *(© Wolfgang Kaehler/Corbis. Reproduced by permission.)*

animals use rituals, a series of behaviors for communication that is performed the same way by all the males or females in a species. They are governed by fixed-action patterns (FAPs), which are virtually identical with instinct (see Instinct and Learning). In courtship, some animals leap and dance, others sing, and still others ruffle their feathers or puff up pouches. The male peacock displays his glorious plumage to the female, and humpback whales advertise their presence under the sea by singing a song that can be heard hundreds of miles away. Courtship behavior enables an animal to find, identify, attract, and arouse a mate. Animals use signals, such as the release of pheromones, or scent signals, as well as visual displays to claim a particular mate or a territory.

CHOOSING A MATE. Usually, the females do the choosing. In some species of birds, males display themselves in a small communal area called a *lek,* where females select a mate from the parading males. Across the animal kingdom, males generally compete with one another for mates, either by fighting or by ritualized displays, and females pick the best quality of

"The Three Graces" by Peter Paul Rubens. A more amply formed female body is an ideal much closer to nature, since our evolutionary lineage has tended to favor women with large hips who are capable of bearing many children. *(© Archivo Iconografico, S.A./Corbis. Reproduced by permission.)*

male available. Several basic factors influence a female in her choice of mate. First, she wants a male who can provide for her offspring, which is why the female tern (a type of bird) selects a good fish catcher. As part of courtship among common terns, the male birds display fish to the females and may even feed them to the females as a way of demonstrating their ability to feed young. Among the long-jawed long-horned beetles that live in the Arizona desert, males battle each other for saguaro cactus fruit, and the females mate in exchange for access to the fruit.

Genetic fitness, or the ability to survive—and to advance the survival of the species—is another important factor in mate selection. This is a large part of the reason why males may fight each other for a female—to show her that they are the most fit. Of course, the female animal does not know that she is choosing on this basis, but she is, and she is likely to select a partner with a striking appearance, capable of energetic displays—both of which are signs of good health.

SECONDARY SEX CHARACTERISTICS, EVOLUTION, AND THE MODEL OF AN IDEAL MATE

Humans, too, have deeply ingrained ideas regarding what makes an attractive potential mate, though to some extent, those ideas are cultural. Most societies regard a muscular male as attractive, for obvious evolutionary reasons: a physically fit male can provide for his mate and offspring, both by acquiring food and other resources and by protecting the nest. On the other hand, the modern American ideal of feminine beauty is more removed from nature. For one thing, this image of an attractive female is a thin body but large breasts—two things that seldom go together in nature but which are possible in the modern world through breast implantation surgery. That is, the achievement of such an ideal is available to a woman with a naturally thin body who also possesses the financial resources (and, of course, the desire) to undergo such surgery.

A woman naturally gifted with large breasts is apt also to have large buttocks, hips, and thighs, and while some men may find such anatomical features (particularly the first) desirable, this more natural version of the female body is not the image usually promoted in advertising or other media. Therefore, if a woman with a full figure wishes to meet societal expectations regarding beauty, she must endure something much more strenuous than a mere operation—a strict low-fat diet and a great deal of exercise. Given these often unnatural expectations from society, it is no wonder that a great many women express frustration with their attempts to conform to them, nor is it any wonder that many women in the modern world simply stop trying to conform. In any case, the American ideal of a thin female is far from universal, in terms of either place or time. Many traditional cultures favor a more amply formed female body, as did Western civilization in the past—a fact exemplified by the paintings of seventeenth-century artists, such as Peter Paul Rubens, famous for his fleshy female subjects. This is an ideal much closer to nature, since our evolutionary lineage has tended to favor women with large hips who are capable of bearing many children.

EVOLUTION AND THE SENSITIVE MAN. As with much else about sexual reproduction and the courtship rituals associated with it, the idea of beauty—itself an expression of secondary sex characteristics—is much simpler for men. As we noted earlier, physical fitness in men almost always is regarded as attractive, but women often value men for other traits, most notably intelligence and sense of humor. Whereas many men place an emphasis on physical appearance in choosing a mate, women's expectations are much more complex and they are also likely to be much more forgiving toward members of the male population who do not look like Tom Cruise or some other Hollywood image of attractiveness.

In defense of modern men, many have complained that modern women do not always want what they say they want. For example, from the early 1970s onward, it often was said in public discussions of sexuality on TV or in magazines that women wanted men to be more sensitive. That is, they wanted men who were more verbal and given to talking about their feelings and who were more aware of the woman and her needs. It is not surprising that men were failing to meet these expectations, since the idea of the "strong, silent type" is a masculine ideal in many cultures. Furthermore, not only civilization but also evolution has tended to favor men who are given more to actions than to words—men who can make war, either on the battlefield or in the business world.

Such aggressive characteristics are all well and good among animals or in a human society just struggling to survive. But in the modern West, where most material needs are easily met, women have sought and expressed a desire for something more from men. Such was the situation that emerged in the wake of the sexual revolution, when people became more open not only

KEY TERMS

CHROMOSOME: A DNA-containing body, located in the cells of most living things, that holds most of the organism's genes.

DIPLOID: A term for a cell that has the basic number of doubled chromosomes.

EGG CELL: A female gamete.

EMBRYO: The stage of animal development in the uterus before the animal is considered a fetus. In humans this is equivalent to the first three months.

FALLOPIAN TUBES: A set of trumpet-like tubes that carries a fertilized egg from the ovary to the uterus.

FAPS: Fixed-action patterns of behavior, or strong responses on the part of an animal to particular stimuli. *FAP* is virtually synonymous with *instinct.*

FERTILIZATION: The process of cellular fusion that takes place in sexual reproduction. The nucleus of a male reproductive cell, or gamete, fuses with the nucleus of an female gamete to produce a zygote.

FETUS: An unborn or unhatched vertebrate that has taken on the shape typical of its kind. An unborn human usually is called a *fetus* during the period from three months after fertilization to the time of birth.

GAMETE: A reproductive cell—that is, a mature male or female germ cell that possesses a haploid set of chromosomes and is prepared to form a new diploid by undergoing fusion with a haploid gamete of the opposite sex. Sperm and egg cells are, respectively, male and female gametes.

GENITALIA: The sex organs, which include the male penis and the female vagina.

GESTATION: The period between fertilization and birth during which the unborn offspring develops in the uterus.

HAPLOID: A term for a cell that has half the number of chromosomes that appear in a diploid or somatic cell.

HORMONE: Molecules produced by living cells, which send signals to spots remote from their point of origin and induce specific effects on the activities of other cells.

about sex itself but about their feelings as well. Although the sexual revolution yielded a number of negative effects, including the loss of mystery associated with sex and an increase in out-of-wedlock pregnancies and cases of sexually transmitted disease, it also made it possible for people to be frank in a way that they had never been before. One outgrowth of this was the revelation, in many women's magazines, that women were no longer satisfied or impressed with strong, silent men.

During the 1970s the model of the sensitive man emerged, and for the first time it became possible for men to talk about their feelings. And they did, in books, in movies such as those of Woody Allen, and in therapist-led encounter groups or men's groups. Gradually, however, a surprising fact emerged: women did not want men to be *too* sensitive and certainly not too nonmasculine. They wanted a man who would talk about his feelings but not a man who wore his heart on his sleeve. They wanted to be treated as equals in the workplace, but if there was a strange noise in the house at night, they wanted the man to go see what it was. They wanted a man who would respect them—but not a man who was so polite and predictable that he possessed no air of mystery.

KEY TERMS CONTINUED

MENSTRUATION: Sloughing off of the lining of the uterus, which occurs monthly in nonpregnant females who have not reached menopause (the point at which menstrual cycles cease) and which is manifested as a discharge of blood.

OVARY: Female reproductive organ that contains the eggs.

OVUM: An egg cell.

PUBERTY: A stage in the maturation of humans and higher primates wherein the person first becomes capable of sexual reproduction. It is marked by the development of secondary sex characteristics, the maturation of the genital organs, and the beginning of menstruation in females. Puberty typically takes place somewhere between the ages of about 10 and 14 years.

SECONDARY SEX CHARACTERISTICS: Those unique traits that mark an individual as a male or female but which are not manifested in the sexual organs themselves. Facial hair and a deep voice in males or breast development as well as hip and buttocks development in females are examples.

SEXUAL REPRODUCTION: One of the two major varieties of reproduction, along with asexual reproduction. In contrast to asexual reproduction, which involves a single organism, sexual reproduction involves two. Sexual reproduction occurs when male and female gametes undergo fusion, a process known as *fertilization,* and produce cells that are genetically different from those of either parent.

SPERM CELL: A male gamete.

TESTES: The pair of male reproductive glands located in the scrotum, a skin-covered sac that hangs from the groin.

UTERUS: A reproductive organ, found in most female mammals, in which an embryo and later a fetus grow and develop.

VAGINA: A passage from the uterus to the outside of the body.

ZYGOTE: A diploid cell formed by the fusion of two gametes.

All of this reveals a great deal about humans and sexual reproduction. First of all, mating, or the interaction between males and females, is about far more than reproduction or even sex. It is an interaction that involves the whole person. Second, human sexuality is surrounded by webs of psychological, social, and spiritual complexity that make it a phenomenon quite different from animal sexuality, which tends to be about little more than procreation. And, third, for all the progress humans have made, and for all the levels of civilization that separate us from our evolutionary roots, there is still something in human sexuality that responds to very basic ideas about sex, the sexes, and the need to propagate the species through mating and reproduction.

WHERE TO LEARN MORE

Avraham, Regina. *The Reproductive System.* Philadelphia: Chelsea House Publishers, 2001.

Cool Nurse (Web site). <http://www.coolnurse.com/>.

Francoeur, Robert T. *Taking Sides: Clashing Views on Controversial Issues in Human Sexuality.* Guilford, CT: Dushkin Publishing Group/Brown and Benchmark, 1996.

"How Human Reproduction Works." *How Stuff Works* (Web site). <http://www.howstuffworks.com/human-reproduction.htm>.

Kimball, Jim. "Sexual Reproduction in Humans." *Kimball's Biology Pages* (Web site). <http://www.ultranet.com/~jkimball/BiologyPages/S/Sexual_Reproduction.html>.

Parker, Steve. *The Reproductive System.* Austin, TX: Raintree Steck-Vaughn, 1997.

"Puberty Guide in Adolescents." *Keep Kids Healthy* (Web site). <http://www.keepkidshealthy.com/adolescent/puberty.html>.

"Puberty Information for Boys and Girls." American Academy of Pediatrics (Web site). <http://www.aap.org/family/puberty.htm>.

Whitfield, Philip. *The Human Body Explained: A Guide to Understanding the Incredible Living Machine.* New York: Henry Holt, 1995.

Winikoff, Beverly, and Suzanne Wymelenberg. *The Whole Truth About Contraception: A Guide to Safe and Effective Choices.* Washington, DC: Joseph Henry Press, 1997.

PREGNANCY AND BIRTH

CONCEPT

One of the greatest dramas in the world of living things is that which takes place in pregnancy and birth. Pregnancy forms a bond between mother and offspring that, in humans at least, lasts throughout life. Humans and many other animals are viviparous, meaning that offspring develop inside the mother's body and are delivered live. By contrast, birds and some other varieties of animal are oviparous, meaning that they deliver offspring in eggs that must develop further before hatching. In the modern world, human females experience birth in several ways—vaginal or cesarean, with anesthetics and without, at home or in a hospital—but just a few hundred years ago, there was little variety in an experience that was almost always painful and dangerous.

HOW IT WORKS

OVIPARITY, VIVIPARITY, AND OVOVIVIPARITY

The birth of live offspring is a reproductive feature shared by mammals, some fishes, and selected invertebrates, such as scorpions, as well as various reptiles and amphibians. Animals who give birth to live offspring are called viviparous, meaning "live birth." In contrast to viviparous animals, other animals—called oviparous, meaning "egg birth"—give birth to eggs that must develop before hatching. Finally, there are ovoviviparous animals, or ones that produce eggs but retain them inside the female body until hatching occurs, so that "live" offspring are born.

Oviparous animals may fertilize their eggs either externally or internally, though all animals that fertilize their eggs externally in nature are oviparous. (See Sexual Reproduction for more about internal and external fertilization.) In cases of internal fertilization, male animals somehow pass their sperm into the female: for example, male salamanders deposit a sperm packet, or spermatophore, onto the bottom of their breeding pond and then induce an egg-bearing (or gravid) female to walk over it. The female picks up the spermatophore and retains it inside her body, where the eggs become fertilized. These fertilized eggs later are laid and develop externally. Oviparous offspring undergoing development before birth obtain all their nourishment from the yolk and the protein-rich albumen, or "white," rather than from direct contact with the mother.

Ovoviviparity is common in a wide range of animals, including certain insects, fish, lizards, and snakes, but it is much less typical than oviparity. Ovoviviparous insects do not supply oxygen or nourishment to their developing eggs; they merely give them a safe brooding chamber for development. Nonetheless, species of ovoviviparous fish, lizards, and snakes appear to provide some nutrition and oxygen to their growing offspring. Because nutrition is provided in these instances, some zoologists consider them examples of true live birth, or viviparity.

VIVIPARITY. Viviparity is the type of birth process that takes place in most mammals and many other species. Viviparous animals give birth to living young that have been nourished in close contact with their mothers' bodies. The offspring of both viviparous and oviparous animals develop from fertilized eggs, but the eggs of viviparous

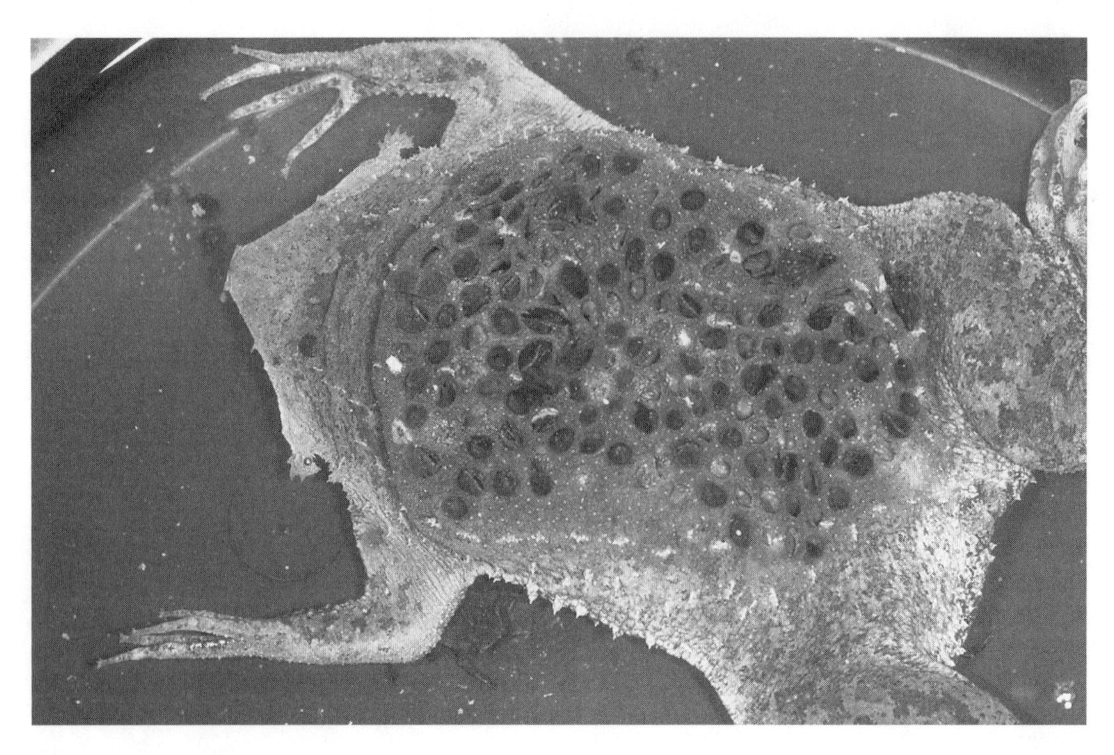

OVIPAROUS ANIMALS GIVE BIRTH TO EGGS THAT MUST DEVELOP BEFORE HATCHING. THE SURINAM TOAD, FOR EXAMPLE, CARRIES HER EGGS ATTACHED TO HER BACK WHILE THEY MATURE. *(© David A. Northcott/Corbis. Reproduced by permission.)*

animals lack a hard outer covering, or shell. Viviparous young grow in the adult female until they are able to survive on their own outside her body. In many cases, the developing fetuses of viviparous animals are connected to a placenta, a special membranous organ with a rich blood supply that lines the uterus in pregnant mammals. It provides nourishment to the fetus through a supply line called an umbilical cord.

All mammals, except for the platypus and the echidnas, are viviparous; only these two unusual mammals, called montremes, lay eggs. (See Speciation for more about mammal species.) Some snakes, such as the garter snake, are viviparous, as are certain lizards and even a few insects. Ocean perch, some sharks, and a few popular aquarium fish are also viviparous. Even certain plants, such as the mangrove and the tiger lily, are described as viviparous because they produce seeds that germinate, or sprout, before they become detached from the parent plant.

FROM ZYGOTE TO FETUS

The essays on Reproduction and Sexual Reproduction discuss the basics of the reproductive process through the point of fertilization. A fertilized egg is called a zygote, but once it begins to develop in the uterus or womb, it is known as an embryo and later, when it begins to assume the shape typical of its species, a fetus. In the uterus, the unborn offspring receives nutrients and oxygen during the period known as gestation, which extends from fertilization to birth. (In humans the gestation period is nine months.)

The zygote forms in one of the mother's fallopian tubes, the tubes that connect the ovaries with the uterus. It then travels to the uterus, where it becomes affixed to the uterine lining. Along the way, the zygote divides a number of times, such that by the time it reaches the uterus it consists of about 100 cells and is called an embryoblast. The exact day on which the embryoblast implants on the uterine wall varies, but it is usually about the sixth day after fertilization. By the end of the first week, a protective sac, known as the amniotic cavity, begins to form around the embryoblast.

EMBRYO AND FETUS. Changes then begin to take place at a rapid rate. As each week passes, the embryo takes on more and more necessary and distinctive features, such as blood vessels in week 3, internal organs in week 5, and finger and thumb buds on the hands in week 7. Unfortunately, miscarriages are not uncommon in the early weeks of pregnancy. The mother's immune system (see Immune System) may react to cells from the embryo that it classifies as "foreign" and begin to attack those cells. The embryo may die and be expelled. The first three months of embryonic development are known as the first trimester, or the first three-month period of growth. At the end of the first trimester, the embryo is about 3 in. (7.5 cm) long and looks like a tiny version of an adult human. Thereafter, the growing organism is no longer an embryo, but a fetus. Fetal development continues through the second and third trimesters until the baby is ready for birth at the end of ninth months.

PREPARING FOR BIRTH

At the end of the gestation period, the mother's uterus begins to contract rhythmically, a process called labor. This is accompanied by the release of hormones, most notably oxytocin. From the time of fertilization, quantities of the hormone progesterone, which keeps the uterus from contracting, are high; but during the last weeks of gestation, maternal progesterone levels begin to drop, while levels of the female hormone estrogen rise. When progesterone levels drop to very low levels and estrogen levels are highest, the uterus begins to contract.

Meanwhile, as birth approaches, the brain's pituitary gland releases oxytocin, a hormone that stimulates uterine contractions and controls the production of milk in the mammary glands (a process called lactation). Synthetic oxytocin sometimes is given to women to induce labor. Scientists believe that the pressure of the fetus's head against the cervix, the opening of the uterus, ultimately initiates the secretion of oxytocin. As the fetus's head presses against the cervix, the uterus stretches and relays a message along nerves to the pituitary gland, which responds by releasing oxytocin. The more the uterus stretches, the more oxytocin is released.

LABOR AND DELIVERY. Rhythmic contractions dilate the cervix, causing the fetus to move down the birth canal and to be expelled together with the placenta, which has supplied the developing fetus with nutrients from the mother during the gestation period. Before delivery, the placenta separates from the wall of the uterus. Since the placenta contains many blood vessels, its separation from the wall of the uterus causes bleeding. This bleeding is normal, assuming that it is not excessive. After the placenta separates from the uterine wall, it moves into the birth canal and is expelled from the vagina. The uterus continues to contract even after the placenta is delivered, and it is thought that these contractions serve to control bleeding.

After the baby is born, the umbilical cord that has attached the fetus to the placenta is clamped. The clamping cuts off the circulation of the cord, which eventually stops pulsing owing to the interruption of its blood supply. The baby now must breathe air through its own lungs, whereas before it has been breathing, fishlike, in the warm, wet environment of the mother's amniotic fluid. The process of labor described here in a very cursory fashion (it is actually much more complicated) can take from less than one hour to 48 hours, but typically the entire birth process takes about 16 hours.

REAL-LIFE APPLICATIONS

CHANGING VIEWS ON CHILDBIRTH

Before modern times, the realm of childbirth was a world exclusive to women, and few men ever entered the birth chamber. It was a place of excruciating pain and serious danger to the mother giving birth, so filled with blood and screaming that few men would have dared enter even if they had wanted to do so. Women had to give birth without anesthesia and any number of other amenities of modern medical care, including sophisticated diagnostic techniques and equipment, such as ultrasound, as well as antiseptic environments and surgical techniques, such as cesarean section.

In those days, birthing assistance was the work of midwives, women who lacked formal schooling in medicine (which was unavailable to most women in any case) but made up in experience for what they lacked in education. By about 1500, however, as medicine began to progress after many centuries of stagnation, male doctors increasingly forced midwives out of a job. In 1540 the European Guild of Surgeons declared that "no carpenter, smith, weaver, or woman shall practice surgery." A major turning point in the male takeover of birthing assistance duties came with the invention of the forceps, tong-like instruments that could be used for extracting a baby during difficult births. The inventor was the

English obstetrician (a physician concerned with childbirth) Peter Chamberlen the Elder (1560–1631), and he and his descendants for a century closely guarded the design of the brilliant invention. Even the mothers on whom it was used never saw the instrument, and midwives were prohibited from using forceps to assist during childbirth.

OBSTETRICIANS TAKE OVER. By the eighteenth century, however, Chamberlen's descendants had released their exclusive claim over the forceps, and use of the instrument spread to other medical professionals. This gave male obstetricians a great technological advantage over female midwives and further ensured the separation of the midwives from the medical profession. By 1750 numerous physicians and surgeons had gained the status of "man-midwives," and the growth of university courses on obstetrics established it as a distinct medical specialty. By the latter part of the 1700s, most women of the upper classes had come to rely on professionally trained doctors rather than midwives, yet in America, where doctors were scarcer than in Europe, the profession of midwife continued to flourish into the 1800s. Still, by the early twentieth century, childbirth had moved out of the home and into the hospital, and at mid-century it had become a completely medical process, attended by physicians and managed with medical equipment and procedures, such as fetal monitors, anesthesia, and surgical interventions.

THE REACTION IN THE LATE TWENTIETH CENTURY. Many women of the late twentieth century found themselves dissatisfied with this clinical approach to childbirth. Some believed that the medical establishment had taken control of a natural biological process, and women who wanted more command over labor and delivery helped popularize new ideas on childbirth that sought to reduce or eliminate medical interventions. Today, some women choose to deliver with the help of a nurse-midwife, who, like her premodern counterparts, is trained to deliver babies but is not a doctor. There are women who even choose home birth, attended by a doctor or midwife or sometimes both. There are even brave souls who, in the face of increasing concern about the effect of anesthesia on the fetus, refuse artificial means of controlling pain and instead rely on breathing and relaxation techniques. For the first time in many years, the screams of women giving birth "naturally" once again filled the halls of hospital maternity wards and home birthing rooms.

MODERN CHILDBIRTH

The last few paragraphs represent an extreme reaction—a view not shared by many women, who have been more than happy to avail themselves of the benefits of childbirth in the modern world. Such benefits include an epidural, a type of anesthetic procedure that serves to alleviate the pain of parturition, or childbirth, while making it possible for the mother to remain conscious. Still, even for women who have no interest in giving birth at home or without the aid of drugs, much has changed in the world of childbirth. Women may choose a happy medium between the medical establishment and more traditional methods, for instance, by opting to consult with an obstetrician *and* a midwife.

Today, many obstetricians are women. This has had an incalculable effect on making childbirth psychologically easier for many women: though some are happy to retain a male obstetrician, many others find themselves much more comfortable being cared for by a physician who, in all likelihood, has given birth herself. The increasingly important role of the female obstetrician, along with other factors, serves to symbolize the fact that the world has progressed beyond the old false dilemma between medical care from a male or a female, between medicine and nature, between hospital and home.

A HOSPITAL AS HOME. Hospital rooms, in fact, are starting to resemble rooms at home. Everywhere one looks in the modern maternity environment, there is evidence that much has changed, not only from the very old days, when male doctors were not involved in childbirth at all, but also from the more recent past, when males took over the process entirely. In a brilliant innovation, many hospitals have created a situation in which the woman gives birth in her own hospital room, which is outfitted with couches, cabinets, curtains, and rocking chairs to make it look like a home rather than a hospital. To emphasize the smooth transition between home life and the delivery room, fathers, once banished from the labor and delivery chambers, now are welcomed as partners in the birth process.

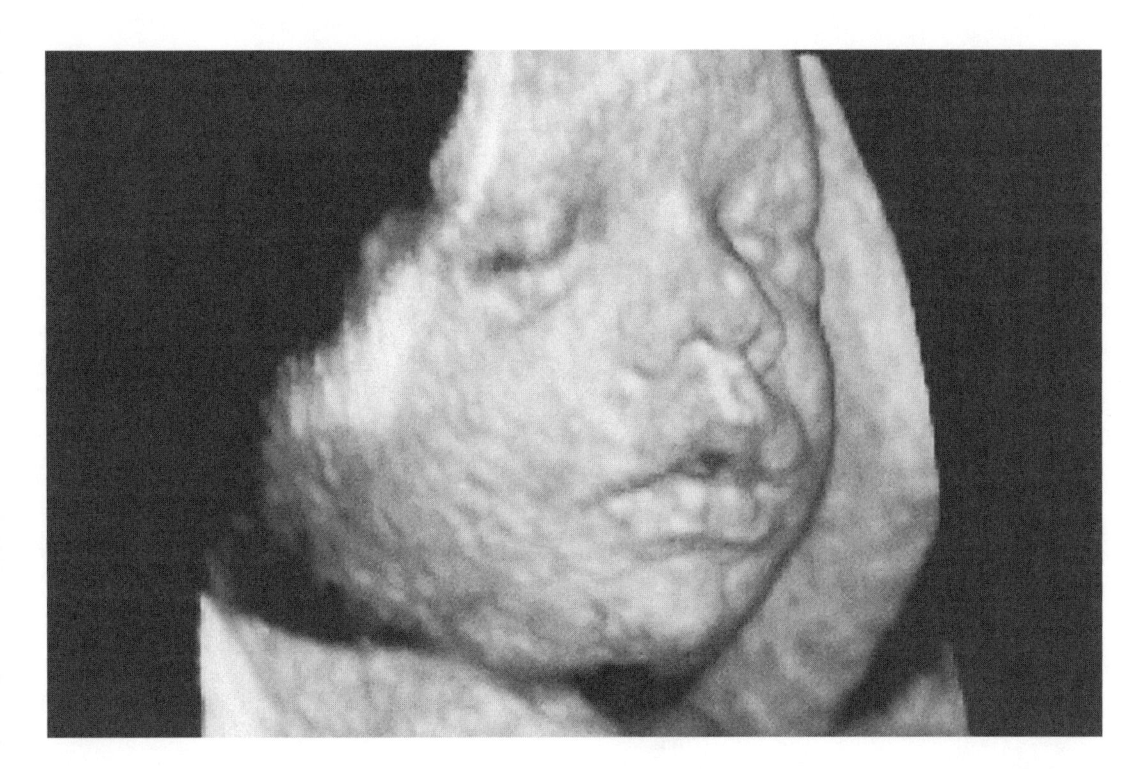

ULTRASOUND IMAGE OF A 30-WEEK-OLD FETUS. BY THE EIGHTEENTH WEEK OF PREGNANCY, ULTRASOUND TECHNOLOGY CAN DETECT MANY STRUCTURAL ABNORMALITIES, SUCH AS SPINA BIFIDA, HEART AND KIDNEY DEFECTS, AND HARELIP. *(© BSIP/Kretz Technik/Photo Researchers. Reproduced by permission.)*

A father may even cut the umbilical cord, and he is certainly likely to be in the delivery room with a video camera, recording the event for posterity—yet another change from the past. Fathers are not the only ones filming in the delivery room. Today, cable television networks, such as the Learning Channel, provide programming that offers a frank view of the delivery process, complete with candid footage that sometimes can be as dramatic as it is revealing. The maternity ward, once a closed place, has increasingly become an open book.

SAVING AND IMPROVING LIVES

Many a mother and father alike can breathe a prayer (or at least a sigh) of thanks for all the innovations that today make birth much safer than it once was. Among them are a variety of techniques for embryo and fetal diagnosis, which help make parents aware of possible problems in the growing embryo. Ultrasound diagnosis, a technique similar to that applied on submarines for locating underwater structures, uses high-pitched sounds that cannot be heard by the human ear. These sounds are bounced off the embryo, and the echoes received are used to identify embryonic size.

By the eighteenth week of pregnancy, ultrasound technology can detect many structural abnormalities, such as spina bifida (various defects of the spine), hydrocephaly (water on the brain), anencephaly (no brain), heart and kidney defects, and harelip (in which the upper lip is divided into two or more parts). On a less dire and much more pleasant note, it can also give future parents an opportunity to gain their first glimpse of their child, and an experienced ultrasound technician usually can tell them the baby's sex if they choose to learn it before the birth.

PRENATAL TESTING. Chorionic villi sampling is the most sophisticated modern technique used to assess possible inherited genetic defects. This test typically is performed between the sixth and eighth week of embryonic development. During the test, a narrow tube is passed through the vagina or the abdomen, and a sample of the chorionic villi (small hairlike projections on the covering of the embryonic sac) is removed while the physician views the baby via ultrasound.

Chorionic villi are rich in both embryonic and maternal blood cells. By studying them, genetic counselors can determine whether the baby will have any of several defects, including

KEY TERMS

EMBRYO: The stage of animal development in the uterus before the point at which the animal is considered a fetus. In humans this is equivalent to the first three months.

FALLOPIAN TUBES: A set of trumpet-like tubes that carries a fertilized egg from the ovary to the uterus.

FERTILIZATION: The process of cellular fusion that takes place in sexual reproduction. The nucleus of a male reproductive cell, or gamete, fuses with the nucleus of a female gamete to produce a zygote.

FETUS: An unborn or unhatched vertebrate that has taken on the shape typical of its kind. An unborn human usually is called a fetus during the period from three months after fertilization to the time of birth.

GESTATION: The time between fertilization and birth, during which the unborn offspring develops in the uterus.

HORMONE: Molecules produced by living cells, which send signals to spots remote from their point of origin and induce specific effects on the activities of other cells.

OVARY: Female reproductive organ that contains the eggs.

OVIPAROUS: A term for an animal that gives birth to eggs that must develop before hatching. Compare with *viviparous.*

OVOVIVIPAROUS: A term for an animal that produces eggs but retains them inside the body until hatching occurs, so that "live" offspring are born. Compare with *oviparous* and *viviparous.*

OVUM: An egg cell.

UTERUS: A reproductive organ, found in most female mammals, in which an embryo and, later, a fetus grows and develops.

VAGINA: A passage from the uterus to the outside of the body.

VIVIPAROUS: A term for an animal that gives birth to live offspring. Compare with *oviparous.*

Down syndrome (characterized by mental retardation, short stature, and a broadened face), cystic fibrosis (which affects the digestive and respiratory systems), and the blood diseases hemophilia, sickle cell anemia, and thalassemia. (Several of these disorders are discussed in different essays throughout this book; for instance, Down syndrome is examined in Mutation.) As with ultrasound, it also can show the baby's gender.

Another important form of prenatal (before the birth of the child) testing is amniocentesis, performed around week 16, in which amniotic fluid is drawn from the uterus by means of a needle inserted through the abdomen. Amniocentesis, too, can reveal the sex of the child, as well as a host of genetic disorders such as Tay-Sachs disease, cystic fibrosis, and Down syndrome. However, amniocentesis involves the risk of fetal loss as a result of disruption of the placenta. Chorionic villi sampling is even more risky, with an even higher possibility of fetal loss than amniocentesis, probably because it is conducted at an earlier stage.

In alpha-fetoprotein screening, which takes place somewhere between the 16th and 18th weeks, proteins from the amniotic sac and the fetal liver are taken as a means of screening for specific defects. Because of uncertainties involved in interpretation of the results, alpha-fetoprotein screening is not a common procedure.

CESAREAN SECTION. Another extremely important technique that has saved the life of many babies and mothers is cesarean section. The normal position for a baby in delivery is head first; when a baby is in the breech position, with its bottom first, it poses grave dangers to both the mother and the child. Not only could the baby fail to emerge in time to begin breathing normally, thus running the risk of brain damage, but it also can become stuck, endangering the life of the mother. Today these dangers are overcome by such techniques as turning the baby and by cesarean section, an operation in which the baby is removed via surgery from the mother's abdomen. Cesarean sections may also be performed due to other complications, including fetal and/or maternal distress.

The term cesarean refers to the Roman emperor Julius Caesar (102–44 B.C.), who supposedly was delivered in this fashion. But the story of Caesar's birth is undoubtedly a legend: until the early modern era, cesarean sections were performed only to save a living baby after the mother had died in childbirth. The reason is that cesareans were likely to be fatal to the mother. Only in the late nineteenth century, by which time doctors had come to understand the importance of providing an antiseptic or germ-free environment, did cesarean sections become practical. Today the C-section, as it is called, has become a routine procedure—one that has saved literally hundreds of thousands, perhaps millions, of lives.

WHERE TO LEARN MORE

Assisted Reproduction Foundation (Web site). <http://www.reproduction.org/>.

Bainbridge, David. *Making Babies: The Science of Pregnancy.* Cambridge, MA: Harvard University Press, 2001.

Facts About Multiples (Web site). <http://mypage.direct.ca/c/csamson/multiples.html>.

Midwifery, Pregnancy, Birth and Breastfeeding (Web site). <http://www.moonlily.com/obc/>.

Pence, Gregory E. *Who's Afraid of Human Cloning?* Lanham, MD: Rowman and Littlefield, 1998.

Pregnancy and Birth (Web site). <http://pregnancy.about.com/mbody.htm>.

Pregnancy and Reproduction Topics. Medline/National Library of Medicine, National Institutes of Health (Web site). <http://www.nlm.nih.gov/medlineplus/pregnancyandreproduction.html>.

Rudy, Kathy. *Beyond Pro-Life and Pro-Choice: Moral Diversity in the Abortion Debate.* Boston: Beacon Press, 1996.

Vaughan, Christopher C. *How Life Begins: The Science of Life in the Womb.* New York: Times Books, 1996.

SCIENCE OF EVERYDAY THINGS

REAL-LIFE BIOLOGY

EVOLUTION

EVOLUTION

PALEONTOLOGY

EVOLUTION

CONCEPT

Among the dominant concepts of the modern world in general, and biology in particular, few are as powerful—or as misunderstood—as evolution. Even the name is something of a misnomer, since it almost implies some sort of striving to reach a goal, as though the "purpose" of evolution were to produce the most intelligent species, human beings. In fact, what drives evolution is not a quest for biological greatness but something much more down to earth: the need for organisms to survive in their environments. Closely tied to evolution are two processes, mutation and natural selection. Natural selection is a process whereby survival is related directly to the ability of an organism to fit in with its environment, while mutation involves changes in the genetic instructions encoded in organisms.

Although the English naturalist Charles Darwin (1809–1882) often is regarded as the father of evolutionary theory, he was not the first thinker to suggest the idea of evolution as such; however, by positing natural selection as a mechanism for evolution, he provided by far the most convincing theory of evolutionary biological change up to his time. In the years since Darwin, evolutionary theory has evolved, but the essential idea remains a sound one, and it is a "theory" only in the sense that it is impossible to subject it to all possible tests. The idea that evolution is somehow still open to question is another pervasive misconception, and it often appears hand in hand with the most pervasive misconception of all—that evolution is in some way anti-Christian, anti-religion, or anti-God.

HOW IT WORKS

THE "WATCH ANALOGY" AND WHAT IT (UNINTENTIONALLY) TEACHES ABOUT EVOLUTION

Throughout this essay, we discuss misconceptions relating to evolution. Such misconceptions have had such a strong impact on modern civilization that it is important to begin by setting aside a few misguided ideas that strike at the very heart of the evolutionary process. Many of these misconceptions are embodied in a popular "argument" against evolution that goes something like this: Suppose you took a watch apart and laid the pieces on the ground. If you came back in a billion years, would you really expect the watch to have assembled itself?

This argument is a virtual museum of all the fallacies associated with evolution. First of all, a watch (or any of the other variations used in similar arguments) is mechanical, not organic or biological, which is the class of objects under discussion within the framework of evolution. In that sense, the answer to this question is easy enough: No, a watch probably never would assemble itself, because it is not made of living material and it has no need for survival.

Another problem with the watch argument is that it starts with impossibly large pieces. Let us assume that the watch *is* a living being; even so, one would not expect its dials and gears to assemble themselves. But evolution does not make such claims: there is nothing in the theory of evolution to lead one to believe that a collection of organs lying around on a beach eventually would piece themselves together to make a whale.

CHARLES DARWIN *(The Library of Congress.)*

According to what paleontologists (see Paleontology) and other scientists can deduce, over the course of three billion years life-forms evolved from extremely simple self-replicating carbon-based molecules to single-cell organisms. This is hardly what one would call breakneck speed. The more visible or "exciting" part of evolution, with the proliferation of species that produced the dinosaurs and (much later) humans, took place in the past billion years. In fact, the pace of change was still very, very slow until about half a billion years ago, and it has been accelerating ever since. For the vast majority of evolutionary history, however, change has been so slow that, by contrast, watching paint dry would be like playing a high-speed video game.

Ironically, for the watch scenario to be truly analogous to anything in evolution, one would have to start with atoms and molecules not whole gears and dials. Opponents of evolutionary theory might take this fact as being favorable to their cause, but if the watch were made of living, organic material rather than metal, it is possible that the molecules would have some reason to join in the formation of organelles and, later, cells. Or perhaps they would not. Therein lies another problem with the watch analogy and, indeed, with many of the attempts to argue against evolution on a religious basis. This might be called the "fallacy of intention," or the idea that evolution is driven by some overall purpose.

THE "FALLACY OF INTENTION." Hidden in the watch analogy is the idea of the watch itself, the finished product, as a "goal." By the same analogy, the single-cell eukaryotes of a billion or two billion years ago were forming themselves for the purpose of later becoming pine trees or raccoons or people. This is not a valid supposition, as can be illustrated by analogies to human history.

The history of human beings, of course, has taken place over a much, much shorter span than evolutionary history. (The Paleontology essay contains several comparisons between the span of human life on Earth and Earth's entire existence.) Moreover, unlike cells, people *do* form goals and act on intentions, so if there were any good example of change with a goal in mind, it would have to come from human beings. Yet even in the few thousand years that humans have existed in organized societies, most trends have occurred not as part of a major plan but as a means of adapting to conditions.

Consider the situation of a group of nomads who lived in what is now southern Russia about 5,000 years ago. At some point, this vast collection of tribes began to migrate outward, some moving into an area that is now central Asia and the Indian subcontinent and others migrating westward. No sane person would argue that the westward-traveling members of this group knew that in moving to the geographically advantageous territory of Europe, they were putting in place conditions that would help give their descendants dominance over most of the planet some 4,500 years later. Rather, they were probably just trying to find better land for grazing their horses.

We cannot say what the Indo-Europeans, as they are known to history, were looking for. Our only evidence that they existed is the similarities between the languages of Europe, India, and Iran, first noted by the German philologist and folklorist Jacob Grimm (1785–1863) at about the same time that Darwin was formulating his theory of evolution. Grimm, in fact, used methods not unlike those of Darwin, but instead of fossils he studied words and linguistic structures. Along the way, he found remarkable links, such as the Sanskrit word *agni,* cousin to the Latin term *ignis* and such modern English words as *ignite.*

In contrast to the Indo-Europeans, we know a great deal about another group of westward-moving nomads, the Huns of around A.D. 300, who were indeed looking for better grazing lands. Dislocated from their native areas by the building of China's Great Wall, the Huns crossed the Danube River, displacing the Ostrogoths. The Ostrogoths, in turn, moved westward, and this migration set in motion a domino effect that would bring an end to the Western Roman Empire in A.D. 476.

Did the Huns intend to destroy the Roman Empire and bring about the Middle Ages? No reasonable person would adopt such a conspiratorial view of history. Even more absurd, did the Chinese build the Great Wall with the idea of precipitating this entire chain of events? Again, no one would assert such a premise. If those trends in the evolution of societies were not goal-directed, why would we assume that cells and organisms would have to be striving toward a particular end to obtain certain results?

CONFUSING EVOLUTION WITH GOD. In fact, there is no driving "purpose" to evolution—no scientifically based substitute for God operating from behind the scenes and manipulating the evolutionary process to achieve its ultimate aims. Evolution is not guided by any one large aim but by a million or a billion small aims—the need for a particular species of mollusk to survive, for instance.

As we discuss in the course of this essay, the idea of an underlying conflict between evolution and Christianity (or any other religion, for that matter) is almost entirely without merit. On the other hand, it is theoretically possible that all the processes of evolution took place *without* a creator—but this still should not pose a threat to anyone's idea of God.

There is nothing in evolution that would lead to the conclusion that there is no God, that the universe is not God's handiwork, or that God does not continue to engage in a personal relationship with each human. Neither is there anything in evolution that would lead to the conclusion that God *does* exist. Rather, the matter of God is simply not relevant to the questions addressed by evolution. In other words, evolution leaves spiritual belief where it should be (at least, according to Christianity): in the realm of individual choice.

NATURAL SELECTION

As we noted earlier, one of the principal mechanisms of evolutionary processes is natural selection. This in itself illustrates the lack of intention, or "goal orientation," in evolution. Like the name *evolution* itself, the term *natural selection* can be deceptive, implying that nature selects certain organisms to survive and condemns others to extinction. In fact, something quite different is at work.

Species tend to overproduce, meaning that the number of field mice, for instance, born in any year is so large that this entire population cannot possibly survive. The reason is that there is never enough of everything—food, water, or living space—for all members of the population to receive what they need. Therefore, only those best adapted to the environment are likely to survive.

FASTER, FURRIER MICE. Suppose, for instance, that the climate in the area where two field mice live is very cold, and suppose that some of the field mice have more protective fur than others; obviously, they are more likely to live. If there are many speedy predators around, judging purely on the basis of that factor alone, it would be easy to predict that the swiftest of the field mice would survive. Thus, faster-running, furrier mice would be "selected" over the slower or less furry mice.

Natural selection is not simply a matter of one particular mouse surviving in an environment. Instead, it involves the survival of specific strains, or lines of descent, that are more suited to the environment in question. Individuals adapted to an environment are more likely to live and reproduce and then pass on their genes to the next generation, while those less adapted are less likely to reproduce and pass on their traits. The genetic strains that survive are not "better" than those that do not—they are only better adapted.

The process of natural selection is ongoing. For example, in generation *A*, the furrier field mice survive and pass on their "furriness" gene to their offspring. Some of the offspring may still not be furry, and these mice will be less likely to survive and reproduce. In addition, since there are almost always several survival factors affecting natural selection, it is likely that other traits also will determine the survivability of certain individual mice and their genes.

For instance, there may be furry but slow mice in generation *B*, which despite their adaptation to temperature conditions are simply not fast enough to get away from predators. Therefore, the mice in generation *C* are likely to be furrier *and* faster than their ancestors. Additional survival factors may come into the picture, to ensure that the average member of generation *D* has sharper teeth in addition to swifter feet and a furrier body.

Although this illustration depicts evolutionary changes as taking place over the course of four generations, they are more likely to occur over the span of 400 or 4,000 or four million generations. In addition, the process is vastly more complicated than it has been portrayed here, because numerous factors are likely to play a part. The essential mechanism outlined here, however, prevails: certain traits are "naturally selected" because individuals possessing those traits are more capable of survival.

THE "SURVIVAL OF THE FITTEST." The concept of natural selection sometimes is rendered popularly as the "survival of the fittest." Scientists are less likely to use this phrase for several reasons, including the fact that it has been associated with distasteful social philosophies or murderous political ideologies—for example, Nazism. Additionally, the word *fittest* is a bit confusing, because it implies "fitness," or the quality of being physically fit.

This implication, in turn, might lead a person to believe that natural selection entails the survival of the *strongest*, which is not the case. Yet this is precisely what proponents of a loosely defined philosophy known as social Darwinism claimed. Popular among a wide range of groups and people in the late nineteenth and early twentieth centuries, social Darwinism could be used in the service of almost any belief. Industrialists and men of wealth asserted that those who succeeded financially did so because they were the fittest, while Marxists claimed that the working class ultimately would triumph for the same reason. Across the political spectrum, social Darwinism confused the meaning of "fittest" with that of other concepts: "strongest," "most advanced," or even "most moral." All of this, it need hardly be said, is misguided, not least because evolutionary theory has nothing to do with race, ethnicity, or social class.

In fact, "survival of the fittest," in a more accurate interpretation, means that individuals that "fit," or "fit in with," their environments are those most likely to survive. This is a far cry from any implication of strength or superiority. Imagine a group of soldiers in combat: Which type of soldier is most likely to survive? Is it the one who scores highest on physical training tests, looks the finest in a uniform, comes from a more socially upper-class home, and has the most advanced education? Or is it the one who keeps his head low, acts prudently, does not rush into dangerous situations without proper reconnaissance, and obeys instruction from qualified leaders?

Clearly, the second set of characteristics has much more to do with survival, even though these qualities may seem less "noble" than the first set. Yet it is by adapting, or proving his or her adaptability, to the environment of war that a soldier survives—not by displays of strength or other types of "fitness" that simply appear impressive. In the same way, the fitness of a species does not necessarily have anything to do with strength: after all, the lion, the "king of beasts," would die out in a polar climate or a desert or an aquatic environment.

MUTATION

Although natural selection is of principal importance in evolution, mutation also plays a pivotal role. Mutation is the process whereby changes take place in the genetic blueprint for an organism as a result of alterations in the physical structure of an organism's DNA (deoxyribonucleic acid). DNA is a molecule in all cells and in many viruses that contains genetic codes for inheritance. DNA carries genetic information that is transmitted from parent to offspring; when a mutation occurs, this new genetic information—often quite different from the genetic code received by the parent from the grandparent—is passed on instead.

Under normal conditions of reproduction, a copy of the DNA from the parent is replicated and transmitted to the offspring. The DNA from the parent normally is copied exactly, but every once in a while errors arise during replication. These errors usually originate in noncoding regions of the DNA and therefore have little effect on the observable traits of the offspring. On the other hand, some mutations may be lethal, and thus the offspring does not survive for the mutation to become apparent. In a very few cases, however, offspring with a slightly modified genetic makeup manage to survive.

CONTRAST WITH ACQUIRED CHARACTERISTICS. Mutation is not to be confused with the inheritance of acquired characteristics, a fallacious doctrine that had its adherents when Darwin was a young man. If acquired characteristics were taken to an extreme, a lumberjack who loses his arm cutting down a tree and later conceives a child with his wife would most likely father a child who is missing an arm. This notion is absurd, and attempts to put forward a workable theory of acquired characteristics in the late eighteenth and early nineteenth centuries involved much greater subtlety. Still, the idea is misguided.

The French natural philosopher Jean Baptiste de Lamarck (1744–1829), one of the leading proponents of acquired characteristics, maintained that giraffes had gained their long necks from the need to stretch and reach leaves at the top of tall trees. In other words, if a giraffe parent had to stretch its neck, a giraffe baby would be born with a stretched neck as well. Later, Darwin's natural selection provided a much more plausible explanation for how the giraffe might have acquired its long neck: assuming that the nutrients it needed were at the highest levels of the local trees, the traits of tallness, long necks, and the ability to stretch would be selected naturally among the giraffe population.

MUTATIONS AND SURVIVAL. Unlike the idea of acquired characteristics, mutation does not entail the inheritance of anatomical traits acquired in the course of an organism's life; rather, it is changes in the DNA that are passed on. For example, when mind-altering drugs became popular among young people in the 1960s, concerns were raised that the offspring of drug takers might suffer birth defects as a result of alterations in their DNA. For the most part, this did not happen. Conditions such as Huntington disease and cystic fibrosis, however, are the result of mutations in DNA; so, too, is albinism, which eliminates skin pigment.

Although mutations often are regarded as undesirable because they can affect the health of individuals adversely, they also can have positive effects for the population in question. Suppose a group of bacteria is exposed to an antibiotic, which rapidly kills off the vast majority of the bacteria. In a fraction of those who survive, however, a mutation may develop that makes them resistant to the medication. Eventually, these mutant bacteria will reproduce, creating more mutants and in time yielding an entire population resistant to the antibiotic.

This is the reason why antibiotics can lose their effectiveness over time: bacteria with mutant genes will render every antibiotic useless eventually. The same often can happen with insect sprays, as roaches and other pests develop into mutant strains that are capable of surviving exposure to these pesticides. Such species, with their short cycles of birth, reproduction, and death, are extremely well equipped for survival as a group, which explains why many an unpleasant "bug" (whether a bacterium or an insect) has long been with us. (See Mutation for more on this subject.)

REAL-LIFE APPLICATIONS

"PROVING" EVOLUTION

Later in this essay, we look at examples of evolution in action and other phenomena that support the ideas of evolutionary theory. But before examining these many "proofs" of evolution, a few words should be said about the very fact that evolution seems to require so much more proof than most other scientific theories.

All scientific ideas must be capable of being proved or disproved, of course, but the demand for proof in the case of evolution goes far beyond the usual rigors of science. In fact, at this point, the people demanding proof are not scientists but certain sectors of the population as a whole—in particular, religious groups or individuals who fear evolution as a challenge to their beliefs.

QUANTUM MECHANICS: A MUCH MORE DIFFICULT IDEA. By contrast, quantum mechanics, though it encompasses ideas completely opposed to common sense, has not sustained anything approaching the same challenge or the demand for proof that evolution has encountered from nonscientists. A theory in physics and chemistry that details the characteristics of energy and matter at a subatomic level, quantum mechanics goes against such common assumptions as the idea that we can know both the location and the speed of an object. It is as though science had proved that down was up and up was down. If there were ever a "dangerous" theory, inasmuch as it undermines all our assumptions about the world, it is quantum mechanics not evolution, which is a fairly straightforward idea by comparison.

Quantum mechanics has gone virtually unchallenged (at least on a social or moral, as opposed to a scientific, basis), whereas even today there are many people who refuse to accept the idea of evolution. Granted, quantum mechanics is a much younger idea, having originated only in the 1920s, and it is vastly more difficult to understand. But the real reason why evolution has come under so much more challenge, of course, has to do with the fact that it is perceived (mistakenly) as challenging the primacy of God.

JUST A THEORY? One of the aspects of evolution often cited by opponents is the fact that it is, after all, the *theory* of evolution. The implication is that if it is still just a theory, it must be open to question. In a sense, this is accurate: for scientific progress to continue, ideas should never be accepted as absolute, unassailable truths. But this is not what opponents of evolution are getting at when they cite its status as a "mere" theory. In fact, their use of this point as a basis for attack only serves to illustrate a misunderstanding with regard to the nature of scientific knowledge.

The word *theory* in "theory of evolution" simply means that evolutionary ideas have not been and, indeed, cannot be tested in every possible circumstance. Most ideas in science are simply theories rather than laws because in few cases is it possible to say with absolute certainty that something always will be the case. One of the few actual scientific laws is the conservation of energy, which holds that for all natural systems the total amount of energy remains the same, though transformations of energy from one form to another take place. This has been tested in such a wide variety of settings and circumstances that there is no reason to believe that would it ever *not* be the case.

By contrast, there probably never will be enough tests on evolution to advance it to the status of a law. The reason is quite simply that evolution takes a long time. Some examples, such as the instances of industrial melanism that we discuss later, unfold within a short enough period of time that humans can observe them. In general, however, evolutionary processes take place over such extraordinarily long spans of time that it would be impossible to subject them to direct observation.

None of this, however, does anything to discredit evolutionary theory. For that matter, the idea that the entire physical world is made of atoms is still technically a theory, though there is no significant movement of people attempting to discredit it. The reason, of course, is that atomic theory does not seem to contradict anyone's idea of God. (This was not always the case, however. Almost 2,500 years ago, a Greek philosopher named Democritus developed the first atomic theory, but because his ideas were associated with atheism, atomic theory was largely rejected for more than two millennia.)

FACING THE FACTS. If people really understood the word theory, they would give it a great deal more respect. Unfortunately, the word so often is misused and applied to anything that has not been proved that it has begun to seem almost like an insult to call evolution a theory. After all, in the present essay, we refer to acquired characteristics as a theory, and in everyday life one often hears much less respectable ideas given the status of theory. For this reason, it is worth taking note of the process, from observation to hypothesis to the formulation of general statements, that goes into the development of a truly scientific theory.

In forming his theory of evolution, Darwin began with several observations about the natural world. Among the things he observed is the fact, which we noted earlier, that for a particular species, more individuals are born than can possibly survive with available resources. On the basis of this observation, he formed a hypothesis, or inference. His inference was that because populations are greater than resources, the members of a population must compete for resources.

A theory is made up of many hypotheses, but to proceed from a collection of hypotheses to a true theory, these inferences must be subjected to rigorous testing. Thus, Darwin, in effect, said to himself, "Is what I have said true? *Are* there more individuals of a species than there are available resources?" Then he began looking for examples, and like a true scientist, he did so with the attitude that if he found examples that contradicted his hypothesis, he would reject the hypothesis *and not the facts.*

As it turns out, of course, there are always more members of a population than there are resources. This can be illustrated in a small way by observing a litter of puppies or piglets struggling to obtain milk from their mother. Chances are that the mother will not have enough teats for all her babies, and the "runt," unless it is able to force its way through the others to the milk source, may die. Only after testing this hypothesis and other hypotheses, such as that of natural selection, did Darwin formulate his theory.

EVOLUTION AND RELIGION

The fact that some puppies or piglets die for lack of milk is not a nice or pleasant thought, but it is the truth. Again, like a true scientist, Darwin accepted reality, without attempting to mold it to fit his personal beliefs about how things should be.

As a great thinker from the generation that preceded Darwin's, the Scottish philosopher David Hume (1711–1776), wrote in his *Enquiry Concerning Human Understanding:* "There is no method of reasoning more common, and yet more blamable, than, in philosophical disputes, to endeavor the refutation of a hypothesis, by a pretense of its dangerous consequences to religion and morality." In other words, there is an understandable, but nonetheless inexcusable, human tendency to evaluate ideas not on the basis of whether they are true but rather on the basis of whether they fit with our ideas about the world.

A scientist may be a Christian, or an adherent of some other religion, and still approach the topic of evolution scientifically—as long as he or she does not allow religious convictions to influence acceptance or nonacceptance of facts. The scientist should start with no preconceived notions and no allegiance to anything other than the truth. If that person's religious conviction is strong enough, it can weather any new scientific idea.

CONFUSING ATHEISM WITH SCIENCE. This brings up an important point regarding the alleged conflict between religion and science. Not all the blame for this belongs with religious groups or individuals who shut their minds to scientific knowledge. Many scientists over the years likewise have adopted the fallacy of maintaining that religion and science are somehow linked, in this case using scientific facts as a basis for rejecting religion.

One such scientist was Darwin himself, who embraced agnosticism because his own findings had proved that the biblical account of creation cannot be *literally* true. In this religious choice, he was following in a family tradition: his grandfather, the physiologist Erasmus Darwin (1731–1802), belonged to the mechanist school, a muddle of atheism, bad theory, and genuine science.

The mechanists claimed that humans were mere machines whose activities could be understood purely in terms of physical and chemical processes. Claims such as these ultimately led to the discrediting of their movement, whose ideas failed to explain such biological processes as growth. At the same time, such mechanist philosophers as the French physician and philosopher Julien de La Mettrie (1709–1751) went far beyond the territory of science, teaching that atheism was the only road to happiness and that the purpose of human life was to experience pleasure.

The thinker who perhaps did the most to confuse science and atheism was one of Darwin's most significant early followers, the German natural scientist and philosopher Ernst Haeckel (1834–1919). It was Haeckel, not Darwin, who first proposed an evolutionary explanation for the origin of human beings, which, of course, was a major step beyond even Darwin's claim that all of life had evolved over millions of years.

In the course of developing this idea, Haeckel, who was a practicing Christian until he read Darwin's *On the Origin of Species by Means of Natural Selection,* renounced his faith and adopted a belief system he called monism, which is based on the idea that there is only a physical realm and no spiritual one. Technically, Haeckel was not an atheist but a pantheist, since his philosophy included the idea of a single spirit that lives in all things, both living and nonliving. Whatever the case, Haeckel's monism is no more scientific than Christianity.

HUMANS AND "MONKEYS." It is interesting that the man who put forward the notorious idea that humans and apes are related also would attempt to turn evolution into a sort of "proof" of atheism. In fact, the evolutionary connection between humans and lower primates, or "monkeys," has long been the most powerful point of contention between religion and evolution.

This, in fact, remains one of the most challenging aspects of evolutionary theory—not because it is hard to see how the human body is similar to an ape's body but because there is such a vast difference between a human mind and that of an ape. Whereas our physical similarity to primates is easy to establish, the fact is that no other animal—ape, dolphin, pig, or dog—comes close to humans in terms of reasoning ability. Nor is it reasoning ability alone that separates humans from other animals. Humans possesses a propensity for conceptualization and a level of self-awareness that sets them completely apart from other creatures, so much so that the brains of apes, cats, birds, and even frogs seem more or less alike compared with that of a human.

Animals are concerned with a few things: eating, sleeping, eliminating waste, and procreating. Some mammals have the ability to engage in play, but there is still no comparison between even the most advanced mammalian brains and that of a human. Other primates have the ability to use sticks or stones as tools, but only humans—practically from the beginning of the species 2.5 million years ago—have the ability to fashion tools. Only humans are gifted, or cursed, with restless minds ever in search of new knowledge.

Does any of this disprove evolution? It does not. Does it pose a significant challenge to the idea that humans and other primates evolved from a common ancestor? Not as it has been stated here. All that has been said in the preceding paragraphs is simply a matter of everyday observation, but it is not a scientific hypothesis, let alone a theory. Clearly, there are some questions still to be answered as to why and how humans developed brains so radically different from those of other primates, but the place for such questioning is *within* the realm of science not outside it.

CREATIONISM. Another thing we can say about the human mind is that it has a tendency to mold ideas toward its own preconceptions as to how things should be. As Hume observed, there is a great temptation, in the minds of all people, to demand that scientific facts conform to a particular set of religious or political beliefs. Such is the case with creationism and "intelligent design theory," two scientific belief systems whose adherents have attempted to challenge evolutionary theory.

Creationism, which sometimes goes by the name of creation science, is based on the belief that God created the universe and did so in a very short period of time. This claim, creationists maintain, can be supported by scientific evidence. Scientific evidence, however, is not really what drives creationism, which is based on a literal reading of the first two chapters of the Book of Genesis. Taken to an extreme, this means that God created the universe about 6,000 years ago in six days of 24 hours each.

Adherents of creationism begin with the premise of a six-day Creation (or at least, a very young Earth) and then look for facts to support the premise—exactly the opposite of the approach taken by true science. The findings of creationists do not change much over the years, unlike evolutionary science, which has continued to develop with new discoveries.

Sometimes creationists attempt to use the findings of evolutionary science against it. For instance, they may interpret industrial melanism (the adaptation of moths to discoloration in the environment caused by pollution, discussed later in this essay) as proof that organisms can change very quickly. This, of course, does not take into account the fact that moths have very short life spans compared with humans, for whom evolutionary change takes much longer. Creationists also point to areas of evolutionary theory where all scientists are not in agreement, citing these as "proof" that the whole theory is unsound.

INTELLIGENT DESIGN THEORY AND THE COURT BATTLE. In contrast to creationism, intelligent design theory is not based on any particular religious position. Instead, it begins with an observation that would find a great deal of agreement among many people, including those who support evolutionary theory. The idea is that evolution alone does not explain fully how life on Earth came to exist as it does, with all its complexity and order. According to intelligent design theory, there must have been some intelligence behind the formation of the universe.

There is another contrast between intelligent design theory and creationism. Whereas it is hard to imagine a genuine scientist embracing creationism, it is not difficult at all to picture a scientific thinker adopting the viewpoint of intelligent design. In fact, this has happened, though long before the "movement" had a name.

Darwin's contemporary, the English naturalist Alfred Russel Wallace (1823–1913), who published his own theory of evolution at about the same time as Darwin's *Origin of Species,* parted ways with Darwin because he maintained that there must be a spiritual force guiding evolution. Only such a force, he maintained, could explain the human soul. From a philosophical and theological standpoint, this idea has a great deal of merit, but because it cannot be tested, it cannot truly be regarded as science.

Neither creationism nor intelligent design has received any support in the scientific community—nor, during court battles over the teaching of creationism in the public schools during the 1980s, did that idea receive the support of the United States justice system. Creationism, the courts ruled, is a religious and not a scientific doctrine. Evolutionary theory is based on an ever increasing body of evidence that is both observable and reproducible. To teach these other doctrines alongside evolution in the public schools would convey the impression that creationism and intelligent design had been subjected to the same kinds of rigorous tests that have been applied to evolution, and this is clearly not the case.

EVIDENCE FOR EVOLUTION

A great deal of evidence for evolution appeared in the seminal text of evolutionary theory (mentioned previously), *On the Origin of Species by Means of Natural Selection,* which Darwin published in 1859. In fact, he had collected much of the evidence he discusses in this volume nearly three decades earlier, from 1831 to 1836, aboard a scientific research vessel off the coast of South America. (He delayed publication because he rightly feared the controversy that would ensue and resolved to present his ideas only when he learned that Wallace had developed his own theory of evolution.)

Just 22 years old, Darwin traveled on the HMS *Beagle,* from which he collected samples of marine life. His most significant work was done on the Galápagos Islands some 563 mi. (900 km) west of Ecuador. As he studied organisms there, Darwin found that they resembled species in other parts of the world, but they were also unique and incapable of interbreeding with similar species on the mainland. He began to suspect that for any particular environment, certain traits came to the forefront, favored for survival by nature.

Back in England, he already had seen such a mechanism at work in the artificial breeding of pigeons, whereby breeders favored certain gene pools—for instance, white-tailed birds—over others. (Breeders of dogs and other animals today still employ artificial-selection techniques to produce desirable strains.) Darwin posited a similar process of selection in nature, only this one was not artificial, directed by a goal-oriented human intelligence, but natural and guided by the need for survival.

THE SPREAD OF SPECIES. Among the phenomena Darwin observed in the Galápagos was the differentiation among the 13 varieties of finch (a type of bird) on the islands as well as the contrasts among these finches and their counterparts on the mainland. As Darwin began to discover, they shared many characteristics, but each variety had its own specific traits (for instance, the ability to crack tough seeds for food) that allowed it to fill a particular niche in its own environment.

From the beginning Darwin was influenced by the recent findings in geology, a newly emerging science whose leading figures maintained that Earth was very, very old. (These scientists included the Scottish geologist Charles Lyell [1797–1875], whose *Principles of Geology,* published between 1830 and 1833, Darwin read aboard the *Beagle*) The relationship between geology and evolution has persisted, and findings in the earth sciences continue to support evolutionary theory.

Among the leading ideas in geology and other geosciences since the mid–twentieth century is plate tectonics, which indicates (among other things) that the continents of Earth are constantly moving. (See Paleontology for further discussion of this topic.) This idea of continental drift provided a mechanism for species differentiation of the kind Darwin had observed.

It appears that in the past, when the landmasses were joined, organisms spread over all available land. Later, this land moved apart, and the organisms became isolated. Eventually, different forms evolved, and in time these distinct organisms became incapable of interbreeding. This is what occurred, for instance, when the Colorado River cut open the Grand Canyon, separating groups of squirrels who lived in the high-

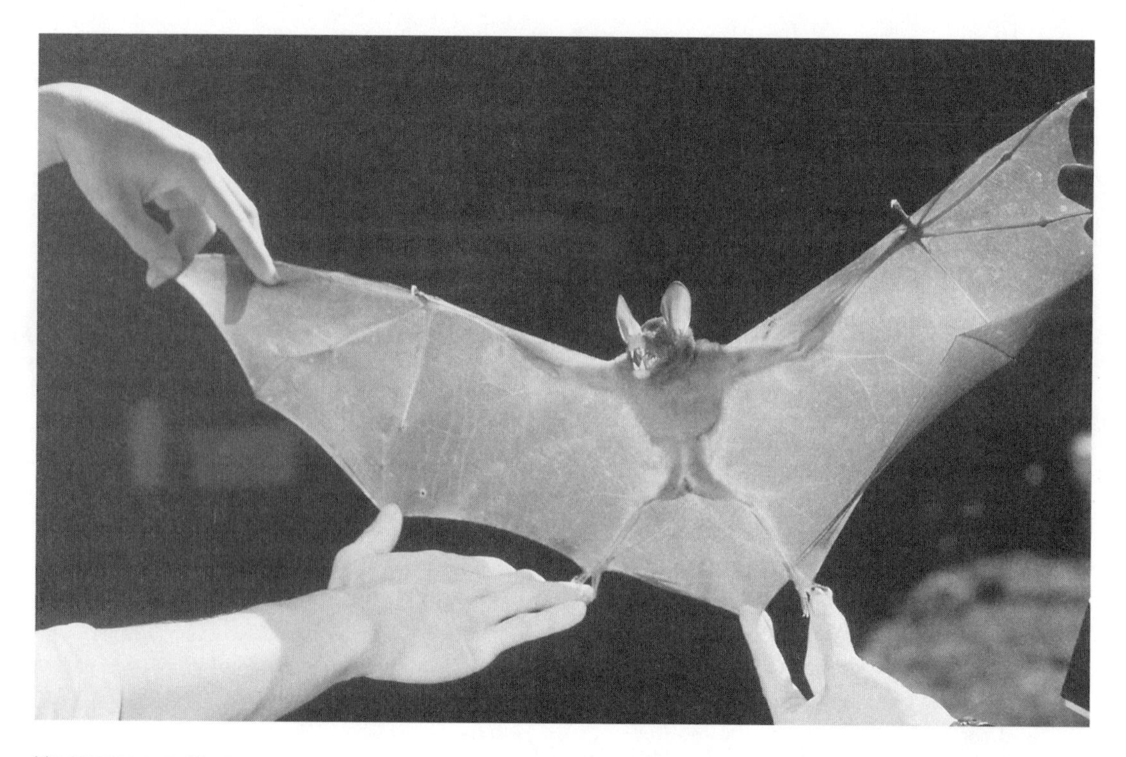

VAMPYRUM BAT. WIDELY DIVERGENT ORGANISMS SOMETIMES POSSESS A COMMON STRUCTURE, ADAPTED TO THEIR INDIVIDUAL NEEDS OVER COUNTLESS GENERATIONS YET REFLECTIVE OF A SHARED ANCESTOR. THE CAT'S PAW, THE DOLPHIN'S FLIPPER, THE BAT'S WING, AND THE HUMAN HAND ARE ALL VERSIONS OF THE SAME ORIGINAL FIVE-DIGIT APPENDAGE, CALLED THE PENDTADACTYL LIMB. *(© Gary Braasch/Corbis. Reproduced by permission.)*

altitude pine forest. Eventually, populations ceased to interbreed, and today the Kaibab squirrel of the northern rim and the Abert squirrel of the south are separate species.

COMMON ANCESTRY. Darwin recognized that some of the best evidence for evolution lies hidden within the bodies of living creatures. If organisms have a history, he reasoned, then vestiges of that history will linger in their bodies—as studies in comparative anatomy show. An example is a phenomenon that sounds as if it is made up, but it is very real: snake hips. Though their ancestors ceased to walk on four legs many millions of years ago, snakes still possess vestigial hind limbs as well as reduced hip and thigh bones.

In some cases widely divergent organisms possess a common structure, adapted to their individual needs over countless generations yet reflective of a shared ancestor. A fascinating example of this is the pentadactyl limb, a five-digit appendage common to mammals and found, in modified form, among birds. The cat's paw, the dolphin's flipper, the bat's wing, and the human hand are all versions of the same original, an indication of a common four-footed ancestor that likewise had limbs with five digits at the end.

The embryonic forms of animals also reflect common traits and shared evolutionary forebears. This is why most mammals look remarkably similar in early stages of development. In some cases animals in fetal form will manifest vestigial features reflective of what were once functional traits of their ancestors. Thus, fetal whales, while still in their mothers' wombs, produce teeth after the manner of all vertebrates (creatures with an internal framework of bones), only to reabsorb those teeth, which they will not need in a lifetime spent filtering plankton through their jaws.

The molecular "language" of DNA also provides evidence of shared evolutionary lineage. When one studies the DNA of humans and chimpanzees, very close similarities rapidly become apparent. Likewise, there are common structures in the hemoglobin, or red blood cells, of different types of organisms. Comparisons of hemoglobin make it possible to pinpoint the date of the last common ancestor of differing species. For example, hemoglobin analysis reveals an ancestor common to humans and frogs dating

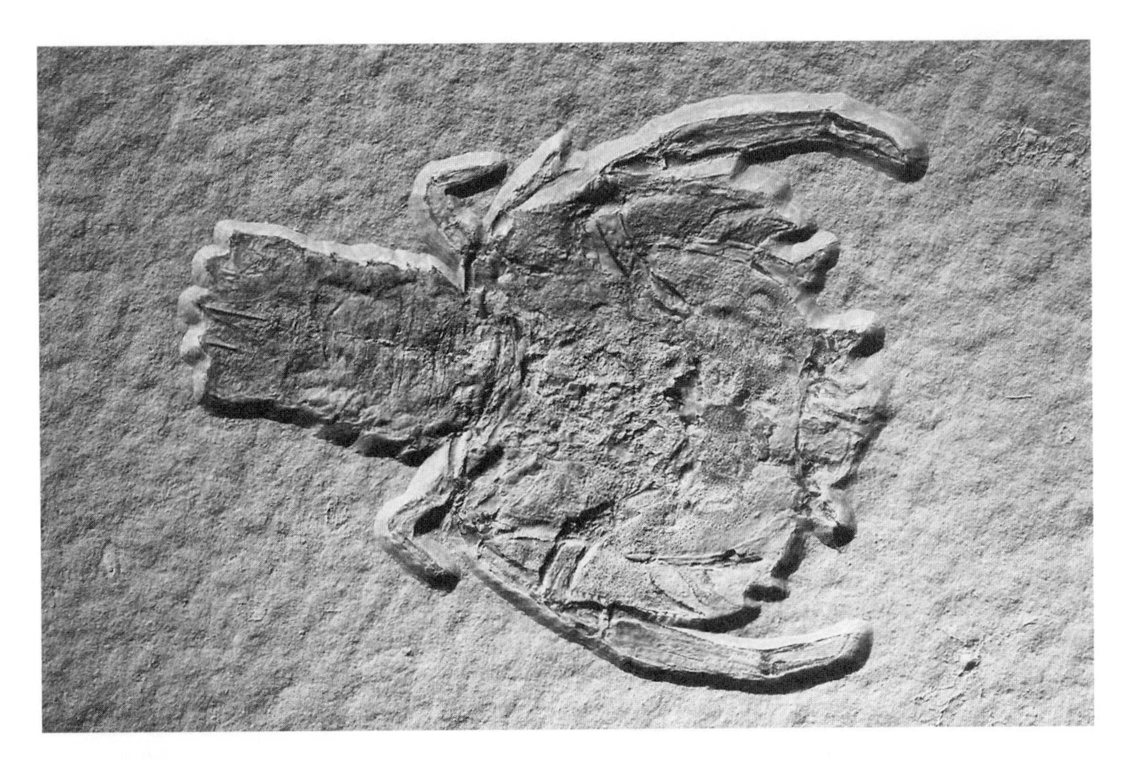

LOBSTER FOSSIL FROM THE LATE JURASSIC PERIOD. THE PRESERVED REMAINS OF SUCH PREHISTORIC LIFE-FORMS APPEAR IN THE ORDER OF THEIR EVOLUTION IN THE STRATA, OR LAYERS, OF EARTH'S SURFACE, WHICH GEOLOGISTS ARE ABLE TO DATE: THE AGE OF A STRATUM ALWAYS CORRELATES WITH THE FOSSILS DISCOVERED THERE. *(© Layne Kennedy/Corbis. Reproduced by permission.)*

back 330 million years, whereas the common human and mouse ancestor lived 80 million years ago, and the ancestor we share with the rhesus monkey walked the earth "only" 26 million years ago.

THE FOSSIL RECORD. The fossil record also provides an amazing amount of evidence concerning common ancestors. Fossilized remains of invertebrates (animals without an internal skeleton), vertebrates, and plants appear in the strata or layers of Earth's surface in the same order that the complexities of their anatomy suggest. The more evolutionarily distant organisms lie deeper, in the older layers, beneath the remains of the more recent organisms. Geologists are able to date rock strata with reasonable accuracy, and the age of a layer always correlates with the fossils discovered there. In other words, there would never be a stratum dating back 400 million years that contained fossils of mastodons, which evolved much later.

A fossil is the remains of any prehistoric life-form, especially those preserved in rock before the end of the last ice age, about 10,000 years ago. The process by which a once living thing becomes a fossil is known as *fossilization.* Generally, fossilization involves changes in the hard portions, including bones, teeth, and shells. This series of changes, in which minerals are replaced by different minerals, is known as *mineralization.*

Fossilized remains of single-cell organisms have been found in rock samples as old as 3.5 billion years, and animal fossils have been located in rocks that date to the latter part of Precambrian time, as long ago as one billion years. Certain fossil types, known as index fossils or indicator species, have been associated strongly with particular intervals of geologic time. An example is the ammonoid, a mollusk that proliferated for about 350 million years, from the late Devonian to the early Cretaceous periods, before experiencing mass extinction.

The fossil record is far from an open book, however, and interpreting fossil evidence requires a great deal of judgment. All manner of natural phenomena such as earthquakes can destroy fossil beds, rendering the evidence unreadable or at least unreliable. Nor is it a foregone conclusion that the animals who left behind fossils are fully representative of the species existing at a given time. Fossils are far more likely to be preserved in certain kinds of protected aquatic environments, for instance, than on land (particularly at higher elevations, where erosion is a significant factor), and therefore paleontologists' knowledge of life forms in the distant past is heavily weighted toward marine creatures.

FAUNAL SUCCESSION AND OTHER FORMS OF DATING. Key to the demonstration of evolution is the age of samples and the idea that many of the processes described took place a long, long time ago. This raises the question of how scientists know the age of things. In fact, they have at their disposal several techniques, both relative and absolute, for dating objects.

One of the earliest ideas of dating in geology was faunal dating, or the use of bones from animals (fauna) to determine age. This was the brainchild of the English engineer and geologist William Smith (1769–1839), whose work is an example of the fact that evolutionary ideas were "in the air" long before Darwin. While excavating land for a set of canals near London, Smith discovered that any given stratum contains the same types of fossils, and therefore strata in two different areas can be correlated. Smith stated this in what became known as the law of faunal succession: all samples of any given fossil species were deposited on Earth, regardless of location, at more or less the same time. As a result, if a geologist finds a stratum in one area that contains a particular fossil and another in a distant area having the same fossil, it is possible to conclude that the strata are the same.

Faunal succession is relative, meaning that it does not provide clues as to the actual age in years of a particular sample. Since the mid–twentieth century, however, scientists have had at their disposal several means for absolute dating, which make it possible to determine the rough age of samples in years. Most of these mechanisms for dating are based on the fact that over time, a particular substance converts to another, mirror substance. By comparing the ratios between them, it is possible to arrive at an estimate as to the amount of time that has elapsed since the organism died.

Chief among the techniques for absolute dating is radiometric dating, which uses ratios between two different kinds of atoms for a given element: stable and radioactive isotopes. Isotopes are atoms that differ in their number of neutrons, or neutrally charged subatomic particles, and radioactive isotopes are ones that spontaneously eject various high-energy particles over time. Because chemists know how long it takes for half the isotopes in a given sample to stabilize (a half-life), they can judge the age of such a sample by examining the ratio of stable to radioactive isotopes. In the case of uranium, one isotopic form, uranium 238, has a half-life of 4,470 million years, which is very close to the age of Earth itself.

Evolution at Work

Every creature that exists today is the result of an incredibly complex, lengthy series of changes brought about by mutation and natural selection, changes that influenced the evolution of that life-form. Take for instance the horse, whose evolutionary background is as well-documented as that of any creature.

The horse family, or Equidae, had its origins at the beginnings of the Eocene epoch about 54 million years ago. This first ancestor, known as *Hyracotherium* or *eohippus* ("dawn horse") was extremely small—only about the size of a dog. In addition, it had four hooves on its front feet and three on each rear foot, with all of its feet being padded, which is quite a contrast with the four unpadded, single-hoofed feet of the modern horse. These and other features, such as head size and shape, constitute such a marked difference from what we know about horses today that many scientists have questioned the status of eohippus as an equine ancestor. However, comparison with fossils from later, also extinct, horses shows a clear line of descent marked by an increase in body size, a decrease in the number of hooves, an elimination of foot pads, lengthening of the legs and fusion of the bones within, development of new teeth suited for eating grass, an increase in the length of the muzzle, and a growth in both the size and development of the brain.

Of course, this was not a clear-cut, neat, and steadily unfolding process, and some features appeared abruptly; still, the progression is there to be observed in the fossil record. Over the course of the many millions of years since eohippus, species have emerged that were distinguished by a particular feature—for example, teeth size and shape—only to disappear if conditions favored species with other traits. Evolutionary lines have branched off, with some dead-ending, and others continuing.

Thus, during the Miocene epoch, which lasted from about 26 million to 7 million years ago, various evolutionary branches competed for a time until the emergence of *Parahippus.* This species had teeth adapted for eating grass, in contrast to those of earlier horse ancestors, which grazed on leaves and other types of vegetation that did not require strong teeth. After Parahippus came *Merychippus,* which resembled a modern pony, and from which came numerous late-Miocene evolutionary lines. Most of these were three-toed, but *Pliohippus* had one toe per foot, and it was from this form that the genus *Equus* (which today includes horses, donkeys, and zebras) began to emerge in the late Pliocene epoch about 3 million years ago.

INDUSTRIAL MELANISM AND THE PEPPER MOTH. Despite the staggering spans of time involved in evolution, one need not look back billions of years to see evolution at work. Both natural selection and mutation play a role in industrial melanism, a phenomenon whereby the processes of evolution can be witnessed within the scale of a human lifetime. Industrial melanism is the high level of occurrence of dark, or melanic, individuals from a particular species (usually insects) within a geographic region noted for its high levels of dark-colored industrial pollution.

With so much pollution in the air, trees tend to be darkened, and thus a dark moth stands a much greater chance of surviving, because predators will be less able to see it. At the same time, there is a mutation that produces dark-colored moths, and in this particular situation, these melanic varieties are selected naturally. On the other hand, in a relatively unpolluted region, the lighter-colored individuals of the same species tend to have the advantage, and therefore natural selection does not favor the mutation.

The best-known example of industrial melanism occurred in a species known as the pepper moth, or *Biston betularia,* which usually lives on trees covered with lichen. (An example of a lichen is reindeer "moss"; see Symbiosis.) Prior to the beginnings of the Industrial Revolution in England during the late eighteenth century, the proportion of light-colored pepper moths was much higher than that of dark-colored ones, both of which were members of the same species differentiated only by appearance.

As the Industrial Revolution got into full swing during the 1800s, factory smokestacks put so much soot into the air in some parts of England that it killed the lichen on the trees, and by the 1950s, most pepper moths were dark-colored. It was at that point that Bernard Kettlewell (1907–1979), a British geneticist and entomologist (a scientist who studies insects), formed the hypothesis that the pepper moths' coloration protected them from predators, namely birds.

Kettlewell therefore reasoned that, before pollution appeared in mass quantities, light-colored moths had been the ones best equipped to protect themselves because they were camouflaged against the lichen on the trees. After the beginnings of the Industrial Revolution, however, the presence of soot on the trees meant that light-colored moths would stand out, and therefore it was best for a moth to be dark in color. This in turn meant that natural selection had favored the dark moths.

In making his hypothesis, Kettlewell predicted that he would find more dark moths than light moths in polluted areas, and more light than dark ones in places that were unpolluted by factory soot. As it turned out, dark moths outnumbered light moths two-to-one in industrialized areas, while the ratios were reversed in unpolluted regions, confirming his predictions. To further test his hypothesis, Kettlewell set up hidden cameras pointed at trees in both polluted and unpolluted areas. The resulting films showed birds preying on light moths in the polluted region, and dark moths in the unpolluted one—again, fitting Kettlewell's predictions.

ANGIOSPERMS AND GYMNOSPERMS. A final interesting example of natural selection at work lies in the comparative success rates of angiosperms and gymnosperms. An angiosperm is a type of plant that produces flowers during sexual reproduction, whereas a gymnosperm reproduces sexually through the use of seeds that are exposed, for instance in a cone. Angiosperms are a beautiful example of how a particular group of organisms can adapt to its environment and do so in a much more efficient way than that of its evolutionary forebears. On the other hand, gymnosperms, with their much less efficient form of reproduction, perhaps one day will go the way of the dinosaur.

Flowering plants evolved only about 130 million years ago, by which time Earth long since

KEY TERMS

ACQUIRED CHARACTERISTICS: Sometimes known as *acquired characters* or *Lamarckism* after one of its leading proponents, the French natural philosopher Jean Baptiste de Lamarck (1744–1829), acquired characteristics is a fallacy that should not be confused with mutation. Acquired characteristics theory maintains that changes that occur in an organism's overall anatomy (as opposed to changes in its DNA) can be passed on to offspring.

DATING: Any effort directed toward finding the age of a particular item or phenomenon. Methods of dating are either relative (that is, comparative and usually based on rock strata, or layers) or absolute. The latter, based on methods such as the study of radioactive isotopes, typically is rendered in terms of actual years or millions of years.

DNA: Deoxyribonucleic acid, a molecule in all cells and in many viruses that contains genetic codes for inheritance.

FOSSIL: The mineralized remains of any prehistoric life-form, especially those preserved in rock before the end of the last ice age.

FOSSILIZATION: The process by which a once living organism becomes a fossil. Generally, fossilization involves mineralization of the organism's hard portions, such as bones, teeth, and shells.

GENE: A unit of information about a particular heritable (capable of being inherited) trait that is passed from parent to offspring, stored in DNA molecules called *chromosomes.*

GEOLOGIC TIME: The vast stretch of time over which Earth's geologic development has occurred. This span (about 4.6 billion years) dwarfs the history of human existence, which is only about 2.5 million years. Much smaller still is the span of human civilization, only about 5,500 years.

had been dominated by another variety of seed-producing plant, the gymnosperm, of which pines and firs are an example. Yet in a relatively short period of time, geologically speaking, angiosperms have become the dominant plants in the world. In fact, about 80% of all living plant species are flowering plants. Why did this happen? It happened because angiosperms developed a means whereby they coexist more favorably than gymnosperms with the insect and animal life in their environments.

Gymnosperms produce their seeds on the surface of leaflike structures, and this makes the seeds vulnerable to physical damage and drying as the wind whips the branches back and forth. Furthermore, insects and other animals view gymnosperm seeds as a source of nutrition. In an angiosperm, by contrast, the seeds are tucked safely away inside the ovary. Furthermore, the evolution of the flower not only has added a great deal of beauty to the world but also has provided a highly successful mechanism for sexual reproduction. This sexual reproduction makes it possible for new genetic variations to develop, as genetic material from two individuals of differing ancestry come together to produce new offspring. (For more about angiosperms and gymnosperms, see Ecosystems and Ecology.)

WHERE TO LEARN MORE

Campbell, Neil A., Lawrence G. Mitchell, and Jane B. Reece. *Biology: Concepts and Connections.* 2nd ed. Menlo Park, CA: Benjamin/Cummings, 1997.

Darwin, Charles, and Richard E. Leakey. *The Illustrated Origin of Species.* New York: Hill and Wang, 1979.

KEY TERMS CONTINUED

HYPOTHESIS: An unproven statement regarding an observed phenomenon.

INDUSTRIAL MELANISM: The high level of occurrence of dark, or melanic, individuals from a particular species (usually insects) within a geographic region noted for its high levels of dark-colored industrial pollution.

INVERTEBRATE: An animal without an internal skeleton.

LAW: A scientific principle that is shown always to be the case and for which no exceptions are deemed possible.

MINERALIZATION: A series of changes experienced by a once living organism during fossilization. In mineralization, minerals in the organism are replaced or augmented by different minerals or the hard portions of the organism dissolve completely.

MUTATION: Alteration in the physical structure of an organism's DNA, resulting in a genetic change that can be inherited.

NATURAL SELECTION: The process whereby some organisms thrive and others perish, depending on their degree of adaptation to a particular environment.

PALEONTOLOGY: The study of life-forms from the distant past, primarily as revealed through the fossilized remains of plants and animals.

SCIENTIFIC METHOD: A set of principles and procedures for systematic study that includes observation; the formation of hypotheses, theories, and laws; and continual testing and reexamination.

THEORY: A general statement derived from a hypothesis that has withstood sufficient testing.

VERTEBRATE: An animal with an internal skeleton.

Dennett, Daniel Clement. *Darwin's Dangerous Idea: Evolution and the Meanings of Life.* New York: Simon and Schuster, 1996.

Evolution and Natural Selection (Web site). <http://www.sprl.umich.edu/GCL/paper_to_html/selection.html>.

Evolution. British Broadcasting Corporation (Web site). <http://www.bbc.co.uk/education/darwin/index.shtml>.

"Evolution FAQs." Talk Origins (Web site). <http://www.talkorigins.org/origins/faqs-evolution.html>.

Evolution. Public Broadcasting System (Web site). <http://www.pbs.org/wgbh/evolution/>.

Evolution. University of California, Berkeley, Museum of Paleontology (Web site). <http://www.ucmp.berkeley.edu/history/evolution.html>.

Levy, Charles K. *Evolutionary Wars: A Three-Billion-Year Arms Race. The Battle of Species on Land, at Sea, and in the Air.* New York: W. H. Freeman, 1999.

Starr, Cecie, and Ralph Taggart. *Biology: The Unity and Diversity of Life.* 7th ed. Belmont, CA: Wadsworth, 1995.

PALEONTOLOGY

CONCEPT

Paleontology is the study of life-forms from the distant past, as revealed primarily through the record of fossils left on and in the earth. It is a subdiscipline of both the biological and the earth sciences, one that brings to bear the techniques of geologic study and several areas of biology, including botany and zoology. But the term *paleontology,* as used in the present context, also provides a convenient means of encompassing, in a single word, the study of the distant biological past. Such study, of course, is related intimately to the investigation of evolutionary processes and phenomena. In the present context, however, our concern is not so much with the theory and principles of evolution, discussed elsewhere in this book, but rather with a relatively short overview of biological history. This area of biological investigation calls upon such concepts as mass extinction and fossilization as well as the study of plant and animal forms that scientists know only from scientific reconstruction of the past rather than from direct experience. Chief among these life-forms are the dinosaurs, which dominated Earth for a period of more than 100 million years, ending about 65 million years ago. This span of time, impressive as it seems on the human scale, is minuscule compared with the entire history of life on Earth.

HOW IT WORKS

PALEONTOLOGY AMONG THE SCIENCES

Paleontology is the investigation of life-forms from the distant past, primarily through the study of fossilized plants and animals. Most people are familiar with fossils, a term that probably calls to mind an image of a flat piece of rock with a shadowy imprint of a leaf or animal on it. In fact, a fossil is the preserved remains of a once living organism that has undergone a process known as *mineralization,* in which the organic materials in hard parts of the organism (for example, teeth or bones) are replaced by minerals, which are inorganic.

The difference between the organic and the inorganic, which we discuss later, is more than a difference between the living and nonliving, or even the living and formerly living. For now, though, it will suffice to say that all living and formerly living things, as well as their parts and products, are organic. The stress on formerly living things is important, since paleontology by definition encompasses plants, animals, and microscopic life-forms that lived a long time ago. (As for just how long "a long time ago" really is, that subject, too, is discussed later in this essay.)

Among the fields related, or subordinate, to paleontology are paleozoology, which focuses on the study of prehistoric animal life; paleobotany, the study of past plant life; and paleoecology, the study of the relationships between prehistoric plants and animals and their environments. Despite the emphasis on past life-forms, however, paleontologists also avail themselves of evidence gleaned from observation of animals living today. In this approach, paleontology shows its relationship to geology, which, along with biology, is one of its two "parents."

GEOLOGY AND PALEONTOLOGY. One of the key concepts in geology is the principle of uniformitarianism, which arose from the early history of the geologic sciences. At a time when the nascent science was embroiled in a debate over Earth's origins and the age of the planet, the Scottish geologist James Hutton (1726–1797) transcended this debate by focusing on the processes at work on Earth in the present day. (In terms of geologic time or even the evolution of species, the eighteenth century and twenty-first century might as well be a few seconds apart, so the expression *present day* applies to Hutton's time as much as to ours.) Rather than simply speculate as to how Earth had come into being, Hutton maintained that scientists could understand the processes that had formed it by studying the geologic phenomena they could see around them.

The reason this approach works is that natural laws do not change over time, nor do the processes that are at work within nature. On the other hand, particular processes may not be in operation at all times, nor can it be assumed that the rate at which the processes take place is the same. The two foregoing sentences encompass some of the observations made by the modern American paleontologist Stephen Jay Gould (1941–2002) with his formulation of the "four types of uniformity" in *Ever Since Darwin: Reflections in Natural History* (1977). It is certainly appropriate that paleontology and geology interface in this concept of uniform change, because both are concerned with piecing together the distant past from the materials available in the present.

In fact, from the standpoint of the earth sciences, paleontology belongs to a branch of geology known as historical geology, or the study of Earth's physical history. (This field of study is contrasted to the other principal branch of the discipline, physical geology, the study of the material components of Earth and of the forces that have shaped the planet.) Other subdisciplines of historical geology are stratigraphy, the study of rock layers, or strata, beneath Earth's surface; geochronology, the study of Earth's age and the dating of specific formations in terms of geologic time; and sedimentology, the study and interpretation of sediments, including sedimentary processes and formations.

BEFORE LIFE

Later in this essay, we consider just what is meant by "geologic time," the grand sweep of several billion years during which Earth evolved from a cloud of gases to its present form. It may be surprising to discover the large proportion of that time during which life existed on Earth; it is equally amazing to learn what a small portion of the biological history of the planet involved anything that modern humans would recognize as "life."

A MUSEUM REPRODUCTION OF NEANDERTHAL MAN, OR HOMO SAPIENS NEANDERTHALENSIS. THE SPAN OF TIME SINCE THE FIRST APPEARANCE OF THE GENUS HOMO (TO WHICH HUMANS, OR HOMO SAPIENS, BELONG) IS MINUSCULE: 2.5 MILLION YEARS COMPARED WITH 4.6 BILLION YEARS, OR ABOUT 0.04% OF THE PLANET'S HISTORY. *(© Bettmann/Corbis. Reproduced by permission.)*

For the moment, let us go even further back, to the very beginning—an almost inconceivably long time ago. Scientists believe that the universe began between 10 billion and 20 billion years ago, with an explosion nicknamed the "big bang," a cataclysm so powerful that it sent galaxies careening outward on a course they maintain even today. Among those galaxies moving outward from the universal point of origin was our Milky Way, which, about six billion years ago, began to develop a rotating cloud of cosmic gas somewhere between its center and its rim. That was the beginning of our solar system.

THE SUN AND EARTH. The center of the cloud, where the greatest amount of gases gathered, was naturally the densest and most massive portion as well as the hottest. There, hydrogen—the lightest of all elements—experienced extraordinary amounts of compression owing to the density of the clouded gases around it, and underwent nuclear fusion, or the bonding of atomic nuclei.

This hot center became the Sun about five billion years ago, but there remained a vast nebula of gas surrounding it. As the fringes of this nebula began to cool, the gases condensed, forming solids around which particles began to accumulate. These were the future planets, including ours. The process of planetary formation took place over a span of about 500 million years. The planets contained various chemical elements, formed by nuclear fusion on the Sun—a fascinating aspect of life on Earth, since it means that the particles that make up the human body came from nuclear reactions on the stars.

THE EARLY ATMOSPHERE. The brand-new Earth possessed an atmosphere that consisted primarily of elemental hydrogen, nitrogen, carbon monoxide, and carbon dioxide. This "air" would have been unbreathable to all but a very small portion of the life-forms that exist on Earth today. Yet for the development of life, it was absolutely essential that no oxygen be present in Earth's atmosphere at that very early time.

The reason is that oxygen is an extremely reactive element, which is why it is involved in an array of chemical reactions (including combustion and rusting) known collectively as *oxidation-reduction reactions.* If oxygen had been present at that time, it likely would have reacted with other elements immediately, rather than permitting the formation of what eventually became organic materials.

WATER. Water appeared as the result of meteorite bombardment from space, which took place in the first half-billion years of Earth's existence. It might seem strange to learn that water, a compound essential to life on Earth, came from the void of space—where, as far as we know, there are no other life-forms. But this is not as strange as it sounds: water itself is inorganic, and even today frozen water exists on several planets in our solar system.

In any case, water eventually accumulated on Earth's surface in quantities sufficient for its condensation, with the result that clouds formed and rain fell on Earth more or less continually for many millions of years. It was then that the beginnings of life made their appearance, in the form of self-replicating molecules of carbon-based matter.

LIFE BEGINS

One famous cliché of science-fiction movies is the use of the phrase "carbon-based life-forms" to describe humans. In fact, *all* living things contain carbon, and if we ever do find life, intelligent or otherwise, on other planets, chances are extremely high that it, too, will be "carbon-based." Carbon, in fact, is almost synonymous with life, and hence the word *organic,* in its scientific meaning, refers to all substances that contain carbon. The only exceptions are the elemental carbon in diamonds or graphite, the carbonate forms that make up many of Earth's rocks, and such oxides as carbon dioxide and monoxide, all of which are considered inorganic.

It may sound as though a huge portion of carbon's possible forms already have been eliminated from the list of organic substances, but, in fact, carbon is capable of forming an almost limitless array of compounds with other elements, particularly hydrogen. A class of molecule known as hydrocarbons, which are nothing but strings of carbon and hydrogen molecules bonded together, is the basis for literally millions of organic compounds, from petroleum to polymer plastics. It may sound odd to hear plastics referred to as *organic,*but this only highlights the difference between the popular and scientific meanings of that term.

ORGANIC MATERIALS. At one time, *organic* referred only to living things, things that were once living, and materials produced by living things (for example, sap, blood, and urine). As recently as the early nineteenth century, scientists believed that organic substances contained a supernatural "life force," but in 1828 the German chemist Friedrich Wöhler (1800–1882) made an amazing discovery.

By heating a sample of ammonium cyanate, a material from a nonliving source, Wöhler converted it to urea, a waste product in the urine of mammals. As he later observed, "without benefit of a kidney, a bladder, or a dog," he had turned an inorganic substance into an organic one. It was almost as though he had created life. Actually, what he had discovered was the distinction between organic and inorganic material, which results from the way in which the carbon chains are arranged.

This explanation of the difference between organic and inorganic is pivotal to understanding how the beginnings of life first formed on Earth. It is an almost inconceivably large step from the nonliving to the living but not nearly so much of a jump from the inorganic to the organic. In fact, it appears that what happened on Earth in its distant past was that organic (but not living) substances underwent chemical reactions with inorganic ones to produce the rudiments of life. The American chemist Stanley Miller (1930–) illustrated this with an experiment that involved a mixture of hydrogen, methane (CH_4), ammonia (NH_3), and water. Subjected to a discharge of electric sparks intended to simulate lightning in Earth's early atmosphere, the mixture eventually yielded amino acids, which are among the chief components of proteins.

EARLY FORMS OF LIFE. The course that early forms of life followed during the first 800 million years of Earth's existence was a lengthy one, and if a person could have glimpsed Earth at any interval of a few million years during this time, it might have seemed as though nothing at all was happening. In fact, however, life-forms were undergoing the most profound changes imaginable.

That span of 0.8 billion years saw a transition from elemental carbon to organic compounds and from organic compounds to organelles, which are discrete components of cells, and finally to cells themselves—the building blocks of life. The processes by which this happened were exceedingly complicated, and modern scientists have little to go on in forming their suppositions as to how these transitions came about. Among many key pieces of information missing from the picture, for instance, is the matter of how and when DNA first appeared in cells. (For more on DNA, see Cells.)

The first cells to form were known as prokaryotic cells, or cells without a nucleus. (These cells, too, are discussed in the essay Cells.) Prokaryotic cells may have been little more than sacs of DNA that were capable of self-replication—much like bacteria today, which are themselves prokaryotic. These early forms of bacteria, which dominated Earth for many millions of years, were apparently anaerobic and eventually split into three branches: archaebacteria, eubacteria, and eukaryotes. Out of the last group grew all other forms of life, including fungi, plants, and animals.

GEOLOGIC TIME MARCHES ON

By about 2.5 billion years ago, bacteria had begun to undergo a form of photosynthesis, as plants do today. As a result, oxygen started to accumulate in Earth's atmosphere, and this had two interesting implications for life on Earth. One of the results of oxygen formation was that formation of "new" cells—that is, spontaneously formed cells that did not come from already living matter—ceased altogether, because they were killed off by reactions with oxygen. Second, aerobic respiration thereafter became the dominant means for releasing energy among living organisms.

The real history of life-forms on Earth—the portion of that history about which paleontologists can learn a great deal from observation of fossils and other materials—dates from the beginning of the Cambrian period, about 550 million years ago. At this point, about 88% of Earth's history already had passed, yet the remaining 12% contains virtually all the really dramatic events in the formation of life on Earth. Later in this essay, we examine a few analogies that help put these time periods into perspective as well as the concept of geologic time itself.

PHASES IN EARTH'S HISTORY. In the course of discussing these topics, it is sometimes necessary to make reference to geologic time divisions—eons, eras, periods, and epochs. These are not units of a specific length in years, like a century or a millennium; instead, they are distinct phases in Earth's history that historical geologists (including paleontologists) have pieced together from fossils and other materials. Their names usually refer to locations where fossils relevant to that phase of geologic time were found: for example, the Jurassic period, whose name became a household word after the release of the 1993 blockbuster movie *Jurassic Park,* is named after the Jura Mountains of Switzerland and France.

The historical juncture mentioned in the paragraph before last—the beginning of the Cambrian period, some 550 million years ago, was also the end of the Proterozoic eon, the third of three eons in what is known as Precambrian time. The fourth and present eon is the Proterozoic, which has included three eras: Paleozoic (about 550–240 million years ago), Mesozoic (about 240–65 million years ago), and Cenozoic (about 65 million years ago to the present).

CATACLYSMS AND CONTINENTAL DRIFT. The divisions between these phases have not been drawn arbitrarily; rather, they are based on evidence suggesting that those points in Earth's history were marked by violent, cataclysmic events. Once life had come into the picture, these cataclysms brought about death on a vast scale, which we discuss later, in the context of mass extinctions. A number of phenomena caused mass extinctions at various points; most notable among them was the impact of meteorites. Also significant was continental drift, which, as its name implies, is the movement of the continents from distant origins.

Just as evolutionary theory informs much of our modern thinking about biology, the theory of plate tectonics holds a dominant position in geology and related earth sciences. Plate tectonics involves the movement of large segments in the crust and the upper mantle of Earth, and one of the outcomes of such movement is continental drift.

Even today, the continents we know are moving slowly away from or toward one another, but that movement is so slow that it would take millions of years for any change to be perceptible. At one time, however, the continents were distributed quite differently than they are today. For example, at the end of the Permian period and the beginning of the Triassic, when the dinosaurs began to appear on the scene, all of Earth's landmasses were joined in a single continent, Pangea, that stretched between the North and South Poles and was surrounded by a vast ocean.

REAL-LIFE APPLICATIONS

OUR PLACE IN THE GRAND SCHEME OF THINGS

All areas of historical geology—including paleontology—are concerned with geologic time, a term that refers to the great sweep of Earth's history. This is a timescale that dwarfs the span of human existence, and to study the history of life on Earth, a mental adjustment of monumental proportions is required. One must discard all notions used in studying the history of human civilization, including such concepts as *modern, medieval,* and *ancient,* which are essentially useless when discussing geologic time.

Human civilization has existed for about 5,500 years, the blink of an eye in geologic terms. Even the span of time since the first appearance of the genus *Homo* (to which humans, or *Homo sapiens,* belong), is minuscule: 2.5 million years compared with 4.6 billion years, or about 0.04% of the planet's history. A couple of analogies, one to a shorter span of time and another to a measurement of space, will help us understand the scope of time encompassed in the evolution of species.

LIFE APPEARS IN JUNE—SO TO SPEAK. Suppose that the entire history of Earth were likened to a single year of 365.25 days, starting with the formation of the planet from a cloud of dust and ending with the present. At what point in the year would the first aerobic life-forms appear? We are not talking about anything approaching a human or even a clam or a fern or a piece of algae—just a single-cell organism that depends on oxygen for energy.

One might guess that such life-forms appeared sometime in January, or at least by the end of winter in late March. In fact, the first aerobic single-cell life-forms arose around June 15—nearly halfway through a year. Even at about the time of Thanksgiving, the most complex organisms would be fish and a few early amphibians.

We tend to associate the dinosaurs with the early phases of Earth's history, but this only illustrates our distorted view of geologic time. In fact, the Jurassic period would be analogous to a period of about five days from December 15–20. By Christmas Day, all remaining dinosaurs would have been headed toward extinction, their dead bodies eventually forming the fossil fuels that power present-day civilization.

By this point we are within a few days of the year's end, and yet nothing remotely resembling a human has appeared. Our own genus would not come on the scene until around 8:00 P.M. on December 31. The New Year's Eve countdown would be nearing by the time human civilization began, at about 42 seconds before midnight. Christ's birth would have occurred at about 14 seconds before midnight, and the final 10-second countdown would begin about the time the Roman Empire fell. The life span of the average person would correspond to approximately half a second or less.

ANOTHER ANALOGY: LOS ANGELES TO NEW YORK. To use another analogy, suppose we are driving from

Los Angeles to New York City, and that this distance corresponds to Earth's history. Once we reach western Nebraska, with about 46% of the distance behind us, we would come to the beginning of the Proterozoic era and the origins of aerobic life-forms. One might wonder why "nothing" happened on all those long miles from L.A. through the deserts, mountains, and plains of the western United States, but as we have seen, a great deal happened: Earth formed from a cloud of gas, was pounded by meteors, and gradually became the home to oceans.

The end of the Proterozoic era (about 545 million years ago) would be at about 88% of the distance from Los Angeles to New York—somewhere around Pittsburgh, Pennsylvania. By this point, the continental plates have formed, oxygen has entered the atmosphere, and soft-bodied organisms have appeared. We are a long way from Los Angeles, and yet almost the entire history of life on Earth, at least in terms of relatively complex organisms, lies ahead of us.

If we skip ahead by about 339 million years—a huge leap not only in time but also in biological development—we come to the point when the dinosaurs appeared. We are now 95% of the way from the beginning of Earth's history to the present, and if it is measured against the distance from Los Angeles to New York, we would be somewhere around Scranton, in eastern Pennsylvania. Another 89 mi. (142 km) from the Scranton area would put us at a point about 65 million years ago, or the time when the dinosaurs became extinct. We then would have less than 40 mi. (64 km) to drive to reach the period where genus *Homo* appeared, by which time we would be in the middle of Manhattan.

Compared with the distance from L.A. to New York, the span of time that *Homo sapiens* has existed would be much shorter than the drive from Central Park to the Empire State Building. The entire sweep of human civilization and history, from about a thousand years before the building of the pyramids to the beginning of the third millennium A.D., would be smaller than a city block. It is *much* smaller, in fact—close to the size of a modest storefront, or 15.54 ft. (5 m).

How Did We Get Here?

On the one hand, the history of relatively complex life-forms is relatively short compared with Earth's history; on the other hand, it is unbelievably long, when one considers the diversity of forms that have evolved in the past half-billion years. What follows is an extremely brief overview, touching on a few points in the development of life on Earth.

The discussion here is far from comprehensive, the purpose being not to provide a detailed overview of evolutionary processes but to illustrate the evolution of life-forms by a few examples. The reader is strongly encouraged to consult a textbook or other reliable information source for greater illumination of these particular topics. Along with this overview, we look at some of the more dramatic aspects of paleontologic study: dinosaurs and mass extinction.

The fossilized history of life on Earth really began in earnest only with the Cambrian period, which saw an explosion of invertebrate (without an internal skeleton) marine forms. These dominated from about 550 to about 435 million years ago. At the latter point, the boundary between the Ordovician and Silurian periods of the Paleozoic era, there occurred the first of five major mass extinctions, as the "supercontinent" of Gondwana crossed the South Pole, freezing most life-forms that were then alive. As always happened with these mass extinctions, some life-forms survived, and within a few million years life again was thriving.

COMING OUT OF THE WATER. One of the favorite subjects of modern cartoonists is the migration of creatures from the water to the land. For example, following the enactment of stiff airport security guidelines in the wake of the September 11 terrorist attacks, a *New Yorker* cartoon in late 2001 showed this evolution from fish to amphibian (a creature that can live both in the water and on the ground) and, ultimately, to the human being. Then, in the last evolutionary sequence, *Homo sapiens* is depicted in a business suit, going through an airport security scanner. Most cartoons based on this event in paleontologic history follow a common theme, which can be summed up thus: "We went through all that evolution just for *this*?"

Certainly, the transition from water to solid ground was a massive step—one of almost inconceivable importance. This invasion of the land, which probably took place about 400 million years ago, perhaps started when fishes of the Devonian period (named after Devonshire in England) began breathing oxygen near the surface of shallow waters. In time they made their way onto the land, and as the number of species grew, their lobed fins developed in different ways for different groups of fish, at length becoming the many limbs and appendages specific to large groups of species, such as birds or mammals. This may seem a bit far-fetched, but even today catfish are known to pull themselves out of the water and onto land if a specific need to do so arises.

Why did the first fish leave the water? Perhaps because the water levels were receding, as they have done periodically in the course of Earth's history as a result of massive freezing of global water supplies—that is, ice ages. Whatever the reason, we can be sure of the fact that the fish were not in any way attempting to reach some "higher" state of development, nor was any force pushing them to "evolve." Therein lies one of the major misconceptions about evolution, even among many of its adherents. The very name *evolution* is somewhat unfortunate, because it suggests that life-forms are moving gradually toward a state of heightened development. Instead, they simply are adapting to circumstances that confront them; hence, the entire phenomenon might well be called by the much less dramatic-sounding name *adaptation*.

Mass Extinction

Let us now pause in our narrative to discuss examples of a phenomenon that has been mentioned several times already: mass extinction. One of the truly amazing things about the history of life on Earth is the way in which forces have seemingly conspired to wipe out virtually all living things, not just once but at least five major times, along with a number of more limited instances.

Over the course of Earth's history—even its very recent history—numerous species have become extinct, usually as a result of their inability to adapt to changes in their natural environment. In the extremely recent past (recent geologically, that is—since the end of the last ice age about 10,000 years ago), some extinctions or endangerments of species have been attributed to human activities, including hunting and the disruption of natural habitats. For the most part, however, extinction is simply a part of Earth's history, a result of the fact that organisms incapable of adaptation soon die out. This is the infamous "survival of the fittest," an aspect of evolutionary theory that many a social or political ideologue has misappropriated for the purpose of insisting that one particular ethnic or social group is naturally superior to another (see Evolution).

Why does mass extinction occur? As we noted earlier, one possible cause of mass extinction is a sudden and dramatic change in ocean levels. Other causes include volcanic eruptions or the effects of events or objects from space—the explosion of a star, perhaps, or the impact of a meteorite on Earth. Although scientists have a reasonable idea of the immediate causes of mass extinction in some cases, their understanding of the root causes still is limited. This fact was expressed by the University of Chicago paleobiologist David M. Raup, who wrote in *Extinction: Bad Genes or Bad Luck?*: "The disturbing reality is that for none of the thousands of well-documented extinctions in the geologic past do we have a solid explanation of why the extinction occurred."

THE GREAT FIVE MASS EXTINCTIONS. The five largest known mass extinctions occurred at intervals of between 50 million and 100 million years over a span of time from about 435 million to 65 million years ago. The first of them we have mentioned as perhaps being associated with the first migrations of organisms to land. This was at the end of the Ordovician period, about 435 million years ago, when a drop in the ocean level wiped out one-fourth of all marine families. No wonder some of the fish adapted to life on land!

Changes in sea level along with climate changes also appear to have brought about the second great mass extinction, which we also have mentioned, near the end of the Devonian period, about 370 million years ago. But no biological cataclysm in Earth's history can compare to the mass extinction that attended the close of the Permian period some 240 million years ago—an event so traumatic to life on Earth that it has been dubbed the *great dying*.

Imagine if 96% of all people alive today were killed—a staggering and terrifying concept. This would be far more fatalities than the combined death toll of both world wars; the Black Death of 1347–1351 and the influenza epidemic of 1918–1920; the various Nazi, Stalinist, and Maoist acts of genocide; and all major earthquakes and other natural disasters in history. Now imagine that along with all those people, 96% of all other life-forms—plants and trees, birds and fishes, single-cell organisms and the

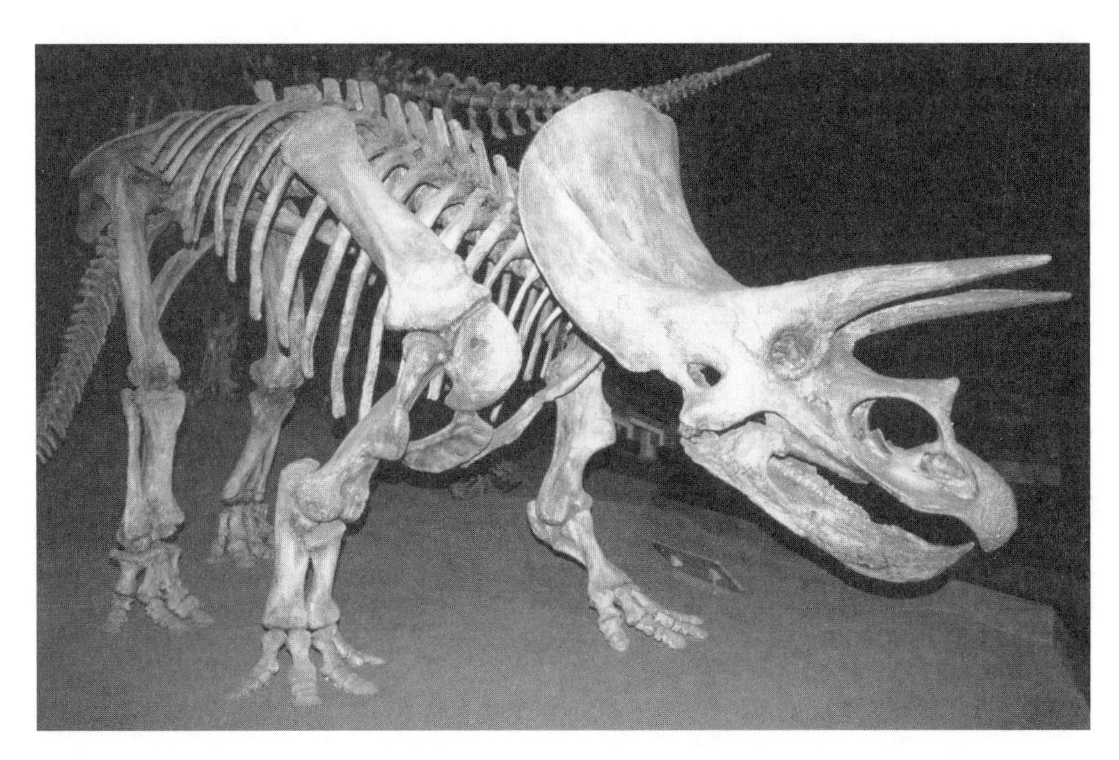

DINOSAURS DOMINATED EARTH FOR A PERIOD OF MORE THAN 100 MILLION YEARS, ENDING ABOUT 65 MILLION YEARS AGO. THE FLYING TRICERATOPS, A CREATURE OF THE LATE CRETACEOUS PERIOD, MAY HAVE BEEN A LINK BETWEEN DINOSAURS AND BIRDS. *(© Kevin Schafer/Corbis. Reproduced by permission.)*

like, which vastly outnumber the relatively small human population of Earth—all ceased to exist as well. Such was the almost incomprehensible scale of the so-called great dying, an event whose cause has not been determined, though it may have resulted from a volcanic eruption in Siberia.

THE AGE OF THE DINOSAURS

Now that we have mentioned only three of the five great mass extinctions, it is appropriate to discuss the creatures that died in the fourth and fifth: the dinosaurs, or "terrible lizards." Though they were the most well-known creatures of the Mesozoic era, which lasted from about 240–65 million years ago, dinosaurs were far from the only notable species on Earth in that phase of geologic history. In fact, it was during the first period of the Mesozoic era, the Triassic, that a group of creatures destined to make a greater impact on Earth—mammals—first appeared. During this time, botanical life included grasses, flowering plants, and trees of both the deciduous (leaf-shedding) and coniferous (cone-bearing) varieties.

Backtracking just a bit, at the beginning of the Devonian period, about 410 million years ago, the first vertebrates (animals with an internal skeleton) made their appearance in the form of jawless fishes. Plant life on land consisted of ferns and mosses. Then came the late-Devonian mass extinction we have mentioned, which removed many of the newly evolved invertebrates, including most fishes. By the end of the Devonian, about 360 million years ago, fish had evolved jaws, and life was thriving on land. Among these life-forms were reptiles in the animal kingdom, and gymnosperms, or plants that reproduce sexually by spreading exposed seeds—a primary example from our own time being pine trees.

MONSTERS OF THE MESOZOIC. Although the existence of dinosaurs has been an established fact for as long as anyone alive can remember, we have had knowledge of them only since the mid–nineteenth century. It was then that early paleontologists began to piece together evidence from fossils, which revealed the record of a time when enormous reptiles had walked the earth. Not all of these dinosaurs were so large, however. Some were as small as chickens, while others—the sauropods, the largest terrestrial animals of all time—were the equivalent of half a dozen stories tall.

Some dinosaurs were awesomely fierce predators, while others were mild-mannered plant eaters. Dinosaurs fell into two groups based on the shape of their hips, which were either lizardlike or birdlike. Though the lizardlike Saurischia emerged first, they lived alongside the birdlike Ornithischia throughout the late Triassic, Jurassic, and Cretaceous periods. Ornithischia were all herbivores, or plant eaters, whereas Saurischia included both herbivores and carnivores, or meat eaters. Naturally, the most fierce of the dinosaurs were carnivores, a group that included the largest carnivore ever to walk the Earth, *Tyrannosaurus rex.*

Most dinosaurs had long tails and necks and walked on four legs, though some were bipedal, meaning either that they had only two legs or that they used only their two rear legs for movement or locomotion. Again, the shape of the creature correlated to some degree with its habits: all four-legged dinosaurs were relatively gentle, slow-moving herbivores, whereas most of the bipedal ones were fast-moving predators. Their teeth also reflected their diet: not surprisingly, carnivores had much sharper cutting surfaces shaped like serrated knives, which made it easy to rip their prey into smaller pieces before swallowing the pieces whole.

As recently as the 1970s, scientists tended to believe that dinosaurs were cold-blooded creatures markedly lacking in brain power. This view is reflected in a number of popular-culture sources, such as the 1983 song by the Police, "Walking in Your Footsteps," which suggests that if humans were to destroy themselves by nuclear warfare, they would seem even more "dumb" than the dinosaurs. By contrast, the film *Jurassic Park,* made a decade later, reflects changes in paleontologists' attitudes toward dinosaurs, gained as a result of continued research. As it turns out, dinosaurs were far from dumb, and they may even have been warm-blooded—unlike most reptiles but like birds, to whom they may have been related.

SOME TYPES OF DINOSAUR. Few creatures have ever captured the imagination of humans as much as *Tyrannosaurus rex*—a particularly impressive feat, given the fact that the last *T. rex* died tens of millions of years before humans ever came on the scene. With a name that means "absolute ruler lizard" in Greek, this terrifying creature reached a maximum length of 45 ft. (14 m) and may have weighed as much as 9 tons (8 metric tons).

T. rex must have made a truly awesome sight running through the forests of its Mesozoic world. Moving on its powerful hind legs with a gait that was at once swift and lumbering, it used its tail as a counterbalance and stabilizer. Once it had come abreast of its prey, *T. rex* probably attacked its victim with powerful head butts and then tore the animal apart with its massive jaws before ripping into it with some 60 dagger-shaped teeth as much as 6 in. (15 cm) long.

Of course, *T. rex* is notable precisely because it was so unusual among dinosaurs in size and, for the most part, power. *Deinonychus,* for example, was much smaller—only about 10 ft. (3 m) and 220 lb. (100 kg)—but these "running lizards" were still fearsome predators, noted for their sickle-like claws. The phrase "sickle-like claws" might call to mind another creature, this one made famous by *Jurassic Park: Velociraptor,* or "swift plunderer," a walking death machine that for all the terror it inspired in moviegoers (and, no doubt, in many a Mesozoic creature) attained a length of only about 6 ft. (2 m).

Among the many lessons about dinosaurs to be learned from *Jurassic Park* was the striking contrast between herbivores and carnivores, a fact emphasized by the scene in which the two children take refuge with a herd of sauropods. These gentle giants possessed four legs the size of large pillars and had extraordinarily long necks and tails. Most famous among the sauropods was the *Apatosaurus,* which achieved a length of 65 ft. (20 m) and a weight of 30 tons (27 metric tons). *Apatosaurus*'s appearance in the scene with the children was not the great creature's first time in the limelight of popular culture: known as *Brontosaurus* until the early 1990s, this dinosaur was a popular fixture of the *Flintstones* cartoon.

In addition to land-based dinosaurs, there were such creatures as the flying *Triceratops,* which may have been a link between dinosaurs and birds. (The name *raptor,* incidentally, usually refers not to a type of dinosaur, but to a bird of prey.) It should be noted that *Triceratops* and *Velociraptor* were creatures of the late Cretaceous period, during which about a hundred different dinosaur genera (plural of *genus*) flourished. This is an important point, because many books picture all manner of dinosaurs coexisting, when, in fact, various ones existed at different times over a period of about 180 million years. Also, it is worth noting that the major nonhuman "star" of *Jurassic Park* did not even live during the Jurassic period!

LATER MASS EXTINCTIONS—AND NEW LIFE

Today, of course, the dinosaurs are long gone—so distant in time, in fact, that the remains of many millions have become petroleum. This has happened in situations where specific conditions prevail: for example, the remains could not be allowed to rot, and decay had to be anaerobic and had to take place within certain types of rock. Yet oil reserves (for as long as they last) are not the only reason why the dinosaurs' disappearance was beneficial from the human standpoint. At the time when dinosaurs controlled the world, mammals were small and hid from predators such as *T. rex.* Without the mass extinction of the dinosaurs, this class of creatures might never have come to the forefront, and human beings might not have evolved.

Actually, two mass extinctions rocked the world of the dinosaurs. The first (fourth among the list of five major mass extinctions) happened at the end of the Triassic, the first period of the Mesozoic era. This was about 205 million years ago, when an asteroid may have hit Earth. Whatever the cause, the result was that creatures in the seas suffered major mass extinction. So, too, did those on the land, but many species of dinosaurs and mammals managed to survive the event, which marked the transition between the Triassic and Jurassic periods.

THEORIES REGARDING THE DINOSAURS' DEMISE. The dinosaurs continued to flourish for another 140 million years, and their descendants might still be walking the earth were it not for the last great mass extinction, which took place about 65 million years ago. What killed the dinosaurs? Paleontologists and other scientists have proposed several theories: a rapid climate change; the emergence of new poisonous botanical species, eaten by herbivorous dinosaurs, that resulted in the passing of toxins along the food web; an inability to compete successfully with the rapidly evolving mammals; and even an epidemic disease to which the dinosaurs possessed no immunity.

Interesting as many of these theories are, none has gained anything like the widespread acceptance achieved by another scenario. According to this highly credible theory, an asteroid hit Earth, hurtling vast quantities of debris into the atmosphere, blocking out the sunlight, and greatly lowering Earth's surface temperature. Around the world, geologists have found traces of iridium deposited at a layer equivalent to the boundary between the Cretaceous and Tertiary periods, the Tertiary being the beginning of the present Cenozoic era. This is significant, because iridium seldom appears on Earth's surface—but it *is* found in asteroids.

THE ASTEROID HITS. It appears that the asteroid smashed into what is now the northern tip of Mexico's Yucatán peninsula. Today that area is home to a crater some 6 mi. (9.6 km) deep and 186 mi. (300 km) in diameter. Note, however, that 65 million years ago, the Yucatán was not exactly where it is now. It had begun to drift in the direction of its present location, but the map of the future continent of North America was quite different from what it is today and included large submerged areas. Among them was the region of impact, which only later rose to the surface as a result of tectonic activity.

The nearest large landmass at the time was a good distance away—equivalent to northern Louisiana. But it hardly mattered that there were no dinosaurs at the point of impact. Traveling at more than 100,000 MPH (160,000 km/h), which is almost four times as fast as the fastest spacecraft built by human beings, the asteroid probably brought about an explosion as intense as many thousands of hydrogen bombs. It may not have raised a mushroom cloud per se, but it probably produced an even more dramatic formation as it sent more than 48,800 cu. mi. (200,000 km^3) of debris and gases into the atmosphere. This would be enough dust to cover the state of Mississippi to a depth of 1 mi. (1.6 km)

The sound of the impact must have been ear-shattering, even far away on what would one day become North America. Then came the tidal waves, some as tall as 394 ft. (120 m), with earthquakes soon to follow. But while these tidal waves undoubtedly caused massive localized death, or mass mortality, what really brought about the mass extinction of the dinosaurs was the aftermath of impact. All that dust in the atmosphere effectively blocked out the Sun's light, bringing about a dramatic cooling on Earth's surface. As a result, plants died, thus depriving herbivorous dinosaurs of an energy source. The herbivores died and then, in a domino effect brought about by the interdependence of components in the food web, the carnivores soon followed.

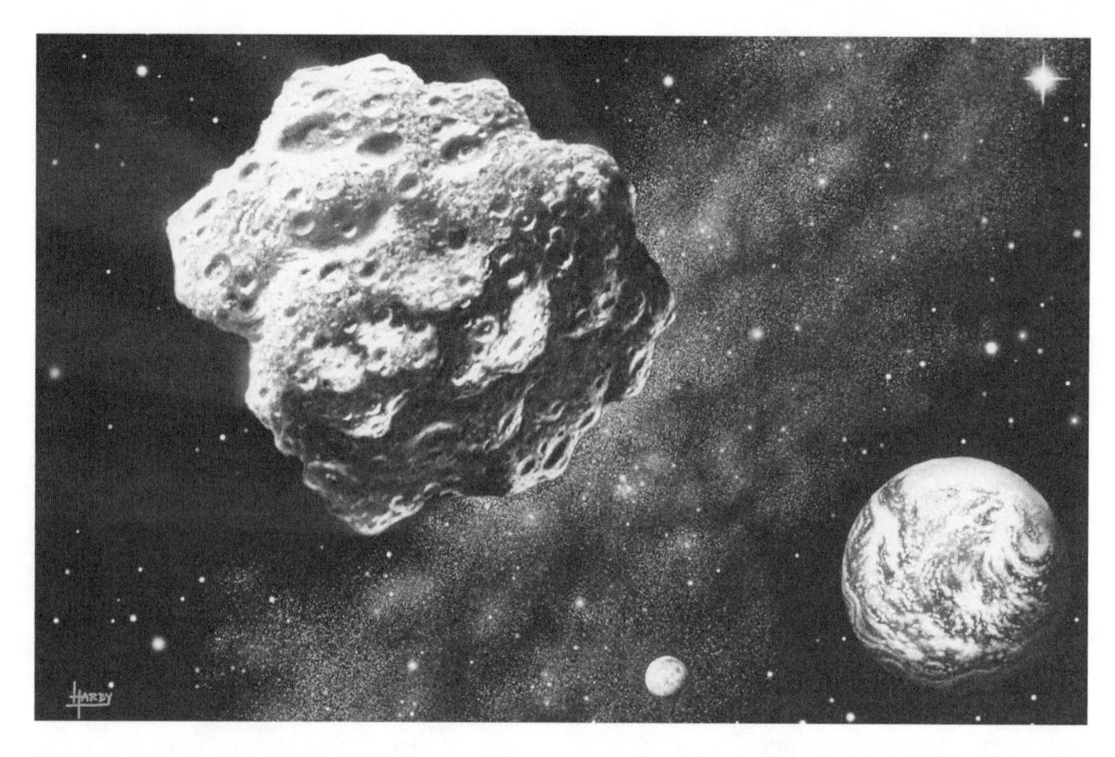

ASTEROID APPROACHING EARTH. ACCORDING TO ONE THEORY, THE MASS EXTINCTION OF THE DINOSAURS RESULTED FROM THE IMPACT OF AN ASTEROID HITTING EARTH, HURTLING VAST QUANTITIES OF DEBRIS INTO THE ATMOSPHERE, BLOCKING OUT THE SUNLIGHT, AND GREATLY LOWERING EARTH'S SURFACE TEMPERATURE. *(© D. Hardy. Photo Researchers. Reproduced by permission.)*

MAMMALS, HUMANS, AND MASS EXTINCTION. When that asteroid hit, it brought an end to the dinosaurs and their Mesozoic world, ushering in the Cenozoic era, which saw the rise of mammals. Some examples of the creatures that proliferated in the course of the past 65 million years are tiny prehistoric horses with four toes as well as a giant rhinoceros-like herbivore, both of which lived in the Eocene epoch about 40 million years ago.

As time went on, the range of species—not only animals and plants but also the much less complex organisms of the other three kingdoms (Protista, including protozoa and algae; Monera such as bacteria; and Fungi)—grew larger and larger. An ice age struck the planet about 1.65 million years ago, bringing about much smaller cases of mass extinction than the ones we have discussed. But other instances of mass extinction would occur at the hands of a creature that appeared about 2.5 million years ago: *Homo sapiens.* The ice age—which was only one of many and thus is referred to often as the "last ice age"—would prove a major turning point in the history of the human species. At the beginning, we were just embarking on the beginnings of the Stone Age, whereas at the end, just 10,000 years ago (or a few seconds ago, in geologic terms), we emerged from the cold, ready to take over the world.

Near the end of the last ice age, peoples from Siberia crossed a land bridge spanning what is today the Bering Strait between Russia and Alaska. Their descendants, of course, are the Native Americans, and *native* is a fitting term, since they arrived in the Americas about 12,000 years ago, or about 7,000 years before the Europeans' ancestors arrived in Europe. What they found, as they poured into the Americas, was a range of species quite different from those known today. There were mammoth and mastodons; giant bears, beaver, and bison; and even saber-toothed "tigers" (which were not directly related to modern-day tigers), camels, and lions. Prehistoric America was also home to horses, but these creatures and many others were wiped out by hunting. Horses did not reappear in the New World until Europeans brought them after A.D. 1500, when they would prove an indispensable aid in European efforts to conquer the Native Americans' lands.

WHERE TO LEARN MORE

Cadbury, Deborah. *Terrible Lizard: The First Dinosaur Hunters and the Birth of a New Science.* New York: Holt, 2001.

KEY TERMS

AEROBIC: Oxygen-breathing.

AMPHIBIAN: A creature capable of living either in the water or on solid ground.

ANAEROBIC: Non-oxygen-breathing.

CARNIVORE: A meat-eating organism.

EON: The longest phase of geologic time. Earth's history has consisted of four eons, the Hadean, Archaean, Proterozoic, and Phanerozoic. The next-smallest subdivision of geologic time is the era.

EPOCH: The fourth-longest phase of geologic time, shorter than an era. The current epoch is the Holocene, which began about 10,000 years ago.

ERA: The second-longest phase of geologic time, after an eon. The current eon, the Phanerozoic, has had three eras, the Paleozoic, Mesozoic, and Cenozoic, which is the current era. The next-smallest subdivision of geologic time is the period.

EUKARYOTE: A cell that has a nucleus and organelles (sections of the cell that perform specific functions) bound by membranes.

FOOD CHAIN: A series of singular organisms in which each plant or animal depends on the organism that precedes it. Food chains rarely exist in nature; therefore, scientists prefer the term *food web.*

FOOD WEB: A term describing the interaction of plants, herbivores, carnivores, omnivores, decomposers, and detritivores in an ecosystem. Each of these organisms consumes nutrients and passes them along to other organisms (or, in the case of the decomposer food web, to the soil and environment). The food web may be thought of as a bundle or network of food chains, but since the latter rarely exist separately, scientists prefer the concept of a food web to that of a food chain.

FOSSIL: The mineralized remains of any prehistoric life-form, especially those preserved in rock before the end of the last ice age.

FOSSILIZATION: The process by which a once living organism becomes a fossil. Generally, fossilization involves mineralization of the organism's hard portions, such as bones, teeth, and shells.

GEOLOGIC TIME: The vast stretch of time over which Earth's geologic development has occurred. This span (about 4.6 billion years) dwarfs the history of human existence, which is only about 2.5 million years. Much smaller still is the span of human civilization, only about 5,500 years.

GYMNOSPERM: A type of plant that reproduces sexually through the use of seeds that are exposed, not hidden in an ovary as with an angiosperm.

HERBIVORE: A plant-eating organism.

ICE AGE: A period of massive and widespread glaciation (the spread of glaciers). Ice ages usually occur in series over stretches of several million or even several hundred million years and have taken place periodically—often in conjunction with mass extinctions—throughout Earth's history.

INVERTEBRATE: An animal without an internal skeleton.

MASS EXTINCTION: A phenomenon in which numerous species cease to exist at

KEY TERMS CONTINUED

or about the same time, usually as the result of a natural calamity.

MINERAL: A naturally occurring, typically inorganic substance with a specific chemical composition and a crystalline structure (that is, a structure in which the constituent parts have a simple and definite geometric arrangement that is repeated in all directions).

MINERALIZATION: A series of changes experienced by a once living organism during fossilization. In mineralization, minerals in the organism are replaced or augmented by different minerals, or the hard portions of the organism dissolve completely.

ORGANIC: At one time chemists used the term *organic* only in reference to living things. Now the word is applied to most compounds containing carbon, with the exception of carbonates (which are minerals) and oxides, such as carbon dioxide.

PALEOBOTANY: An area of paleontology involving the study of past plant life.

PALEOECOLOGY: An area of paleontology devoted to studying the relationships between prehistoric plants and animals and their environments.

PALEONTOLOGY: The study of life-forms from the distant past, primarily as revealed through the fossilized remains of plants and animals.

PALEOZOOLOGY: An area of paleontology devoted to the study of prehistoric animal life.

PERIOD: The third-longest phase of geologic time, after an era. The current eon, the Phanerozoic, has had 11 periods, and the current era, the Cenozoic, has consisted of three periods, of which the most recent is the Quaternary. The next-smallest subdivision of geologic time is the epoch.

PHOTOSYNTHESIS: The biological conversion of light energy (that is, electromagnetic energy) from the Sun to chemical energy in plants.

PROKARYOTE: A cell without a nucleus or organelles bound by membranes.

VERTEBRATE: An animal with an internal skeleton.

Gould, Stephen Jay. *Ever Since Darwin: Reflections in Natural History.* New York: W. W. Norton, 1977.

K–12: Paleontology—Dinos (Web site). <http://www.ceismc.gatech.edu/busyt/paleo.html>.

Morris, S. Conway. *The Crucible of Creation: The Burgess Shale and the Rise of Animals.* New York: Oxford University Press, 1998.

Munro, Margaret, and Karen Reczuch. *The Story of Life on Earth.* Toronto: Douglas and McIntyre, 2000.

Oceans of Kansas Paleontology (Web site). <http://www.oceansofkansas.com/>.

Paleontology and Fossils Resources. University of Arizona Library, Tucson (Web site). <http://www.library.arizona.edu/users/mount/paleont.html>.

Palmer, Douglas. *Atlas of the Prehistoric World.* Bethesda, MD: Discovery Communications, 1999.

Raup, David M. *Extinction: Bad Genes or Bad Luck?* New York: W. W. Norton, 1991.

Singer, Ronald. *Encyclopedia of Paleontology.* Chicago: Fitzroy Dearborn Publishers, 1999.

Starr, Cecie, and Ralph Taggart. *Biology: The Unity and Diversity of Life.* 7th ed. Belmont, CA: Wadsworth, 1995.

University of California, Berkeley Museum of Paleontology (Web site). <http://www.ucmp.berkeley.edu/>.

USGS (United States Geological Survey) Paleontology Home Page (Web site). <http://geology.er.usgs.gov/paleo/>.

BIODIVERSITY AND TAXONOMY

TAXONOMY

SPECIES

SPECIATION

TAXONOMY

CONCEPT

Taxonomy is the area of the biological sciences devoted to the identification, naming, and classification of living things according to apparent common characteristics. It is far from a simple subject, particularly owing to many disputes over the rules for classifying plants and animals. In terms of real-life application, taxonomy, on the one hand, is related to the entire world of life on Earth, but on the other hand, it might seem an ivory-tower discipline that it has nothing to do with the lives of ordinary people. Nonetheless, to understand the very science of life, which is biology, it is essential to understand taxonomy. Each discipline has its own form of taxonomy: people cannot really grasp politics, for instance, without knowing such basics of political classification as the difference between a dictatorship and a democracy or a representative government and one with an absolute ruler. In the biological sciences, before one can begin to appreciate the many varieties of organisms on Earth, it is essential to comprehend the fundamental ideas about how those organisms are related—or, in areas of dispute, *may be* related—to one another.

HOW IT WORKS

TAXONOMY IN CONTEXT

The term taxonomy is actually just one of several related words describing various aspects of classification in the biological sciences. In keeping with the spirit of order and intellectual tidiness that governs all efforts to classify, let us start with the most general concept, which happens to be classification itself. Classification is a very broad term, with applications far beyond the biological sciences, that simply refers to the act of systematically arranging ideas or objects into categories according to specific criteria.

While its meaning is narrower than that of classification, even taxonomy still has broader applications than the way in which it is used in the biological sciences. In a general sense, taxonomy refers to the study of classification or to methods of classification—for example, "political taxonomy," as we used it in the introduction to this essay. Literary critics sometimes refer to a writer's taxonomy of characters. Within the biological sciences, however, the term designates specifically a subdiscipline involving the process and study of the identification, naming, and classification of organisms according to apparent common characteristics.

PHYLOGENY AND NOMENCLATURE. Two other terms that one is likely to run across in the study of taxonomy are phylogeny and nomenclature. Phylogeny is the evolutionary history of organisms, particularly as that history refers to the relationships between life-forms and the broad lines of descent that unite them. Taxonomy is less fundamental a concept than phylogeny. Whereas taxonomy is a human effort to give order to all the data, phylogeny is the true evolutionary relationship between living organisms. Some scientists call phylogeny the *tree of life,* meaning that it represents the underlying hierarchical structure by which life-forms evolved and are related to one another.

The word naming was used earlier in the definition of taxonomy because it is a familiar, easily understandable word. However, a more accurate term, and one that helps illuminate the distinction between taxonomy and systematics, is

THE GREEK PHILOSOPHER ARISTOTLE IS REGARDED AS THE FATHER OF TAXONOMY. THE ARISTOTELIAN PRINCIPLES OF CLASSIFICATION WERE GOVERNED BY THE IDEA THAT THERE ARE CONSTANT, UNCHANGING "ESSENCES" THAT UNITE CLASSES OF ORGANISMS, WHICH IS COMPLETELY AT ODDS WITH THE EMPIRICAL MENTALITY THAT GOVERNS TAXONOMY TODAY. *(The Library of Congress.)*

nomenclature. The latter can be defined as the act or process of naming as a well as a system of names, particularly one used in a specific science or discipline.

HOMOLOGOUS AND ANALOGOUS FEATURES

Before going on to discuss methods of classification, it is important to note just which characteristics of an organism's morphological aspect (i.e., structure or form) are important to scientists working in the field of taxonomy. In theorizing relationships between species, taxonomists are not interested in what are known as analogous features, those characteristics that are superficially similar but not as a result of any common evolutionary origin. Rather, they are interested in homologous features, or features that have a common evolutionary origin, even though they may differ in terms of morphological form.

One example of a shared evolutionary characteristic, discussed briefly in the essay Evolution, is the pentadactyl limb, a five-digit appendage common to mammals and found, in modified form, among birds. This is a homologous feature, indicating a common ancestor that likewise had limbs with five digits at the end. By contrast, there is no indication of a close evolutionary relationship in the fact that birds, butterflies, and bats all have wings that are similar in shape. Rather, the laws of physics require that a wing be of a certain shape in order to hold an object aloft, which is why the contour of an airplane wing, when viewed from the side, is remarkably like that of a bird's wing where it joins the animal's body.

CLADISTICS AND NUMERICAL TAXONOMY

Cladistics is a system of taxonomy that distinguishes taxonomic groups or entities on the basis of shared derived characteristics, hypothesizing evolutionary relationships to arrange them in a treelike, branching hierarchy. The expression *derived characteristics* in this definition means that the characteristics that unite two types of organism are not necessarily present in a shared evolutionary ancestor. Rather, they have developed over the course of evolutionary history since the time of that shared ancestor.

In explaining cladistics to the ordinary human being, the vast majority of science writers seem to be at a loss as to how to make the topic comprehensible. Thus, such terms as *derived characteristics* and its opposite, *primitive characteristics,* usually are left undefined. A welcome exception is Paul Willis, who, in an on-line article for the Australian Broadcasting Corporation (see Where to Learn More) gave a wonderful illustration that was an attempt to analyze the relationships between a mouse, a lizard, and a fish.

"They've all got backbones," Willis wrote, "so the feature 'backbone' is useless [as an indication of evolutionary branching]; it's a 'primitive' character that tells you nothing. But the [derived] feature 'four legs' is useful because it's an evolutionary novelty shared only between the lizard and the mouse. This implies that the lizard and mouse are more closely related to each other than either is to the fish. Put another way, the lizard and the mouse share a common ancestor that had four legs." Willis went on to note that "the more evolutionary novelties we can find that support a particular relationship, the greater our confidence that the relationship is correct. 'Air breathing,' 'neck' and 'amniotic egg' are another three evolutionary novelties that tie the lizard and the mouse together and leave the fish as a more distant relative."

NUMERICAL TAXONOMY. Cladistics, the most widely applied approach to taxonomy, has undergone considerable change since it was introduced by the German zoologist Willi Hennig (1913–1976) in the 1950s. Particularly important has been the marriage of cladistics with another taxonomic idea born in the mid–twentieth century, phenetics, or numerical taxonomy. Introduced by the Austrian biologist Robert Reuven Sokal (1926–) and the English microbiologist Peter Henry Andrews Sneath (1923–), numerical taxonomy is an approach in which specific morphological characteristics of an organism are measured and assigned numerical value, so that similarities between taxa (taxonomic groups or entities) can be compared mathematically. These mathematical comparisons are performed through the use of algorithms, or specific step-by-step mathematical procedures for computing the answer to a particular problem. The aim of numerical taxonomy is to remove all subjectivity (such as the taxonomist's "intuition") from the process of classification. Initially, many traditional taxonomists rejected numerical taxonomy, because its results sometimes contradicted their own decades-long studies of comparative morphological features. Nearly all modern taxonomists apply numerical methods in taxonomy, although there is often heated debate as to which particular algorithms should be used.

IDENTIFICATION, CLASSIFICATION, AND NOMENCLATURE

Earlier, taxonomy was defined in terms of its relationship to the identification, classification, and nomenclature of taxa. Let us now briefly consider each in turn, with the understanding that they are exceedingly complex, technical subjects that can be treated here in the most cursory fashion. The process of identification is a particularly complex one. When an apparently new taxon is discovered, a taxonomist prepares an organized written description of the characteristics of similar species, which are referred to as a taxonomic key. Instead of using pictures, which often poorly convey the natural variations in morphological features, taxonomists prefer to use a taxonomic key in written form, which provides much more detail and exactitude.

To put it in colloquial terms, by referring to a taxonomic key, a taxonomist may determine that if an organism "looks like a duck and quacks like a duck, it must be a duck"—only, in this instance, the taxa being compared are much more specific than the common term *duck* and the characteristics much more precisely described. (For one thing, there are several dozen species in the genus *Anas,* which includes all "proper" ducks, and many more species in the family Anatidae, or waterfowl, that are commonly called by "duck names"—including such amusingly named species as the ruddy duck, lack duck, freckled duck, and comb duck.) If there is no already established "duck" that the species in question resembles, the taxonomist may have discovered an entirely new genus, family, order, class, or even phylum.

A taxonomist may use what is called a dichotomous key, which presents series of alternatives much like a flow chart. For example, if the flowers of a sample in question are white and the stem is woody, then (depending on additional alternatives) it could be either species *A* or species *B.* If the flowers are not white and the stem is herbaceous (non-woody), then, presented with another set of additional alternatives, it is possible that the plant is either species *C* or species *D.*

CLASSIFICATION. In discussing cladistics and phenetics, we touched briefly on the *process* of classification. Suffice it to say that this process is far more complex and technically elaborate than these few paragraphs can begin to suggest. We return later to specifics of classification as they relate to systems and innovations introduced by the Greek philosopher Aristotle (384–322 B.C.), the Swedish botanist Carolus Linnaeus (1707–1778), and the English naturalist Charles Darwin (1809–1882), the three most important men in the history of taxonomy before the twentieth century. For the present, our focus is on the overall ranking system.

There are many possible ranks of classification but only seven that are part of what is known as the obligatory taxonomy, or obligatory hierarchy. These ranks are kingdom, phylum, class, order, family, genus, and species. Listed here are all possible ranks, with obligatory ranks in italics.

- *Kingdom*
- Subkingdom

- *Phylum*
- Subphylum
- Superclass
- *Class*
- Subclass
- Infraclass
- Cohort
- Superorder
- *Order*
- Suborder
- Superfamily
- *Family*
- Subfamily
- Tribe
- *Genus*
- Subgenus
- *Species*
- Subspecies

The reader occasionally may come across nonobligatory ranks, most notably subphylum, but for the most part the only ranks referred to in this book are the obligatory ones.

NOMENCLATURE. In accordance with a tradition established by Linnaeus, all group names are in Latin, thus facilitating ease of communication. There are some rules concerning names of groups: for instance, those of families use the suffix *-idae.* In the world of taxonomy, however, few rules are accepted universally. Even as basic a term as *phylum* is not universal, since botanists prefer the word *division.*

The proper name of any ranking more general than species is capitalized (e.g., phylum Chordata), with species and subspecies names in lowercase. Genus, species, and subspecies names are rendered in italics (e.g., *Homo sapiens,* or "man the wise"), while proper names of the more general groupings are presented in ordinary type (e.g., class Mammalia). If the same name appears a second time in the same article, the genus name usually is abbreviated: thus, *H. sapiens.*

Just as most people (with such rare exceptions as Cher and Madonna) are identified by two names, a personal and a family name, taxonomy makes use of a system called binomial nomenclature, in which each type of plant or animal is given a two-word name, with the first name identifying the genus and the second the species. In binomial nomenclature, the genus name is analogous to the family name, inasmuch as there are many species within a genus, and the species name is like a personal name. The difference is that whereas there may be thousands of boys and men named John Smith, there is only one species called *Homo sapiens.* Beyond the species name, there may be subspecies names: humans are subspecies *sapiens,* so our full species name with subspecies is *Homo sapiens sapiens.* Additional rules govern the inclusion of a name or an abbreviation, at the end of the species or subspecies name, to recognize the individual who first identified it.

REAL-LIFE APPLICATIONS

THE URGE TO CLASSIFY

One might ask what all the fuss is about. Why is classification so important? We attempt to answer that question from a few angles, including a brief look at the lengthy historical quest to develop a workable taxonomic system. But what was the original impulse that motivated that quest? One clue can be found in the Greek roots of the word *taxonomy: taxis,* or "arrangement," and *nomos,* or "law." The search for a taxonomic system represents humankind's desire to make order out of the complexities with which nature presents us. When it comes to the organization of ideas (including ideas about the varieties of life-forms), this desire for order is more than a mere preference. It is a necessity.

THE ANALOGY TO A LIBRARY. Imagine a library without any organizational system, with books simply crammed willy-nilly on the shelves. Such a place would be totally chaotic, and if one happened to find a book one was looking for, it would be a case of pure luck. The odds would be weighted heavily against such luck, especially in a university library or a large municipal or regional one. Just as a good-size university library has upward of a million volumes, and many large university libraries have several million, so there are at least a couple of million identified species, and the total may be much larger. Some entomologists (scientists who study insects) speculate that there may be ten million species of insect alone.

THE LURE OF A NEW SPECIES. When a zoologist or botanist discovers what he or she believes to be a new species, the taxonomic system provides a standard against which to check it—rather as you would do if you thought you had discovered a book that was not in the library. If the "new" species matches an established one, that may be the end of the story—unless the scientist has discovered a new aspect of the species or a new subspecies. And if there is no match in the taxonomic "library," the scientist has discovered an entirely new life-form, with all the grand and terrifying ramifications that may ensue.

The new species might be an herb from which a cure can be synthesized for a devastating disease, or it could be a parasite that carries a new and previously unknown malady. Whatever it is, it is better to know about it than not to know, and though the vast majority of "new" species are not nearly as exciting as the preceding paragraph would imply, each has its part to play in the overall balance of life. Discovery of new species is particularly important when those species are endangered or might be in the process of disappearing even as they are identified.

NONSCIENTIFIC TAXONOMY

Without knowing anything about scientific taxonomy, almost anyone can begin to classify animals and perhaps plants. If we limit the discussion purely to animals, there are many basic parameters according to which we could classify them, just off the tops of our heads, as it were. For example, there are aquatic and terrestrial animals, and these general groupings can be broken down further according to biome or habitat (see Biomes). There are animals that walk, fly, swim, slither, or move by some other means. Animals can be divided according to their forms of reproduction, whether asexual or sexual, oviparous or viviparous (expelling or retaining a fertilized egg, respectively), and so on. As discussed in Food Webs, animals may be classified as herbivores, carnivores, omnivores, or detritivores or as primary, secondary, or tertiary consumers. They may be endothermic or ectothermic (warm-blooded or cold-blooded), and they may be covered with scales, feathers, fur, or skin. (In the last case, that skin may be protected by either mucus or hair.)

On and on go the categories, and if one is inclined toward a classifying mind, this kind of mental exercise can be fun. Certainly, little children enjoy it, and many educational programs and games call on the child to group animals thus. Although these kinds of groupings, and the efforts to place animals into one group or another, constitute a form of classification, there is a great difference between this and scientific taxonomy.

SCIENCE VERSUS "COMMON SENSE." Taxonomy is tied closely to evolutionary study, and Darwin's theory of evolution was a turning point in the history of scientific classification. Thus, taxonomists are concerned more with the evolutionary patterns that link organisms than they are with what may be only superficial similarities. Habitat, for instance, is significant in studying biomes, but it seldom plays a role in taxonomy. Nor is the ability to fly, as we have noted, necessarily an indicator of taxonomic similarities.

A striking example of the difference between scientific taxonomy and "common sense" classification is the fact that whales and dolphins are grouped along with other mammals (class Mammalia) rather than with fish and other creatures that most readily come to mind when thinking of aquatic organisms. In fact, whales and dolphins share not only a wide array of primitive characteristics with mammals (for example, the pentadactyl limb described earlier) but also the derived characteristic that defines *mammal*: the secreting of milk from mammary glands, by which a mother feeds her young. Not only is it impossible to get milk from a fish (even family Chanidae, known by the common name "milkfish"), but fish lack even that primitive characteristic, the pentadactyl limb, that links mammals, at least distantly, with nonmammalian creatures, such as birds (class Aves).

COMMON TERMS AND FOLK TAXONOMY. For the sake of convenience, in many places throughout this book, common terms such as *bird, horse, fish,* and so forth are used. But common terms are far from adequate in a scientific context, because such terminology can be deceptive, as exemplified by the nonduck "ducks" mentioned earlier. Likewise, shellfish and starfish are not "fish" as that term is usually understood. But while common terminology can be misleading, sometimes correlations with scientific taxonomy can be found in what is known as folk taxonomy. The latter is a term for the taxonomic systems applied in relatively isolated non-Western societies. For example, the folk taxonomy of native peoples in New Guinea identified 136 bird species in the mountains of that island, a figure that came amazingly close to the 137 species identified by the German-born

ILLUSTRATION OF A SEA MONSTER FROM "DE NATURA RERUM" BY ALBERTUS MAGNUS. DURING THE MIDDLE AGES, TAXONOMIC WRITINGS CONSISTED PRIMARILY OF BOOKS OF IMAGINARY CREATURES; THE FIRST SIGNS OF SCIENTIFIC REAWAKENING IN TAXONOMY CAME WITH PLANT AND ANIMAL CATALOGUES BY SUCH MEDIEVAL SCHOLARS AS MAGNUS. *(© Gianni Dagli Orti/Corbis. Reproduced by permission.)*

American evolutionary biologist Ernst Mayr (1904–) when he studied New Guinea's birds using scientific methods.

ARISTOTLE, LINNAEUS, DARWIN, AND BEYOND

Among his many other accomplishments as a thinker, Aristotle is regarded as the father of the biological sciences and of taxonomy. Among the dominant ideas in his work as a philosopher are the concepts of hierarchy and classification, and thus he took readily to the idea of classifying things. At his school in Athens, he put his students to work on all sorts of taxonomic pursuits, from listing the champions at the Pythian Games (a festival like the Olympics) to classifying the constitutions of various Greek city-states to analyzing the body parts of animals. Aristotle himself dissected hundreds of animals to understand what made them tick, and he proved to be some 2,000 years ahead of his time in recognizing that the dolphin is a mammal and not a fish. His system of classification, however, was a far cry from the ideas that developed in nineteenth-century taxonomy; rather than searching for evolutionary lines of descent, he ranked animals in order of their physical complexity.

In most aspects of his other work, Aristotle established sharp distinctions between his own ideas and those of his teacher, Plato (427?–347 B.C.). For example, Aristotle rejected Plato's position that every idea we can conceive is but a dim reflection of an essential concept—for example, that our idea of "red" is only a shadowy copy of the perfect notion of "redness." Yet in his taxonomy, Aristotle seemed to hark back to his days as Plato's star pupil. The Aristotelian principles of classification were governed by the idea that there are constant, unchanging "essences" that unite classes of organisms. This idea of essences is completely at odds with the empirical (experience-based) mentality that governs taxonomy today. Nonetheless, for two millennia, Aristotelian ideas represented the cutting edge in taxonomy and much else.

THE MIDDLE AGES AND RENAISSANCE. After Aristotle and his brilliant student Theophrastus (371?–287? B.C.), the father of botany, there would be no Western biological theorists of remotely comparable stature until the time of the Renaissance. In the meantime, taxonomy, as with so many other areas of learning in Europe, declined badly. During the Middle Ages, what passed for taxonomic writings consisted primarily of bestiaries, books full of fanciful and imaginary creatures, such as the unicorn. The first signs of scientific reawakening in the biological sciences in general, and taxonomy in particular, came with plant and animal catalogues by such great medieval scholars as Peter Abelard (1079–1142) and Albertus Magnus (*ca.* 1200–1280). Even so, their work consisted primarily of summations of existing Aristotelian knowledge rather than new contributions.

In the sixteenth century, the Swiss scientist Konrad von Gessner (1516–1565) wrote *Historia animalium* (1551–1558), a groundbreaking work that included descriptions of many animals never before seen by most Europeans. Gesner also denounced the practice of including fictitious animals in bestiaries. Around the same time, the discoveries of new plant and animal species in the New World began to point up the need for a taxonomy that went beyond Aristotle's. The first scholar of the modern era to attack this problem was the Italian botanist Andrea Cesalpino (1519–1603), but nearly two centuries would pass before the development of a workable classification system.

LINNAEUS AND OTHERS. The man who revolutionized taxonomy was born Carl von Linné but adopted the Latinized name Carolus Linnaeus. Even that late in scientific history, scholars still wrote chiefly in Latin, not because they were trying to adhere to tradition but because it remained a common language between educated people of different countries. Thus, Linnaeus's great work, which he first published in 1737 but revised numerous times, was named *Systema naturae,* or "The Natural System." Thanks to Linnaeus, Latin became enshrined permanently as the language of taxonomy the world over, but this was far from his only accomplishment.

It was Linnaeus who introduced binomial nomenclature, in a 1758 revision of his *Systema,* and also Linnaeus who established several of the obligatory rankings. Moreover, he instituted the first taxonomic keys, and his system, first applied in botany, became accepted in the zoological community as well. Others, including Baron Georges Cuvier (1769–1832), Michel Adanson (1727–1806), and Comte Georges Buffon (1707–1788), refined Linnaeus's system, but he stands as a towering figure in the discipline.

Later, the French natural philosopher Jean Baptiste de Lamarck (1744–1829) proposed a distinction between vertebrates, or animals with spinal columns, and invertebrates. Today this distinction is not considered as useful as it once was, since it is lopsided—that is, there are nine times as many invertebrates as vertebrates in the animal kingdom—but at the time, it represented an advancement. Less questionable were the distinctions introduced in 1866 by the German biologist Ernst Haeckel (1834–1919) between plants, animals, and single-cell organisms. As Haeckel reasoned, at the level of unicellular organisms, distinctions between plant and animal really make no sense.

DARWIN AND THE TWENTIETH CENTURY. By far the most influential figure in taxonomy during the nineteenth century was the man also recognized as the most influential figure in all of biology during that era: Darwin. Whereas Linnaeus had retained the Aristotelian focus on the "essence" of the animal's features, Darwin swept away such notions and, in his *Origin of Species* (1859), proposed that the "community of descent" is "the one known cause of close similarity in organic beings" and therefore the only reasonable basis for taxonomic classification systems. As result of Darwin's work, taxonomists became much more oriented toward the representation of phylogeny in their classification systems. Therefore, instead of simply naming and cataloguing species, modern taxonomists also try to construct evolutionary trees showing the relationships between different species.

Since Darwin's time, taxonomy has seen numerous innovations, including the introduction of cladistics by Hennig and of numerical taxonomy by Sokal and Sneath. Taxonomists today make use of something unknown at the time of Darwin: DNA (deoxyribonucleic acid, a molecule that contains genetic codes for inheritance), which provides a wealth of evidence showing relationships between creatures. For example, a comparison of human and chimpanzee DNA reveals that we share more than 98% of the same genetic material, indicating that the two lines of descent are related more closely than either is to apes.

THE FIVE KINGDOMS

There are several taxonomic systems, distinguished in part by the number of different kingdoms that each system recognizes. The system used in this book is that of five kingdoms, listed here, which is the result of modifications by the American biologists Lynn Margulis (1938–) and Karlene V. Schwartz (1936–) to the work of earlier taxonomists. (It should be noted that biologists are increasingly using a system of six kingdoms under three domains: eubacteria, arachaea, and eukaryotes. For the sake of simplicity, however, the five-kingdom system is used here.) These five kingdoms are as follows:

Monera: bacteria, blue-green algae, and spirochetes (spiral-shaped, undulating bacteria). Members of this kingdom, consisting of some 10,000 or more known species, are single-cell prokaryotes, meaning that the cell has no distinct nucleus. Some researchers have divided Monera into Eubacteria, or "true" bacteria, and Archaebacteria, which are bacteria-like organisms capable of living in extremely harsh and sometimes anaerobic (oxygen-lacking) environments, such as in acids, saltwater, or sewage.

Protista (or Protoctista): protozoans, slime molds (which resemble fungi), and algae other than the blue-green variety. Made up of more than 250,000 species, this kingdom is distinguished by the fact that its members are single-cell organisms, like the Monera. These organisms, however, are eukaryotes, or cells with a nucleus as well as organelles (sections of the cell that perform specific functions).

Fungi: fungi, molds, mushrooms, yeasts, mildews, and smuts (a type of fungus that afflicts certain plants). Fungi are multicellular, consisting of specialized eukaryotic cells arranged in a filamentous form (that is, a long, thin series of cells attached either to one another or to a long, thin cylindrical cell). There are some 100,000 varieties of fungi.

Plantae: plants, of which there are upward of 250,000 species. Although *plant* is a common term, there is no universally accepted definition that includes all plants and excludes all nonplants. One of the most important characteristics of plants is the fact that they receive their nutrition almost purely through photosynthesis. Beyond the plant kingdom, this is true only of a few protests and bacteria. (For the most part, the three lower kingdoms obtain nutrition through absorption.) Other characteristics of plants include the fact that they are incapable of locomotion; have cells that contain a form of carbohydrate called cellulose, making their cell walls more or less rigid; are capable of nearly unlimited growth at certain localized regions (unlike most animals, which have set numbers of limbs and so forth); and have no sensory or nervous system.

Animalia: animals, of which there are more than 1,000,000 species. Like plants, animals are characterized by specialized eukaryotic cells, but also like plants, the comprehensive definition of *animal* is not as obvious as one might imagine. Mobility, or a means of locomotion, is not a defining characteristic, since sponges and corals are considered animals. The principal difference between animals and plants is at the cellular level: animals either lack cells walls entirely or have highly permeable walls, unlike the cellulose cell walls in plants. Another defining characteristic of *animal* is that they obtain nutrition by feeding on other organisms. Additionally, animals usually have more or less fixed morphological characteristics and possess a nervous system. The fact that most animals are mobile helps account for the large number of animal species compared with those of other kingdoms; over the course of evolutionary history, mobility brought about the introduction of animals to a wide range of environments, which required a wide range of adaptations.

A DIVER INSIDE A BARREL SPONGE ALONG A CORAL REEF IN THE CARIBBEAN. MOBILITY, OR A MEANS OF LOCOMOTION, IS NOT A DEFINING CHARACTERISTIC OF KINGDOM ANIMALIA; INDEED, SPONGES AND CORALS ARE CONSIDERED ANIMALS. *(© Jeffrey L. Rotman/Corbis. Reproduced by permission.)*

Space does not permit a discussion of the various phyla, let alone the smaller divisions, in anything like the detail we have accorded to kingdoms. Furthermore, the distinctions among most phyla, apart from higher animals and some plants, makes for rather dry reading to a nonscientist. These divisions are discussed in further detail, however, within the essays Species and

Speciation. The latter essays also address the definition of species, a great and continuing challenge that faces taxonomists.

TAXONOMY IN ACTION

Two stories reported in *National Geographic News* online (see "Where to Learn More") in 2001 and 2002 illustrate the fact that scientific classification is an ongoing process, and that the world of taxonomy is frequently home to controversies and surprises. Lee R. Berger of the *Geographic* reported the first story, on December 17, 2001, under the heading "How Do You Miss a Whole Elephant Species?" As it turns out, there are not just two species of elephant, as had long been believed, but three.

In addition to the Asian elephant (*Elephas maximus*) scientists had long recognized the African savanna elephant, or *Loxodonta africana,* as a second species. However, DNA testing (see Genetics and Genetic Engineering) in 2001 revealed a second African variety, *Loxodonta cyclotisare* or the African forest elephant, formerly believed to constitute merely a subspecies.

The news was not entirely new: as early as a century prior to the announcement of the "new" species, zoologists had begun to suspect that the forest elephant was a separate grouping distinguished by a number of characteristics. For example, the forest elephant is physically smaller, with males seldom measuring more than 8 ft. (2.5 m) at the shoulder, as compared to 13 ft. (4 m) for a large savanna male. Additionally, ivory samples confiscated from poachers or illegal hunters have revealed that the material in the tusks of the forest variety is pinker and harder than that of its savanna counterpart.

Recognition of the third elephant species followed years of argument as to whether the two African varieties are capable of interbreeding, which would indicate that they are not separate species. That debate was rendered moot by the DNA studies, which showed that the African forest and savanna elephants are less closely related genetically than are lions and tigers, or horses and zebras.

A "NEW" INSECT ORDER. The identification of the forest elephant in 2001 was a major taxonomic event, inasmuch as the elephant itself is a large and commonly known creature. However, it was still a matter only of identifying a new species, whereas in 2002, for the first time in 87 years, taxonomists identified an entirely new insect order. Actually, the order consists of a single known species, but this one is so different from others that it must be grouped separately. Discovered in Namibia, in southwestern Africa, the creature was given the nickname "the gladiator" in honor of the Academy Award-winning 2000 film of that name.

Entomologist Oliver Zompro of the Max Planck Institute of Limnology in Plön, Germany, described the creature as "a cross between a stick insect, a mantid, and a grasshopper," according to the *Geographic.* Because its first body segment is the largest, it is distinguished from a stick insect, whereas it differs from a mantid inasmuch as it uses both fore and mid-legs to capture prey. And while it looks like a grasshopper, "the gladiator" cannot jump.

Measuring as much as 1.6 in. (4 cm) long, the insect, whose order is designated as *Mantophasmatodea,* is a carnivorous, nocturnal creature. Its discovery raised the number of known insect orders to 31, a discovery that Piotr Naskrecki, director of the Conservation International Invertebrate Diversity Initiative, compared to finding a mastodon or saber-toothed tiger. Colorado State University ecologist Diana Wall described the discovery as "tremendously exciting" and told the *Geographic,* "This new order could be a missing link to determining relationships between insects and other groups. ... Every textbook discussing the orders of insects will now need to be rewritten."

WHERE TO LEARN MORE

Classification—The Dinosaur FAQ (Web site). <http://www.miketaylor.org.uk/dino/faq/index.html#4>.

The Germplasm Resources Information Network (GRIN), Agricultural Research Service (Web site). <http://www.ars-grin.gov/npgs/tax/>.

Goto, H. E. *Animal Taxonomy.* London: Arnold, 1982.

Lacey, Elizabeth A., and Robert Shetterley. *What's the Difference?: A Guide to Some Familiar Animal Look-Alikes.* New York: Clarion Books, 1993.

Margulis, Lynn, and Karlene V. Schwartz. *Five Kingdoms: An Illustrated Guide to the Phyla of Life on Earth.* New York: W. H. Freeman, 1988.

National Geographic News (Web site). <http://news.nationalgeographic.com/news/>.

O'Neil, Dennis. *Classification of Living Things/Palomar College* (Web site). <http://anthro.palomar.edu/animal/>.

Parker, Steve. *Eyewitness Natural World.* New York: Dorling Kindersley, 1994.

KEY TERMS

ALGORITHM: A specific set of step-by-step procedures for computing answers to a mathematical problem.

ANALOGOUS FEATURES: Morphological characteristics of two or more taxa that are superficially similar but not as a result of any common evolutionary origin. For example, birds, bats, and butterflies all have wings, but this is not because they are closely related. Compare with *homologous features.*

BINOMIAL NOMENCLATURE: A system of nomenclature in biological taxonomy whereby each type of plant or animal is given a two-word name, with the first name identifying the genus and the second the species. Genus name is always capitalized and abbreviated after the first use, and species name is lowercased. Both are always shown in italics; thus, *Homo sapiens* and, later in the same document, *H. sapiens.*

BIOME: A large ecosystem (community of interdependent organisms and their inorganic environment) characterized by its dominant life-forms. There are two basic varieties of biome: terrestrial, or land-based, and aquatic.

CLADISTICS: A system of taxonomy that distinguishes taxonomic groups or entities (i.e., taxa) on the basis of shared derived characteristics, hypothesizing evolutionary relationships to arrange these in a treelike, branching hierarchy. Cladistics is one of several competing approaches to taxonomic study.

CLASS: The third most general obligatory of the taxonomic classification ranks, after *phylum* but before *order.*

CLASSIFICATION: A very broad term, with application far beyond the biological sciences, that refers to the act of systematically arranging ideas or objects into categories according to specific criteria. A more specific term is *taxonomy.*

EUKARYOTE: A cell that has a nucleus as well as organelles (sections of the cell that perform specific functions) bound by membranes.

FAMILY: The third most specific of the seven obligatory ranks in taxonomy, after *order* but before *genus.*

FUNGI: One of the five kingdoms of living things, consisting of multicellular eukaryotic cells arranged in a filamentous form (that is, a long, thin series of cells attached either to one another or to a long, thin cylindrical cell.) Fungi include "true" fungi, molds, mushrooms, yeasts, mildews, and smuts (a type of fungus that afflicts certain plants).

GENUS: The second most specific of the obligatory ranks in taxonomy, after *family* but before *species.*

HOMOLOGOUS FEATURES: Morphological characteristics of two or more taxa that indicate a common evolutionary origin, even though the organisms may differ in terms of other morphological features. An example is the pentadactyl limb, common to many birds and most mammals (e.g., the human's four fingers and thumb), which indicates a common ancestor. Compare with *analogous features.*

KINGDOM: The highest or most general ranking in the obligatory taxonomic system. In the system used in this book, there

KEY TERMS CONTINUED

are five kingdoms: Monera, Protista, Fungi, Plantae, and Animalia.

MONERA: One of the five kingdoms of living things, consisting of single-cell prokaryotes, including bacteria, blue-green algae, and spirochetes (spiral-shaped undulating bacteria that may cause such diseases as syphilis).

MORPHOLOGY: Structure or form, or the study thereof.

NOMENCLATURE: The act or process of naming, or a system of names—particularly one used in a specific science or discipline. See also *binomial nomenclature.*

NUMERICAL TAXONOMY: An approach to taxonomy in which specific morphological characteristics of an organism are measured and assigned numerical value, so that similarities between two types of organism can be compared mathematically by means of an algorithm. Numerical taxonomy also is called *phenetics.*

OBLIGATORY TAXONOMY (OR OBLIGATORY HIERARCHY): The seven taxonomic ranks by which all species must be identified, whether or not they also are identified according to nonobligatory categories, such as subphylum, cohort, or tribe. These ranks are *kingdom, phylum, class, order, family, genus,* and *species.*

ORDER: The middle of the seven obligatory ranks in taxonomy, more specific than *class* but more general than *family.*

PHENETICS: Another name for *numerical taxonomy.*

PHOTOSYNTHESIS: The biological conversion of light energy (that is, electromagnetic energy) from the Sun to chemical energy in plants. In this process carbon dioxide and water are converted to sugars.

PHYLOGENY: The evolutionary history of organisms, particularly as that history refers to the relationships between lifeforms, and the broad lines of descent that unite them.

PHYLUM: The second most general of the obligatory taxonomic classification ranks, after *kingdom* and before *class.*

PROKARYOTE: A cell without a nucleus.

PROTISTA (OR PROTOCTISTA): One of the five kingdoms of living things, consisting of single-cell eukaryotes. Protista include protozoans, slime molds (which resemble fungi), and algae other than the blue-green variety.

SPECIES: The most specific of the seven obligatory ranks in taxonomy.

SYSTEMATICS: The science of classifying and studying organisms with regard to their natural relationships.

TAXON: A taxonomic group or entity.

TAXONOMY: The area of the biological sciences devoted to the identification, nomenclature, and classification of organisms according to apparent common characteristics. The word *taxonomy* also can be used more generally to refer to the study of classification or to methods of classification (e.g., "the taxonomy of Dickens's characters.")

Simpson, George Gaylord. *Principles of Animal Taxonomy.* New York: Columbia University Press, 1961.

Taxonomy Browser, National Center for Biotechnology Information, National Library of Medicine, National Institutes of Health (Web site). <http://www.ncbi.nlm.nih.gov/Taxonomy/>.

The Tree of Life Web Project (Web site). <http://beta.tolweb.org/tree/>.

Tudge, Colin. *The Variety of Life: A Survey and a Celebration of All the Creatures That Have Ever Lived.* London: Oxford University Press, 2000.

Whyman, Kathryn. *The Animal Kingdom: A Guide to Vertebrate Classification and Biodiversity.* Austin, TX: Raintree Steck-Vaughn, 1999.

Willis, Paul. "Dinosaurs and Birds: The Story." Australian Broadcasting Corporation (Web site). <http://www.abc.net.au/science/slab/dinobird/story.htm>.

SPECIES

CONCEPT

One of the challenges that faces a student of the biological sciences is the seemingly endless array of unfamiliar terms that one must learn. It is a relief to come across a relatively familiar one, such as species. Although it has a scientific sound to it, the word has entered everyday language, such that when people use it, most everyone understands what is meant. Or do they? As it turns out, there is no hard and fast definition for the word. Nonetheless, it is easy enough to find examples of species, since there are many millions of them in five kingdoms of living things—a product of a phenomenon know as *speciation,* whereby evolutionary lines of descent diverge and new species are created. In the world today, there are many interesting groups of species, distinguished neither by evolutionary line nor by taxonomy but instead by the ways in which they interact with their environments. Among these groups are endangered species, of whose existence most people are aware, owing to the spread of the environmentalist message through media and entertainment outlets since the early 1970s. Less familiar is another broad group that in many cases threaten humans: introduced or invasive species.

HOW IT WORKS

TAXONOMY IN BRIEF

The concept of species falls under the heading of taxonomy, the area of the biological sciences devoted to the identification, naming, and classification of living things according to apparent common characteristics. Taxonomy is discussed in detail within the essay on that subject, but to appreciate the topic of species in context, it is helpful to have at least some knowledge of the larger subject. At one time taxonomists were concerned most with the morphological characteristics (i.e., the structure or form) of organisms as a basis for classifying many species within a larger grouping. Today, however, shared evolutionary lineage is much more important than morphological features in determining whether taxa (plural of *taxon,* meaning a taxonomic group or entity) can be classed together. Organisms may be linked closely in terms of evolutionary lines of descent but differ in a particular morphological aspect as a result of the adaptive changes that accompany natural selection. The latter, a key concept in the theory of evolution put forward by the English naturalist Charles Darwin (1809–1882), is a process whereby some organisms thrive and others perish, depending on their degree of adaptation to a particular environment.

It is therefore possible for organisms in a particular environment to develop a common adaptive mechanism through generations of natural selection, even though those organisms themselves are not related to fish closely in terms of evolutionary line of descent. Thus, whales and dolphins, mammals that live underwater, evolved the ability to swim just as well as fish, but that does not mean they are connected closely. Conversely, organisms may be close, or relatively close, in terms of evolutionary lines of descent yet differ in significant morphological features. To use the whale and dolphin example again, these creatures are classified as mammals owing to certain particulars (discussed later), but they differ from the vast majority of mammals in that they have no legs. They do, however, have four appendages, just like the rest of the mammalian class; as a result of natural selection, however, theirs ceased to operate as legs (an encumbrance for life in the water) a long time ago, and today they function instead as fins.

OBLIGATORY RANKS

The classification system used today is an outgrowth of a system developed by the Swedish botanist Carolus Linnaeus (1707–1778) in the 1730s. The realms of zoology and botany, areas of biology devoted to the study of animal and plant life, respectively, differ somewhat with regard to their classification systems, but both use international codes of nomenclature with roots in the Linnaean system. There are many possible ranks of classification, but only seven are obligatory, meaning that all species must be assigned a place in these groupings. The obligatory ranks are listed here. The entire list of rankings, including versions of obligatory ranks with such prefixes as *sub-*, *super-*, and *infra-*, as well as such additional ranks as cohort or tribe, are given in Taxonomy. Note the difference between the zoological and botanical names for the second rank.

Obligatory Taxonomic Ranks

- Kingdom
- Phylum (*Division* in botany)
- Class
- Order
- Family
- Genus
- Species

As discussed in Taxonomy, this book uses a system of five kingdoms, whose characteristics are defined in that essay. Even at the level of kingdom, not everything is delineated precisely (see the discussion in Taxonomy), and there are significant areas of dispute. For example, some taxonomic systems include viruses. Because viruses are not cellular in structure and are not universally regarded as true organisms, however, they are not included in the five-kingdom system used here.

Below the level of kingdom, definitions become even more difficult. Organisms are grouped into phyla on the basis of body plan or organization, but there is no regular pattern for grouping within the smaller categories. For example (as noted later herein), humans are placed within their particular phylum and subphylum on the basis of their spinal columns and overall internal bone structure, but those specifics play no significant role in categorizing them within any of the more specific groupings to which they belong. Furthermore, the generic definitions of the categories—for example, *class* as opposed to class Mammalia, class Insecta, or some other class in the taxonomic system—are purely relative. In other words, class is simply the obligatory rank that is more specific than phylum but more general than order.

DESIGNATING A SINGLE SPECIES WITHIN THE RANKS. When preparing an outline for a paper, students are taught that no topic should have only one subheading; instead, that solitary subheading should be moved up one level. Such rules do not apply in taxonomy, and it is not necessary that there be more than one subgroup within a larger group. For example, there might be only one class in a phylum. Taxonomists use detailed definitions to single out particular groups, such as class Mammalia. The following list shows the placement of humans within the larger taxonomic universe, along with brief explanations of a few (though far from all) characteristics that define each group.

- Kingdom Animalia: Multicell eukaryotic (that is, possessing cells with a nucleus and specialized compartments called organelles) organisms that obtain their nutrition solely by feeding on other organisms. (Other defining characteristics of Animalia are discussed in Taxonomy.)
- Phylum Chordata: Animals whose bodies, at least at some point in their life cycles, include a rudimentary internal skeleton with a stiff supporting rod known as a notochord. All chordates at some point also breathe through gills (in the case of a human, while still in the womb). Other characteristics set apart chordates, including a tail or the remnants of one. Humans belong to the subphylum Vertebrata, or chordates with a spinal column.
- Class Mammalia: Vertebrates that feed their young from special milk-secreting glands, known as mammae, located on the mother's body. Mammals have other distinguishing characteristics, such as a hinged lower jaw attached to the skull.
- Order Primates: A group of mammals whose characteristics may include some version of an opposable digit (e.g., the human thumb) and other features that,

while they are prevalent among most primates, are not universal to them. Not every one of these traits is exclusive to primates, a group that includes prosimians (e.g., lemurs), monkeys, apes, and humans.

- Family Hominidae: Primates noted for their erect posture, large brains, rounded skulls, small teeth, bipedal locomotion (i.e., they walk on two legs), and tendency to use language for communication. Humans are the only surviving species in the family, but extinct hominids include *Homo habilis* (about 1.6 million years ago) and *H. erectus* (about two million years ago) as well as the more distant *Australopithecus* (about eight million years ago).
- Genus *Homo*: Hominids with especially large skulls as well as the features that characterize family Hominidae. Members of this genus, which included *H. erectus* and *H. habilis* as well as *H. sapiens,* also are known for their ability to fashion precise tools.
- Species *Homo sapiens*: Members of the genus *Homo* ("man") noted for, among other things, the ability to use symbols and writing. This category includes modern humans and the extinct Cro-Magnon and Neanderthal man.

Note that the proper name of any ranking more general than species is capitalized (e.g., phylum Chordata), with species (and subspecies) names in lowercase. Genus, species, and subspecies names are rendered in italics (e.g., *Homo sapiens,* or "man the wise"), whereas proper names of the more general groupings are presented in ordinary type (e.g., class Mammalia). If the same name appears a second time in the same article, the genus name usually is abbreviated: thus, *H. sapiens.* Another important abbreviation is spp., implying several species within a genus—for example, *Quercus* spp. refers to more than one species of oak.

Taxonomy makes use of a system called binomial nomenclature, in which each species is identified by a two-word name, designating genus and species proper. Beyond the species name, there may be subspecies names: humans are subspecies *sapiens,* so our full species name with subspecies is *Homo sapiens sapiens.* Additional rules govern the inclusion of a name or an abbreviation at the end of the species or subspecies name, to recognize the person who first identified it—in this case, Linnaeus. Hence the proper full name of our species is *Homo sapiens sapiens* Linneaus, 1758.

THE MYSTERY OF SPECIES

If one studies the delineation of humans' place in the overall taxonomic structure, one may notice that for several groupings, the defining characteristics are a bit "fuzzy around the edges." This is true even of the animal kingdom, as noted in Taxonomy: mobility and locomotion, seemingly so integral to the definition of *animal,* are not prevalent among all animal species. Given the many gray areas and areas of dispute in the larger taxonomic categories, it should come as no surprise that the smallest of the obligatory rankings, that of species, lacks a precise definition.

The most widely accepted definition of species is the one put forward by the German-born American evolutionary biologist Ernst Mayr (1904–) in the 1940s. Mayr's idea, known as the biological species concept, defines a species as a population of individual organisms capable of mating with one another and producing fertile offspring in a natural setting. Members of two different, but closely related species in some cases can mate with one another to produce *infertile* offspring, the most well-known example being the mule, a sterile hybrid produced by the union of a male donkey and a female horse.

The definition offered by the biological species concept requires qualification. While many plants and animals reproduce sexually, many more do not; *no* single-cell life-forms reproduce in this way, yet there are certainly many different and distinct species of bacteria and protozoa. Thus, a further qualification typically is added to the definition of species*:* members of the same species share a gene pool, or a total sum of genes. Genes carry information about heritable traits, which are passed from parent to offspring. Whereas the gene pool is shared by members of a species, nonmembers of that species have genes that do not belong to that gene pool. To use a rudimentary example, let's say that there is a gene pool containing genes *x, y,* and *z.* Individuals that have these genes fit within the gene pool, but an individual with gene *w* does not.

The definition of species remains challenging, with special problems raised in the area of botany. It is also sometimes possible to confuse

INUIT VILLAGERS BUTCHER A WHALE. AMONG ENDANGERED SPECIES IS THE RIGHT WHALE, SO CALLED BECAUSE WHALERS OF THE NINETEENTH CENTURY CONSIDERED IT THE "RIGHT" WHALE TO HUNT: IT SWIMS SLOWLY AND CLOSE TO SHORE AND CAN BE FOUND AND SLAUGHTERED EASILY. *(© Lowell Georgia/Corbis. Reproduced by permission.)*

species and race, a grouping that applies not only in the world of humans but also that of other animal and even plant species. Race is different from species inasmuch as races are not isolated genetically from one another; in other words, there are no biological barriers to interbreeding between races. (See Speciation for a discussion of the process whereby single species develop over time into more than one reproductively isolated species.)

REAL-LIFE APPLICATIONS

ENDANGERED SPECIES

An endangered species is any plant, animal, or microorganism that is at risk of becoming extinct or at least of disappearing from a particular local habitat. Over the course of Earth's geological history, species have become extinct naturally—sometimes in large proportions, as discussed in the context of mass extinction in Paleontology. In modern times, however, species and their natural communities are threatened mostly by human activities.

The number of endangered species worldwide is not known. In the United States—a country that, unlike most, expends considerable effort on keeping track of its endangered species—there were more than 750 species and subspecies listed by the late 1990s as endangered under the federal Endangered Species Act. Additional endangered species are being added at a rate of about 50 per year, and there is a "waiting list" of an estimated 3,500 candidate species.

Efforts at monitoring endangered species in the United States have directed a disproportionate amount of attention toward larger organisms; consequently, smaller endangered species from such groups as arthropods, mosses, and lichens have received insufficient attention. The regions of the United States with the largest numbers of endangered species are in the humid southeast and the arid southwest. These areas tend to have the unfortunate combination of unique ecological communities alongside runaway urbanization and resource development.

Overdevelopment and destruction of habitats is perhaps the most well-known ways that humans endanger the survival of species. For example, the habitat of the northern spotted owl is under threat from loggers in the Pacific Northwest (see Succession and Climax). Another threat is the introduction of new species, particularly predators, to an area that is not their natural habitat—a topic we discuss in more depth later in this essay.

HUNTING THE ESKIMO CURLEW. Another way humans threaten species is by excessive hunting. An example of a species thus threatened is the Eskimo curlew (*Numenius borealis*), a sandpiper (a type of bird) that was still abundant in North America during the nineteenth century. A large, friendly creature, it was hunted in vast numbers during its seasonal migrations over the prairies and coasts of Canada and the United States and during its winter seasons in South America. (See Migration and Navigation for more about birds' winter migrations.) The Eskimo curlew became very rare by the end of the nineteenth century, and the last time an Eskimo curlew nest was seen (1866), the guns of the Civil War were practically still smoking. The last time a scientific team collected an Eskimo curlew specimen was in 1922. It might seem that the bird is extinct, but this is not the case. Although it is extremely rare, there have been a few reliable sightings of individuals and small flocks of this species, mostly during migration in Texas and elsewhere but also in its breeding habitat in the Canadian Arctic. Once abundant, the Eskimo curlew now hangs on by a thread.

RIGHT WHALES AND BLUE WHALES. More familiar is the endangerment of whales, a cause made popular by many a "Save the Whales" bumper sticker. Among endangered animals of this group are the blue whale (*Balaenoptera musculus*) and various species from the genus *Balaena,* or right whale. The latter species gained its common name because whalers of the nineteenth century considered it the "right" whale to hunt: it swims slowly and close to shore and so could be found and slaughtered easily. In addition, it yields a large amount of oil, used for lighting lamps in the era when Herman Melville's *Moby Dick* (1851) was written. The estimated world population of right whales is currently about 2,000 individuals, much depleted from the historical high numbers; though it is now protected from whaling, it suffers an excessive mortality rate from ship collisions.

As for the blue whale, it occurs virtually worldwide, and with a typical weight of 150 tons (136 tonnes) and a length of 100 ft. (30 m), it is the largest animal ever to have lived on Earth. Because it is such a fast swimmer, it could not be hunted effectively by whalers in sailing ships. Once steam-powered ships were invented, however, these whales were taken in tremendous numbers and became endangered. Because of its precarious status, this species has not been hunted for several decades, but it remains rare and endangered.

The Fate of the Dodo

When a species becomes extinct, it is gone forever. It is like a family whose last member has died without leaving an heir, but in this case the impact is potentially much more profound. Several thousand species have become extinct as the result of human activities, mostly hunting, in the past few hundred years, and of these species perhaps none is more well known than *Raphus cucullatus,* or the dodo.

Long before the application of the term clueless in the 1990s, a person out of touch or out of step was called a *dodo.* How did the bird's name come to be a synonym for stupidity? Perhaps it is just the funny sound of the name, or perhaps it is the fact that the dodo looked a bit like a turkey, another bird name used for someone of less than exemplary capabilities. Or perhaps the application of the name *dodo* in this way carries a hint of blaming the victim—the implication that the dodo somehow played a part in its own extinction.

In fact, the dodo's only shortcoming was its inability to overcome the threat posed by an extremely dangerous predator: the human. A member of the dove or pigeon family, the dodo was flightless and lacked natural enemies until humans discovered its homeland, the Indian Ocean island of Mauritius, in the early sixteenth century. First came the Portuguese and then, in 1598, the Dutch, who made the island a colony in 1644. By 1681 the dodo had ceased to exist. Not only did sailors collect the birds for food, but introduced species, including dogs, cats, pigs, monkeys, and rats, also preyed on dodos. They were subjected to regular slaughter by sailors, but the species managed to breed and survive on the remote areas of the island for a time. After the establishment of their colony, however, Dutch

A MEMBER OF THE DOVE OR PIGEON FAMILY, THE DODO WAS FLIGHTLESS AND LACKED NATURAL ENEMIES UNTIL HUMANS DISCOVERED ITS HOMELAND, THE ISLAND OF MAURITIUS, IN THE EARLY SIXTEENTH CENTURY. BY 1681, THROUGH THE EFFECTS OF PREDATION, THE DODO HAD CEASED TO EXIST. *(© Bettmann/Corbis. Reproduced by permission.)*

settlers launched what amounted to an extermination campaign.

No part of Earth's living environment can be removed without repercussions, and the destruction of the dodo illustrates the ripple effect that occurs when one species is eliminated. As it turned out, the bird had a symbiotic relationship (see Symbiosis) with the dodo tree, or *Calvaria major,* whose fruit it ate, thus releasing the seeds to germinate. With the dodo gone, the dodo tree stopped being able to reproduce. Fortunately, it is a species with a long life, and some specimens of *C. major* continue to survive after some 300 years; when those die, however, this species, too, will be extinct.

EXOTIC, INTRODUCED, AND INVASIVE SPECIES

An introduced species is one that has been spread to a new environment or habitat as a result of human activity. An invasive species may or may not have been spread by humans (the ones we discuss were), but as the name suggests, it threatens an aspect of the habitat to which it has been introduced. Both introduced and invasive species are examples of exotic species, or species that have been introduced to a region or continent, usually but not always through human activity.

In the case of species introduced by humans, some were introduced deliberately and were intended to improve conditions for some human activity (for example, in agriculture) or to achieve desired aesthetic results—for example, when colonists wanted to plant a flower or tree that reminded them of home. Other introductions have been accidental, as when plants were brought with soil transported as ballast in ships or insects were conveyed with timber or food.

BENEFICIAL AND HARMFUL INTRODUCTIONS. Some introduced species have been wildly successful. In fact, most agricultural plants and animals are introduced species: for example, wheat (*Triticum aestivum*) was originally native only to a small region of the Middle East, but it now grows virtually anywhere conditions are suitable for its cultivation. Likewise, corn, or maize (*Zea mays*), has spread far beyond its home in Central America. The domestic cow (*Bos taurus*) once lived only in Eurasia and the turkey (*Meleagris gallopavo*) only in North America, but today these species can be found throughout the world. If all introduced species were like cows and corn, or turkeys and wheat, there would not be much cause for alarm. Many introduced species are invasive, however, and pose a wide variety of threats—threats to their environments or, in some cases, to human well-being. All manner of weeds and pests are among the nefarious roll-call of invasive species, a broad grouping that ranges from nuisances to serious dangers.

ACCIDENTAL AND DELIBERATE INTRODUCTIONS. There are more than 30,000 introduced species in the United States, and most of them enhance rather than diminish the quality of life. For example, there are the many species introduced by colonists to make them feel more comfortable in their new homes, among them, the Norway maple (*Acer platanoides*), linden (*Tilia cordata*), horse chestnut (*Aesculus hippocastanum*), and other trees as well as many exotic species of shrubs and herbaceous plants. The European settlers also introduced some species of birds and other animals with which they were familiar, such as the starling (*Sturnus vulgaris*), house sparrow (*Passer domesticus*), and pigeon, or rock dove (*Columba livia*).

These are all deliberate introductions; on the other hand, accidental introductions are more likely to be undesirable. When cargo ships from Europe did not have a full load of goods, they had to carry other heavy material as ballast, to help the vessel maintain its stability on the ocean. Early ships to the New World often used soil as ballast, and upon arriving, sailors dumped this soil near the port. In this way, many European weeds and other soil-dwelling organisms arrived in the Americas. In addition, ships have used water as ballast since the late nineteenth century, and many aquatic species have become widely distributed by this practice. This is how two major pests, the zebra mussel (*Dreissena polymorpha,* discussed later) and the spiny water flea (*Bythothrepes cederstroemii*) were introduced to the Great Lakes from European waters.

Several European weeds are toxic to cattle if eaten in large quantities, and when these plants become abundant in pastures, they represent a significant potential problem. Some examples of toxic introduced weeds in the pastures of North America include common Saint-John's-wort (*Hypericum perforatum*), ragwort (*Senecio jacobaea*), and common milkweed (*Asclepias syriaca*). Several introduced insects have become troublesome pests in forests, as is the case with the gypsy moth (*Lymantria dispar*), which has defoliated many trees since its introduction to North America from Europe in 1869.

Similarly, the introduced elm bark beetle (*Scolytus multistriatus*) has helped spread Dutch elm disease, itself caused by an introduced fungus, *Ceratocystis ulmi.* It would be interesting to note the irony inherent in this affliction, which at first glance seems to involve another apparently introduced species, the "Dutch elm." There is no such tree, however; the name refers to the fact that the disease arrived in America from Holland, probably some time after World War I. Its principal victim is the American elm, or *Ulmus americana.*

DELIBERATE AND HARMFUL INTRODUCTIONS. Not all harmful introduced species were introduced accidentally. Settlers from Europe deliberately brought pets, such as the domestic dog (*Canis familiaris*) and cat (*Felis catus*); while these pets may add greatly to the quality of human life, they can cause problems, because they are wide-feeding predators. Such creatures threaten vulnerable animals in many places, especially isolated oceanic islands. Among other predators are mongooses (family Viverridae), often introduced to get rid of snakes, as well as omnivores, such as pigs (*Sus scrofa*) and rats (*Rattus* spp.) Meat-eating animals are not the only threat: herbivores such as sheep (*Ovis aries*) and goats (*Capra hircus*) also endanger plant life in some areas as a result of overgrazing.

A particularly striking example of harmful, deliberate species introduction is the Nile perch (*Lates niloticus*). First introduced to Africa's Lake Victoria in the 1950s, it has proved an economically important food source, with a large worldwide market. The problem is that the Nile perch is an extraordinarily active predator and has brought about a tragic mass extinction of native fishes. Until the 1980s, Lake Victoria supported an extremely diverse community of more than 400 species of fish, mostly cichlids (family Cichlidae), with 90% of those species being endemic, meaning that they exist only in one area. About one-half of the endemic species are now extinct in Lake Victoria because of predation by the Nile perch, although some species survive in captivity, and a few are still in the lake.

KILLER BEES, ZEBRA MUSSELS, AND KUDZU. Three notable examples of invasive species in America are

THE AFRICANIZED, OR "KILLER," BEE IS ONE EXAMPLE OF AN INVASIVE SPECIES. THESE BEES HAVE NO MORE VENOM THAN DOMESTICATED HONEYBEES, BUT THEY ATTACK MORE QUICKLY AND IN GREAT NUMBERS AND HAVE SPREAD THROUGHOUT MUCH OF SOUTH AND CENTRAL AMERICA, TEXAS, AND CALIFORNIA. *(© S. Camazine/Photo Researchers. Reproduced by permission.)*

Africanized honeybees (*Apis mellifera scutellata*), better known as "killer bees"; the zebra mussel (*Dreissena polymorpha*); and kudzu (*Pueraria lobata*). The first "killer" bees were released accidentally by a Brazilian bee breeder in 1957. These aggressive insects have no more venom than domesticated honeybees (another *A. mellifera* subspecies, which is also an Old World import), but they attack more quickly and in great numbers. Interbreeding with resident bees and sometimes traveling with cargo shipments, Africanized bees have spread at a rate of up to 200 mi. (320 km) a year and now threaten humans, fruit orchards, and domestic bees throughout much of South and Central America and north to Texas and California.

The zebra mussel was introduced to the Great Lakes in about 1985 in ballast water dumped by a ship or ships arriving from Europe. It colonizes any hard surface, including rocks, wharves, industrial water-intake pipes, and the shells of native bivalve mollusks. A bean-sized female zebra mussel can produce 50,000 larvae (an immature form of an animal) in a single year. Growing in masses with up to 70,000 individuals per square foot, zebra mussels clog pipes, suffocate native clams, and destroy the breeding habitats of other aquatic animals. These invaders have placed a great burden not only on the environment but also on the economy of the Great Lakes region: area industries spend hundreds of millions of dollars annually to unclog pipes and equipment.

Kudzu is an integral part of culture in the southern United States, but it originated in Japan and did not arrive on American shores until 1876. In that year, numerous foreign governments sent exhibits to the Centennial Exposition, held in Philadelphia to honor the country's 100th birthday. Two generations later, during the Great Depression, the U.S. Soil Conservation Service began promoting the use of kudzu for erosion control.

At a time when work was scarce, young men in the government-sponsored Civilian Conservation Corps (CCC) earned a living by planting kudzu throughout the South. The federal government paid farmers as much as $8.00 an acre—a fabulous sum at the time—to plant kudzu fields. Before another generation had passed, in 1953, the federal government stopped promoting the use of kudzu. In 1972, just four years shy of a

KEY TERMS

BINOMIAL NOMENCLATURE: A system of nomenclature in biological taxonomy whereby each type of plant or animal is given a two-word name, with the first name identifying the genus and the second the species. The genus name is always capitalized and abbreviated after the first use, and the species name is lowercased. Both are always shown in italics—thus, *Homo sapiens* and, later in the same document, *H. sapiens.*

CLASS: The third most general of the obligatory taxonomic classification ranks, after *phylum* but before *order.*

DNA: Deoxyribonucleic acid, a molecule in all cells, and many viruses, containing genetic codes for inheritance.

ENDANGERED SPECIES: Any plant, animal, or microorganism that is at risk of becoming extinct or at least of disappearing from a particular local habitat.

ENDEMIC SPECIES: Species that exist in only one geographic region.

EUKARYOTE: A cell that has a nucleus as well as organelles (sections of the cell that perform specific functions) bound by membranes.

EXOTIC SPECIES: Species that have been introduced to a region or continent, usually but not always through human activity. See also *introduced species* and *invasive species.*

EXTINCTION: A condition in which all members of a taxon have ceased to exist.

FAMILY: The third most specific of the seven obligatory ranks in taxonomy, after *order* but before *genus.*

GENE: A unit of information about a particular heritable (capable of being inherited) trait that is passed from parent to offspring, stored in DNA molecules called *chromosomes.*

GENE POOL: The sum of all the genes shared by a population, such as that of a species.

GENUS: The second most specific of the obligatory ranks in taxonomy, after *family* but before *species.*

HYBRID: The product of sexual union between members of two species or other smaller and less genetically separate groups, such as two races. In the case of species hybrids, the process of hybridization involves genetic abnormalities that lead in most cases to sterility.

century after its first introduction, kudzu was officially declared a weed by the U.S. Department of Agriculture.

Obviously, something had gone wrong. The problem was that kudzu grew *too* well—so fast, in fact, that in the minds of many southerners, it began to possess some sort of mystical significance. This preoccupation with kudzu is reflected in the work of several Georgians, whose state has been particularly afflicted with the vine. There is the poem "Kudzu" by James Dickey as well as the cover of *Murmur,* the music group R.E.M.'s 1983 debut, which features a photograph of a kudzu-covered railroad trestle near the group's hometown of Athens.

Kudzu covered more than railroad tracks, and in the mid–twentieth century, it began to seem as though it would cover the entire South with its tangled vines. The plant is capable of growing by as much as 1 ft. (0.3 m) per day during the summer and can cover virtually anything that is not moving. Over the course of a good

KEY TERMS CONTINUED

INTRODUCED SPECIES: A species that has been spread to a new environment or habitat, whether deliberately or accidentally, as a result of human activity. Introduced species, like invasive species, are considered exotic species.

INVASIVE SPECIES: An exotic species that threatens some aspect of the habitat to which it has been introduced.

KINGDOM: The highest or most general ranking in the obligatory taxonomic system. In the system used in this book there are five kingdoms: Monera, Protista, Fungi, Plantae, and Animalia.

MORPHOLOGY: Structure or form, or the study thereof.

NATURAL SELECTION: The process whereby some organisms thrive and others perish, depending on their degree of adaptation to a particular environment.

NOMENCLATURE: The act or process of naming or a system of names—particularly one used in a specific science or discipline. See also *binomial nomenclature.*

ORDER: The middle of the seven obligatory ranks in taxonomy, more specific than *class* but more general than *family.*

PHYLUM: The second most general of the obligatory taxonomic classification ranks, after *kingdom* and before *class.*

PROKARYOTE: A cell without a nucleus.

RNA: Ribonucleic acid, the molecule translated from DNA in the cell nucleus, the control center of the cell, that directs protein synthesis in the cytoplasm, or the space between cells.

SPECIATION: The divergence of evolutionary lineages and creation of new species.

SPECIES: The most specific of the seven obligatory ranks in taxonomy. Species often are defined as a population of individual organisms capable of mating with one another and producing fertile offspring in a natural setting. Also, members of the same species share a gene pool.

TAXON: A taxonomic group or entity.

TAXONOMY: The area of the biological sciences devoted to the identification, nomenclature, and classification of organisms according to apparent common characteristics.

year, kudzu can grow by as much as 60 ft. (20 m), and it has proved impervious to many herbicides. One herbicide used in Auburn, Alabama, actually made it grow better! Thanks to the development of better chemical treatments, and the use of grazing animals, such as goats, kudzu no longer is perceived as such a great threat. Additionally, various entrepreneurs and scientists have set out to make use of the vine in weaving baskets or in preparing foods and medicines. Ground kudzu root, called *kuru,* has long been used in foods and medications in China and Japan.

One might wonder why Japan is not covered in kudzu and why kudzu is not crawling up the Great Wall of China. The answer is more than a little interesting from a biological standpoint. When kudzu was transplanted to America, it was taken out of its native environment and thus away from the local insects that threatened its growth. In its new home there were no threats to its spread, and with no obstacles in its way, it began to take over the South. (For more about the development of species, see Speciation. See also the discussion of keystone and indicator species in Food Webs.)

WHERE TO LEARN MORE

All Species Foundation (Web site). <http://www.all-species.org/>.

"Endangered Species on EE-Link." EE-Link (Environmental Education Link), North American Association for Environmental Education (Web site). <http://eelink.net/EndSpp/>.

Integrated Taxonomic Information System (ITIS), United States Department of Agriculture (Web site). <http://www.itis.usda.gov/>.

Invasive Species, National Agricultural Library, U.S. Department of Agriculture (Web site). <http://www.invasivespecies.gov/>.

Levy, Charles K. *Evolutionary Wars: A Three-Billion-Year Arms Race—The Battle of Species on Land, at Sea, and in the Air.* New York: W. H. Freeman, 1999.

Schilthuizen, Menno. *Frogs, Flies, and Dandelions: Speciation—The Evolution of New Species.* New York: Oxford University Press, 2001.

Schwartz, Jeffrey H. *Sudden Origins: Fossils, Genes, and the Emergence of Species.* New York: John Wiley and Sons, 1999.

Species 2000 (Web site). <http://www.sp2000.org/>.

Van Driesche, Jason, and Roy Van Driesche. *Nature out of Place: Biological Invasions in the Global Age.* Washington, DC: Island Press, 2000.

Vergoth, Karin, and Christopher Lampton. *Endangered Species.* New York: F. Watts, 1999.

SPECIATION

CONCEPT

One of the defining characteristics of a species is its reproductive isolation: the fact that among animals and plants that reproduce sexually, it is impossible for members of two different species to mate and produce fertile offspring. Speciation is the process whereby a single species develops over time into two distinct, reproductively isolated species. It is one of the key evolutionary processes and is responsible for the diversity of life that exists on Earth. In the following essay we explore not only the basic facts of speciation and biological diversity but also an example of adaptive radiation, in the form of the wide range of species within the mammalian order.

HOW IT WORKS

SPECIES AND SPECIATION

The concept of species, discussed in the article devoted to that subject, is an extraordinarily complex one. Owing to limitations of space, that essay only hints at the many details, the competing schools of thought, and the varying definitions of *species.* Likewise, in the present context, it is possible to examine the concept of speciation only in the most cursory fashion. In addition to consulting the essay on Species for more information, the reader is encouraged to review the article on Taxonomy.

Taxonomy is the area of the biological sciences devoted to the identification, nomenclature, and classification of organisms according to apparent common characteristics. It uses a wide array of specialized rankings for grouping animals, but only seven of them are essential to most biology students. These seven, known as the obligatory hierarchy, are kingdom, phylum, class, order, family, genus, and species. In the case of mammals, it is also useful to refer to subphylum, which in this case is Vertebrata (see the classification of humans in Species), but for the most part it is enough for the beginning student to attain at least some mastery of the obligatory ranks.

Note that species is the most specific of these ranks, which is fitting, because *species* and *specific* come from the same Latin root, *specie,* or "kind." Nonetheless, it is difficult to define species beyond a reference to its place among the categories of the obligatory taxonomy. According to the biological species concept, discussed briefly in Species, a species is any population of individual organisms capable of mating with one another and producing fertile offspring in a natural setting. This is far from the only definition, however.

INTERSPECIFIC MATING. Occasionally, it is possible to produce an infertile hybrid, such as a mule, which is created by the mating of a male donkey and a female horse, or a hinny, the product of the less common union between a male horse and a female donkey. The infertility is due to genetic disorders that arise when mating takes place between distinct species, and even this imperfect product is possible only by mating two species that are very closely related. Donkeys and horses, for instance, both belong to family Equidae, which makes them very closely connected.

In the taxonomic ranking of humans, this would be equivalent to a human mating with a fellow hominid, or member of family Hominidae. If the long-extinct genus *Australopithecus* were

A SPECIES IS A POPULATION OF INDIVIDUAL ORGANISMS CAPABLE OF MATING WITH ONE ANOTHER AND PRODUCING FERTILE OFFSPRING. OCCASIONALLY, IT IS POSSIBLE TO PRODUCE AN INFERTILE HYBRID, SUCH AS A MULE, THROUGH THE MATING OF A MALE DONKEY AND A FEMALE HORSE. *(© D. Robert & Lori Franz/Corbis. Reproduced by permission.)*

still around, it is not inconceivable that humans could mate with them and produce at least sterile offspring. Of course, it is unlikely that many humans would want to mate with *Australopithecus,* the most famous example of which was named "Lucy" after the Beatles' song "Lucy in the Sky with Diamonds." Standing about 3.5-5 ft. (1–1.5 m) tall, *Australopithecus* was very close in appearance to a modern ape who lived about four million years ago.

All of humans' close relatives are extinct, and today our nearest relatives are members of the order Primates: apes, monkeys, and marsupials. It is impossible to imagine a human mating with one of these animals and producing offspring of any kind. Likewise, it is extremely unlikely that a horse or donkey could mate with a tapir or rhinoceros, which are about as distant in relation to them as other primates are to us. (These species all belong to the order Perissodactyla, herbivorous mammals possessing either one or three hoofed toes on each hind foot. We discuss this group, along with all other mammalian orders, later in this essay.)

THE PROBLEM OF DEFINING SPECIES. Although the biological species concept is accepted widely, it has its shortcomings, not least of which is the fact that not all species reproduce sexually. Although sexual reproduction is the case with a wide array of animals and even plants, quite a few organisms reproduce by some asexual means: for example, single-cell organisms reproduce by splitting.

Among the competing definitions of species is the phenetic (or morphological) species concept, which relies in part on common sense. According to the phenetic species concept, a species is the smallest possible population of organisms that consistently and continually remains distinct and distinguishable by ordinary methods of observation. There are also a variety of definitions that fall under the heading "phylogenetic species concepts," all of which maintain in one way or another that taxonomic classifications should incorporate the most widely recognized hypotheses regarding the evolutionary lines of descent that produced the organisms in question.

THE PROCESS OF SPECIATION

Clearly, there is no hard and fast definition of species, but in general terms, everyone who has some familiarity with the concept has at least a basic knowledge of what does and does not qualify as a species. We will leave finer distinctions to trained taxonomists and other biologists and move on to a fact regarding which there is no disagreement: a wide array of species exists in the world today. Some estimates calculate the number of species in the five kingdoms—animals, plants, monerans, protista, and fungi (see Taxonomy for a very brief identification of each)—at about 1.5 million.

This is only the number of identified species, however. Other figures, based on the probable numbers of unidentified species in the world, put the sum total in the tens of millions. Whatever the case, it is obvious that over the course of evolutionary history (discussed in Evolution and Paleontology), there has been a widespread adaptive radiation—that is, a diversification of species as a result of specialized adaptations by particular populations of organisms.

Speciation events are described as either allopatric or sympatric. Allopatric ("different places") speciation occurs when a population of organisms is divided by a geographic barrier, a great example being the division of squirrel species caused by the formation of the Grand Canyon (see Evolution). Another example is the speciation of the black-throated green warbler, which today consists of one species in the eastern United States, along with three others in the western part of the country. Some scientists speculate that there may once have been a single species of black-throated green warbler, whose population was split by the formation of a glacier during the Pleistocene epoch. The latter was the period of the last ice age, which ended about 10,000 years ago, but the end of the ice age was a slow process. It may be that glaciers, formed in the latter part of that time, helped to separate what became three different western species.

Species share the same gene pool, or the sum of all genetic codes possessed by members of that species. The isolation of two populations slowly results in differences between gene pools, until the two populations are unable to interbreed either because of changes in mating behavior or because of incompatibility of the DNA between the two populations. (Deoxyribonucleic acid, or DNA, contains genetic codes for inheritance. See Genetics for more on this subject.) More rare than allopatric speciation, sympatric ("same place") speciation happens when a group of individuals becomes reproductively isolated from the larger population of the original species. This type of speciation typically results from mutation, or alterations in DNA that result in a genetic change.

Studies of three-spined sticklebacks, a variety of freshwater fish, in British Columbia have revealed what appears to be a fascinating example of sympatric speciation. Evolutionary biologist Dolph Schluter and others have discovered that the region contains two species of stickleback, one with a large mouth that feeds on large prey close to shore, the other with a small mouth that feeds on plankton in open water. Both species jointly inhabit five different lakes. Through DNA analysis, scientists have determined that the lakes were colonized independently by common marine ancestors, meaning that the process of sympatric speciation between the two varieties had to have occurred independently at least five times. This seems to indicate a situation of competition for resources that favored stickleback species at either extreme of size, as opposed to those of medium size and medium-sized mouths.

RATE OF EVOLUTIONARY CHANGE. Closely tied to speciation is the rate of evolutionary change, or the speed at which new species arise. This is a long process, one that is usually not observable within a human lifetime or even the span of many lifetimes, though bacteria at least have shown some evolutionary change in their growing resistance to antiobiotics (see Infection). DNA analysis (see Genetics and Genetic Engineering for more about DNA) has been used to examine the rate of evolutionary change. To perform such analysis, it is necessary first to determine the percentage of similarity between the organisms under study: the greater the similarity, the more recently the organisms probably diverged from a common stock. Data obtained in this manner then must be corroborated by information obtained from other sources, such as the fossil record and comparative anatomy studies.

At certain times the rate of evolutionary change can be very rapid, leaving little fossil evidence of intermediate forms, a phenomenon known as *punctuated equilibrium.* This is contrasted with phyletic (that is, evolutionary) gradualism. Of course, the term *rapid* in this context is relative, since we are talking about vast spans of time. Life on Earth has existed for about 3,000 million years, and the fossil record goes back some 1,000 million years. This is the case, in part, because to leave fossilized remains, an organism must have "hard parts" that can become mineralized to turn into fossils. (See Paleontology for more on these subjects.)

REAL-LIFE APPLICATIONS

THE DIVERSITY OF MAMMALS

One of the most interesting examples of speciation is that which has produced the vast array of species, including humans, that fall within the mammalian class. Mammals began evolving before the dawn of the Cenozoic era about 65 million years ago. The Cenozoic era, which started with a catastrophic event that brought about the mass extinction of the dinosaurs and the end of the Mesozoic era (see Paleontology), is truly the age of the mammal. Just as dinosaurs dominated the Mesozoic, today the world belongs to mammals as to no other class of creature.

Since its humble beginnings in the shadow of the dinosaurs, class Mammalia has undergone a massive radiation to the point that today some 4,625 species of mammal, in about 125 families and 24 orders, are recognized. (That number is changing, as noted later in the context of elephants.) This diversity is tied closely to mammals' enormous mobility, which facilitated their spread throughout the world. Aside from much less complex life-forms, such as arachnids and insects (see Parasites and Parasitology), mammals are believed to be distributed more widely throughout the world than any other comparable taxonomic grouping. Insects may be the most diverse of all animal classes, with numbers of species that may be many times greater than the number of mammals, but considering mammals' much-greater level of physical development and complexity, the diversity of their species is astounding.

MAMMALS' EARLY EVOLUTION. In the next section we list the orders of mammals and give very brief descriptions of each. The purpose here is not to provide anything like a comprehensive discussion but rather to illustrate the enormous range of species in a class that includes anteaters, dolphins, humans, elephants, and bats. The fact that all these diverse creatures, and many more, emerged from a common evolutionary lineage is almost as amazing as the fact that this common ancestor was a reptile.

Mammals are believed to have come from the reptilian order Therapsida, which emerged during the Triassic period (from about 245 to 208 million years ago) in the early part of the Mesozoic era. Over the course of many millions of years, these creatures began to develop a number of mammal-like qualities—in particular, endothermy, or the ability to maintain internal temperature regardless of environmental conditions. In other words, these cold-blooded creatures became warm-blooded. This evolutionary process was as complex as it was lengthy. Nor was there a clean break with the past—no moment when the therapsids faded away or when it would have been clear that mammals had taken the place of their reptilian ancestors. Rather, in what must have been a fascinating taxonomic situation, for many millions of years, species that combined aspects of both reptiles and mammals walked the earth.

MAMMALIAN ORDERS

The listing of the 20 orders of living mammals that follows is arranged not alphabetically but in the probable order in which these groups evolved. (This is not to imply that the process was orderly or linear; it was not.) Very few dates are given, simply because there is much dispute in most cases. Numbers of species within each order are also a subject of debate among taxonomists, and therefore these numbers are not always precise.

In the essay, Species, there is a taxonomic listing of the obligatory ranks for humans; included within that listing is a short description of the kingdom (Animalia), phylum (Chordata), and subphylum (Vertebrata) to which mammals belong. *Mammal* itself is defined as a vertebrate (an animal with a spinal column) that feeds its young from special milk-secreting glands, termed *mammae,* located on the mother's body. Mammals are warm-blooded or endothermic, meaning that their internal temperatures remain relatively stable, and their bodies usually are covered with hair. They have other distinguishing characteristics as well, such as a relatively large cranium (skull) with a hinged lower jaw attached to it.

MONOTREMES. Order Monotrema consists of primitive, egg-laying mammals spread throughout parts of the region known as Oceania, which includes Australia, New Zealand, and islands of the southeastern Pacific. The habitat of this order lies specifically in Australia, Tasmania, and New Guinea. Monotremes, as they are called, are distinguished further by the fact that their mammary glands are without nipples, that teeth are present only in the young, and that adults have horny beaks.

The monotremes illustrate the fact that to be constituted as an order or family, a taxon, or taxonomic group, need not have large numbers. The entire order consists of a single existing species, the duck-billed platypus, which constitutes a family of its own, and two species of echidnas, creatures that look like a cross between a platypus and a porcupine. (The "porcupine" look comes from the fact that their bodies are covered in spines, or spiky protrusions.)

MARSUPIALS. The marsupials, or order Marsupialia, include two other famous animal citizens of Oceania: the kangaroo and its close relative, the wallaby. Marsupials' young are poorly developed at birth and must continue to grow while attached to their mothers' nipples. For this reason, they must remain close to the mother, and therefore natural selection for marsupials favored those strains in which females possess a pouch bearing four teats.

Immediately after birth, the young marsupial (a kangaroo baby is called a joey) installs itself in the mother's pouch. Given this situation, marsupials can support only one offspring a year and thus are not given to the large litters that characterize another order, Carnivora, which we discuss later. Kangaroo offspring remain in the pouch until the age of about 7-10 months, by which time the mother has conceived again; the female kangaroo goes into heat just a few days after giving birth. The embryonic kangaroo remains in a state of dormancy, or arrested development, until the older sibling has left the pouch.

The marsupial order (some authorities call it a *superorder,* with numerous subordinate orders) consists of some 240 species. The greatest number of these, including many species of kangaroo, wallaby, wombat, and koala, are found exclusively in Australia. Some 70 additional species are scattered across parts of Oceania, including Australia, Tasmania, New Guinea, and smaller islands. There are an additional 70 species in the Americas, including four species of the genus *Didelphis*—the large American opossum, better known in the southern United States as possums.

Why the preponderance of marsupials in Australia and Oceania as a whole? The reason lies in Earth's geologic history, which has seen regular collisions and divisions of the continents, which are even today shifting slowly under our feet. It appears that prior to about 70 million years ago, at a time when most of Earth was united in a single supercontinent called Pangaea, marsupials originated in what is now North America and migrated to the land masses that became Australia and Oceania. Because the bulk of marsupial species remained on Australia and nearby areas when Pangaea began to break apart, marsupials underwent much greater speciation there than in North America.

XENARTHRANS, INSECTIVORES, SCANDENTIA, AND DERMOPTERA. Known variously as xenarthrans and edentates, members of order Xenarthra

MARSUPIALS' YOUNG ARE POORLY DEVELOPED AT BIRTH AND MUST CONTINUE TO GROW WHILE ATTACHED TO THEIR MOTHERS' NIPPLES. IMMEDIATELY AFTER BIRTH, THE YOUNG KANGAROO (CALLED A JOEY) INSTALLS ITSELF IN THE MOTHER'S POUCH AND REMAINS THERE UNTIL THE AGE OF 7-10 MONTHS. *(© Michael S. Yamashita/Corbis. Reproduced by permission.)*

either lack teeth or have very small ones. Evolutionary development has adapted the forward limbs of these creatures for digging or for holding on to the branches of trees. Included in this order are some 30 species of sloth, anteater, and armadillo. Sloths are herbivores, armadillos are omnivores (i.e., they eat plants and small animals), and anteaters, as their name would suggest, are hard-core insectivores.

The term insectivore can refer to any organism that lives by eating insects, but it is also the name for members of the order Insectivora, which includes shrews, hedgehogs, moles, and various other, less well known groups. Some 400 species, of which about 300 are shrews in a single family (Soricidae), make up this order. Not only their diet but also their pointed snouts and rodent-like appearance distinguishes this group. Many, but not all, are diggers, like the xenarthrans. Like all mammals, they have the pentadactyl limb (an appendage with five digits, like the human arm and hand)—in their case, a foot with five toes.

Order Scandentia, which is identical with the family Tupaiidae, or tree shrews, is sometimes grouped with order Insectivora. Despite their name, tree shrews, of which there are five genera and between 15 and 19 species, may live either on the ground or in the trees. Squirrel-like in appearance, they have strong claws on all their toes and are excellent climbers.

Another very small order of tree-dwellers is Dermoptera, which consists of just two species of flying lemur. Found in Indonesia and the Philippines, these creatures are equipped with skin flaps adapted for gliding. This aspect of their morphological makeup calls to mind the "flying" squirrel, but their nocturnal habits are more like those of lemurs (discussed later, with other primates); hence their common name.

CHIROPTERANS. Among the most fascinating of mammalian orders is Chiroptera, better known as bats. This order, which consists of about 900 species in some 175 genera, is the only group of truly flying mammals, as opposed to the "flying" lemurs we just discussed. Yet they, too, have the pentadactyl limb, only in their case the forelimb has been adapted as a wing. Among the intriguing features of bats is their use of acoustic orientation, or echolocation, to find their way through the dark caves and nocturnal exteriors that make up their world. Contrary to popular belief, they are not blind, but they do have very small eyes, simply because vision is not important for bat navigation. (See Migration and Navigation for more about this subject.)

As befits an order with such a wide array of species, bats run the gamut with respect to their eating habits. The majority of bat species are insectivores that consume many thousands times their weight in insects each year. Many others are fruit bats, important members of the ecosystems they occupy, because they consume fruit and spread seeds, helping assist in seed dispersal. (Some bats also aid in pollination; see Reproduction). Then there are the three species of vampire bat, which are largely responsible for bats' unfortunate reputation with humans.

Members of subfamily *Desmodontinae,* species of vampire bat include *Diaemus youngi* and *Diphylla ecaudata.* However, the best-known variety is the common vampire bat, *Desmodus rotundus,* which is native to an area that stretches from the southwestern United States to the northwestern third of South America. As one would suspect of a creature named vampire, they live off blood, which they suck at night from birds or other mammals, including humans. These creatures are livestock pests and strike fear in humans both because of the imaginary association with vampires and for the quite real threat of contracting rabies from them. The vast majority of bats, however, are creatures that cause no harm to humans and often are unfairly persecuted as the result of human prejudices. Even in the case of the vampire bat, there is a strong possibility that it may one day help save human lives. Scientists have discovered that the saliva of *Desmodus rotundus* is better than any other known substance for keeping blood from clotting; therefore, vampire bat saliva may one day be adapted for use in treating heart attacks and strokes.

PRIMATES. The order of humans, Primates, falls approximately in the middle of the mammalian class in terms of evolutionary order. This is an interesting aspect of speciation, evolution, and taxonomy: even though humans themselves are the most advanced of all creatures, it is not a logical necessity that we should come from the most recently evolved order. In fact, the opposite would seem to be the case. To produce a species whose intelligence dwarfs that of all other animals, the line of descent should be a long one. Where primates are concerned, that is certainly the case. The oldest primate samples date back some 75 million years, or long before the end of the Mesozoic.

Because it is from primates that humans draw their lineage, more has been written about primate evolution than on that of all other mammalian orders combined. The subject is such a vast one that we will not attempt to approach it here, except to encourage the reader to study in more detail elsewhere the process by which the human lineage emerged from order Primates, family Hominidae, and genus *Homo.*

Primates consist of two broad groups, suborders Prosimii and Anthropoidea. The first, the prosimians, includes five families (or six, since tree shrews are sometimes included) of lemurs, lorises, and tarsiers. The other suborder, known as the *higher* primates, encompasses another six families: marmosets and tamarins; South American monkeys other than marmosets; African and Asian monkeys; lesser apes, or siamangs and gibbons; great apes, or orangutans, gorillas, and chimpanzees; and humans, both living and extinct. (Most orders contain extinct members, but for the most part they are not discussed here.)

Most primates are tree dwellers, and among the approximately 230 species, there is enormous variation in eating habits. Many lemurs are insectivores, while great apes tend to be fruit eaters. Quite a few are omnivores, though no primate other than humans is known for eating large mammals, such as cows, sheep, and pigs. The pentadactyl limb (an appendage with five digits) is a significant feature for primates, which alone have the advantage of the opposable thumb for grasping. Humans and a few other primate species are also the only animals with four limbs who are not only capable of standing upright but also function best in this way.

CARNIVORES. As with insectivore, *carnivore* is a name both for an eating preference—in this case, meat—and for members of a primate order, Carnivora. Most members of this extraordinarily varied group eat meat, including, in some cases, the "meat" of insects. Bears and some other species are omnivorous, meaning that they also eat plants, and hyenas and jackals are classic examples of detritivores, or animals who feed on the remains of other creatures.

The distinction between detritivore and carnivore relates not to the materials each consumes but to their place in the food web. Rather than consume live creatures, hyenas and jackals feed on the carcasses of dead ones. Usually these creatures are artiodactyls (discussed later), such as antelopes, which have been killed by other carnivores—big cats, such as the lion or cheetah. After the big cats have fed on the fleshy parts of the prey, hyenas come to consume the flesh that remains, and they are followed by jackals and vultures, swoop in to pick the bones. These detritivores help process the remains of formerly living things, which ultimately return to the soil. (See Food Webs for more on this subject.)

Clearly, all members of order Carnivora eat meat, though in different ways and sometimes in combination with fruit or other vegetation. Natural selection has equipped them for this purpose with sharp claws and teeth. Carnivora includes some 270 species grouped into ten families, listed here:

- Canidae (dogs, wolves, jackals, and foxes)
- Felidae (cats)
- Hyaenidae (hyenas)
- Mustelidae (skunks, mink, weasels, badgers, and otters)
- Otariidae (eared seals)
- Odobenidae (walrus)
- Phocidae (earless seals)
- Procyonidae (raccoons)
- Ursidae (bears)
- Viverridae (mongooses and civets).

Note that Felidae is a particularly varied family of some 36 species: lions, lynxes, tigers, leopards, and even ordinary domesticated cats. Thirty-five species belong to a single subfamily, Felinae, which is native to most parts of the world other than Australia, Madagascar, most oceanic islands, and, of course, Antarctica. The last species, the cheetah (*Acinonyx jubatus*), is segregated into another subfamily, Acinonychinae, primarily because this cat, native to Africa and southwest Asia, is a daytime hunter, unlike its nocturnal cousins.

CETACEANS AND SIRENIANS. Orders Cetacea and Sirenia include the majority of aquatic mammals, as opposed to the many amphibious mammals, such as seals, sea lions, sea elephants, and walruses, that belong variously to families Otariidae, Odobenidae, and Phocidae of the order Carnivora. Cetaceans include whales, dolphins, and porpoises, while sirenians are made up of just three species of manatee and one of dugong. Sirenians are large, friendly creatures that inhabit the Atlantic coast and tributary rivers (manatee) or the Indian and Pacific coastlines (dugong), but cetaceans are much more familiar.

With cetaceans, two questions, one specific and one general, often arise. The answer to the first of these, "What is the difference between a dolphin and a porpoise?," is that a porpoise is smaller and more chubby and has a blunt snout, whereas a dolphin has a beaklike snout. Some taxonomists and marine biologists put porpoises in the same family as dolphins, whereas others treat them as two different families. The more general, and much more important, question is "Why are these mammals living in the water?" In fact, life itself first appeared in the sea, so perhaps the question should be "Why or how did anything start living on land?" The transition from water to land took place long before the age of the dinosaurs, much less the emergence of mammals, but later, some mammals began to return to the water, probably about 70 million years ago.

Certainly, there is no question that cetaceans are mammals, a fact first recognized by Aristotle (384–322 B.C.), a Greek thinker who is noted not only as one of the greatest philosophers of all time but also as the father of the biological sciences. (See Taxonomy for more about Aristotle's contributions.) As Aristotle observed, whales and dolphins bear live young and suckle them with milk-producing glands; their bodies have hair, albeit only very small strands; and they possess lungs, breathing air through a blowhole.

As evidence of their terrestrial, or land-based, origins, consider that a whale fetus possesses the remnants of four limbs, each with five fingers (the pentadactyl limb), like any land mammal. Adult whales and dolphins have the streamlined, fishlike morphological appearance that is necessary for life underwater, but their resemblance to fishes is of the superficial, analogous variety discussed in Taxonomy. They have maintained and modified key terrestrial features; for example, a blowhole atop the head—one in dolphins, two in whales—replaces the nostrils, and thus the passageways for food and air are completely separate. This differs from the situation with most terrestrial mammals, which take in food and air through the same opening.

PROBOSCIDEANS, PERISSODACTYLS, AND HYRAXES. Moving from the largest aquatic mammals, the whales, to the largest terrestrial variety, we come to the order Proboscidea, which includes elephants. Our discussions of most orders in class Mammalia have illustrated particular aspects of taxonomy and speciation, and so it is with proboscideans, which give evidence of the many species from the past that are gone forever. The order is a large one, with three suborders and some 300 species, but anyone who searches for most of those species will search in vain. All but three species are extinct. These three are the Asian elephant (*Elephas maximus*) and two varieties of African elephant (*Loxodonta africana* or African savanna elephant and *Loxodonta cyclotisare,* the African forest elephant). Until 2001, taxonomists believed that there were only two living species of elephant. (See Taxonomy for more on this subject.)

Another 16 species belong to the order Perissodactyla, herbivores whose hind feet bear either one or three hoofed toes. Included among perissodactyls is another large animal from the grasslands of Africa and tropical Asia: the rhinoceros. The group also encompasses donkeys, zebras, and tapirs, but by far the most important in human terms is *Equus caballus,* the domesticated horse. Described by the French zoologist Comte Georges de Buffon (1707–1788) as "the proudest conquest of man," the horse was domesticated (adapted so as to be useful and advantageous for humans) some 6,000 years ago. Nonetheless, feral or wild horses remain an important subspecies.

Order Hyracoidea also consists of hoofed mammals: seven species of hyrax, a primarily herbivorous creature native to Africa and the far southwestern extremities of Asia. Hyraxes are sometimes lumped in with pikas under the term rock rabbit, but, in fact, pikas are lagomorphs, a group we discuss later. To further the confusion, hyraxes are probably the animal called a coney in the Bible, even though there is an animal called a *cony* (no "e") that is actually a lagomorph. This is just one of many examples of confusion resulting from the complexities of the animal world and humans' attempts to name and classify its members.

TUBULIDENTATES. For most orders, the lowercase adjectival name (e.g., primates, carnivores, insectivores, and so on) is commonly used. On the other hand, the names hyracoidean and tubulidentate are seldom used for more obscure groups, such as Hyracoidea and Tubulidentata. If any order of mammal is obscure, it is Tubulidentata, which emerged some 60 million years ago and which consists of a single species: the African ant bear (*Orycteropus afer*). The latter creature is better known by the name *aardvark,* which in the Afrikaans language means "earth pig."

Aardvarks at first glance might seem to belong with anteaters, sloths, and armadillos in the order Edentata, and that is what taxonomists thought for a long time. The latter half of the name tubulidentate, however, suggests the area of differentiation: the teeth. Aardvarks' teeth are unique among those of all mammals. Viewed from the top, the aardvark's jawbone is V-shaped, with the teeth midway along either side of the *V.* The teeth themselves are not fixed to the jaw but rest in the flesh attached to it, and instead of being covered with enamel, they are protected with a cementlike substance. The substance comes from tubules that run under the teeth.

ORDER PHOLIDOTA CONSISTS OF SEVEN SPECIES OF SCALY ANTEATER, OR PANGOLIN. THE WORD PANGOLIN COMES FROM A MALAY TERM MEANING "ROLLING OVER," A REFERENCE TO THE FACT THAT WHEN IT IS THREATENED, THE ANIMAL CURLS INTO A LITTLE BALL. *(© Keren Su/Corbis. Reproduced by permission.)*

ARTIODACTYLS. Like many other mammalian orders we have discussed, members of order Artiodactyla are ungulates, or hoofed animals. Whereas perissodactyls have odd-numbered toes, artiodactyls have even-numbered toes—either two or four—on each foot. This large group, consisting of some 220 species, comprises a wide variety of well-known species in nine families. Among them are cows, pigs, sheep, goats, deer, antelope, bison, camels, giraffes, hippopotamuses, and numerous less well known varieties, such as okapi, pronghorn, peccaries, and deerlike chevrotains.

The camel family is particularly widespread geographically, including as it does many varieties—the llama, alpaca, and vicuña of South America—whose home is far from the habitats in the Near East that are associated with the camel. In this family is what may be a previously undiscovered species, whose existence the British Broadcasting Corporation (BBC) reported in early 2001. Living on a former nuclear weapons testing range in a remote region of Chinese central Asia, these creatures drink saltwater, which in itself is an unusual characteristic.

Although it is extraordinarily hardy, even by the standards of camels, the central Asian camels are threatened; fewer than 1,000 remain. As the BBC reported, this makes them more endangered than the more well known giant panda. The creatures survived nuclear testing in the area, which ceased in 1996, but they continue to be threatened by much less spectacular varieties of explosive: dynamite and land mines, planted by hungry locals. John Hare of the Wild Camel Protection Foundation told the BBC, "We found land mines put by the saltwater springs. So when the camels come to drink, they step on them. Bang! They are blown to pieces and picked up as meat."

As to whether the camels constitute a separate species, the molecular geneticist Olivier Hanotte told the BBC: "There are two possibilities here. One is that the domestic camel was bred from these wild ones some time back in history. The second is that the domestic camel we see today was bred from another species that has disappeared. This would mean that these wild camels are a totally separate species." As of early 2002 the camels' fate, both practically and taxonomically, remained undecided.

PHOLIDOTA/PANGOLINS. Order Pholidota consists of seven species of scaly anteater, or pangolin. Members of this order are normally called *pangolins,* rather than an adjecti-

KEY TERMS

ADAPTIVE RADIATION: A diversification of species over time as a result of specialized adaptations by particular populations of organisms.

ALLOPATRIC SPECIATION: A type of speciation that occurs when a population of organisms is divided by a geographic barrier.

ANALOGOUS FEATURES: Morphologic characteristics of two or more taxa that are superficially similar but not as a result of any common evolutionary origin.

CARNIVORE: A meat-eating organism, or an organism that eats *only* meat (as distinguished from an *omnivore*).

CLASS: The third most general of the obligatory taxonomic classification ranks, after *phylum* but before *order.*

DETRITIVORES: Organisms that feed on waste matter, breaking organic material down into inorganic substances that then can become available to the biosphere in the form of nutrients for plants. Their function is similar to that of decomposers; however, unlike decomposers—which tend to be bacteria or fungi—detritivores are relatively complex organisms, such as earthworms or maggots.

DNA: Deoxyribonucleic acid, a molecule in all cells, and many viruses, containing genetic codes for inheritance.

DOMESTICATE: To adapt an organism, whether plant or animal, so as to be useful and advantageous for humans.

FAMILY: The third most specific of the seven obligatory ranks in taxonomy, after *order* but before *genus.*

GENE: A unit of information about a particular heritable (capable of being inherited) trait that is passed from parent to offspring, stored in DNA molecules called *chromosomes.*

GENE POOL: The sum of all the genes shared by a population, such as that of a species.

GENUS: The second most specific of the obligatory ranks in taxonomy, after *family* but before *species.*

HERBIVORE: A plant-eating organism.

HYBRID: The product of sexual union between members of two species or other smaller and less genetically separate groups, such as two races. In the case of species hybrids, the process of hybridization involves genetic abnormalities that lead in most cases to sterility.

INSECTIVORE: An insect-eating organism.

KINGDOM: The highest or most general ranking in the obligatory taxonomic system. In the system used in this book there are five kingdoms: Monera, Protista, Fungi, Plantae, and Animalia.

val form of Pholidota. The word *pangolin* comes from a Malay term meaning "rolling over," a reference to the fact that when it is threatened, the animal curls into a little ball. As with the aardvark, members of this order once were grouped with Edentata but now are considered a separate order. Pangolins are also like aardvarks in the sense that their evolutionary relationship to other mammals is not clear.

RODENTS, LAGOMORPHS, AND MACROSCELIDEANS. Rodents, or mem-

KEY TERMS CONTINUED

MORPHOLOGY: Structure or form, or the study thereof.

MUTATION: Alteration in the physical structure of an organism's DNA, resulting in a genetic change that can be inherited.

NATURAL SELECTION: The process whereby some organisms thrive and others perish, depending on their degree of adaptation to a particular environment.

NUMERICAL TAXONOMY: An approach to taxonomy in which specific morphological characteristics of an organism are measured and assigned numerical value, so that similarities between two types of organism can be compared mathematically by means of an algorithm. Numerical taxonomy also is called *phenetics.*

OMNIVORE: An organism that eats both plants and other animals.

ORDER: The middle of the seven obligatory ranks in taxonomy, more specific than *class* but more general than *family.*

PENTADACTYL LIMB: An appendage with five digits, like the human arm and hand. This appendage is common to all mammals, though it may take very different forms—for example, the dolphin's flipper.

PHENETICS: Another name for *numerical taxonomy.*

PHYLOGENY: The evolutionary history of organisms, particularly as that history refers to the relationships between lifeforms, and the broad lines of descent that unite them.

PHYLUM: The second most general of the obligatory taxonomic classification ranks, after *kingdom* and before *class.*

SPECIATION: The divergence of evolutionary lineages and creation of new species. See *allopatric speciation* and *sympatric speciation.*

SPECIES: The most specific of the seven obligatory ranks in taxonomy. Species often are defined as a population of individual organisms capable of mating with one another and producing fertile offspring in a natural setting. Also, members of the same species share a gene pool.

SYMPATRIC SPECIATION: A type of speciation that occurs when a group of individuals becomes reproductively isolated from the larger population of the original species. This type of speciation typically results from mutation.

TAXON: A taxonomic group or entity.

TAXONOMY: The area of the biological sciences devoted to the identification, nomenclature, and classification of organisms according to apparent common characteristics.

VERTEBRATE: An animal with a spinal column.

bers of order Rodentia, are familiar to us as both pests and pets as well as aids to research through their use as test subjects in laboratories. They are also the most abundant of all mammalian orders: about one-fourth of all families, 35% of all genera, and 50% of all living species of mammal are rodents. The group consists of some 2,205 species, among them mice, rats, squirrels, beavers, gophers, and porcupines. The name *rodent* comes from the Latin *rodere,* meaning "to gnaw," and, indeed, the defining characteristic of rodents is their chisel-like upper front teeth.

Whereas rats typically are despised creatures, mice (distinguished from rats simply because they are smaller) often are considered cute—that is, if the mouse in question is a pet or a laboratory mouse, rather than a pest chewing up the insulation or electrical wiring in someone's house. The fact that rodents are so often pests and pets arises in part from rodents' close association with humans. This is a distinction in itself, since few mammal orders manage to live successfully in such close proximity to humans. Not only do squirrels often live around human dwellings, but other species (for better or worse) often enter structures where humans live or work. Particularly notorious in this regard are black rats (*Rattus rattus*) and Norway rats (*R. norvegicus*), which are just two of some 500 rat species.

The two remaining orders of mammal also are composed of small, furry creatures. Lagomorphs, or members of the order Lagomorpha, are small mammals with large upper incisors (front teeth) but no canines or eyeteeth and with molars that lack roots. The 80-odd species of lagomorphs include rabbits, hares, and their lesser-known cousin the pika, or mouse-hare. The difference between rabbits and hares relates to their conditions at birth: rabbits are furless, blind, and helpless, whereas hares are furry, have open eyes, and are capable of hopping within minutes.

Finally, 28 species of elephant shrew, or jumping shrew, make up the order Macroscelidea, a collection of species known for their long, flexible, sensitive snouts. Some authorities group macroscelideans with order Insectivora, whereas others place them in another order, Mentophyla, with tree shrews. The latter often have been placed variously in orders Scandentia, Insectivora, or Primates, indicating that many areas of mammalian taxonomy remain in dispute.

WHERE TO LEARN MORE

Boxhorn, Joseph. "Observed Instances of Speciation." Talk. Origins (Web site). <http://www.talkorigins.org/faqs/faq-speciation.html>.

Kirby, Alex. "'New' Camel Lives on Salty Water." British Broadcasting Corporation (Web site). <http://news.bbc.co.uk/hi/english/sci/tech/newsid_1156000/1156212.stm>.

"Mammalia." *Animal Diversity Web,* The University of Michigan Museum of Zoology http://animaldiversity.ummz.umich.edu/chordata/mammalia.html>.

Mammal Species of the World (MSW). Smithsonian National Museum of Natural History, Department of Systematic Biology—Vertebrate Zoology (Web site). <http://www.nmnh.si.edu/msw/>.

Marks, Jonathan. *Human Biodiversity: Genes, Race, and History.* New York: Aldine de Gruyter, 1995.

Norton, Bryan G. *The Preservation of Species: The Value of Biological Diversity.* Princeton, NJ: Princeton University Press, 1986.

Patent, Dorothy Hinshaw. *The Challenge of Extinction.* Hillside, NJ: Enslow Publishers, 1991.

———. *Biodiversity.* Illus. William Muñoz. New York: Clarion Books, 1996.

Schilthuizen, Menno. *Frogs, Flies, and Dandelions: Speciation—The Evolution of New Species.* New York: Oxford University Press, 2001.

Speciation (Web site). <http://www.ultranet.com/~jkimball/BiologyPages/S/Speciation.html>.

UCMP Hall of Mammals, University of California, Berkeley, Museum of Paleontology (Web site). <http://www.ucmp.berkeley.edu/mammal/mammal.html>.

DISEASE

DISEASE
NONINFECTIOUS DISEASES
INFECTIOUS DISEASES

DISEASE

CONCEPT

Disease is a term for any condition that impairs the normal functioning of an organism or body. Although plants and animals also contract diseases, by far the most significant disease-related areas of interest are those conditions that afflict human beings. They can be divided into three categories: intrinsic, or coming from within the body; extrinsic, or emerging from outside it; and of unknown origin. Until the twentieth century brought changes in the living standards and health care of industrialized societies, extrinsic diseases were the greater threat; today, however, diseases of intrinsic origin are much more familiar. Among them are stress-related diseases, autoimmune disorders, cancers, hereditary diseases, glandular conditions, and conditions resulting from malnutrition. There are also illnesses, such as Alzheimer's disease, whose causes remain essentially unknown.

HOW IT WORKS

CLASSIFYING DISEASES

Any condition that impairs the normal functioning of an organism can be called a disease. In the human organism, as in all others, there are certain basic requirements, which in the human body include the need for a certain proper amount of oxygen, acidity, salinity (salt content), nutrients, and so on. These conditions must all be maintained within a very narrow range, and any deviation can bring about disease.

Diseases can be classified into three general groups. There are conditions that are infectious, or extrinsic, meaning that they are caused by an infection through which a virus, bacterium, or other parasite enters the body. Infectious diseases, infections, and the immune system that usually protects us against them are discussed elsewhere in this book. Our attention in the present context will be devoted to the other two broad categories—noninfectious, or intrinsic, diseases and diseases of unknown origin.

CLASSIFYING INTRINSIC DISEASES. There are several basic varieties of intrinsic disease, or conditions that are neither contagious nor communicable. These varieties are listed in the next few paragraphs. The essay Noninfectious Diseases includes a discussion of other systems for classifying diseases of either the intrinsic or the extrinsic variety.

Hereditary diseases: diseases that are genetic, meaning that they are passed down from generation to generation. An example, discussed in Noninfectious Diseases, is hemophilia. Heredity is not a "cause," and some of the diseases of unknown origin may be transmitted from parent to offspring. Some forms of cancer are hereditary as well, as are other conditions discussed elsewhere in this book. (See Nonifectious Diseases, Mutation, and Heredity.)

Glandular diseases: Conditions involving a gland—that is, a cell or group of cells that filters material from the blood, processes that material, and secretes it either for use again in the body or to be eliminated as waste. Examples include diabetes mellitus, examined in Noninfectious Diseases, as well as various kidney and liver diseases, among them, hepatitis and jaundice. Goiter, a swelling in the neck area caused by a diet poor in iodine, is both a glandular and a dietary condition, a fact that illustrates the overlap between disease types.

Dietary diseases: These are all illnesses that relate to nutrient deficiencies—either an overall lack of adequate nutrition (i.e., malnutrition) or the absence of a key nutrient. Examples include pellagra, scurvy, and rickets, all of which are vitamin deficiencies, as well as kwashiorkor, which brings about a swollen belly and is caused by a lack of protein. Vitamin deficiencies are discussed in Vitamins, and kwashiorkor and other varieties of malnutrition are examined in Nutrients and Nutrition.

Cancers: Cancer is not just one disease but some 100 conditions. Its two main characteristics are uncontrolled growth of diseased cells in the human body and migration of the disease from the original site to distant sites within the body. If the spread is not controlled, cancer can result in death. (See Noninfectious Diseases for more.)

Stress-related diseases: Some heart conditions are hereditary or glandular, but quite a few diseases of the heart and circulatory system are exacerbated by stress. Examples include heart murmurs, hardening of the arteries, and varicose veins. We will examine heart disease and the general effects of stress shortly.

Autoimmune diseases: This is a particularly terrifying category of disease, because it involves a rejection of the body itself by the body's own immune system. Autoimmune diseases, examples of which include lupus and rheumatoid arthritis, are discussed in The Immune System.

DISEASES OF UNKNOWN ORIGIN. Finally, there are diseases for which there is no known cause. In some cases, it is possible that heredity, diet, or some other aspect of human existence has a role, but it is not certain. And even if, say, heredity plays a part, the exact hereditary factors are not established. In any case, many of the categories of disease we have listed do not amount to "causes," but rather are types of disease. Moreover, some diseases classifiable in one of the listed categories also belong in the ranks of the diseases with unknown causes. For instance, many autoimmune diseases are mysterious to scientists. Likewise, chronic fatigue syndrome, considered a disease of unknown origin, is obviously a stress-related disorder, while fibromyalgia, characterized by sore muscles and tissues, may be stress-related as well. Two brain diseases of unknown origin, Creutzfeldt-Jakob and Alzheimer's disease, are discussed near the conclusion of this essay.

REAL-LIFE APPLICATIONS

A CHANGING THREAT

At one time the diseases that posed the greatest threat to human survival were infectious ones, such as the Black Death (actually a combination of bubonic and pneumonic plague), which killed about a third of Europe's population during the period from 1347 to 1351. Plagues or epidemics, in fact, are among the persistent themes in history, punctuating the fall of empires and the rise of others.

A plague that struck the eastern Roman (Byzantine) Empire in the sixth century, for instance, brought an end to a plan by the great Justinian I to reconquer the Italian peninsula and restore Roman rule in western Europe. It also spelled the beginning of the end of Byzantine glory (though the empire hung on until 1453) and opened the way for the rise of Islam and Muslim influence over the Mediterranean. Thus, the course of history up to the present day, including the events of the European Middle Ages, the Crusades, and even the modern-day conflict between the West and Islamic terrorists, can be traced in part back to a plague in about A.D. 540.

Wherever people have gathered in large numbers, infectious diseases have arisen. Smallpox and chicken pox, cholera and malaria, diphtheria and scarlet fever, influenza and polio—these and many other diseases have threatened the very survival of whole populations, bringing about a collective death toll that dwarfs that of twentieth-century wars and genocide. Yet it was in the twentieth century—ironically, the era when humans discovered the capacity to kill themselves in truly frightening numbers through world wars, nuclear weaponry, and totalitarian social experiments—that the threat of infectious diseases began to recede.

Thanks to successful vaccination programs, many infectious diseases are largely a thing of the past. This is true even of smallpox, a scourge that effectively ended in 1978 thanks to a United Nations inoculation program, but which reemerged as a potential threat of biological ter-

ILLUSTRATION FROM THE FIFTEENTH-CENTURY TOGGENBERG BIBLE OF PEOPLE AFFLICTED WITH BUBONIC PLAGUE. AT ONE TIME SUCH INFECTIOUS DISEASES POSED THE GREATEST THREAT TO HUMAN SURVIVAL; IN FACT, PLAGUES AND EPIDEMICS ARE AMONG THE PERSISTENT THEMES IN HISTORY, PUNCTUATING THE FALL OF EMPIRES AND THE RISE OF OTHERS. *(© Bettmann/Corbis. Reproduced by permission.)*

rorism in the hands of political terrorists, such as the Islamic terrorist Osama bin Laden. It would be difficult for bin Laden's al-Qaeda organization to acquire samples of the virus, however; they are stored in only four or five laboratories worldwide (kept there for the purpose of making more vaccine if needed) and remain under heavy guard.

THE RISE OF INTRINSIC DISEASES. Rather than infectious diseases, the much greater threat today is in the form of intrinsic diseases, or ones that are neither communicable nor contagious. The leading causes of death in the United States are as follows:

- Heart disease
- Cancer
- Stroke
- Chronic obstructive lung diseases (e.g., emphysema)
- Accidents (motor vehicle or other)
- Pneumonia and influenza
- Diabetes mellitus
- Suicide
- Kidney disease
- Chronic liver disease and cirrhosis

Note that the one item on the list that is not an intrinsic disease and is not disease-related at all: accidents. After that is the first extrinsic disease entry on the list, number 6, pneumonia and the closely related condition influenza. Number 8, of course, is not related to disease—at least not physical disease. The high incidence of suicide, with 11.1 such deaths per 100,000 population, probably reflects the fact that the United States is an industrialized, wealthy nation. Ironically, people who are eking out a living, struggling for survival, are far less likely to end it all voluntarily.

Also reflective of America's high level of development is the overwhelming preponderance of intrinsic, noninfectious diseases on the list. Unquestionably, the greatest threat to human health today takes the form of noninfectious diseases, such as heart disease, cancer, and diseases of the circulatory system. This is true only in the industrialized world, however: whereas only about 25% of all patients who visit doctors in the United States do so because of infectious diseases, more than two-thirds of all deaths worldwide are caused by infectious diseases, such as malaria.

STRESS AND HEART DISEASE

Stress, simply put, is a condition of mental or physical tension brought about by internal or external pressures. Many events can cause stress: something as simple as taking a test or driving through rush-hour traffic or as traumatic as the death of a loved one or contracting a serious illness. Stress may be short-lived, as when facing a particular deadline, or it may be the ongoing, crippling stress related to a job that is slowly killing the victim.

People who experience severe traumas, such as soldiers in combat, may experience a condition called post-traumatic stress disorder (PTSD). This condition first came to public attention after World War I, a war that completely dwarfed all preceding conflicts in its intensity and brutality. Formerly bright-eyed, optimistic youths came home behaving like madmen or nervous wrecks, and soon the condition gained the nickname *shell shock.* (Actually, shell shock dated back more than 50 years, to what might be regarded as the first modern war in the West—the first "total war" involving relatively sophisticated weaponry and a fully engaged citizenry: America's Civil War, from which combatants returned home with a condition known as "soldier's heart.")

EFFECTS OF STRESS. Whereas PTSD has a distinct psychological dimension, in many stress-related diseases there is not as obvious a link between mental states and bodily disorders. Nonetheless, it is clear that stress kills. Some of the physical signs of stress are a dry mouth and throat, headaches, indigestion, tremors, muscle tics, insomnia, and a tightness of the muscles in the shoulders, neck, and back. Emotional signs of stress include tension, anxiety, and depression. During stress, heart rate quickens, blood pressure increases, and the body releases the hormone adrenaline, which speeds up the body's metabolism. Stress may disrupt homeostasis, an internal bodily system of checks and balances, leading to a weakening of immunity.

Diseases and conditions associated with stress include adult-onset diabetes (see Noninfectious Diseases), ulcers, high blood pressure, asthma, migraine headaches, cancer, and even the common cold. The last, of course, is an infectious illness, but because stress impairs the immune system, it can leave a person highly susceptible to infection. Furthermore, medical researchers have determined that long-term stress causes the accumulation of fat, starch, calcium, and other substances in the linings of the blood vessels. This condition ultimately results in heart disease.

HEART DISEASES. The human heart weighs just 10.5 oz. (300 g), but it contracts more than 100,000 times a day to drive blood through about 60,000 mi. (96,000 km) of vessels. An average heart will pump about 1,800 gal. (6,800 l) of blood each day. With exercise, that amount may increase as much as six times. In an average lifetime the heart will pump about 100 million gal. (380 million l) of blood. The heart is divided into four chambers: the two upper atria and the two lower ventricles. The wall that divides the right and left sides of the heart is the septum. Movement of blood between chambers and in and out of the heart is controlled by valves that allow transit in only one direction.

Given its importance to human life, it follows that heart disease is an extremely serious condition. Among the many illnesses that fall under the general heading of heart disease is congenital heart disease, a term for any defect in the heart that is present at birth. About one of every 100 infants is born with some sort of heart abnormality, the most common form being the atrial septal defect, in which an opening in the septum allows blood from the right and left atria to mix.

Coronary heart disease, also known as coronary artery disease, is the most common form of heart disease. A condition termed arteriosclerosis, in which there is a thickening of the artery walls, or a variety of arteriosclerosis known as atherosclerosis results when fatty material, such as cholesterol, accumulates on an artery wall. This forms plaque, which obstructs blood flow. When the obstruction occurs in one of the main arteries leading to the heart, the heart does not receive enough blood and oxygen, and its muscle cells begin to die.

CREUTZFELDT-JAKOB DISEASE

A particularly frightening category of unexplained diseases includes those that attack and destroy the brain. Among them are two conditions named after German scientists: the psychiatrists Alfons Maria Jakob (1884–1931) and Hans Gerhard Creutzfeldt (1885–1964) and the neurologist Alois Alzheimer (1864–1915). Creutzfeldt-Jakob dis-

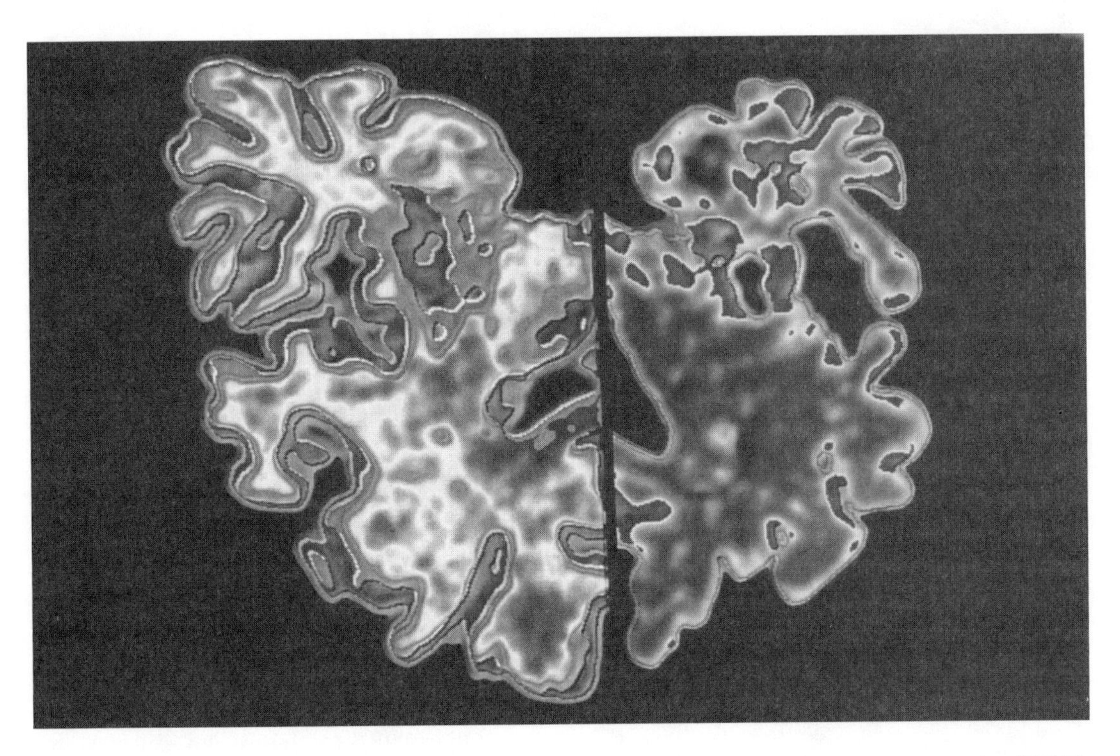

COMPUTER GRAPHIC OF THE BRAIN OF AN ALZHEIMER PATIENT (LEFT) COMPARED WITH A NORMAL BRAIN (RIGHT). ALZHEIMER'S DISEASE SHRINKS THE BRAIN, WHICH SHOWS SIGNS OF THE DEGENERATION OF NERVE CELLS, TANGLED PROTEIN FILAMENTS, AND LESIONS CAUSED BY ACCUMULATION OF BETA-AMYLOID PROTEIN. *(Photograph by Alfred Pasieka. Photo Researchers, Inc. Reproduced by permission.)*

ease, fortunately, is a rare condition. The disease, first described by the two doctors in the 1920s, initially shows itself with the loss of memory, and within a few weeks it progresses to visual problems, loss of coordination, and seizure-like muscular jerking. Death usually follows within a year.

It appears that Creutzfeldt-Jakob disease ensues when a certain protein in the brain, known as prion protein, changes into an abnormal form. As to what causes that change, scientists remain in the dark. The disease attacks about one of a million people worldwide, and victims are typically about 50-75 years of age. During the 1990s something strange happened: the disease began affecting relatively large numbers of young people in the United Kingdom. A 1996 report of British medical experts, however, linked the surge in Creutzfeldt-Jakob cases to what might be considered a dietary condition: bovine spongiform encephalopathy, or mad cow disease, contracted from eating cattle with a form of prion disease. The only way to contract such a condition, however, is by eating the brain or spinal cord of an affected cow, something that could only happen in the case of hamburger or sausage, in which one does not always know what one is getting. The cows themselves got the disease from eating feed tainted with by-products of other cows, and as a result of the outbreak, Great Britain issued wide-ranging controls prohibiting the production of feed containing any materials from cows. (These particular feed-production practices were never common in the United States.)

ALZHEIMER'S DISEASE

Whereas Creutzfeldt-Jakob disease is a little-known condition, Alzheimer's disease is all too familiar to the families of the more than four million sufferers in America today. Note the reference to the families rather than the victims themselves: one of the most devastating aspects of Alzheimer's disease is the patient's progressive loss of contact with reality, such that a patient in an advanced stage does not even know that he or she has the disease. A progressive brain disease that brings about mental deterioration, Alzheimer's disease is signaled by symptoms that include increasingly poor memory, personality changes, and a loss of concentration and judgment. Although most victims are older 65 years, Alzheimer's is not a normal result of aging. Up until the 1970s people assumed that physical and mental decline were normal and unavoidable

KEY TERMS

DISEASE: A general term for any condition that impairs the normal functioning of an organism.

EXTRINSIC: A term for a disease that is communicable or contagious and comes from outside the body. Compare with *intrinsic.*

GLAND: A cell or group of cells that filters material from the blood, processes that material, and secretes it either for use again in the body or to be eliminated as waste.

INTRINSIC: A term for a disease that is not communicable or contagious and comes from inside the body. Compare with *extrinsic.*

features of old age and dismissed such cases of deterioration as "senility." Yet as early as 1906, Alzheimer himself discovered evidence that pointed in a different direction.

In that year Alzheimer was studying a 51-year-old woman whose personality and mental abilities were obviously deteriorating. She forgot things, became paranoid, acted strangely, and just over four years after he began working with her, she died. Following an autopsy, Alzheimer examined sections of her brain under a microscope and noted deposits of an unusual substance in her cerebral cortex—the outer, wrinkled layer of the brain, where many of the higher brain functions, such as memory, speech, and thought, originate. The substance Alzheimer saw under the microscope is now known to be a protein called beta-amyloid. About 75 years later scientists and physicians began to recognize a strong link between "senility" and the condition Alzheimer had identified. Since then, the public has become more aware of the disease, especially since Alzheimer's disease has stricken such well-known figures as the former president Ronald Reagan (1911–) and the actress Rita Hayworth (1918–1987).

THE IMPACT OF ALZHEIMER'S DISEASE. A slight decline in short-term memory (as opposed to long-term memories of childhood and the like) is typical even in healthy elderly adults, but the memory loss seen in Alzheimer's disease is much more severe. As years pass, memory loss becomes greater, and personality and behavioral changes occur. Later symptoms include disorientation, confusion, speech impairment, restlessness, irritability, and the inability to care for oneself. Although victims may remain physically healthy for years, the progressive decline of their mental faculties is ultimately fatal: eventually, the brain loses the ability to control basic physical functions, such as swallowing. Persons with Alzheimer's disease typically live between five and ten years after diagnosis, although improvements in health care in recent years have enabled some victims to survive for 15 years or even longer.

Improvements in health care also may help explain the fact that the numbers of Alzheimer victims are growing. Medical discoveries of the twentieth century served to prolong life greatly, such that there are far more people alive today who are 65 years of age or older than there were in 1900. More accurate reporting no doubt plays a part as well. Whereas about 2.5 million cases were reported throughout the 1970s, by the end of the twentieth century there were some four million living Alzheimer victims, and by the mid-twenty-first century that number is expected to climb to the range of 13 million if physicians do not find a cure. Meanwhile, Alzheimer's causes the deaths of more than 100,000 American adults each year and costs $80–90 billion annually in health-care expenses.

UNDERSTANDING ALZHEIMER. It is not a simple procedure to diagnose Alzheimer's disease, and despite all the medical progress since the time of Alois Alzheimer, the "best" method for determining whether someone has the condition is hardly a good one. The only possible physical procedure for definitively diagnosing Alzheimer's disease is to open the skull and remove a sample of brain tissue for microscopic examination. This is rarely done, of course, because brain surgery is far too drastic a procedure for simply obtaining a sample of tissue.

The immediate cause of Alzheimer's is the death of brain cells and a decrease in the connections between those cells that survive. But what causes *that?* Many scientists today believe that the presence of beta-amyloid protein is a cause in itself, while others maintain that the appearance of the protein is simply a response to some other, still unknown phenomenon. Researchers have found that a small percentage of Alzheimer cases apparently are induced by genetic mutations, but most cases result from unknown factors. Various risk factors have been identified, but they are not the same as causes; rather, a risk factor simply means that if a person has *x,* he or she is more likely to have *y.* Risk factors for Alzheimer's include exposure to toxins, head trauma (former president Reagan suffered a serious head injury before the onset of Alzheimer's disease), Down syndrome (a genetic disorder that causes mental retardation), age, and even gender (women are more likely than men to suffer from Alzheimer's disease).

Familial Alzheimer's disease, an inherited form, accounts for about 10% of cases. Approximately 100 families in the world are known to have rare genetic mutations that are linked with early onset of symptoms, and some of these families have an aggressive form of the disease in which symptoms appear before age 40. The remaining 90% of cases may be caused by various combinations of genetic and as yet undefined environmental factors.

WHERE TO LEARN MORE

Centers for Disease Control and Prevention (Web site). <http://www.cdc.gov/>.

DeSalle, Rob. *Epidemic!: The World of Infectious Disease.* New York: New Press, 1999.

Diseases, Disorders and Related Topics. Karolinska Institutet/Sweden (Web site). <http://www.mic.ki.se/Diseases/>.

Environmental Diseases from A to Z. National Institute of Environmental Health Sciences/National Institutes of Health (Web site). <http://www.niehs.nih.gov/external/a2z/home.htm>.

Epidemiology. University of Minnesota, Crookston (Web site). <http://sunny.crk.umn.edu/courses/biolknut/1020/micro3>.

Ewald, Paul W. *Plague Time: How Stealth Infections Cause Cancers, Heart Disease, and Other Deadly Ailments.* New York: Free Press, 2000.

Garrett, Laurie. *The Coming Plague: Newly Emerging Diseases in a World out of Balance.* New York: Farrar, Straus, Giroux, 1994.

Moore, Pete. *Killer Germs: Rogue Diseases of the Twenty-First Century.* London: Carlton Books, 2001.

National Cancer Institute, National Institutes of Health (Web site). <http://www.nci.nih.gov/>.

Oncolink: University of Pennsylvania Cancer Center (Web site). <http://oncolink.upenn.edu/>.

Oldstone, Michael B. A. *Viruses, Plagues, and History.* New York: Oxford University Press, 1998.

"Plant and Animal Bacteria Diseases." University of Texas Institute for Cellular and Molecular Biology (Web site). <http://biotech.icmb.utexas.edu/pages/science/bacteria.html#disease>.

World Health Organization (Web site). <http://www.who.int/home-page/>.

NONINFECTIOUS DISEASES

CONCEPT

In contrast to infectious, or extrinsic, diseases, noninfectious, or intrinsic, conditions are neither contagious nor communicable. They arise from inside the body as a result of hereditary conditions or other causes, such as dietary deficiencies. Although infectious forms historically have been the most life-threatening varieties of disease and remain so even today in much of the world, noninfectious disease is a far more serious concern in industrialized nations, such as the United States. Some categories of intrinsic diseases include stress-related, dietary, and autoimmune conditions as well as diseases of unknown origin. Additionally, there are several other categories we examine in this essay: hereditary diseases, such as hemophilia; glandular conditions, of which diabetes mellitus is a powerful example; and cancer, of which there are approximately 100 different varieties.

HOW IT WORKS

THE THREAT OF NONINFECTIOUS DISEASES

The world has long suffered under the threat of infectious diseases, some of which include smallpox, chicken pox, cholera, malaria, diphtheria, scarlet fever, influenza, polio, pneumonia, and even the common cold. Except for the last one, which is rarely fatal, such conditions have racked up a considerable death toll. For centuries people attributed these diseases to all manner of false causes, ranging from divine curses to an imbalance of bodily fluids. Only with the development of the microscope in the 1600s did scientists begin to identify the real cause behind most infectious diseases: bacteria, viruses, and other parasites. Infectious diseases, discussed elsewhere in this book (see Infection, Infectious Diseases, and Parasites and Parasitology), continue to pose a threat to life in underdeveloped nations. In North America, Europe, the emerging capitalist democracies of eastern Asia, and a handful of other materially and technologically advanced lands, however, infectious diseases have taken a back seat to noninfectious ones as well as to other threats.

CAUSES OF DEATH IN AMERICA. In the United States today, the four leading causes of death (see list in Disease) are noninfectious conditions: heart disease, cancer, stroke, and chronic obstructive lung diseases. The first of these conditions is discussed in Disease, and the second is examined later in this essay. Stroke, which occurs when a clot obstructs the flow of blood to the brain, may be classified along with heart disease as a stress-related ailment of the circulatory system. Chronic obstructive lung diseases include emphysema, cystic fibrosis, and other conditions with widely differing causes that nonetheless all affect the same organ in the same way.

The remainder of the list includes two non-disease-related conditions (accidents at no. 5 and suicide at no. 8), three more noninfectious diseases (diabetes mellitus at no. 7, kidney disease at no. 9, and chronic liver disease and cirrhosis at no. 10); and just one infectious-disease-related set of causes. The latter, no. 6, consists of two diseases, pneumonia and influenza, which may be the result of another type of infection, the only infectious condition that poses a serious threat in the Western world today: AIDS, or acquired immunodeficiency syndrome.

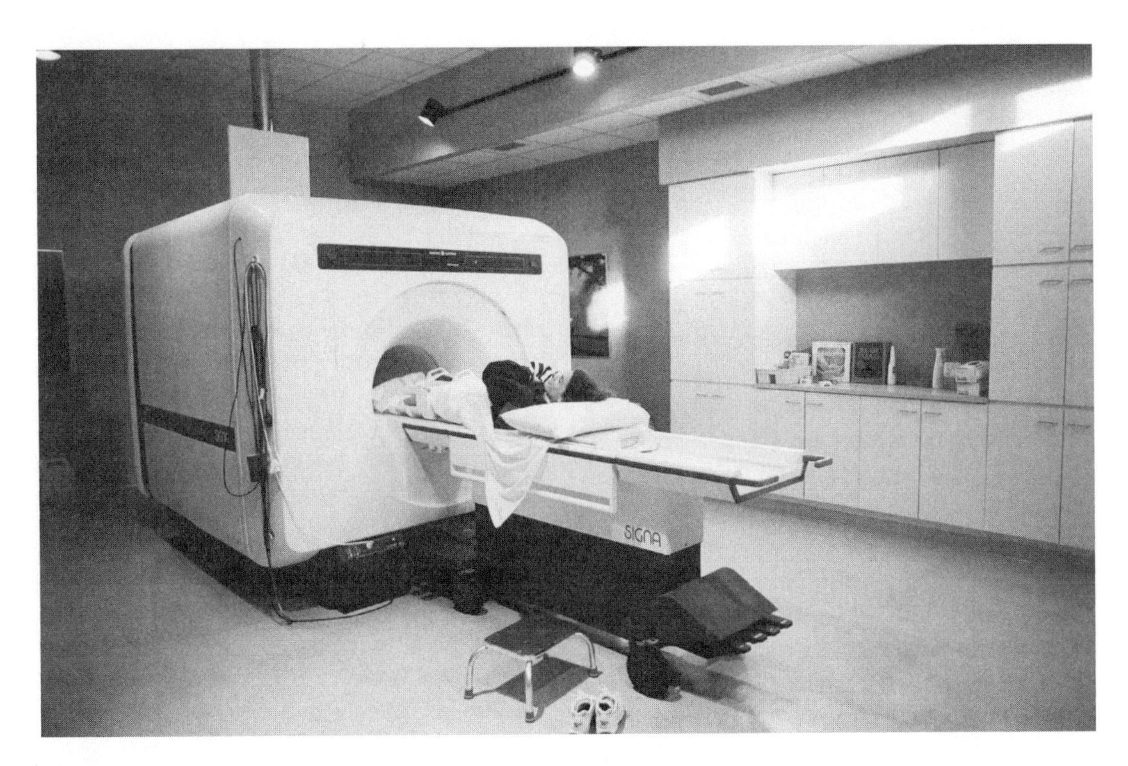

MAGNETIC RESONANCE IMAGING IS ONE TYPE OF DIAGNOSTIC TEST USED TO LOCATE TUMORS. *(© Neal Preston/Corbis. Reproduced by permission.)*

VARIETIES OF DISEASE

The essay Disease classifies noninfectious diseases as follows: hereditary or genetic diseases (e.g., hemophilia, discussed later in this essay); glandular diseases, or conditions involving a group of cells that filters material from the blood (e.g., diabetes mellitus, also included in the present essay); dietary diseases (see the essays Vitamins and Nutrients and Nutrition); cancers (discussed here); stress-related diseases (see Disease); autoimmune diseases (see Immunity and Immunology); and diseases of unknown origin (see Immunity and Immunology).

This is one way of dividing up ailments, and it happens to be the method applied in the essays of this book that deal with diseases. This method has the advantage of illustrating the wide variation in noninfectious diseases, but it would not necessarily be the best model to use for an in-depth professional study of disease. Scientists who study illnesses typically use one of several methods of classification that, while less broadly based than the one used here—and perhaps less interesting as well—are more efficient, because they group all diseases according to the same characteristics.

SOME OTHER CLASSIFICATION SYSTEMS. Among the systems doctors and scientists use for grouping diseases are topographic; anatomic; physiological; pathologic; etiologic, or causal; and epidemiological classifications. Topographic classification refers to bodily region or system: for instance, the circulatory system, the neurological system, and so on. The second method, anatomic (by organ or tissue), also uses parts of the body as a criterion for classification. The designation of heart and lung diseases, used earlier in discussing leading causes of mortality in the United States, is an example of this system.

Physiological classifications divide diseases in terms of function or effect (for example, metabolic disorders, some of which are discussed in Metabolism), while pathologic classifications separate diseases by the nature of the process that the disease takes: for example, inflammatory diseases. Etiologic, or causal, classifications are used most commonly in discussing infectious diseases, where broad types of causes can include viruses, bacteria, or other types of parasites. Likewise, epidemiological classifications usually refer to infectious diseases. Epidemiology is an area of the medical sciences devoted to the study of disease, including its incidence, distribution, and control within a population.

REAL-LIFE APPLICATIONS

CANCER

Although we are accustomed to hearing of "cancer" as though it were one disease, it is actually many diseases, close to 100 in number. Some of the most common varieties include skin, lung, and colon cancer, as well as breast cancer in women and prostate cancer in men. Blood and lymph node cancers, known as *leukemias* and *lymphomas,* respectively, are widespread, whereas cancer of the kidneys, ovaries, uterus, pancreas, bladder, and rectum are included among the cancers that most often affect Americans.

As this listing suggests, most cancers attack either body parts or systems and therefore are often classified anatomically or topographically; yet several characteristics unite these conditions. Cancer strikes the genes, which are carriers of genetic information that make up part of DNA (deoxyribonucleic acid), a molecule that appears in all cells. By gaining control at this level, the cancer is like a terrorist who has established a grip on all the communication or transportation systems in a country.

Many genes produce proteins that play a part in controlling the processes of cell growth and division. An alteration, or mutation, to the DNA molecule can disrupt the genes and produce faulty proteins, causing the cells to become abnormal and multiply. The abnormal cell begins to divide uncontrollably and eventually forms a new growth, known as a *tumor,* or neoplasm. In a healthy person, the immune system can recognize the neoplastic cells and destroy them before they have a chance to divide. Some mutant cells may escape immune detection, however, and survive to become tumors or cancers. (The immune system is discussed in Immunity and Immunology.)

Tumors can be either benign or malignant. A benign tumor is slow growing, does not spread or invade surrounding tissue, and, once removed, usually does not recur. A malignant tumor, on the other hand, invades surrounding tissue and spreads to other parts of the body. Therefore, even if the malignant tumor is removed, if the cancer cells have spread to the surrounding tissues, cancer will return. If the cancer cells are allowed to keep growing in number, migrating from the site of origin and spreading throughout the body, they eventually will kill the patient.

CAUSES AND TYPES. A majority of cancers are caused by changes in the cell's DNA because of damage from the environment. Environmental factors that are responsible for the initial mutation in DNA are called *carcinogens,* and there are many types, which we discuss shortly. Additionally, some cancers have a genetic basis: in other words, a person can inherit faulty DNA from his or her parents, which can predispose the patient to cancer of one kind or another. While there is scientific evidence that both factors (environmental and genetic) play a role, less than 10% of all cancers are purely hereditary.

There are several different types of cancers. In addition to leukemias and lymphomas, mentioned earlier, there are carcinomas, or cancers that arise in the epithelium (the layers of cells covering the body's surface and lining the internal organs and various glands). These types alone account for about 90% of all cancers. Some forms of skin cancer are melanomas, which typically originate in the pigment cells. Other forms of cancer include sarcomas (cancers of the supporting tissues of the body, e.g., bone, muscle, and blood vessels), and gliomas, or cancers of the nerve tissue.

DIAGNOSING CANCER. Many signs indicate the onset of cancer, among them, changes in the size, color, or shape of a wart or a mole; a sore that does not heal; or persistent cough, hoarseness, or sore throat. Many other diseases can produce similar symptoms, however, and for this reason it is important for a person to visit a doctor for regular checkups and diagnosis. Usually, diagnosis calls for fairly routine physical examination, though in the case of cancers of the reproductive organs, "routine" can still be plenty invasive.

Doctors examining women for cancers of the ovaries, uterus, cervix, and vagina must palpate the internal organs—that is, examine them by touch. For males, inspection of the rectum and the prostate is included in the physical examination. The doctor inserts a gloved finger into the rectum and rotates it slowly to feel for any growths, tumors, or other abnormalities. The doctor also palpates the testicles to identify any lumps, thickening, or differences in the size, weight, or firmness. Such examinations, as well as diagnoses for certain other types of cancer in private parts (namely, colon cancer), can be less than pleasant, but they are certainly preferable to an early and painful death.

If the patient has an abnormality that could be indicative of cancer, the doctor may order diagnostic tests. These tests may include laboratory studies of sputum or saliva, blood, urine, and stool (feces). To locate tumors, such imaging tests as computerized tomography (CT) scans, magnetic resonance imaging (MRI), ultrasound, or fiber-optic scope examinations may be used. The most definitive diagnostic test, however, is the biopsy, in which a piece of tissue is surgically removed for examination under a microscope. Besides confirming whether a patient has cancer, the biopsy also provides information about the type of cancer, the stage it has reached, the aggressiveness of the cancer, and the extent of its spread.

Screening examinations, conducted regularly by health care professionals, can result in the detection of cancers at an early stage. In addition, advances in molecular biology (an area of biology concerned with the physical and chemical basis of living matter) and cancer genetics have led to the development of several tests for assessing one's risk of developing cancers. These new techniques include genetic testing, in which molecular probes are used to identify mutations in certain genes that have been linked to particular cancers. At present, however, there are limitations to genetic testing, a fact that emphasizes the need for better strategies of early detection.

TREATING CANCER. Although there is as yet no cure for cancer, there are treatments designed to remove as much of the tumor or tumors as possible and to prevent the recurrence or spread of the cancer. Cancer treatment can take many different forms, including surgery, radiation, chemotherapy, immunotherapy, hormone therapy, and bone-marrow transplantation. Physicians recommend specific treatments based on the needs, condition, and illness of the particular patient.

Surgery, the most frequently used form of cancer treatment, involves the removal of the visible tumor. It is most effective when a cancer is small and confined to one area of the body. Radiation, which kills tumor cells by bombarding them with high-energy waves or particles, may be used alone in cases where a tumor is unsuitable for surgery. More often, however, it is used in conjunction with surgery and chemotherapy or with drugs to kill cancer cells. While it can be highly effective, chemotherapy may be physically difficult for the patient and may have side effects, including temporary hair loss.

Immunotherapy uses the body's own immune system to destroy cancer cells. The various immunological agents being tested include substances produced by the body as well as vaccines. Unlike traditional vaccines, cancer vaccines do not prevent cancer; instead, they are designed to train the immune system of the cancer patient's body to attack and destroy cancer cells. Hormone therapy is standard treatment for some types of cancers that are hormone-dependent and which grow faster in the presence of particular hormones. Among them are cancers of the prostate, breast, and uterus, and the therapy is designed to block the production or action of the hormones involved.

A particularly aggressive form of treatment is bone-marrow transplantation, which involves taking tissue from within a donor's bone cavities, where blood-forming cells are located, and transplanting it into the patient. In addition, cancer patients may use massage, reflexology, herbal remedies, and other forms of treatment dubbed "alternative," meaning that they usually are not recognized by the mainstream of the medical profession. (Lack of official recognition does not necessarily mean anything: many people have benefited from alternative cancer treatments.)

WHO IS AT RISK?

One of every four deaths in the United States is from cancer, and each year more than a million Americans are diagnosed with some form of the disease. Of these people, about half will die of the disease. Cancer can attack anyone, even children, though, fortunately, cases of cancer in very young patients are the exception. Most cases are seen in middle-aged or older adults. Although scientists are a long way from being able to predict who will get cancer (much less effectively prevent it), they have identified numerous risk factors. A risk factor is not necessarily a cause per se (though it may be); rather, if there seems to be a link between a particular behavior and a specific disease, that behavior is referred to as a *risk factor* for that disease. Major risk factors for cancer are tobacco use, alcohol consumption, diet, certain types of sexual and reproductive behavior, infectious agents, family history, occupation, environment, and pollution.

SPECIFIC RISK FACTORS. Approximately two-fifths of all cancer deaths annually are linked to tobacco use and excessive alcohol consumption. In addition to the relationship between smoking and lung cancer (80–90% of all lung cancer patients are smokers), smoking also has been shown to be a contributory factor to a whole host of other cancers. Moreover, scientists have established that second-hand smoke (or passive smoking) can increase one's risk of cancer. Nor is "smokeless tobacco" a safe alternative to cigarettes: snuff and chewing tobacco have been associated with countless cases of mouth cancer.

Excessive drinking is a risk factor in liver cancer and other illnesses, and the deadly combination of tobacco *and* excessive alcohol use significantly increases the chances that a person will contract mouth, pharynx, larynx, or esophageal cancer. Tobacco and alcohol are not the only cancer-related agents that people take into their bodies: about one-third of all cancer deaths annually are related to the things people eat. For example, immoderate intake of fat, leading to obesity, has been associated with cancers of the breast, colon, rectum, pancreas, prostate, gallbladder, ovaries, and uterus.

There are even varieties of cancer that are linked to contagious diseases. Since the mid-1970s, scientists have obtained evidence showing that approximately 15% of all cancer deaths worldwide can be traced to viruses, bacteria, or parasites. One such pathogen, the human papilloma virus, is sexually transmitted. Having too many sex partners and becoming sexually active too early have been shown to increase one's chances of contracting cancer of the cervix. (On the other hand, women who do not have children or those who have them late in life have a higher risk of both ovarian and breast cancer.)

Certain cancers, including those of the breast, colon, ovaries, and uterus, recur generation after generation in some families, and therefore family history and genetics cannot be ruled out as risk factors. In addition, less well-known cancers, such as the eye condition known as retinoblastoma, have been traced to certain genes that can be tracked within a family. Thus, it is possible that inheriting particular genes makes a person susceptible to certain cancers.

OCCUPATION, ENVIRONMENT, AND POLLUTION. Cancers arising from occupational hazards make up a particularly significant group. They account for 4% of all cancer deaths, and there is a certain poignancy in the fact that these cancer victims contracted their illness not by drinking or smoking or bad eating but simply by earning a living. Nowhere was this poignancy more evident than in some of the first documented cases of cancer arising from occupational causes.

In 1775 the English surgeon Percivall Pott (1714–1788) described the high incidence of cancer of the scrotum among former chimney sweeps, most of them men in their twenties. This was bad enough, but the fact that they had contracted the cancer much earlier hinted at a reality even more grim. In the harsh early days of the Industrial Revolution, before child labor laws had even been imagined, boys as young as four—orphans or children of desperately poor families—were put to work cleaning the insides of chimneys. The chimneys were so narrow that only children could clean them, and because of the tight fit, even small boys were unable to wear any clothes while doing their job. As a result, soot became embedded in their skin, and, since few people bathed, it stayed there, a silent killer whose effects became apparent only many years later.

In addition to helping make English society aware of the injustices put upon members of its lowest classes, Pott's research introduced the medical world to the concept of occupational health. Still, until the last third of the twentieth century, large numbers of industrial workers in the West labored at occupations with built-in cancer hazards. Thus, asbestos workers contracted lung cancer at a high rate, and a link became apparent between bladder cancer and dye and rubber industries. Likewise, connections were established between skin or lung cancer and the jobs of smelters, gold miners, and arsenic workers; between leukemia and glue and varnish workers' occupations; between liver cancer and the business of PVC (polyvinyl chloride) manufacturers; and between lung, bone, and bone marrow cancer and the work of radiologists and uranium miners.

Radiation itself is an environmental hazard that affects not only workers in those specialized industries just named but also anyone who has been exposed to radioactive materials. Fortunately, this is less of a hazard today, thanks to numerous bans on nuclear testing; nonetheless, radiation—including ultraviolet radiation from

PORTRAIT OF YOUNG CHIMNEY SWEEPERS FROM THE 1890S. BEFORE CHILD LABOR LAWS, BOYS AS YOUNG AS FOUR WERE PUT TO WORK CLEANING THE INSIDES OF CHIMNEYS; MANY LATER CONTRACTED CANCER OF THE SCROTUM FROM EXPOSURE TO SOOT. *(© Bettmann/Corbis. Reproduced by permission.)*

the Sun—causes 1–2% of all cancer deaths. Additionally, it has been estimated that 1% of cancer deaths are due to air, land, and water pollution, particularly as a result of chemical dumping in water supplies.

HEMOPHILIA

In contrast to the boys Pott treated, sufferers from hemophilia in the past often came from the highest echelons of society. A hereditary disease that primarily affects males, hemophilia was passed down through many royal lines, with the females acting as carriers and some of the males in the bloodline becoming victims of the disease. England's Queen Victoria (1819–1901) had several sons with hemophilia who died before they had the opportunity to become king. Her nephew, Russia's Czar Nicholas II, had a son who was hemophiliac as well. The boy's affliction caused his mother to seek the help of the charismatic "healer" Grigori Rasputin, whose close involvement with the royal family fed Russian discontent and helped contribute to the overthrow of the czar in 1917. (Ironically, the boy died not from hemophilia but perished, along with his family, before a Bolshevik firing squad.) Thus, once again (as noted in Disease), disease affected the course of history.

Its name taken from Greek and Latin words that together mean "love of bleeding," hemophilia

ALEXIS, SON OF CZAR NICHOLAS II OF RUSSIA, HAD HEMOPHILIA, A HEREDITARY DISEASE THAT IMPAIRS THE CLOTTING OF BLOOD. THE DISEASE PRIMARILY AFFECTS MALES AND HAS BEEN PASSED DOWN THROUGH MANY ROYAL BLOODLINES. *(© Bettmann/Corbis. Reproduced by permission.)*

is caused by a genetic defect that prevents the body from developing proteins needed to help the blood clot. What would be a minor bruise or scratch for an ordinary person is therefore a life-threatening situation to a hemophiliac, who runs a severe risk of bleeding to death from even the most minor cut. A hemophiliac must therefore live in a state of constant fearfulness: as a youngster, for instance, he cannot run and jump and get into mischief like other boys, for fear that he might skin his knee—a minor pain for most boys but a serious injury to him. For hemophiliacs today, there is good news and bad news. The good news is that the condition can be treated with transfusions containing the necessary proteins, and this has extended the life expectancy of some victims. The bad news, however, is that in this day and age the treatment itself might kill them: the transfusions originate from donated blood, which may contain the virus that causes AIDS, as well as pathogens linked to other diseases.

DIABETES MELLITUS

Much more prevalent than the genetic disorder hemophilia, or any one form of cancer, is diabetes mellitus, a serious condition caused by an absence of, or an insufficient amount of, insulin. Insulin, a hormone produced by the pancreas in varying amounts, helps maintain a normal concentration of glucose, or blood sugar (see Carbohydrates). Diabetes in general is a glandular disease, the gland in this case being the pancreas, but type 1 diabetes also is considered an autoimmune disorder because the body's own immune system destroys its insulin-producing cells. (See Immunity and Immunology and The Immune System for more about autoimmune disorders.)

Diabetes prevents the body from putting glucose to use, and instead large amounts of it are excreted in the urine. The word *diabetes* means "siphon," a reference to one of its major symptoms: frequent urination in an attempt to expel glucose. The urine itself is full of sugar; hence the term *mellitus,* meaning "honey." (In this vein, it is worth mentioning Chen Chuan, an eighth-century Chinese physician who surely qualifies as one of history's most dedicated scientists. He was the first to describe the sweetness of urine in patients suffering from diabetes, presumably as the result of firsthand research that went above and beyond the call of duty.)

EXTENT AND TYPES OF DIABETES. More than 12 million Americans, and some 100 million people worldwide, are affected by diabetes. That number is increasing by 5-6% annually, primarily the result of the population's increased longevity, combined with other factors such as increasing obesity and consumption of rich, processed, and carbohydrate-loaded foods (and drinks such as beer). Approximately 300,000 deaths each year are attributed to diabetes, which is of two principal varieties. Type 1, or insulin-dependent diabetes, is present at birth, is characterized by insulin deficiency, and normally is treated by taking insulin injections. Type 2, or non-insulin-dependent diabetes, arises not at birth, but somewhat later (though it can occur in childhood) and then among people who have normal insulin levels. (With type 2 diabetes, the problem is the body's inability to use its insulin efficiently.) Type 2, which typically stems from dietary causes—is preventable, but it is not treatable by insulin injections. Type 2 diabetes also may temporarily affect pregnant women, who may experience heightened glucose levels in a condition known as gestational diabetes. (There is another kind of diabetes altogether, a rare condition known as diabetes insipidus, which involves inadequate production of another hormone, vasopressin.)

EFFECTS OF DIABETES. One might wonder why problems with blood sugar could be so serious as to kill a million people every three years and to qualify diabetes as one of the leading killers in America. The reason is that the body depends on glucose as a source of immediate energy, and in the absence of usable glucose, it begins instead to use its fat cells. This rapid burning of fat produces a surplus of organic compounds known as ketones, and ketone accumulation brings about an accumulation of acids in the blood, a condition known as ketoacidosis. Severe ketoacidosis can cause nausea, vomiting, and a loss of consciousness, or diabetic coma. If the patient does not receive a shot of insulin, he or she can die. Thus, the symptoms displayed by Julia Roberts's diabetic character in her first major movie, *Steel Magnolias* (1989), were not overdone. Diabetes can also bring about other conditions, including blindness, kidney diseases, and long-term organ damage.

THE TRIUMPH OF BANTING AND BEST. As difficult as life is for diabetics today, it is infinitely better than it was before 1921. That was the year when the Canadians Frederick Banting (1891–1941), a surgeon, and Charles Herbert Best (1899–1978), a physiologist, isolated insulin. Thanks to their work, and the subsequent development of insulin therapy—typically using insulin harvested from cows or pigs—deaths from ketoacidosis and diabetic coma declined. A person with type 1 diabetes, formerly consigned to a dramatically shortened life of misery, could hope to have something approaching a normal existence. Today diabetes remains a life-shortening illness, and doctors and scientists continue to search for a cure, but ever since 1921, the lot of those with diabetes has been improving steadily. (Banting was later knighted and shared the 1923 Nobel Prize in physiology and medicine for his achievement. Best, however, in one of history's great snubs, received neither honor, because he had not earned his doctorate at the time of the discovery.)

KEY TERMS

CARCINOGEN: A substance or agent that induces the development of cancer.

DISEASE: A general term for any condition that impairs the normal functioning of an organism.

DNA: Deoxyribonucleic acid, a molecule in all cells, and many viruses, containing genetic codes for inheritance.

EXTRINSIC: A term for a disease that is communicable or contagious and comes from outside the body. Compare with *intrinsic.*

GENE: A unit of information about a particular heritable (capable of being inherited) trait that is passed from parent to offspring and stored in DNA molecules called *chromosomes.*

GENETICS: The study of hereditary traits passed down from one generation to the next through the genes.

GLAND: A cell or group of cells that filters material from the blood, processes that material, and secretes it either for use again in the body or to be eliminated as waste.

INTRINSIC: A term for a disease that is not communicable or contagious and comes from inside the body. Compare with *extrinsic.*

WHERE TO LEARN MORE

American Cancer Society (Web site). <http://www.cancer.org/>.

American Cancer Society's Guide to Complementary and Alternative Cancer Methods. Atlanta, GA: American Cancer Society, 2000.

American Institute for Cancer Research (Web site). <http://www.aicr.org>.

CancerCare (Web site). <http://www.cancercare.org/>.

Centers for Disease Control and Prevention (Web site). <http://www.cdc.gov/>.

Izenberg, Neil, ed. *Human Diseases and Conditions.* New York: Scribner, 2000.

Steen, R. Grant, and Joseph Mirro. *Childhood Cancer: A Handbook from St. Jude Children's Research Hospital with Contributions from St. Jude Clinicians and Scientists.* Cambridge, MA: Perseus Publishing, 2000.

INFECTIOUS DISEASES

CONCEPT

The history of the human species, it has been said, is the history of infectious disease. Over the centuries, humans have been exposed to a vast amount and array of contagious conditions, including the Black Death and other forms of plague, typhoid fever, cholera, malaria, influenza, and the acquired immunodeficiency syndrome, or AIDS. Only in the past few hundred years have scientists begun to have any sort of accurate idea concerning the origin of such diseases, through the action of microorganisms and other parasites. Such understanding has led to the development of vaccines and methods of inoculation, yet even before they made these great strides in medicine, humans had an unseen protector: their own immune systems.

HOW IT WORKS

INFECTION AND IMMUNITY

There are two basic types of disease: ones that are infectious, or extrinsic, meaning that they are contagious or communicable and can be spread by contact between people, and ones that are intrinsic, or not infectious. Diseases in general and noninfectious diseases in particular are discussed in essays devoted to those subjects. So, too, is infection itself, a subject separate from infectious diseases: a person can get an infection, such as tetanus or salmonella, without necessarily having a disease that can be passed on through contact with others in the same way that colds, malaria, or syphilis is spread.

The background on scientists' progressive understanding of the microorganisms that cause disease and the means of fighting these microorganisms are discussed in Infection. Among the leading figures in that history were the French chemist and microbiologist Louis Pasteur (1822–1895) and the German bacteriologist Robert Koch (1843–1910), who contributed greatly to what is known today as germ theory—the idea that infection and infectious diseases are brought about by microorganisms. In most cases, the organisms are too small to be seen with the naked eye. They include varieties of amoeba and worm, discussed in the essay Parasites and Parasitology, as well as viruses and some forms of bacteria and fungi, which together are known as pathogens, or disease-carrying parasites. Other terms related to infectious diseases, their agents, and the prevention and study of them are defined in the essay Infection.

IMMUNE MECHANISMS. The human body has numerous mechanisms for protecting itself from infectious disease, the first line of defense being the skin. Skin shields us all the time from unseen attackers and generally is able to prevent pathogens from entering the body; however, any break in the skin, such as a cut or scrape, provides an opening for microorganisms to invade the body. Germs that normally would be prevented from entering the body are able to invade the bloodstream through such openings. This is why it is so very important, in any situation involving potential contact with infection, to protect the skin. With the advent of AIDS, doctors and members of other professions who are likely to touch people carrying diseases—including officers arresting addicts or prostitutes—are much more likely to do their work wearing heavy plastic gloves.

Suppose that a microorganism makes it through the barrier of skin, thanks to a cut or other opening; if so, the body puts into action a second defensive mechanism, the immune system. This system is a network of organs, glands, and tissues that protects the body from foreign substances. Without a properly functioning immune system, a person could die simply by walking out the door in the morning and coming into contact with an airborne infectant. Even in relatively healthy people, the immune system may be unable to react adequately to an invasion of microorganisms. In such cases, disease develops.

TRANSMISSION OF DISEASES

Infectious diseases, by definition, are transmitted easily from one person to another. We have all been told, for instance, not to drink after someone who has a cold. On a much more serious level, persons who are sexually active or potentially sexually active, but not settled in a monogamous (one-partner) relationship, are advised to avoid unprotected sexual contact so as not to contract AIDS or some other sexually transmitted disease (STD). In these and many other cases, microorganisms travel from the carrier of the disease to the uninfected person. (Actually, in the case of AIDS, the pathogen is a virus, which is not, strictly speaking, an organism or even a living thing; however, viruses usually are lumped in with bacteria, amoeba, and some fungi as microorganisms.)

Pathogens can be spread by many methods other than direct contact, including through water, food, air, and bodily fluids—blood, semen, saliva, and so on. For instance, any time a person with an infection coughs or sneezes, they may be transmitting illness. This is how diseases such as measles and tuberculosis are passed from person to person. AIDS and various STDs, as well as many other conditions, such as hepatitis, are transferred when one person comes into contact with the bodily fluids of another. This is the case not only with sexual intercourse but also with blood transfusions and any number of other interactions, including possibly drinking after someone. (Contrary to rumors that circulated in the early 1980s, when AIDS first made itself known, that particular syndrome cannot be transferred by saliva, but the common cold and other viral infections can be.)

Cholera, caused by a bacterium found in dirty wells and rivers from India to England (in the 1800s, at least), is an example of a waterborne disease. Many foodborne pathogens tend to bring about what would be more commonly thought of as an illness than a disease, since in everyday language the latter term implies a long-term affliction, whereas food poisoning usually lasts for a week or so. (Still, some forms of food poisoning can be fatal.) Bacterial contamination may occur when food is not cooked thoroughly, is left unrefrigerated, is prepared by an infected food handler, or otherwise is handled in an unsanitary or improper fashion. (The case of Typhoid Mary, discussed near the conclusion of this essay, is an extreme example of this form of transmission.)

Additionally, diseases may be transferred by vectors—animals (usually insects) that carry microorganisms from one person to another. Vectors may spread a disease either by mechanical or by biological means. Mechanical transmission occurs, for example, when flies transfer the germs for typhoid fever from the feces (stool) of infected people to food eaten by healthy people. Biological transmission takes place when an insect bites a person and takes infected blood into its own system. Once inside the insect's gut, the disease-causing organisms may reproduce, increasing the number of parasites that can be transmitted to the next victim. This is how the *Anopheles* mosquito vector, for instance, transfers malaria.

REAL-LIFE APPLICATIONS

A TOUR OF DISEASES

The range of infectious diseases, from conditions that merely cause discomfort to those that bring about death, is truly staggering. Some have brought about vast epidemics that have wiped out huge populations, and many have changed the course of history, while others are hardly known to anyone outside the ranks of epidemiologists and the victims of the disease. Some, such as smallpox, have been eradicated or largely eradicated through inoculation campaigns, while others, most notably AIDS, continue to elude efforts to defeat them.

Diseases can be classified according to the systems or body parts affected. Some of those systems and parts, with examples of diseases relating to each, include the following.

- Upper respiratory tract: common cold, sinusitis, croup
- Lower respiratory tract: pneumonia, bronchitis
- Cardiovascular system: rheumatic fever
- Central nervous system: meningitis, encephalitis
- Genitourinary tract: sexually transmitted diseases (i.e., venereal diseases, such as syphilis, gonorrhea, and the herpes simplex viral infection)
- Gastrointestinal tract: cholera, salmonella, hepatitis
- Bones and joints: septic arthritis
- Skin: warts, candida
- Eyes: conjunctivitis (pink eye)

Another way to classify diseases is according to the types of organism that cause them: bacteria, viruses, or other forms of parasite, particularly worms, amoeba, and insects. The first two groups are discussed in further detail within Infection and the other varieties of parasite in Parasites and Parasitology.

Bacterial infections include anthrax, botulism, tetanus (lockjaw), leprosy, tuberculosis, diphtheria, whooping cough, plague, and a variety of pneumococcal, staphylococcal, and streptococcal illnesses. Among viral illnesses and diseases are the common cold, influenza, infectious mononucleosis, smallpox, chicken pox, measles, mumps, rubella (or German measles), yellow fever, poliomyelitis (i.e., polio), rabies, herpes simplex, and AIDS. Diseases related to other varieties of parasite include malaria, Rocky Mountain spotted fever, trichinosis, scabies, and river blindness. Nonmicroscopic parasites, particularly such worms as hookworm and pinworm, bring about disease-like forms of parasitic infestation within the body.

PLAGUES

From earliest times infectious diseases have wreaked havoc on the human species, and this was particularly so with the various plagues that struck Europe in ancient and medieval times. As noted in Infection, a plague in the fifth century B.C. helped bring an end to the golden age of Greek civilization. A thousand years later, another plague befell Greece, which by then dominated what remained of the Roman Empire. Based in Byzantium (Constantinople) this realm became known to history as the Byzantine (Eastern Roman), Empire, though its citizens saw themselves simply as "Romans" and thus as the inheritors of Roman civilization. Italy itself had fallen under the control of nomadic invaders, the Visigoths, but Emperor Justinian I (483–565) undertook a vast and costly campaign to wrest control of the Italian peninsula from the barbarians. Had he succeeded, the entire course of medieval history in Western Europe might have been different; he did not, largely because of a plague that swept Constantinople in 541.

Through a series of interconnected events, the plague permanently weakened Byzantium and left the Mediterranean world ripe for conquest by a new power: Islam. Both directly and indirectly, the plague of 541 served to divide Eastern and Western Europe. Not only was the Roman Empire never truly reunited, meaning that the two halves of the continent grew increasingly separate, but the rise of Islam made possible the Crusades (1095–1291). The latter sowed further discord between the East and the West, owing to the fact that Western European crusaders overran Byzantium and incited trouble between the Byzantines and Arabs. Ultimately, the split between Eastern and Western Europe, which became particularly pronounced during the years of Communism and the Iron Curtain (1945–1990), can be traced to the plague of 541.

THE BLACK DEATH

The Byzantine plagues (there were several, occurring at intervals of a few generations), killed millions of people, yet for sheer scope of destruction—and, perhaps, historical impact—they were dwarfed by the plague that devastated Europe in the years 1347–1351. This one became known as the Plague (with a capital *P*) or by another name that gave some hint of the terror that was as much a part of the epidemic as the ghastly physical symptoms it brought on: the Black Death.

It began in Asia and quickly made its way to the shores of the Black Sea, where it erupted in September 1346. The first outbreak in Western Europe occurred 13 months later, in October 1347, at the Sicilian port of Messina, from whence it was an easy jump to the Italian mainland. By the following April all of Italy was infected; meanwhile, the Plague had reached Paris in

A POLITICAL CARTOON FROM ABOUT 1870 ILLUSTRATES THE UNSANITARY CONDITIONS IN NEW YORK CITY, WITH THE POLITICIAN AND PUBLIC WORKS COMMISSIONER BOSS TWEED WELCOMING A CHOLERA EPIDEMIC. *(© Bettmann/Corbis. Reproduced by permission.)*

January 1348, and within a year, 800 people a day were dying in that city alone. Quickly it penetrated the entire European continent and beyond, from North Africa to Scandinavia and from England to the hinterlands of Russia. By 1351 it had spread so far and wide that sailors arriving in Greenland found its ports deserted.

The only merciful thing about the Black Death was that death came quickly. Victims typically died within four days—a hundred hours of agony. If they caught a strain of bubonic plague, their lymph glands swelled; if it was pneumonic plague, the lungs succumbed first. Either way, as the end approached, the victim turned purplish-black from respiratory failure—hence the name Black Death.

SOCIAL IMPACT OF THE PLAGUE. Lacking any modern concept of what causes disease, people looked for spiritual explanations. Some believed that the world was coming to an end, while others joined sects of flagellants, religious enthusiasts who wandered the countryside, beating themselves with lashes as a way of doing penance. The flagellants were tied closely tied to a rising trend toward anti-Semitism: searching for someone to blame, Europeans found a convenient scapegoat in the Jews, who, they claimed, had started the Plague by poisoning the wells of Europe.

The Black Death aptly illustrates how infectious diseases can have an impact on history in ways both big and small. In just five years the disease killed about 30% of Europe's population, which had been 100 million in 1300 but which would not reach that level again until 1500. All over the continent, farms were emptied and villages abandoned, leading to scarcity and higher prices. In the short run, these economic conditions spurred peasant revolts, but in the long run, the shortage of workers brought about higher wages and contributed to the emergence of the working and middle classes. Neither popes nor priests, neither kings nor noblemen, were any more equipped than the common people to confront the fearsome disease, and this, too, helped provoke the rise of competing classes and new centers of power in European society.

THE ETIOLOGY OF THE PLAGUE. The Black Death, in short, may be regarded as the beginning of the end of the Middle Ages—a hideously painful event that nevertheless carried positive consequences, which might hardly have been achieved without it. The irony was that the force at the center of all this devastation and change was too small to be seen by the naked eye. Although the disease was carried by rats, the cause of the Black Death was actually a bacillus known today as *Pastuerella pestis* or *Yersinia pestis,* which uses fleas as a vector. Modern medicines such as streptomycin, a variety of antibiotic developed after World War II, would have stopped the Plague, but such concepts were a long time in coming. Although the worst phase of the epidemic ended in 1351, it continued to spread, reaching Moscow by 1353; the next five centuries saw occasional outbreaks of the disease. As late as 1894 a strain of plague killed more than six million people in Asia over the course of 14 years.

THE CHANGING FACE OF DISEASE

The many biblical passages dealing with leprosy illustrate the role that infectious disease has played in human life from the earliest times. The fact that leprosy causes the victim's skin to turn ghostly white and brings about a gradual withering away of body parts must certainly have seemed like a curse from God. In fact, leprosy, also known as Hansen disease, is caused by the bacillus *Mycobacterium leprae,* and despite the many fears throughout the ages associated with touching lepers, it is not very contagious. A scene in the 1973 blockbuster *Papillon* illustrates this fact. The title character, a prison escapee played by Steve McQueen, takes a drag from a cigar offered to him by a leper, who then asks him if he knew that leprosy is not contagious. Papillon says no, indicating that he simply intended to build a sense of shared risk with someone who he hoped would aid his escape.

The example of leprosy shows something about the many curiosities involved in diseases and their study: for example, the fact that a disease can be infectious without being significantly contagious. Leprosy is by definition infectious, inasmuch as it is caused by a pathogen known as *Mycobacterium leprae,* but the latter is unusual for a number of reasons, including the fact that it is extremely slow in dividing, unlike most bacteria. After years of study, researchers are still not clear as to how leprosy is transmitted, and many believe that genetics may play a role. Thanks to increased understanding of the disease, the stigma that used to go with leprosy—including the reference to people with the disease as "lepers"—has largely been lifted. Yet places such as the leprosy facilities at Carville, Louisiana, and Molokai, Hawaii, continued to exist for many years, if only because the disfigurement associated with the disease influenced the separation of leprosy sufferers from the rest of society. In 1998, with only about 6,000 victims of the disease left in the entire country, the federal government closed the facilities at Carville and Molokai.

Leprosy remains a threat, with some two million cases of the disease worldwide, primarily in nations of Asia, Africa, and Latin America that are both underdeveloped and located in tropical zones. It has, however, ceased to be the worldwide danger that it once was, and as such it joins ranks with numerous other afflictions that formerly held all of humankind in the grip of terror. For example, tuberculosis, caused by a bacillus that attacks the lungs, afflicted a huge population in the nineteenth century, bringing an end to the careers of figures that ranged from the great English poet John Keats to the American gunslinger Doc Holliday. Holliday, in fact, traveled to Tombstone, Arizona, where he and Wyatt Earp participated in the infamous shootout at the O.K. Corral, because he thought the climate would help his condition. Their story has been portrayed in countless films; for example, in *Tombstone* (1993), Val Kilmer gives an extremely convincing

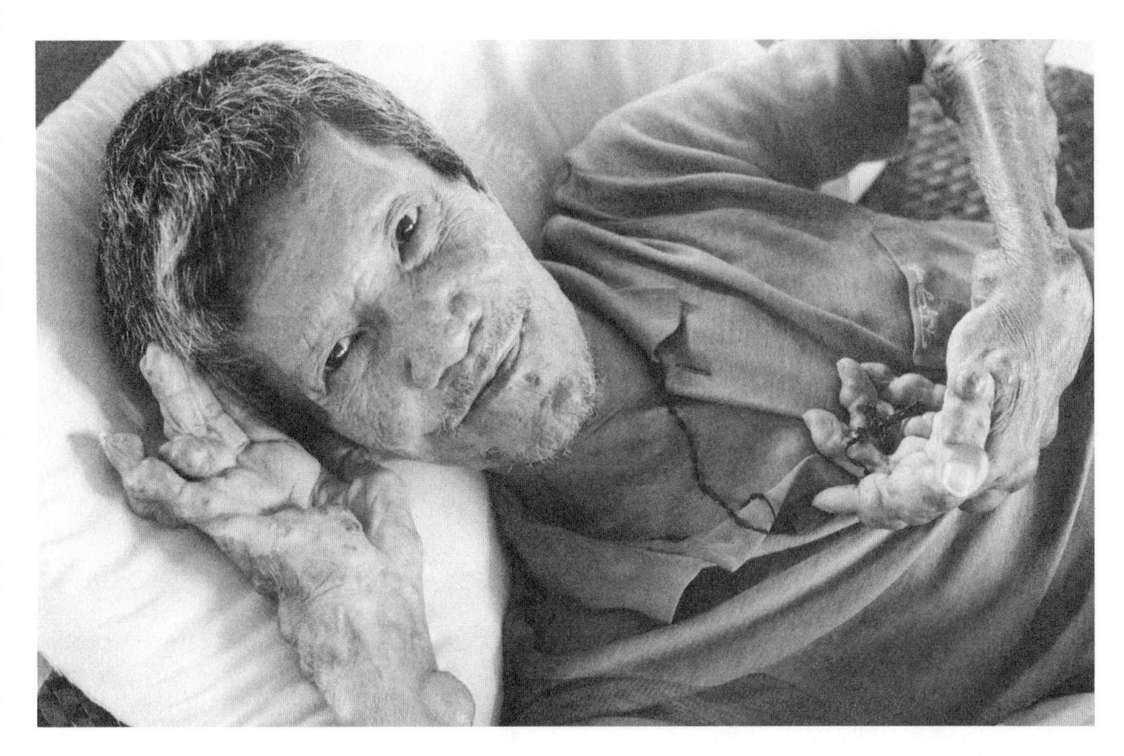

LEPROSY, CAUSED BY THE BACILLUS MYCOBACTERIUM LEPRAE, BRINGS ABOUT A GRADUAL WITHERING AWAY OF BODY PARTS. THERE ARE SOME TWO MILLION CASES OF THE DISEASE WORLDWIDE, PRIMARILY IN THE UNDERDEVELOPED NATIONS OF ASIA, AFRICA, AND LATIN AMERICA. *(© Paul A. Souders/Corbis. Reproduced by permission.)*

portrayal of the debilitating effects that Holliday's tuberculosis (aggravated by his lifestyle) must have had on him. Today, tuberculosis is not nearly the scourge that it once was, though it remains a problem, particularly because of patients' increasing resistance to the antibiotics used to treat it. (See Infection for more about antibiotics.)

VACCINATION AND CONTINUING THREATS. When Europeans invaded the lands of Native Americans, they brought with them a host of microorganisms to which they had developed an immunity but to which the Indians were completely vulnerable. Although Europeans and their descendants had developed immunities to various diseases, thanks to generations of exposure to pathogens, they and the rest of the world remained vulnerable to a host of contagious disease, including cholera, smallpox, chicken pox, measles, mumps, yellow fever, polio, malaria, and many others. Today, vaccines have virtually eradicated many of these contagious diseases and keep others at bay. (Anyone who has ever had a cholera vaccine, which causes the patient's body to become miserably sore, achy, and tender for about 48 hours, has some idea of just how awful the disease itself must be.) Polio, which once posed an enormous threat to American children and crippled one of America's greatest leaders, President Franklin D. Roosevelt, is an artifact of history, thanks to vaccines developed after World War II.

Yet some killers never really die. For instance, malaria, caused by a protozoan parasitic genus known as *Plasmodium* and spread by mosquito biological vectors, infects from 300 to 500 million people annually and kills up to 2.7 million people every year. Although the substance known as quinine showed some promise as a treatment during most of the nineteenth and twentieth centuries, Plasmodium has become increasingly resistant to it. In the search for a cure for what has been called "the most devastating disease in history," some 100,000 drugs have been tested.

SOME OTHER KILLERS

The twentieth century saw its own version of the Plague, in the form of the 1918–1920 influenza epidemic. Carried to all corners of the globe by soldiers returning from World War I, "the Influenza," as it came to be known (again with a capital letter to distinguish it as the greatest outbreak of a particular disease), killed 20 million

FOLLOWING THE SEPTEMBER 11, 2001, TERRORIST ATTACKS ON THE UNITED STATES, A SERIES OF LETTERS CONTAINING ANTHRAX SPORES SHOWED UP AROUND THE COUNTRY, AND EXPOSURE TO THE DISEASE LED TO A HANDFUL OF DEATHS. IN ONE INCIDENT HAZARDOUS MATERIALS EXPERTS WERE CALLED TO CAPITOL HILL TO INVESTIGATE. *(© AFP/Corbis. Reproduced by permission.)*

people—more than the war itself. Then there is the greatest epidemic of the latter part of the twentieth century and the early twenty-first century: AIDS. This disease is linked to the human immunodeficiency virus (HIV), a retrovirus (see Infection for an explanation of retrovirus) that causes a gradual breakdown of the victim's immune system.

People do not die of AIDS per se but of the illnesses—particularly pneumonia or Kaposi's sarcoma, a cancer of the tissues—to which AIDS makes them susceptible. The disease is transmitted primarily by sexual contact and intravenous drug use. A smaller number of particularly tragic cases result from no actions on the part of the victim, who in this case is either the recipient of infected blood or the child of a mother with AIDS. Since the disease first came to public attention in 1981, 21.8 million people worldwide (and about 750,000 in the United States) have died from it. The vast majority of deaths have been in sub-Saharan Africa, and 90% of all AIDS cases are in developing countries. Worldwide, approximately 36.1 million people have either HIV or AIDS. (For more about AIDS, see Immunity and Immunology.)

THE EBOLA VIRUS. AIDS was not the only infectious condition to come out of central Africa and terrorize the world in the late twentieth century. Beginning in about 1975, numerous viruses, previously unknown and terrifyingly lethal, emerged from tropical regions of Africa, South America, and Asia. So great was the rise of new infectious diseases that some epidemiologists believed this was tied with economic development: as humans cultivated previously undeveloped lands and delved into more isolated parts of the world, they might be exposing new viruses.

Few of these inspired as much terror as the Ebola virus, and the fear is understandable, given the effects of the disease. Three to nine days after the illness enters the body, the victim begins to experience fever and other flu-like symptoms, sudden exhaustion, sore throat, muscle pain, and headache. Vomiting and diarrhea soon follow, and the vomit and stools are black with blood. Soon hemorrhaging occurs, with blood flowing from the nose, ears, and even the eyes. Internal organs begin to liquefy, and within three weeks of contracting the virus, the victim is usually dead.

An almost unbelievably hideous condition, Ebola might seem at first glance a great deal like the Black Death. Why, then, has it not ravaged whole populations the way the Plague did? It is certainly not because scientists have a cure for Ebola; the best doctors can hope to do, if they detect the disease early enough, is to provide supportive care, such as blood transfusions, that may save the patient's life. Yet even the worst outbreaks of the disease have not occurred on anything like the scale of the Plague: the worst known outbreak of Ebola, in Uganda in 2000–2001, killed 425 people.

Part of the reason Ebola is not capable of spreading rapidly is, ironically, because it is such an efficient killer: it kills its human victims before they have a chance to spread it to many other victims. Other than nonfatal incidents in laboratories in the United States, England, and Italy, as well as one case in a monkey export facility in the Philippines—various primates are carriers—all Ebola cases and outbreaks have been in Africa, primarily in Zaire (now Democratic Republic of the Congo), Sudan, and Gabon. Many times, local conditions, situations, and practices have exacerbated the spread of the disease. For example, in 1996, a group of people in Gabon found a dead chimpanzee in the forest and ate it; as a result, 37 people died. The Uganda outbreak became much worse than it might have been because locals, lacking education as to antiseptic procedures, failed to take proper precautions. Many died as a result of attending funerals of Ebola victims at which bodies were not disposed of properly.

TYPHOID MARY. Sometimes a single person can be a walking epidemic, as in the case of the Irish cook Mary Mallon (1869–1938), better known as "Typhoid Mary." Mallon was an example of the fact that some people, because of genetic characteristics or other specifics, can act as carriers of a disease without ever contracting it themselves. Even though Typhoid Mary had *Salmonella typhosa* bacteria in her system, she did not get sick; still, she was highly contagious, and her profession as cook made her particularly dangerous. At least three deaths and 53 cases of typhoid fever were linked directly to her, with thousands of other probable cases of infection indirectly caused by this human vector.

Part of what made her so notorious—hence her nickname, given to her by the press—was the fact that Mallon did not seem to care how many people she infected. In the first decade of the twentieth century, authorities tracked her down as the cause of, or at least a contributing factor in, an outbreak of typhoid in the New York City area. Instead of cooperating with officials, Mallon repeatedly escaped before being caught and confined to Riverside Hospital on New York's North Brother Island in 1910. She served three years in isolation there before her release, after which she promptly went back to work as a cook—despite explicit orders not to do so. It was this (and an outbreak of typhoid fever at her place of work, which happened to be a hospital) that earned her the nickname by which she became known to history. She was caught again in 1915 and spent the remainder of her life on North Brother Island.

THE THREAT OF BIOLOGICAL WARFARE. Infinitely more despicable than Typhoid Mary are terrorists and rogue nations that would willingly unleash infectious disease on large, unsuspecting civilian populations. One such pathogen is *Bacillus anthracis*, the cause of anthrax, a deadly bacterial disease of cattle and other grazing animals. Under the right circumstances, anthrax can kill a human in about 36 hours, though a number of antibiotic treatments are effective in the early stages of the disease.

During the late twentieth century, the United States and Soviet Union experimented with

KEY TERMS

EXTRINSIC: A term for a disease that is communicable or contagious and comes from outside the body. Compare with *intrinsic*.

GERM THEORY: A theory in medicine, widely accepted today, that infections, contagious diseases, and other conditions are caused by the actions of microorganisms.

IMMUNE SYSTEM: A network of organs, glands, and tissues that protects the body from foreign substances.

IMMUNITY: The condition of being able to resist a specific disease, particularly through means that prevent the growth and development or counteract the effects of pathogens.

INFECTION: A state or condition in which parasitic organisms attach themselves to the body or to the inside of the body of another organism, causing contamination and disease in the host.

INTRINSIC: A term for a disease that is not communicable or contagious and comes from inside the body. Compare with *extrinsic*.

PATHOGEN: A disease-carrying parasite, usually a microorganism.

STD: Sexually transmitted disease.

VECTOR: An organism, such as an insect, that transmits a pathogen to the body of a host.

the use of anthrax in biological warfare, and an accidental release of anthrax spores at a Soviet lab in 1979 led to some 68 deaths. Following the September 11, 2001, terrorist attacks on the World Trade Center in New York City and on the Pentagon, a series of letters containing anthrax spores showed up around the United States, and exposure to the disease led to a handful of deaths. Although the attacks were linked initially to Osama bin Laden and his al-Qaeda organization, authorities increasingly began to suspect that home-grown terrorists were simply exploiting the September 11 attacks as cover for their own deeds.

Still, there was little doubt that bin Laden, the Iraqi dictator Saddam Hussein, or North Korea's ruling clique would use biological agents if the opportunity arose. One threat that loomed in the aftermath of September 11 was the possibility that bin Laden's followers would reintroduce the smallpox virus, which had been eradicated by worldwide vaccinations during the 1970s. The reason why smallpox could pose such a great threat is precisely that it has been eliminated, and few Americans born after 1973 have received vaccines. Unless they gained access to one of the two labs worldwide (one in the United States and one in Russia) where smallpox virus is stored for the purpose of making vaccines, however, terrorists would be unable to obtain a sample. (It is this matter of access that led authorities to suspect that the anthrax attacks were an "inside job.")

Another biological agent that poses a threat is *Clostridium botulinum,* which causes botulism, a toxic condition that can result in paralysis. Members of the fanatic Japanese cult Aum Shinrikyo attempted unsuccessfully to launch botulism attacks in Tokyo on three occasions in 1995. The Japanese government itself—that is, the Axis Japanese government of World War II—experimented with another biological agent, tularemia, or *Francisella tularensis.* The pathogen, which causes lung inflammation and death, is considered one of the most dangerous forms of biological weapon, because it is extremely efficient and easy to spread. America's military, borrowing an idea from its former enemy, developed its own *F. tularensis* strain in the late 1960s but destroyed its stockpile in 1973.

WHERE TO LEARN MORE

Centers for Disease Control and Prevention (Web site). <http://www.cdc.gov/>.

Cranmer, Hilarie. *Anthrax Infection. Emedicine.com* (Web site). <http://www.emedicine.com/emerg/topic864.htm>.

DeSalle, Rob. *Epidemic!: The World of Infectious Disease.* New York: New Press, 1999.

Everything You Need to Know About Diseases. Springhouse, PA: Springhouse Corporation, 1996.

Ewald, Paul W. *Plague Time: How Stealth Infections Cause Cancers, Heart Disease, and Other Deadly Ailments.* New York: Free Press, 2000.

Hoff, Brent H., Carter Smith, and Charles H. Calisher. *Mapping Epidemics: A Historical Atlas of Disease.* New York: Franklin Watts, 2000.

Infection and Immunity. University of Leicester Microbiology and Immunology (Web site). <http://www-micro.msb.le.ac.uk/MBChB/MBChB.html>.

Marr, Lisa. *Sexually Transmitted Diseases: A Physician Tells You What You Need to Know.* Baltimore, MD: Johns Hopkins University Press, 1998.

Oldstone, Michael B. A. *Viruses, Plagues, and History.* New York: Oxford University Press, 1998.

Shein, Lori. *AIDS.* San Diego: Lucent Books, 1998.

IMMUNITY

IMMUNITY AND IMMUNOLOGY

THE IMMUNE SYSTEM

IMMUNITY AND IMMUNOLOGY

CONCEPT

Immunity is the condition of being able to resist a specific disease, particularly through means that prevent the growth and development of disease-carrying organisms or counteract their effects. It is regulated by the immune system, a network of organs, glands, and tissues that protects the body from foreign substances. Immunology is the study of the immune system, immunity, and immune responses. Progress in immunology over the past two centuries has made inoculation—the prevention of a disease by the introduction to the body, in small quantities, of the virus or other microorganism that causes the disease—widely accepted and practiced. Despite such progress, however, some diseases evade human efforts to counteract them through medicine or other forms of treatment. This is particularly the case with a disease in which the immune system shuts down entirely: a condition known as acquired immunodeficiency syndrome, or AIDS.

HOW IT WORKS

IMMUNITY AND THE IMMUNE SYSTEM

The functioning of the immune system is considered in a separate essay, along with the means by which that system responds to foreign invasion. Also included in that essay is a discussion of allergies, which arise when the body responds to ordinary substances as though they were pathogens, or disease-carrying parasites. The body cannot know in advance what a pathogen will look like and how to fight it, so it creates millions and millions of different lymphocytes, a type of white blood cell. The principal types of lymphocyte are B cells and T cells. These cells recognize random antigens, or substances capable of requiring an immune response.

Certain researchers believe that while some B cells and T cells are directed toward fighting an infection, others remain in the bloodstream for months or even years, primed to respond to another invasion of the body. Such "memory" cells may be the basis for immunities that allow humans to survive such plagues as the Black Death of 1347–1351 (see Infectious Diseases). Other immunologists, however, maintain that trace amounts of a pathogen persist in the body and that their continued presence keeps the immune response strong over time.

IMMUNOLOGY

Immunology is the study of how the body responds to foreign substances and fights off infection and other disease-causing agents. Immunologists are concerned with the parts of the body that participate in this response, and this investigation takes them beyond looking merely at tissues and organs to studying specific types of cells or even molecules.

From ancient times, humans have recognized that some people survive epidemics, when the majority are dying. About 1,500 years ago in India, physicians even practiced a form of inoculation, as we discuss later. The modern science of immunology, however, had its beginnings only in 1798, when the English physician Edward Jenner (1749–1823) published a paper in which he maintained that people could be protected from the deadly disease smallpox by the prick of a needle dipped in the pus from a cowpox boil. (Cowpox is a related, less-lethal disease that, as its name suggests, primarily affects cattle.)

Later, the great French biologist and chemist Louis Pasteur (1822–1895) theorized that inoculation protects people against disease by exposing them to a version of the pathogen that is harmless enough not to kill them but sufficiently like the disease-causing organism that the immune system learns to fight it. Modern vaccines against such diseases as measles, polio, and chicken pox are based on this principle.

HUMORAL AND CELLULAR IMMUNITY. In the late nineteenth century, a scientific debate raged between the German physician Paul Ehrlich (1854–1915) and the Russian zoologist Élie Metchnikoff (1845–1916) concerning the means by which the body protects against diseases. Ehrlich and his followers maintained that proteins in the blood, called antibodies, eliminate pathogens by sticking to them. This phenomenon and the theory surrounding it became known as humoral immunity. Metchnikoff and his students, on the other hand, had noted that certain white blood cells could swallow and digest foreign materials. This cellular immunity, they claimed, was the real way that the body fights infection. In fact, as modern immunologists have shown, both the humoral and cellular responses identified by Ehrlich and Metchnikoff, respectively, play a role in fighting disease.

REAL-LIFE APPLICATIONS

Inoculation and Vaccines

Inoculation is the prevention of a disease by the introduction to the body, in small quantities, of the virus or other microorganism that causes that particular ailment. It is a brilliant idea, yet one that seems to go against common sense. For that reason, it was a long time in coming: not until the time of Jenner, in about 1800, did the concept of inoculation become widely accepted in the West. Nonetheless, it had been applied more than 13 centuries earlier in India.

In the period between about 500 B.C. and A.D. 500, Hindu physicians made extraordinary strides in a number of areas, pioneering such techniques as plastic surgery and the use of tourniquets to stop bleeding. Most impressive of all was their method of treating smallpox, which remained one of the world's most deadly diseases until its eradication in the late 1970s. Indian physicians apparently took pus or scabs from the sores of a mildly infected patient and rubbed the material into a small cut made in the skin of a healthy person. The Indians' method was risky, and there was always a chance that the patient would become deathly ill, but the idea survived and gradually made its way west over the ensuing centuries.

SMALLPOX VACCINATION. Smallpox, or variola, is carried by a virus that causes the victim's body to break out in erupting, pus-filled sores. Eventually, these sores dry up, leaving behind scars that may alter the appearance of the victim permanently, depending on the intensity of the disease. Such was the case with Lady Mary Wortley Montagu (1689–1762), a celebrated English writer and noblewoman. Known for her passionate relationships, romantic and otherwise, Lady Montagu had been scarred from youth by smallpox, and no doubt this experience gave her heightened concern for the victims of the disease. While she was in Turkey with her husband, Edward, an ambassador, she became aware of an inoculation method, probably based on the Hindu practice of many centuries before, used by local women. Lady Montagu arranged for her three-year-old son to be inoculated against smallpox in 1717, and after returning home, initiated smallpox inoculations in England.

Nonetheless, the problem remained that the inoculated person contracted a serious case of the disease and died, at least some of the time. More than 80 years later, in 1796, during a smallpox epidemic, Jenner decided to test a piece of folk wisdom to the effect that anyone who contracted cowpox became immune to human smallpox. He took cowpox fluid from the sores of a milkmaid named Sarah Nelmes and rubbed it into cuts on the arm of an eight-year-old boy, James Phipps, who promptly came down with a mild case of cowpox. Soon, however, James recovered, and six weeks later, when Jenner injected him with samples of the smallpox virus, the boy was unaffected.

Jenner, who published his findings after conducting additional tests, coined a new term for the type of inoculation he had used: *vaccination,* from the Latin word for cowpox, *vaccinia.* (The

THE MODERN SCIENCE OF IMMUNOLOGY HAD ITS BEGINNINGS IN 1798, WHEN THE ENGLISH PHYSICIAN EDWARD JENNER PUBLISHED A PAPER IN WHICH HE MAINTAINED THAT PEOPLE COULD BE PROTECTED FROM THE DEADLY DISEASE SMALLPOX BY THE PRICK OF A NEEDLE DIPPED IN THE PUS FROM A COWPOX BOIL. *(© Bettmann/Corbis. Reproduced by permission.)*

latter term comes from the Latin *vacca,* or "cow," the source of such terms as the French *vache.*) With the success of his vaccine, Jenner was awarded a sum of money to continue his work, and he soon oversaw the vaccination of thousands of English citizens, including the royal family. The practice spread to Germany and Russia and then to the United States. In Lady Montagu's time, the American clergyman Cotton Mather (1663–1728) had been an advocate of vaccination, and now President Thomas Jefferson (1743–1826) became an ardent proponent of Jenner's methods.

RABIES AND POLIO INOCULATION. The next advancement in the study of vaccines came almost 100 years after Jenner's discovery. In 1885 Pasteur saved the life of Joseph Meister, a nine-year-old boy who had been attacked by a rabid dog, by using a series of experimental rabies vaccinations. Pasteur's rabies vaccine, the first human vaccine created in a laboratory, was made from a version of the live virus that had been weakened by drying it over potash (sodium carbonate—burnt wood ashes).

Exactly 70 years later, the American microbiologist Jonas Salk (1914–1995) created a vaccine for poliomyelitis (more commonly known as polio), in which the skeletal muscles waste away and paralysis and often permanent disability and deformity ensue. Although polio had been known for ages, the first half of the twentieth century had seen an enormous epidemic in the United States.

The most famous victim of this scourge was the future president Franklin D. Roosevelt (1882–1945), who contracted it while on vacation in 1921. Throughout the 1930s and 1940s, polio remained a threat, especially to children; at the peak of the epidemic, in 1952, it killed some 3,000 Americans in one year, while 58,000 new cases were reported. At the same time, Salk was working on his vaccine, which finally was declared safe after massive testing on schoolchildren. In 1961 an oral polio vaccine developed by the Polish-born American virologist Albert Sabin (1906–1993) was licensed in the United States. Whereas the Salk vaccine contained the killed versions of the three types of poliovirus that had been identified in the 1940s, the Sabin vaccine used weakened live poliovirus. Because it was taken by mouth, the Sabin vaccine proved more convenient and less expensive to administer than the Salk vaccine, and it soon overtook the latter in popularity. By the early 1990s health organizations reported that polio was close to extinction in the Western Hemisphere.

TRIUMPHS AND CONTINUING CHALLENGES. Thanks to these and other vaccines, many life-threatening infectious diseases have been forced into retreat. In the United States, children starting kindergarten typically immunized against polio, diphtheria, tetanus, measles, and several other diseases. Other vaccinations are used only by people who are at risk of contracting a disease, are exposed to a disease, or are traveling to an area (usually in the Third World) where particular diseases are common. Such vaccinations include those for influenza, yellow fever, typhoid, cholera, and hepatitis A.

Internationally, 80% of the world's children had been inoculated as of 1990 for six of the primary infectious diseases: polio, whooping cough, measles, tetanus, diphtheria, and tuberculosis. Smallpox was no longer on the list, because efforts against it had proved overwhelmingly successful. (See Infectious Diseases for more on the threat, or nonthreat, of smallpox as a form of biological warfare.) Despite these successes, however, each year more than two million children who have not received any vaccinations die of infectious diseases. Even polio has continued to be a threat in some parts of the world: as many as 120,000 cases are reported around the world each year, most in developing regions. And as if the threat from age-old diseases were not enough, in the last quarter of the twentieth century a new killer entered the fray: AIDS.

AIDS

A viral disease that is almost invariably fatal, AIDS destroys the immune systems of its victims, leaving them vulnerable to a variety of illnesses. No cure has been found and no vaccine ever developed. The virus that causes AIDS has proved to be one of the most elusive pathogens in history, and so far the only effective way not to contract the disease is to avoid sharing bodily fluid with anyone who has it. This means not having sex without condoms (and, to be on the truly safe side, not having sex outside a committed, fully monogamous relationship) and not engaging in intravenous drug use. But there are some people who have contracted the AIDS virus through no actions or fault of their own: people who have received it in blood transfusions or, even worse, babies whose AIDS-infected mothers have passed the disease on to them.

Within two to four weeks of being infected with the virus that causes AIDS (HIV, human immunodeficiency virus), a patient will experience what at first seems like flu: high fever, headaches, sore throat, muscle and joint pains, nausea and vomiting, open ulcers in the mouth, swollen lymph nodes, and perhaps a rash. As the immune system begins to fight the invasion, some cells produce antibodies to neutralize the viruses that are floating free in the bloodstream. Killer T cells destroy many other cells infected with the AIDS virus, and the patient enters a phase of the disease in which no symptoms are evident.

Although at this point it seems as though the worst is over, in fact, the AIDS virus is at work on the immune system, quietly destroying the body's protection by infecting those T cells that would protect it. With an immune system that gradually becomes more and more unresponsive, the patient is made vulnerable to any number of infections. Normally, the body would be able to fight off these attacks with ease, but with the immune system itself no longer functioning properly, infectious diseases and cancers are free to take over. The result is a long period of increasing misery and suffering, sometimes accompanied by dementia or mental deterioration caused by the ravaging of the brain by disease. Whatever the course it takes, the end result of AIDS is always the same: not just death but a miserable, excruciatingly painful death.

BIRTH OF A KILLER. Believed to have originated in Africa, where the majority of AIDS cases still are found (see Infectious Diseases for statistics on AIDS), the disease first appeared in the United States in 1981. In that year two patients were diagnosed with an unusual form of pneumonia and with Kaposi's sarcoma, a type of cancer that previously had struck only people of Mediterranean origin aged 60 years and older. The appearance of that condition in younger persons of non-Mediterranean origin prompted an investigation by the United States Centers for Disease Control and Prevention (CDC).

Through the efforts of physicians both inside and outside the CDC, understanding of AIDS—the name and acronym appeared in 1982—gradually emerged. In 1983 scientists at the Pasteur Institute in Paris, as well as a separate team in the United States, identified the virus that causes AIDS, a pathogen that in 1986 was given the name human immunodeficiency virus (HIV). Further research showed that HIV, a retrovirus (see Infectious Diseases for an explanation of retrovirus), is subdivided into two types: HIV-1 and HIV-2. In people who have HIV-2, AIDS seems to take longer to develop; however, neither form of HIV carries with it a guarantee that a person will contract the disease. At first it was believed that if someone were HIV-positive, meaning that the person had the virus, it was a virtual death sentence. Therefore in 1991, when the basketball superstar Earvin "Magic" Johnson (1959–) announced that he was HIV-positive, it was an extremely melancholy event. Fans and admirers all over the world assumed that Johnson shortly would contract AIDS and begin to wither away in the process of suffering an exceedingly panful, dehumanizing death.

The fact that Johnson was alive and healthy more than ten years after the diagnosis of his infection with HIV serves to indicate that there is a great deal of difference between being HIV-positive and having AIDS. It also says much about people's emerging understanding of the disease and the virus that causes it. So, too, does Johnson's experience as he attempted, twice, to make a return to the court after retiring in the wake of his HIV announcement. Before examining his experiences, let us look at the social climate engendered by this politically volatile immunodeficiency syndrome.

CHANGING VIEWS ON AIDS. AIDS first was associated almost exclusively with the male homosexual community, which contracted the disease in large numbers. This had a great deal to do with the fact that male homosexuals were apt to have far more sexual partners than their heterosexual counterparts and because anal intercourse is more likely to involve bleeding and hence penetration of the skin shield that protects the body from infection. The association of AIDS with homosexuality led many who considered themselves part of the societal mainstream to dismiss AIDS as a "gay disease," and the fact that intravenous drug users also contracted the disease seemed only to confirm the prejudice that AIDS had nothing to do with heterosexual non-junkies. Some so-called Christian ministers even went so far as to assert, sometimes with no

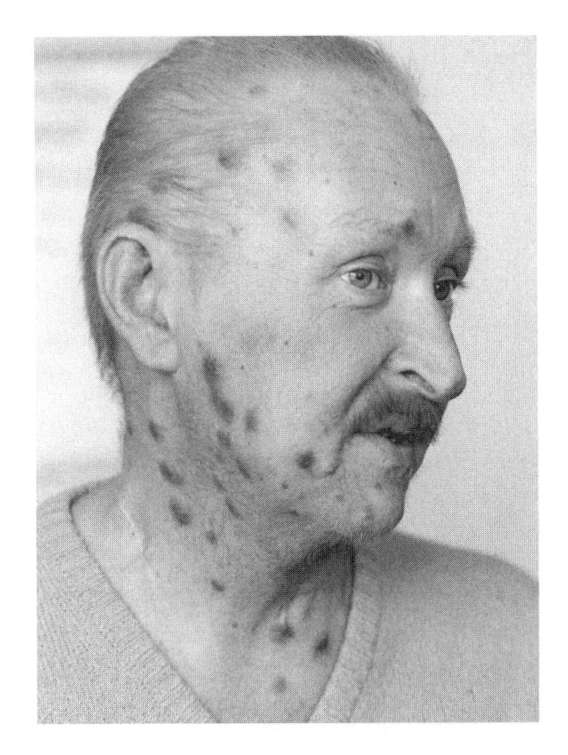

A viral disease that is almost invariably fatal, AIDS destroys the immune systems of its victims, leaving them vulnerable to a variety of illnesses, among them, Kaposi's sarcoma (shown here), a type of cancer that before the 1980s struck only people of Mediterranean origin aged 60 years and older. *(© Roger Ressmeyer/Corbis. Reproduced by*

small amount of satisfaction, that AIDS was God's punishment for homosexuality.

Then, during the mid-1980s, AIDS began spreading throughout much of society: to heterosexuals, hemophiliacs (see Noninfectious Diseases) and others who received blood, and even babies. The fact that AIDS could be transferred through heterosexual intercourse proved that it was not just a disease of homosexuals. Nor were all homosexuals necessarily susceptible to it. In fact, the safest of all sexual groups was homosexual women, who often tended toward monogamy and whose form of sexual contact was least invasive.

As AIDS spread throughout society, so did paranoia. Rumors circulated that a person could catch the disease from a mosquito bite or from *any* contact with the bodily fluids of another person—not just semen or blood but even sweat or saliva. People with AIDS began to acquire the status lepers once had held (see Infectious Diseases). By the mid-1990s views had changed considerably, and society as a whole had a much more realistic view of AIDS. This came about to

KEY TERMS

ALLERGY: A change in bodily reactivity to an antigen as a result of a first exposure. Allergies bring about an exaggerated reaction to substances or physical states that normally would have little significant effect on a healthy person.

ANTIBODIES: Proteins in the human immune system that help the body fight foreign invaders, especially pathogens and toxins.

ANTIGEN: A substance capable of stimulating an immune response or reaction.

APC: An antigen-presenting cell—a macrophage that has ingested a foreign cell and displays the antigen on its surface.

B CELL: A type of white blood cell that gives rise to antibodies. Also known as a *B lymphocyte.*

EPIDEMIC: Affecting or potentially affecting a large proportion of a population (*adj.*) or an epidemic disease (*n.*)

HUMORAL: Of or relating to the antibodies secreted by B cells that circulate in bodily fluids.

IMMUNE SYSTEM: A network of organs, glands, and tissues that protects the body from foreign substances.

IMMUNITY: The condition of being able to resist a particular disease, particularly through means that prevent the growth and development or counteract the effects of pathogens.

IMMUNOLOGY: The study of the immune system, immunity, and immune responses.

INOCULATION: The prevention of a disease by the introduction to the body, in small quantities, of the virus or other microorganism that causes the disease.

LYMPHOCYTE: A type of white blood cell, varieties of which include B cells and T cells, or B lymphocytes and T lymphocytes.

MACROPHAGE: A type of phagocytic cell derived from monocytes.

MONOCYTE: A type of white blood cell that phagocytizes (engulfs and digests) foreign microorganisms.

MONOGAMOUS: Having only one mate.

PATHOGEN: A disease-carrying parasite, usually a microorganism.

PHAGOCYTE: A cell that engulfs and digests another cell.

T CELL: A type of lymphocyte, also known as a T lymphocyte, that plays a key role in the immune response. T cells include cytotoxic T cells, which destroy virus-infected cells in the cell-mediated immune response; helper T cells, which are key participants in specific immune responses that bind to APCs, activating both the antibody and cell-mediated immune responses; and suppressor T cells, which deactivate T cells and B cells.

VACCINE: A preparation containing microorganisms, usually either weakened or dead, which are administered as a means of increasing immunity to the disease caused by those microorganisms.

some extent because of increased education and awareness—and in no small part because of Johnson, who was by far the most widely known and admired HIV-positive celebrity.

MAGIC COMES BACK. After playing on the United States "Dream Team" that trounced all opponents at the 1992 Summer Olympics in Barcelona, Spain, Johnson attempted a comeback with the Lakers the following year. Owing to fears on the part of many other players that they might contract AIDS by coming into close contact with him on the court, however, he decided again to retire. In December 1991 Johnson had established the Magic Johnson Foundation to promote AIDS awareness, and he devoted himself to this and other AIDS-related causes as well as to other ventures. Raising money for AIDS led him out onto the basketball court again in October 1995, when he and the American All Stars faced an Italian team in a benefit game, with an unsurprisingly lopsided score of 135–81.

Then, in February 1996, Johnson made his second attempted comeback with the Lakers. He ended up retiring again four months later, this time for good, but because he had chosen to and not because he had been forced to do so. Thanks in part to his AIDS education programs, in his second comeback Johnson discovered that players realized that they were not likely to catch the virus on the court. As the New Jersey Nets' player Jayson Williams told one reporter, "You've got a better chance of Ed McMahon knocking on your door with $1 million than you have of catching AIDS in a basketball game."

WHERE TO LEARN MORE

Aaseng, Nathan. *Autoimmune Diseases.* New York: Franklin Watts, 1995.

American Autoimmune Related Diseases Association, Inc. (AARDA) (Web site). <http://www.aarda.org/>.

Benjamini, Eli, and Sidney Leskowitz. *Immunology: A Short Course.* New York: Liss, 1988.

Clark, William R. *At War Within: The Double-Edged Sword of Immunity.* New York: Oxford University Press, 1995.

Dwyer, John M. *The Body at War: The Miracle of the Immune System.* New York: New American Library, 1989.

Edelson, Edward. *The Immune System.* New York: Chelsea House, 1989.

How Your Immune System Works. How Stuff Works (Web site). <http://www.howstuffworks.com/immune-system.htm>.

"Infection and Immunity." University of Leicester Microbiology and Immunology (Web site). <http://www-micro.msb.le.ac.uk/MBChB/MBChB.html>.

"The Lymphatic System and Immunity." Estrella Mountain Community College (Web site). <http://gened.emc.maricopa.edu/bio/bio181/BIOBK/BioBookIMMUN.html>.

"Magic Johnson Retires Again, Saying It's on His Own Terms This Time." *Jet*, June 3, 1996, p. 46.

UNAids: The Joint UN Programme on HIV/AIDS (Web site). <http://www.unaids.org/>.

THE IMMUNE SYSTEM

CONCEPT

The immune system is a network of organs, glands, and tissues that protects the body from foreign substances. These substances include bacteria, viruses, and other infection-causing parasites and pathogens. Usually, the immune system is extremely effective in performing its work of defending the body, but sometimes an error occurs in this highly complex system, and it can lead to terrible mistakes. The result can be an allergic reaction, which can be as simple as a case of the sniffles and as serious as a fatal condition. Or the error can manifest as an autoimmune disorder, such as lupus, in which the body rejects its own constituents as foreign invaders.

HOW IT WORKS

THE IMMUNE SYSTEM IN GENERAL

The human body is under near constant attack from pathogens, or disease-carrying parasites, of the type discussed in Infection, Infectious Diseases, and Parasites and Parasitology. No human would live very long without the immune system, which includes two levels or layers of protection, the nonspecific and the specific defenses. The nonspecific defenses, including the skin and mucous membranes, serve as a first defensive line for preventing pathogens from entering the body. The specific defenses are activated when these microorganisms get past the nonspecific defenses and invade the body.

For the immune system to work properly, two things must happen: first, the body must recognize that it has been invaded, either by pathogens or toxins or by some other outside threat. Second, the immune response must be activated quickly, before the invaders destroy many body tissue cells. For the immune system to respond effectively, several conditions must be in place, including the proper interaction of nonspecific and specific defenses. The nonspecific defenses on the skin do not identify the antigen (a substance capable of stimulating an immune response or reaction) that is attacking or potentially attacking the body; instead, these defenses simply react to the presence of what it identifies as something foreign. Often, the nonspecific defenses effectively destroy microorganisms, but if these defenses prove ineffective and the microorganisms manage to infect tissues, the specific defenses go into action. The specific defenses function by detecting the antigen in question and mounting a response that targets it for destruction.

THE MAJOR HISTOCOMPATIBILITY COMPLEX. How does the specific system "know" what is foreign and what is part of the body? The cell membrane of every cell is studded with various proteins, which together are known as the major histocompatibility complex, or MHC. The MHC is a kind of pass code, since all cells in the body must possess an identical pattern so that the body will identify those cells as belonging to the "self." An invading microorganism, such as a bacterium, does not have the same MHC, and when the immune system encounters it, it alerts the body that it has been invaded by a foreign cell.

Every person has his or her nearly unique MHC, and the response of the immune system to foreign MHC can pose a problem where organ transplants are concerned. Because the immune

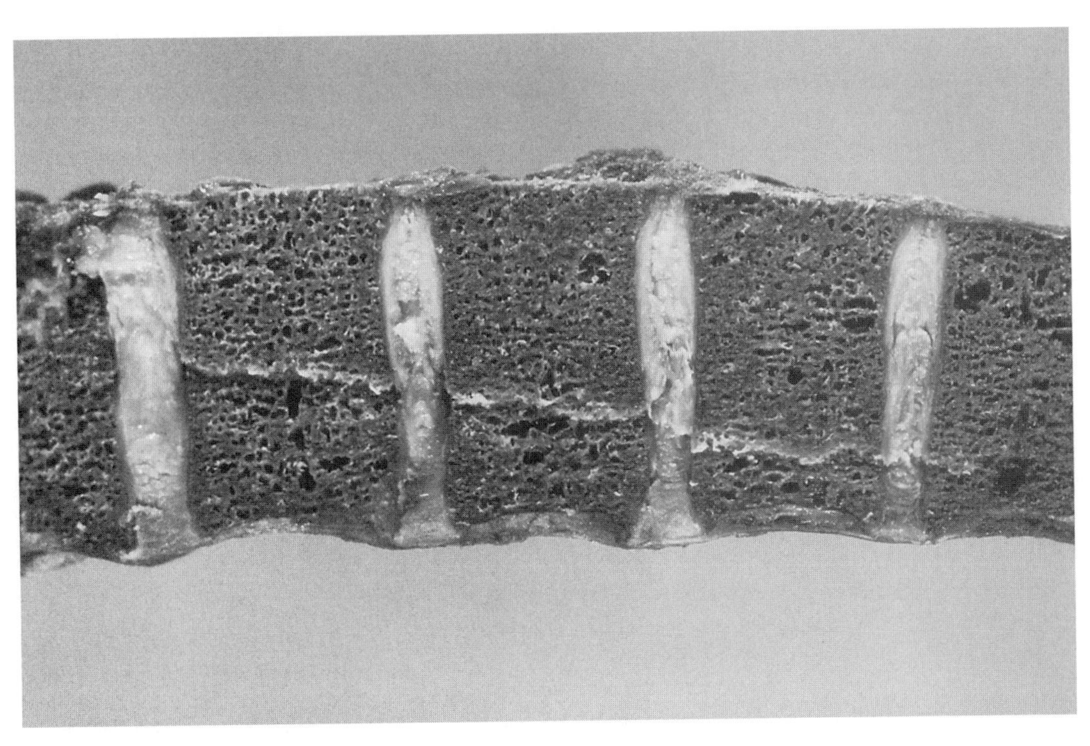

MARROW, THE SOFT TISSUE AT THE CORE OF BONES, IS A KEY PRODUCER BOTH OF LYMPHOCYTES AND OF ANOTHER COMPONENT OF BLOOD, THE HEMOGLOBIN-CONTAINING RED BLOOD CELLS. *(© Lester V. Bergman/Corbis. Reproduced by permission.)*

system interprets the transplanted organ, with its foreign MHC, as an invader, the body may reject the transplant, and therefore organ recipients usually take immunosuppressant drugs to quell the immune response. Furthermore, doctors often attempt transplants only between close relatives, who are likely to have genetically similar MHCs, or try to find organs that match in the major histocompatibility antigens.

PARTS OF THE IMMUNE SYSTEM

The organs of the immune system include the lymphatic vessels, lymph nodes, tonsils, thymus, Peyer's patch, and spleen. Each of these organs either produces the cells that participate in the immune response or serves as a site for immune function. Lymphocytes, a type of white blood cell, are concentrated in the lymph nodes, which are masses of tissue that act as filters for blood at various places throughout the body-most notably the neck, under the arms, and in the groin. As the lymph (white blood cells plus plasma) filters through the lymph nodes, foreign cells are detected and overpowered.

The tonsils, located at the back of the throat and under the tongue, contain large numbers of lymphocytes and filter out potentially harmful bacteria that might enter the body via the nose and mouth. Peyer's patches, scattered throughout the small intestine and appendix, are lymphatic tissues that perform this same function in the digestive system. The thymus gland, located within the upper chest region, is another site of lymphocyte production, though it is most active during childhood. The thymus gland continues to grow until puberty, protecting a child through the critical years of early development, but in adulthood it shrinks almost to the point of vanishing.

Marrow, the soft tissue at the core of bones, is a key producer both of lymphocytes and of another component of blood, the hemoglobin-containing red blood cells. Because of its critical role in the immune system, it is a very serious decision to allow marrow to be extracted (itself an extremely serious operation, of course) for use in a cancer treatment, as described in Noninfectious Diseases. The spleen, in addition to containing lymphatic tissue and producing lymphocytes, acts as a reservoir for blood and destroys worn-out red blood cells.

ANTIBODIES, B CELLS, AND T CELLS. The functioning of the immune system also calls into play a wide array of substances, most notably antibodies and the two significant varieties of lymphocyte: B cells and T cells. Antibodies, the most well known of the three, are proteins in the human immune system that help fight foreign invaders. B cells (B lymphocytes) are a type of white blood cell that gives rise to antibodies, whereas T cells (T lymphocytes), are a type of white blood cell that plays an important role in the immune response. T cells are a key component in the cell-mediated response, the specific immune response that utilizes T cells to neutralize cells that have been infected with viruses and certain bacteria. There are three types of T cells: cytotoxic, helper, and suppressor T cells. Cytotoxic T cells destroy virus-infected cells in the cell-mediated immune response, whereas helper T cells play a part in activating both the antibody and the cell-mediated immune responses. Suppressor T cells deactivate T cells and B cells when needed, and thus prevent the immune response from becoming too intense.

The intricacies of the immune system's functioning are far beyond the scope of this essay. The reader interested in a more in-depth review of the substances, organs, glands, and processes is encouraged to seek clarification from a textbook. On the other hand, a very basic and nontechnical example of how the body resists infection can help clarify, in general terms, how the immune system does its work.

REAL-LIFE APPLICATIONS

PROTECTING THE BODY

As discussed in Infection, not all bacteria are bad; in fact, many are helpful or even essential to humans. When the word bacteria is mentioned, however, most of us think of the "bad" bacteria, which is understandable, since there are so many of them and their effects can be so dramatic. Suppose such a bacterium enters the body, which is an easy situation to imagine—it happens all the time. Indeed, even as you are reading these words, literally trillions of bacteria the world over are attempting to invade human bodies, including your own. Their chances of success are determined by the immune system and response.

Most of the time, the skin provides us with sufficient protection from invaders, but if the skin is broken, it creates a pathway for invasion. Even a minor cut on a finger can serve as an opening for a microorganism that, once inside the body, will flourish in the body's warm, blood-washed interior. When it is established, the bacterium begins to divide rapidly, but already the specific immune system has begun to mount its resistance, and sometimes evidence of the battle can be seen on the outside—for example, in the form of a red, pus-exuding welt. In the bloodstream, lymphocytes engulf bacteria and carry them toward the lymph nodes. For this reason, when the body is under attack, it begins producing white blood cells at an accelerated rate, and for this reason doctors sometimes measure a patient's white blood cell count. If the number is high, the physician knows that an infection is active somewhere in the patient's body.

Killer blood cells, known by the generic name phagocyte, engulf the bacteria and digest them, but even as this is occurring, the rapid reproduction of the bacterium provides a challenge to the immune system. If the infectious agent reproduces at a rate beyond the control of the immune system, the physician may provide help in the form of an antibiotic. Alternatively, he or she may lance (cut open) a superficial infection to allow it to drain and to provide access for an antiseptic agent. If the bacterial invasion is minor, the immune system soon dispatches the invader, and the system returns to normal.

Often, some of the white blood cells form antibodies against such invading bacteria, so that the immune system will be better armed to combat any future invasions by the same microorganism. The white blood cell count returns to its normal level, but still with the capability of mobilizing the immune defense on short notice. It is this response that is the basis for inoculations against certain infections, a topic discussed in Immunity. Sometimes, however, something goes wrong in the production of antibodies, and instead of properly protecting the body against invaders, the immune system creates an allergy.

ALLERGIES

An allergy is a change in bodily reactivity to an antigen as a result of a first exposure. Allergies bring about an exaggerated reaction to substances or physical states that would normally have little significant effect on a healthy person. Although the immune system behaves as if it is

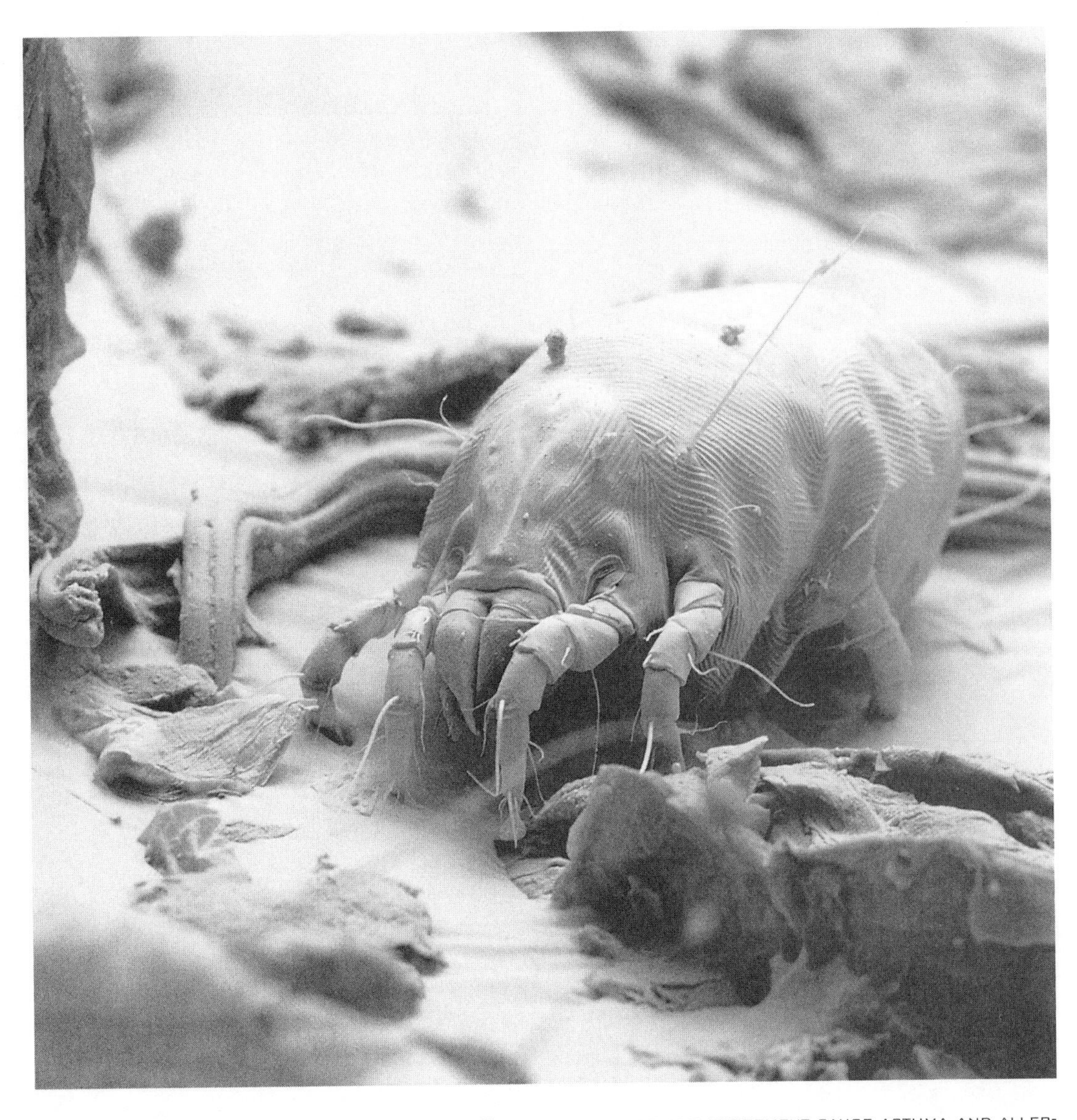

Electron micrograph of a single dust mite. Their dead bodies and excrement cause asthma and allergic reactions to household dust. *(© Eye of Science/Photo Researchers. Reproduced by permission.)*

fighting off a pathogen, in fact, it is launching a complex series of reactions against an irritant. The irritant, or allergen, may well be an otherwise innocuous substance that hardly bothers a person without the allergy. It could even be something that other people enjoy—for example, peanuts or bananas—or at least something, such as animal hair, that does not typically cause people undue discomfort. Allergies also may involve a substance, such as venom from a bee sting, that most people consider far from pleasant but which does not pose a serious threat to someone who is not allergic to it.

In extreme cases of allergic reaction, the situation that follows exposure to an allergen truly is one of life and death. The immune response may be accompanied by a number of stressful symptoms, ranging from mild reactions, such as hives (the formation of red, swollen areas on the surface of the skin) to a life-threatening situation known as anaphylactic shock. The latter, a condition characterized by a sudden drop in blood pressure and difficulty in breathing, can be accompanied by acute skin irritation in the form of angry red boils all over the body. Collapse or coma can ensue and may result in death.

CAUSES OF ALLERGY. Pollens from grasses, trees, and weeds produce such allergic reactions as sneezing, runny nose, swollen nasal tissues, headaches, blocked sinuses, and watery, irritated eyes. Of the 46 million allergy sufferers in the United States, about 25 million

KEY TERMS

ALLERGY: A change in bodily reactivity to an antigen as a result of a first exposure. Allergies bring about an exaggerated reaction to substances or physical states that normally would have little significant effect on a healthy person.

ANTIBODIES: Proteins in the human immune system that help fight foreign invaders, especially pathogens and toxins.

ANTIGEN: A substance capable of stimulating an immune response or reaction.

ANTISEPSIS: The practice of inhibiting the growth and multiplication of microorganisms, usually by ensuring the cleanliness of the environment.

APC: An antigen-presenting cell—a macrophage that has ingested a foreign cell and displays the antigen on its surface.

B CELL: A type of white blood cell that gives rise to antibodies. Also known as a *B lymphocyte.*

CELL-MEDIATED RESPONSE: The specific immune response that utilizes T cells to neutralize cells that have been infected with viruses and certain bacteria.

GLAND: A cell or group of cells that filters material from the blood, processes that material, and secretes it either for use again in the body or to be eliminated as waste.

HEMOGLOBIN: An iron-containing protein in red blood cells that is responsible for transporting oxygen to the tissues and removing carbon dioxide from them. Hemoglobin is known for its deep red color.

HUMORAL: Of or relating to the antibodies secreted by B cells that circulate in bodily fluids.

IMMUNE SYSTEM: A network of organs, glands, and tissues that protects the body from foreign substances.

IMMUNITY: The condition of being able to resist a particular disease, particularly through means that prevent the growth and development or counteract the effects of pathogens.

have this form of allergy, known to scientists as rhinitis but to the populace as hay fever. Other common allergens are dust and dust mites, pet hair and fur, insect bites, certain foods or drugs, and skin contact with specific chemical substances. About 12 million Americans are allergic to a variety of chemicals.

Some people are allergic to a wide range of substances, while others are affected by only a few or none. Why the difference? The reasons can be found in the makeup of an individual's immune system, which may produce several chemical agents that cause allergic reactions. The main immune system substances responsible for the symptoms of allergy are the histamines that are produced after exposure to an allergen. When an allergen first enters the body, the lymphocytes make what are known as E antibodies. These antibodies attach to mast cells, large cells that are found in connective tissue and contain histamines. The histamines are chemicals released by basophils, a type of lymphocyte, during the inflammatory response.

The second time a given allergen enters the body of a person who has an allergy, it becomes attached to the E antibodies. They stimulate the mast cells to discharge their histamines and other anti-allergen substances. One type of histamine travels to various receptor sites in the nasal passages, respiratory system, and skin, dilating smaller blood vessels and constricting airways. The results include some of the reactions associ-

KEY TERMS CONTINUED

IMMUNOLOGY: The study of the immune system, immunity, and immune responses.

LYMPH: That portion of the blood that includes white blood cells and plasma but not red blood cells.

LYMPH NODES: Masses of tissue at certain places in the body that act as filters for blood.

LYMPHOCYTE: A type of white blood cell, varieties of which include B cells and T cells, or B lymphocytes and T lymphocytes.

MACROPHAGE: A type of phagocytic cell derived from monocytes.

MHC: Major histocompatibility complex, a group of proteins found on the membrane of each cell in an organism's body. Since all cells in a particular body have the same MHC pattern, the MHC is a kind of pass code, identifying cells in the body as belonging to the "self."

MONOCYTE: A type of white blood cell that phagocytizes (engulfs and digests) foreign microorganisms.

PATHOGEN: A disease-carrying parasite, usually a microorganism.

PHAGOCYTE: A cell that engulfs and digests another cell.

T CELL: A type of white blood cell, also known as a *T lymphocyte,* that plays a key role in the immune response. T cells include cytotoxic T cells, which destroy virus-infected cells in the cell-mediated immune response; helper T cells, which are key participants in specific immune responses that bind to APCs, activating both the antibody and cell-mediated immune responses; and suppressor T cells, which deactivate T cells and B cells.

TISSUE: A group of cells, along with the substances that join them, which form part of the structural materials in plants or animals.

WHITE BLOOD CELLS: Blood cells that are colorless, lack hemoglobin, have a nucleus, and include the lymphocytes and other varieties.

ated with allergies, for instance, sneezing or the formation of hives. Another type of histamine constricts the larger blood vessels and travels to the receptor sites found in the salivary and tear glands and in the stomach's mucosal lining. These histamines stimulate the release of stomach acid, thus creating a stomach ulcer condition.

TREATMENTS. There are many treatments for allergy, including (obviously) avoidance of the substance to which the patient is allergic. Among these treatments are the administration of antihistamines, which either inhibit the production of histamine or block histamines at receptor sites. After the administration of antihistamines, E antibody receptor sites on the mast cells are blocked, thereby preventing the release of the histamines that cause the allergic reactions. The allergens are still there, but the body's allergic reactions are suspended for the period of time that the antihistamines are active. Antihistamines, sold both in prescription and over-the-counter forms, also constrict the smaller blood vessels and capillaries, thereby removing excess fluids. Decongestants can bring relief as well, but they can be used for only a short time, since their continued use can irritate and intensify the allergic reaction.

In cases of extreme allergic reaction leading to anaphylactic shock, the patient may require an injection of epinephrine (also sometimes called *adrenaline*), a hormone that the body produces for responding to situations of fear and danger.

In the case of anaphylactic shock, which involves such severe constriction of the breathing passages that the patient runs a risk of suffocation, epinephrine causes the passages to open, making it possible to breathe again. It also constricts the blood vessels, increasing the pressure and making the blood move more rapidly throughout the body. The body's own supply of epinephrine is not enough to counteract anaphylactic shock, however, and therefore a person experiencing that condition must receive an emergency injection containing many times the amount of the hormone naturally supplied by the body. It may be administered at a hospital, though doctors usually advise people with severe allergies to keep an emergency supply on hand.

AUTOIMMUNE DISORDERS

Allergies are one example of an immune system gone awry, and though they can be fatal, they typically are a reaction to only one or two substances. An autoimmune disorder, on the other hand, is an entirely different class of phenomenon: it a condition in which a person's body actually rejects itself. This condition comes about when the ability of the immune system to read MHCs becomes scrambled, such that it fails to recognize cells from within the body and instead rejects them as though they came from outside. As a result, the body sets in motion the same destructive operation against its own cells that it normally would carry out against bacteria, viruses, and other such harmful invaders.

The reasons why the immune system becomes dysfunctional are not well understood, but most researchers agree that a combination of genetic, environmental, and hormonal factors plays into autoimmunity. They also speculate that certain mechanisms may trigger it. First, a substance normally restricted to one part of the body, and therefore not usually exposed to the immune system, is released into other areas, where it is attacked. Second, the immune system may mistake a component of the body for a similar foreign component. Third, cells of the body may be altered in some way, by drugs, infection, or some other environmental factor, so that they are no longer recognizable as "self" to the immune system. Fourth, the immune system itself may be dysfunctional, for instance, because of a genetic mutation.

SOME AUTOIMMUNE DISEASES. Examples of autoimmune disorders include lupus, rheumatoid arthritis, autoimmune hemolytic anemia, pernicious anemia, and type 1 diabetes mellitus. (The last of these diseases is discussed in Noninfectious Diseases.) Lupus, or systemic lupus erythematosus, is seen mainly in young and middle-aged women, and its symptoms include fever, chills, fatigue, weight loss, skin rashes (especially a "butterfly" rash on the face), patchy hair loss, sores in the mouth or nose, enlargement of the lymph nodes, stomach problems, and irregular menstrual cycles. Lupus also may induce problems in the cardiopulmonary, urinary, and central nervous systems and can cause seizures, depression, and psychosis.

Rheumatoid arthritis, as its name suggests, is a type of both rheumatism and arthritis, which are general names for diseases associated with inflammation of connective tissue. Rheumatoid arthritis occurs when the immune system attacks and destroys the tissues that line bone joints and cartilage. The disease can affect any part of the body, although some joints may be more susceptible than others are. As it progresses, joint function diminishes sharply, and deformities arise.

Like rheumatism and arthritis, anemia is a general term for several conditions. Forms of it are marked either by a lack of red blood cells (hemoglobin) or by a shortage in total blood volume, and these deficiencies can produce effects that range from lethargy or sluggishness to death. Autoimmune hemolytic anemia occurs when the body makes antibodies that coat red blood cells. Patients have been known to experience a variety of symptoms, including jaundice, characterized by a yellowish coloration, before dying—sometimes just a few weeks after showing the first signs of the disease.

Pernicious anemia was so named at a time when it, too, was almost always fatal (*pernicious* means "deadly"), though treatments developed in the twentieth century have changed that situation. A disorder in which the immune system attacks the lining of the stomach in such a way that the body cannot metabolize vitamin B_{12} (see Vitamins), pernicious anemia manifests symptoms that include weakness, sore tongue, bleeding gums, and tingling in the extremities. Because the disease leads to a decrease in stomach acid, nausea, vomiting, loss of appetite, weight loss, diarrhea, and constipation are also possible. Furthermore, since B_{12} is essential to the functioning of the nervous system, a deficiency can result in a host of neurological problems, including weakness, lack of coordination, blurred vision, loss of fine motor skills, impaired sense of taste, ringing in the ears, and loss of bladder control.

WHERE TO LEARN MORE

All About Allergies. About.com (Web site). <http://gened.emc.maricopa.edu/bio/bio181/BIOBK/BioBookIMMUN.html>.

Clark, William R. *At War Within: The Double-Edged Sword of Immunity.* New York: Oxford University Press, 1995.

Davis, Joel. *Defending the Body: Unraveling the Mysteries of Immunology.* New York: Atheneum, 1989.

Deane, Peter M. G., and Robert H. Schwartz. *Coping with Allergies.* New York: Rosen Publishing Group, 1999.

Focus on Allergies (Web site). <http://www.focusonallergies.com/script/main/hp.asp>.

"Infection and Immunity." University of Leicester Microbiology and Immunology (Web site). <http://www-micro.msb.le.ac.uk/MBChB/MBChB.html>.

Joneja, Janice M. Vickerstaff, and Leonard Bielory. *Understanding Allergy, Sensitivity, and Immunity: A Comprehensive Guide.* New Brunswick, NJ: Rutgers University Press, 1990.

The Vaccine Page: Vaccine News and Database (Web site). <http://vaccines.org/>.

Vaccine Safety (Web site). <http://www.vaccines.net/>.

Young, Stuart H., Bruce S. Dobozin, and Margaret Miner. *Allergies: The Complete Guide to Diagnosis, Treatment and Daily Management.* New York: Plume, 1999.

INFECTION

PARASITES AND PARASITOLOGY
INFECTION

PARASITES AND PARASITOLOGY

CONCEPT

When people hear the word *parasite,* one of the first ideas or images that probably comes to mind is that of disease. Though many parasites do carry diseases, including some extremely deadly ones, "disease-carrying" is not necessarily a defining characteristic of a parasite. Rather, a parasite can be identified as any organism that depends on another organism, the host, for food, shelter, or some other benefit and which receives these benefits in such a way that the host experiences detrimental effects as a consequence. Theoretically, organisms from all across the kingdoms of living things can be characterized as parasites; in practice, however, the realm of organisms studied by parasitologists is confined to protozoa and various species within the animal kingdom, mostly worms and arthropods. Included among these organisms are countless varieties of tapeworm and roundworm as well as a parade of insects that have plagued humankind since the dawn of time: cockroaches, lice, bedbugs, flies, fleas, ticks, mites, and mosquitoes.

HOW IT WORKS

SYMBIOSIS

Parasitism belongs within the context of symbiosis, a term for a biological relationship in which two species live in close proximity to each other and interact regularly in such a way as to benefit one or both of the organisms. In addition to the symbiosis of two species (that is, at least two individuals representing two different species), it is also possible for a symbiotic relationship to exist between two organisms of the same species. Organisms engaging in symbiotic relationships are called *symbionts.*

There are three basic types of symbiosis, differentiated as to how the benefits (and the detriments, if any) are distributed. In the case of parasitism, of course, the arrangement benefits only one of the two: the parasite, an organism that obtains nourishment or other life support at the expense of the host. Though predation (the relationship of predator to prey) is technically a form of symbiosis, it usually is not considered in the context of symbiotic relationships; therefore, parasitism is really the only variety of symbiosis that is detrimental to one of the organisms. Unlike the predator killing its prey, however, the parasite allows its host to live as long as possible, since it depends on the host for support and sustenance.

A second type of symbiosis, commensalism, likewise involves a benefit to only one of the two organisms, but in commensalism the beneficiary, or commensal, manages to receive its benefits without causing any detriment to the host. The other variety of symbiosis is mutualism, an arrangement from which both partners receive benefits. (Commensalism and mutualism, along with the overall concept of symbiosis, are discussed at much greater length in Symbiosis.)

In addition to the distinctions among parasitism, commensalism, and mutualism, symbiotic relationships are distinguished according to the participants' ability to live without each other. In a facultative relationship, the partners can live apart successfully, whereas in an obligate one, the interacting species are incapable of living separately. Needless to say, for most parasites the relationship is obligate, whereas from the host's viewpoint, the arrangement is more than facultative: not only *could* the host live without the parasite, but it actually would be better off without the tiny hanger-on.

DEFINING THE LIMITS OF PARASITOLOGY

The realm of parasitism and parasitic creatures is far larger and more varied than that of parasitology. In other words, the range of organisms that exhibit parasitic behavior is much greater than the variety of species that are considered within the realm of parasitological study. Among the parasitic species excluded from ordinary parasitology are two named in the title of a book in the bibliography at the end of this essay: *Despicable Species: On Cowbirds, Kudzu, Hornworms, and Other Scourges,* by Janet Lembke.

The cowbird is a species that exploits the instinctive tendency of other birds to care for their young. Rather than raise its offspring, it leaves its eggs with other birds, which mistake the eggs for their own and provide them with food and care. Meanwhile, the adult cowbird goes on about its business, freed from the responsibility of raising its own progeny. (See Instinct and Learning for more on this subject.)

Kudzu is a plant that grows at an amazing rate, covering hillsides and virtually every surface to which it can attach itself. During the 1930s agricultural officials in the American South advocated the planting of kudzu, imported from China, as a means of controlling erosion on hillsides. The plant did help control erosion, but it virtually took over parts of Georgia, Alabama, Mississippi, and neighboring states. Only through successful, or at least partially successful, eradication campaigns did county and state governments, as well as private landowners, manage to stem the tide of kudzu's onslaught.

Cowbirds are clearly parasitic in their behavior, and kudzu is unquestionably a pest in plant form (i.e., a weed), but this does not truly make them parasites in the sense that parasitologists use the term. Nor, as we noted earlier, does the capacity to cause or carry disease necessarily define a parasite, though many parasites (known collectively as pathogens) do carry diseases. And even though many species of viruses, bacteria, and fungi exhibit parasitic behavior and can be transmitted by parasites, scientists usually study them separately in the context of infectious diseases.

PROTOZOA, WORMS, AND ARTHROPODS. After making all the exclusions indicated in the foregoing paragraph, the species studied within the realm of parasitology are only those creatures within certain phyla from two kingdoms: animals and protists. The latter category, which includes algae (other than blue-green algae), slime molds, and protozoa, is made up primarily of unicellular, eukaryotic organisms. (A eukaryote is a cell that has a nucleus, as well as organelles, or sections of the cell that perform specific functions, enclosed in membranes.)

Among protists, of principal concern is protozoa, which, unlike other protists, are capable of moving on their own. All the species we discuss in the remainder of this essay fall into one of three general categories: protozoa, worms, and arthropods, the latter two being groups of creatures within the animal kingdom.

THE TAXONOMY OF WORMS AND ARTHROPODS

Taxonomy, or the area of the biological sciences devoted to the classification of species, is an exceedingly complex area of study. Within that realm, there are numerous matters that either are or have been areas of dispute, among them the classification of protozoa as protists rather than animals. Owing to differences between taxonomic systems and the complications involved in explaining the characteristics that unify members of particular phyla (the next-largest major taxonomic ranking after kingdom), we will dispense with any effort to delineate classifications of species rigorously.

In other words, we will take a much simpler approach, a fact signaled by the use of the very general word *worm.* Whereas arthropods are a genuine phylum whose characteristics we examine shortly, *worms* is simply a broad term that encompasses numerous phyla: Platyhelminthes (flatworms), Nemertea (or Rhynchocoela, ribbon worms), Acanthocephala (spiny-headed worms), Aschelminthes, Priapulida (priapulids), and Annelida (annelid worms).

The classification of the many species of worm serves to illustrate just how complex a subject taxonomy can be. The group Aschelminthes, for example, is divided into five classes—Rotifera, Gastrotricha, Kinorhyncha, Nematoda, and Nematomorpha—that sometimes are treated as separate phyla. Yet another classification system groups the worms into phyla completely different from the ones named here. Therefore, rather than hopelessly confusing the issue, in the present context we will call them all simply *worms* and not attempt to map the complex relationships between the species that fall under that heading.

ARTHROPODS. In contrast to worms, arthropods, or members of the phylum Arthropoda, are much easier to identify. Arthropoda is the largest phylum in the animal kingdom, accounting for some 84% of all known animal species. Nearly 900,000 arthropod species have been identified, the vast majority of them being insects, but some zoologists maintain that this number represents but a tiny fraction of all species within this enormous phylum. In fact, there may be as many as ten million species of insect alone.

Arthropods are identified by a nonliving exoskeleton (an external skeleton), by segmented bodies, and by jointed appendages in pairs. The four subphyla of phylum Arthropoda are Trilobita, Crustacea, Chelicerata, and Uniramia. The first of these subphyla became extinct during the "Great Dying" that marked the end of the Permian period some 245 million years ago. (See Paleontology for more on this subject.) The second includes numerous organisms that humans eat, such as crabs and shrimp, as well as a few parasitic species. The most familiar parasites, however, belong to the other two subphyla. Within Chelicerata, by far the largest class is Arachnida, whose name might be familiar from that of the 1990 horror film about spiders, *Arachnophobia.* In addition to spiders, Arachnida includes scorpions, ticks, and mites, among which are many parasitic creatures.

Finally, there is subphylum Uniramia, which includes several classes, such as Chilopoda and Diplopoda (centipedes and millipedes, respectively); the largest class is unquestionably Insecta. In the latter group are some of the most obnoxious creatures known, including the genera *Anopheles* (mosquitoes), *Glossina* (tsetse fly), and *Climex* (bedbugs) as well as the species *Periplaneta americana* (American cockroach), *Phthirus pubis* (pubic or crab louse), and *Tunga penetrans* (sand flea).

PARASITE HOMES AND LIFE CYCLES

Most insects are ectoparasites, or parasites that live outside a host's body, whereas protozoa and worms are endoparasites, or parasites that live inside a host's body. As we shall see, insects are often vectors, or organisms that transmit pathogens, meaning that they serve as a delivery vehicle for the disease-causing protozoa and worms. Because endoparasites interact with the bodies of their hosts in much more complex ways, we devote more space to those species.

In general, the life cycle of most worms begins in the body of a definitive host, a host that provides a setting for the sexual reproduction of parasites. (For most of the parasites we consider, humans are the definitive host.) The parasite's eggs pass from the body of the definitive host, usually through the feces, and hatch in an inorganic medium, typically water. From there, the parasites enter the body of either a vector or an intermediate host. The latter is defined by the fact that sexual reproduction of parasites does not take place in its body; for the parasites to reproduce sexually, they must enter the body of a definitive host, where the cycle begins again. There may be more than one intermediate host, and they are identified, respectively, as first intermediate host, second intermediate host, and so on.

For protozoa the life cycle is much as we have described; however, the concept of a definitive host, as we have defined it, does not usually enter in, since protozoa reproduce asexually. Nonetheless, they typically begin life in the intestinal tracts or other organs of a host, pass through the feces into the water supply, and enter the body either of a vector or a new host.

REAL-LIFE APPLICATIONS

PROTOZOA

They may be single-cell organisms, compared with the many millions of cells that make up the vastly more complex human body, but protozoa have been the cause of more human suffering and death than almost any category of disease-carrying organism. Intestinal protozoa are common throughout the world, particularly in areas where food and water sources are subject to contamination from animal and human waste. Like soldiers sneaking into a city disguised as civilians, protozoa typically enter the body while still in an inactive state, thus getting around any defenses the immune system might put up against them.

Protozoa in this inactive state, encased in a protective outer membrane, are known as cysts. As cysts, they enter the gastrointestinal tract of the host before developing into a mature form that feeds and reproduces.

Other types of parasitic protozoa infect the blood or tissues of their hosts. These protozoa typically enter the host through a vector. Often the vector is an invertebrate (an animal without an internal skeleton), such as an insect, that feeds on the host and passes the protozoan on through the bite wound. The effects that protozoan parasites can have on human bodies range from no symptoms to extreme ones. Some victims of *Trichomonas vaginalis,* known colloquially as trick, are asymptomatic, meaning that they show no obvious symptoms of their condition. It is an STD, or sexually transmitted disease, for which men are typically carriers, passing it on to female sexual partners. Males do exhibit some symptoms, but usually it is the female who experiences the worst symptoms of the infection, including a burning sensation when urinating, genital itching, and a white discharge. When such pain is not present, the disease can be detected only by finding the trophozoites (protozoa in a feeding stage, as opposed to a reproductive or resting stage) in secretions within the victim's genital tract.

INTESTINAL PROTOZOA. A classic intestinal protozoan is *Giardia lamblia.* Like all unicellular organisms, it is extremely small, with a length of about 15 microns, or micrometers, equal to about 0.00059 in. It has a teardrop shape with two nuclei on either side of a rod called an axostyle, and under a scanning electron microscope the nuclei look like eyes separated by a nose. For this reason, people who have glimpsed *G. lamblia* under such magnification have reported the decidedly unsettling sensation of a microorganism "staring back at them."

These protozoa carry a condition known as *giardiasis.* As cysts, they pass through the feces of one host and enter the body of the next through contaminated food or water. The resulting symptoms can range from nothing to severe diarrhea. Not surprisingly, giardiasis is fairly common in developing countries, where open sewers commonly run through the city streets. Also, there are places in the third world where farmers use human feces as a fertilizer, another significant source of infection. Yet giardiasis is not simply an affliction of people in developing nations; rather, the ease with which the pathogen can contaminate water supplies makes it a condition known the world over. Even in the United States and other industrialized nations, *G. lamblia* may find its way into water supplies when waste-disposal systems are placed too close to wells. Campers who unwisely drink water from mountain streams also may contract giardiasis, which may come from beavers—hence the nickname "beaver fever," which is sometimes given to giardiasis contracted in the wild.

Far more notorious and destructive is *Entamoeba histolytica,* carrier of amoebic dysentery. The process of infection is much the same as with *G. lamblia,* but once *E. histolytica* enters the host's body, it can wreak considerably more damage. In most cases, the parasite causes only diarrhea and relatively minor gastrointestinal problems, but it can bring about a rupture of the gastrointestinal tract (the stomach and intestines). Trophozoites may even pass into the circulatory system or other organs, such as the liver, and amoebic dysentery can be fatal.

BLOOD INFECTIONS. Among the most significant protozoan blood infections is African trypanosomiasis, or "sleeping sickness." The parasite enters humans through the bite of the tsetse fly, a bloodsucking pest that serves as vector for the *Trypanosoma* genus of parasites. The result is fever, inflammation of the lymph nodes (masses of tissue at certain places in the body that filter blood), and various negative effects on the brain and spinal cord that bring about extreme lethargy or tiredness—hence the name. Sleeping sickness, prevalent throughout central Africa, is frequently fatal.

Still worse is the disease carried by the *Plasmodium* genera: malaria, which has been described at the leading health problem in the world today. It is estimated that more than two billion people live in regions where malaria is endemic (native), that the number of persons infected may be as high at 750 million, and that as many as three million people die of the disease each year. Four species of *Plasmodium,* borne by mosquitoes, are capable of infecting humans and causing malaria. *P. falciparum* is the most dangerous of the four strains; it can kill a healthy adult in 48 hours.

Inside the human host, the parasite finds a home in the red blood cells, where it reproduces until the cell bursts, spreading more parasites throughout the blood. These new parasites infect and destroy more red blood cells. As a new generation of parasites bursts from the blood cells, the body tries to defend itself by causing a fever, which does destroy many (but not all) of the pathogens. Scientists have developed several chemical antidotes to malaria; unfortunately, strains of pathogen resistant to these antidotes are on the rise in various parts of the world.

WORMS

The protozoa we have discussed up to this point, as well as the conditions that they carry, are far from pleasant. But the many species we examine under the general heading of "worms" are vastly more disgusting, if that is to be believed. It should be stated that much of what follows is not for those of weak stomach and probably should not be read while eating.

In contrast to protozoa, worms are relatively complex, large creatures with fairly long life spans. For example, the genera *Schistosoma,* or blood flukes, average 0.4 in. (10 mm) in length, while *Clonorchis sinensis* (Chinese, or oriental, liver fluke) may attain lengths of about 1 in. (25 mm). Additionally, they can live for very long spans of time: a blood fluke (fluke is simply a more common way of designating a nematode, a type of worm) may last as long as 30 years.

During that time, a female blood fluke can produce several hundred eggs *a day.* Blood flukes are one of the few worm species in which the sexes are separate; most worms are hermaphrodites, or both male and female. This brings up another distinction between worms and protozoa: the fact of sexual reproduction. From this follows yet another distinction, the involvement (typically) of one or more intermediate hosts as well as a definitive host, in whose body sexual reproduction takes place.

Flukes and other worms can be as destructive as many of their protozoan counterparts. Blood flukes, for instance, kill some one million people a year by inflaming tissues and causing organs, including the liver and small intestine, to cease functioning. Oriental liver flukes, so named because they are common throughout eastern Asia, pass through the bodies of snails (first intermediate host) to fish (second intermediate host) and ultimately to humans (definitive host), who are infected by eating raw or undercooked fish. They block the bile ducts in the liver, a condition that can be fatal.

FLUKES AND OTHER WORMS CAN BE AS DESTRUCTIVE AS MANY OF THEIR PROTOZOAN COUNTERPARTS; ORIENTAL LIVER FLUKES, FOR EXAMPLE, PASS THROUGH THE BODIES OF SNAILS TO FISH AND THEN TO HUMANS, WHO ARE INFECTED BY EATING RAW OR UNDERCOOKED FISH. THEY BLOCK THE BILE DUCTS IN THE LIVER, A CONDITION THAT CAN BE FATAL. *(Photo Researchers. Reproduced by permission.)*

TAPEWORMS AND THE DANGERS OF UNDERCOOKED MEAT. There are numerous species of tapeworm, which may be as long as 50 ft. (15 m). Their bodies are made up of segments called proglottids, which contain male and female sexual organs and which house their eggs. Proglottids break off, usually into the feces of the host, thus enabling the spread of the tapeworm. One variety, *Dipylidium caninum* (cucumber tapeworm), uses dogs or cats as a definitive host, entering the body when the animal ingests a flea or louse, which serves as an intermediate host. Several varieties of worm can enter the body through improperly cooked pork. This is a testament to the wisdom of the injunctions against eating pork in the Old Testament and Koran, which were written for peoples in a world without refrigeration or sophisticated medical knowledge.

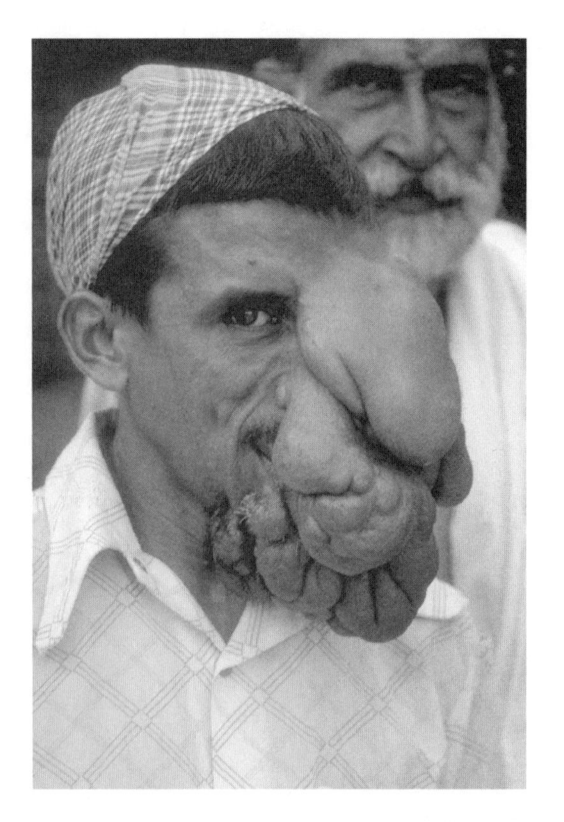

FOUND IN CENTRAL AFRICA AND PARTS OF ASIA, WUCHERERIA BANCROFTI IS CARRIED BY MOSQUITOES AND CAUSES A CONDITION KNOWN AS ELEPHANTIASIS. THIS PARASITE MAKES ITS WAY TO THE LYMPH NODES AND BRINGS ABOUT GROTESQUE SWELLING. *(© Sheldan Collins/Corbis. Reproduced by permission.)*

The *Taenia* genus of tapeworm includes *T. solium,* the pork tapeworm, which has been known to crawl out of the anus of an infected human. Another worm often found in undercooked pork is *Trichinella spiralis,* which brings about the condition known as trichinosis. The latter disease is rarely fatal, but it can cause extreme discomfort, characterized by sore, tender muscles. Thanks to enhanced efforts at meat inspection, the incidence of trichinosis in U.S. pigs has dropped to less than 1%. Nonetheless, it is still quite possible to ingest the parasite from eating undercooked game, particularly bear.

WORMS THAT AFFLICT LARGE POPULATIONS. Hookworms such as *Necator americanus* afflict more than a billion people as well as untold numbers of dogs and cats. In fact, one of the ways that the parasites spread is through human exposure to dog and cat feces. Some hookworms can produce as many as 25,000 eggs a day, wreaking havoc on the host, on whose blood the hookworms depend for sustenance. As its name implies, *N. americanus* is found in the New World, and was a widespread affliction throughout the southeastern United States during the early part of the twentieth century. Another common species, *Ancylostoma duodenale,* is found in southern Europe, northern Africa, northern Asia, and parts of South America.

Pinworms, carriers of a condition known as enterobiasis, are also common in the United States, and in much of the world as a whole. They primarily afflict preschool and school-age children who live in crowded conditions, though adults can often contract enterobiasis as well. Pinworms infect about half as many people as hookworms, but the effects can be extremely painful, especially because they are felt in the perianal skin, or the skin around the anus, and (for adult women) in the genitals. The eggs hatch on the perianal skin, which causes it to itch, but if the victim scratches the area, this can cause additional bacterial infections. The affliction of pinworms can be especially severe for children, who often experience irritability, insomnia, restlessness, and other behavioral changes. They are highly contagious and can contaminate nonliving surfaces such as bed linens and carpets; therefore, if one family member is infected, the whole family usually must receive treatment. Drugs for treating pinworm are sold either as prescription or over-the-counter drugs, and treatment usually involves two doses over the course of two weeks.

Ascaris lumbricoides, or intestinal roundworms, may infect as much as one-fourth of the world's population. Affected areas are mostly in the third world, though in the United States, an estimated 10,000 cases of roundworm infection—frequent in dogs and cats—occur in humans. The roundworm pathogen goes through a strange life cycle inside the human body, hatching in the small intestine and making its way into the air passages, only to be swallowed and returned to the small intestine. It can cause a fluid buildup in the lungs, resulting in what is known as *ascaris pneumonia,* an often fatal disease. As with *T. solium,* the intestinal roundworm has been known to escape the body, though in an even more shocking way. Because it is very sensitive to anesthetics, the parasite may evacuate the body of a patient in a surgical recovery room by moving from the small intestine to the stomach and out the nose or mouth.

OTHER WORMS. Several other worms have been associated with various dis-

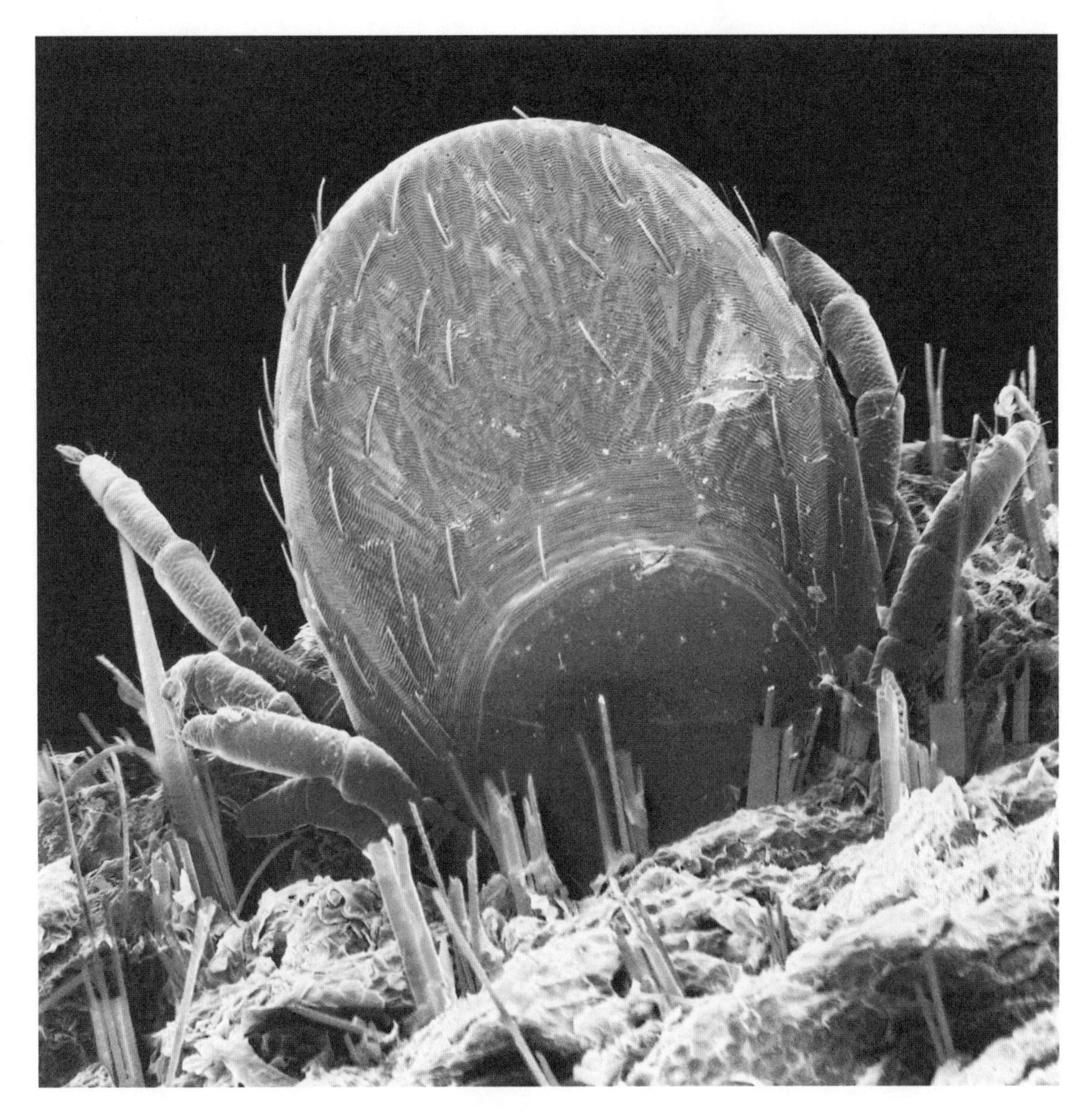

SCANNING ELECTRON MICROGRAPH OF A TICK FEEDING IN HUMAN SKIN. A TICK IS A PARASITIC ARTHROPOD THAT CAN SERVE AS A VECTOR FOR BACTERIAL INFECTIONS, SUCH AS ROCKY MOUNTAIN SPOTTED FEVER. *(Photo Researchers. Reproduced by permission.)*

eases in parts of the third world. Among them are *Loa loa,* or the "eye worm," found primarily in equatorial Africa. These worms live just under the skin and, as their name indicates, just under the surface of the eyeball. *Onchocerca volvulus,* also common in Africa as well as Arabia and Latin America, afflicts an estimated 30 million people. When it enters the human eye, it dies, eventually bringing about blindness. Because the vector, a black fly, breeds in rivers, the disease is commonly called *river blindness,* and in some towns it has blinded as much as 40% of the adult population.

Found in central Africa and parts of Asia, *Wuchereria bancrofti* is carried by mosquitoes and causes a condition known as *elephantiasis.* The parasite makes its way to the lymph nodes, particularly those that drain lower parts of the body, and brings about grotesque swelling, usually in the legs and genitals. (Visitors to the "Parasites and Parasitological Resources" Web site of Ohio State University's College of Biological Sciences, will find links to photographs of elephantiasis victims. It should be noted that these images include grotesquely swollen male genitalia and may be too graphic for some viewers.)

ARTHROPODS

After the preceding discussion of worms, mere insects and other arthropods probably will

KEY TERMS

ARTHROPOD: A term for members of the phylum Arthropoda, largest in the animal kingdom. Arthropods are identified by a nonliving exoskeleton (an external skeleton), by segmented bodies, and by jointed appendages that appear in pairs. Among the classes within this phylum are Arachnida (spiders, scorpions, ticks, and mites) and Insecta.

ASYMPTOMATIC: Displaying no symptoms of a disease or other condition.

CLASS: The third-largest major, or obligatory, taxonomic classification rank, smaller than *phylum* but larger than *order.*

COMMENSALISM: A symbiotic relationship in which one organism, the commensal, benefits without causing any detriment to the other organism, the host.

CYST: A protozoan in an inactive state, encased in a protective outer membrane. Protozoa usually enter the bodies of hosts in the form of cysts.

DEFINITIVE HOST: The host in whose body a parasite reproduces sexually. A parasite enters a definitive host either through a vector, an intermediate host, or through some other source, such as a water supply.

ECTOPARASITES: Parasites that live outside a host's body.

ENDOPARASITES: Parasites that live inside a host's body.

EUKARYOTE: A cell that has a nucleus, as well as organelles (sections of the cell that perform specific functions) that are bound by membranes.

FACULTATIVE: A term for a symbiotic relationship in which partners are capable of living apart.

GASTROINTESTINAL TRACT: The stomach and intestines.

GENUS: The second-smallest major taxonomic classification rank, smaller than *family* and larger than *species.*

HOST: The term for an organism that provides a benefit or benefits to another organism in a symbiotic relationship of commensalism or parasitism. See also *intermediate host* and *definitive host.*

INTERMEDIATE HOST: A creature that serves as host to a parasite, which it receives either directly (typically through the water supply) or from a vector before passing it on to a definitive host. Parasites do not reproduce sexually inside an intermediate host.

LYMPH NODES: Masses of tissue at certain places in the body that act as filters for blood.

not have the same power to disgust the reader as they otherwise might have had. This is particularly so because most arthropods are ectoparasites, and not as many of them are associated with diseases. This is not to say that arthropod parasites are anything pleasant or that they could be described as upstanding citizens of the biological world—only that, compared with worms, they are not quite so revolting.

Certainly many insect parasites, such as mosquitoes (genus *Anopheles*), are notorious vectors. Of the 400 or so mosquito species, about a dozen carry malaria, and many others transmit canine heartworm. Some parasites are known as *mechanical,* as opposed to biological, vectors. This means that they may happen to carry cysts of protozoa, or worm eggs, on their exteriors, but they do not serve as the biological carrier of any

KEY TERMS CONTINUED

OBLIGATIVE: A term for a symbiotic relationship in which the partners, if they were separated, would be incapable of continuing to live.

PARASITE: A general term for any organism that depends on another organism for support, which it receives at the expense of the other organism.

PARASITISM: A symbiotic relationship in which one organism, the parasite, benefits at the expense of the other organism, the host.

PARASITOLOGY: A biological discipline devoted to the study of parasites but primarily those among the animal and protist kingdoms. Parasitic bacteria, fungi, and viruses usually are studied within the context of infectious diseases.

PATHOGEN: A disease-carrying parasite, typically a microorganism.

PHYLUM: The second-largest major taxonomic classification rank, smaller than *kingdom* but larger than *class.*

PROTISTS: A term for members of the kingdom Protista, which includes algae (other than blue-green algae, which are monerans), slime molds (a group of about 500 species that resemble fungi), and protozoa.

PROTOZOAN: A term that refers to some 50,000 species of mobile protists.

SUBPHYLUM: A classification rank, smaller than that of phylum but larger than that of class, that is sometimes used to indicate a level between those two major ranks.

SYMBIOSIS: A biological relationship in which (usually) two species live in close proximity to each other and interact regularly in such a way as to benefit one or both of the organisms. Symbiosis may exist between two or more individuals of the same species as well as between two or more individuals representing two different species. The three principal varieties of symbiosis are *mutualism, commensalism,* and *parasitism.*

TAXONOMY: The area of the biological sciences devoted to the classification of organisms according to apparent common characteristics.

TROPHOZOITE: A protozoan in a feeding stage, as opposed to a reproductive or resting stage.

VECTOR: An organism, such as an insect, that transmits a pathogen to the body of a host.

disease. Such is the case with the American cockroach (*Periplaneta americana*). But one thing unites all insect parasites: they are pests, and if any of their species ever became endangered (something that is not likely to happen), very few people would be upset.

ARTHROPODS THAT CAUSE ITCHING. The bedbug (genus *Cimex*) is an insect parasite that is not known to serve as a vector, either mechanical or biological, for any disease—yet it certainly gives rise to more than its share of misery. Like many another arthropod parasite, it causes its hosts to itch. Most familiar from the old saying "don't let the bedbugs bite," bedbugs live not only in beds but also in clothing, furniture, laundry, and numerous other areas around a house. And though they depend on human blood, it does no good simply to vacate a house for a few days in the hope of starving them; they can survive for up to year on the blood supplies they have stored in their bodies.

Then there are body lice (*Pediculus humanus*) and pubic, or crab, lice (*Phthirus pubis*). Although people may tend to associate the problem of lice with the poor, lice afflict all socioeconomic classes. Body lice, as their name indicates, spread all over the body and particularly the head, while crab lice nest either in the pubic area or anywhere else they can find short, thick hair: armpits, eyebrows, eyelashes, beards, sideburns, and mustaches. Other than inducing itching, most lice do not pose a serious threat—unless they happen to be carrying typhus or other diseases.

Other insect parasites known to cause itching are cat and dog fleas (*Ctenocephalides felis* and *C. canis,* respectively), which may carry the cucumber tapeworm we discussed earlier. Sand fleas (*Tunga penetrans*) are not associated with any disease, but by causing hosts to scratch, they can bring about secondary infections, and they become so firmly attached to the host that they usually require surgical removal.

Like the insects we have described, most arachnid parasites also produce itching. Among these parasites are several varieties of tick, which can serve as vectors for bacterial infections, such as Rocky Mountain spotted fever. Tick species such as *Ixodes dammini* in the northeastern United States carry Lyme disease, named after the town in Connecticut were it was first identified in 1975. Effects of Lyme disease can range from itching to severe neurological disorders. There are also several varieties of mange mites, among them *Sarcoptes scabiei,* which afflicts humans and animals with scabies, characterized by severe skin inflammation and itching.

WHERE TO LEARN MORE

Berenbaum, May. *Bugs in the System: Insects and Their Impact on Human Affairs.* Reading, MA: Addison-Wesley, 1995.

"Biology 160, Animal Behavior: Symbiosis and Social Parasitism." Department of Biology, University of California at Riverside (Web site). <http://www.biology.ucr.edu/Bio160/lecture25.html>.

Knutson, Roger M. *Furtive Fauna: A Field Guide to the Creatures Who Live on You.* New York: Penguin Books, 1992.

Lembke, Janet. *Despicable Species: On Cowbirds, Kudzu, Hornworms, and Other Scourges.* New York: Lyons Press, 1999.

Parasites and Parasitism. University of Wales, Aberystwyth (Web site). <http://www.aber.ac.uk/parasitology/Edu/Para_ism/PaIsmTxt.html>.

Parasites and Parasitological Resources. Ohio State University College of Biological Sciences (Web site). <http://www.biosci.ohio-state.edu/~parasite/home.html>.

Trager, William. *Living Together: The Biology of Animal Parasitism.* New York: Plenum Press, 1986.

The World of Parasites (Web site). <http://martin.parasitology.mcgill.ca/jimspage/worldof.htm>.

The World Wide Web Virtual Library: Parasitology (Web site). <http://www.aan18.dial.pipex.com/urls.htm>.

Zimmer, Carl. *Parasite Rex: Inside the Bizarre World of Nature's Most Dangerous Creatures.* New York: Free Press, 2000.

INFECTION

CONCEPT

Humans may hold dominance over most other life-forms on Earth, but a few varieties of organism have long held mastery over us. Ironically, these life-forms, including bacteria and viruses, are so small that they cannot be seen, and this, in fact, has contributed to their disproportionate influence in human history. For thousands of years, people attributed infection to spiritual causes or, at the very least, to imbalances of "humors," or fluids, in the human body. Today germ theory and antisepsis—the ideas that microbes cause infection and that a clean body and environment can prevent infections—are ingrained so deeply that we almost take them for granted. Yet these concepts are very recent in origin, and for a much longer span of human history people quite literally wallowed in filth—with predictable consequences.

HOW IT WORKS

WHAT IS INFECTION?

The term infection refers to a state in which parasitic organisms attach themselves to the body, or to the inside of the body, of another organism, causing contamination and disease in the host organism. Parasite refers generally to any organism that lives at the expense of another organism, on which it depends for support. Numerous parasites and the diseases they cause are discussed in the essay Parasites and Parasitology; in the present context, we are concerned primarily with infections that relate to bacteria and viruses.

Almost all infections contracted by humans are passed along by other humans or animals. Infections fall into two general categories: exogenous, or those that originate outside the body, and endogenous, which occur when the body's resistance is lowered. Examples of exogenous infection include catching a cold by drinking after someone else from the same glass; coming down with salmonella after ingesting undercooked eggs, meat, or poultry; getting rabies from a dog bite; or contracting syphilis, AIDS (acquired immunodeficiency syndrome), or some other sexually transmitted disease from an infected partner.

Any number of factors—lack of sleep, prolonged exposure to extreme cold or moisture, and so on—can lower the body's resistance, opening the way for an endogenous infection. Malnutrition, illness, and trauma also can be factors in endogenous infection. Substance abuse, whether it be the use of tobacco in its many forms, excessive drinking, or drug use, lowers the body's resistance. Furthermore, all of these behaviors tend to be coupled with poor eating habits, which invite infection by denying the body the nutrients it needs.

SOME TERMS

A whole array of terminology attends the study of infection and infectious diseases, a subject that is touched upon in the present context but explored at length in its own essay as well. Among these terms are the names for the different branches of study relating to infection, its agents, and the resulting diseases. Although germ theory is a term (defined later) that is used widely in the context of infection, *germ* itself—a common word in everyday life—is not used as much as microorganism or pathogen. The latter word refers to disease-carrying parasites, which are usually microorganisms. Two of the principal types of pathogen, bacteria and viruses, are discussed later in this essay.

Words relating to the effects of infectious agents include *epidemic,* an adjective meaning "affecting or potentially affecting a large proportion of a population"; as a noun, the word refers to an epidemic disease. *Pandemic* also doubles as an adjective, meaning "affecting an extremely high proportion of a population over a wide geographic area," and a noun, referring to a disease of pandemic proportions. Areas of study relating to pathogens, their effects, and the prevention of those effects include the following.

- Bacteriology: An area of the biological sciences concerned with bacteria, including their importance in medicine, industry, and agriculture
- Epidemiology: An area of the medical sciences devoted to the study of disease, including its incidence, distribution, and control within a population
- Etiology: A branch of medical study concerned with the causes and origins of disease. Also, a general term referring to all the causes of a particular disease or condition
- Immunology: The study of the immune system, immunity, and immune responses
- Pathology: The study of the essential nature of diseases
- Virology: The study of viruses

In addition, there are several terms relating to the prevention of infection.

- Antibiotic: A substance produced by, or derived from, a microorganism, which in diluted form is capable of killing or at least inhibiting the action of another microorganism. Antibiotics typically are not effective against viruses.
- Antisepsis: The practice of inhibiting the growth and multiplication of microorganisms
- Germ theory: A theory in medicine, widely accepted today, that infections, contagious diseases, and other conditions are caused by the actions of microorganisms
- Immunity: A condition of being able to resist a particular disease, particularly through means that prevent the growth and development of pathogens or counteract their effects
- Inoculation: The prevention of a disease by the introduction to the body, in small quantities, of the virus or other microorganism that causes the disease
- Vaccine: A preparation containing microorganisms, usually either weakened or dead, which are administered as a means of increasing immunity to the disease caused by those microorganisms

Some of these words appear in this essay and others in related essays on infectious diseases and immunity.

BACTERIA

Five major groups of microorganisms are responsible for the majority of infections. They include protozoa and helminths, or worms—both of which are considered in Parasites and Parasitology—as well as bacteria and viruses. Bacteria and viruses often are discussed, along with fungi (the fifth major group), in the context of infection and infectious diseases. In the present context, however, we limit our inquiry to viruses and bacteria.

Bacteria are very small organisms, typically consisting of one cell. They are prokaryotes, a term referring to a type of cell that has no nucleus. In eukaryotic cells, such as those of plants and animals, the nucleus controls the cell's functions and contains its genes. Genes carry deoxyribonucleic acid (DNA), which determines the characteristics that are passed on from one generation to the next. The genetic material of bacteria is contained instead within a single, circular chain of DNA.

Members of kingdom Monera, which also includes blue-green algae (see Taxonomy), bacteria generally are classified into three groups based on their shape: spherical (coccus), rodlike (bacillus), or spiral- or corkscrew-shaped (spirochete). Some bacteria also have a shape like that of a comma and are known as *vibrio.* Spirochetes, which are linked to such diseases as syphilis, sometimes are considered a separate type of creature; hence, Monera occasionally is defined as consisting of blue-green algae, bacteria, *and* spirochetes.

The cytoplasm (material in the cell interior) of all bacteria is enclosed within a cell membrane that itself is surrounded by a rigid cell wall. Bacteria produce a thick, jellylike material on the surface of the cell wall, and when that material forms a distinct outer layer, it is known as a capsule. Many rod, spiral, and comma-shaped bacteria have whiplike limbs, known as flagella, attached to the outside of their cells. They use these flagella for movement by waving them back and forth. Other bacteria move simply by wiggling the whole cell back and forth, whereas still others are unable to move at all.

Bacteria most commonly reproduce by fission, the process by which a single cell divides to produce two new cells. The process of fission may take anywhere from 15 minutes to 16 hours, depending on the type of bacterium. Several factors influence the rate at which bacterial growth occurs, the most important being moisture, temperature, and pH, or the relative acidity or alkalinity of the substance in which they are placed.

Bacterial preferences in all of these areas vary: for example, there are bacteria that live in hydrothermal vents, or cracks in the ocean floor, where the temperature is about 660°F (350°C), and some species survive at a pH more severe than that of battery acid. Most bacteria, however, favor temperatures close to that of the human body—98.6°F (37°C)—and pH levels only slightly more or less acidic than water. Since they are composed primarily of water, they thrive in a moist environment.

VIRUSES

One of the interesting things about bacteria is their simplicity, coupled with the extraordinary complexity of their interactions with other organisms. As simple as bacteria are, however, viruses are vastly more simple. Furthermore, the diseases they can cause in other organisms are at least as complex as those of bacteria, and usually much more difficult to defeat. Whereas there are "good" bacteria, as we shall see, scientists have yet to discover a virus whose impact on the world of living things is beneficial. There is something downright creepy about viruses, which are not exactly classifiable as living things; in fact, a virus is really nothing more than a core of either DNA or RNA (ribonucleic acid), surrounded by a shell of protein.

Two facts separate viruses from the world of the truly living. First, unlike all living things (even bacteria), viruses are not composed of even a single cell, and, second, a virus has no life if it cannot infect a host cell. When we say "no life" in this context, we truly mean *no life.* Although parasites, including bacteria and those species discussed in Parasites and Parasitology, depend on other organisms to serve as hosts, they can live when they are between hosts. They are rather like a person between jobs: without other means of support, the person eventually will go broke or starve, but typically such a person can hang on for a few months until he or she finds a new job. A virus without a host, on the other hand, is simply not alive—not dead, like a formerly living thing, but more like a machine that has been switched off.

Once a virus enters the body of a host, it switches on, and the result is truly terrifying. In order to produce new copies of itself, a virus must use the host cell's reproductive "machinery"—that is, the DNA. The newly made viruses then leave the host cell, sometimes killing it in the process, and proceed to infect other cells within the organism. As for the organisms that viruses target, their potential victims include the whole world of living things: plants, animals, *and* bacteria. Viruses that affect bacteria are called bacteriophages, or simply phages. Phages are of special importance, because they have been studied much more thoroughly than most viruses; in fact, much of what virologists now know about viruses is based on the study of phages.

REAL-LIFE APPLICATIONS

BACTERIA AND HUMANS

Not all bacteria are harmful; in fact, some even are involved in the production of foods consumed by humans. For example, bacteria that cause milk to become sour are used in making cottage cheese, buttermilk, and yogurt. Vinegar and sauerkraut also are produced by the action of bacteria on ethyl alcohol and cabbage, respectively. Other bacteria, most notably *Escherichia coli* (*E. coli*) in the human intestines, make it possible for animals to digest foods and even form vitamins in the course of their work. (See Digestion for more on these subjects.) Others function as decomposers (see Food Webs), aiding in the chemical breakdown of organic materials, while still others help keep the world a cleaner place by consuming waste materials, such as feces.

Despite its helpful role in the body, certain strains of *E. coli* are dangerous pathogens that can cause diarrhea, bloody stools, and severe abdominal cramping and pain. The affliction is rarely fatal, though in late 1992 and 1993 four people died during the course of an *E. coli* outbreak in Washington, Idaho, California, and Nevada. More often the outcome is severe illness that may bring on other conditions; for example, two teenagers among a group of 11 who became sick while attending a Texas cheerleading camp had to receive emergency appendectomies. The pathogen is usually transmitted through undercooked foods, and sometimes through other means; for example, a small outbreak in the Atlanta area in the late 1990s occurred in a recreational water park.

BACTERIAL INFECTIONS. Many bacteria attack the skin, eyes, ears, and various systems in the body, including the nervous, cardiovascular, respiratory, digestive, and genitourinary (i.e., reproductive and urinary) systems. The skin is the body's first line of defense against infection by bacteria and other microorganisms, although it supports enormous numbers of bacteria itself. Bacteria play a major role in a skin condition that is the bane of many a young man's (and, less frequently, a young woman's) existence: acne. Pimples or "zits," known scientifically as *Acne vulgaris,* constitute one of about 50 varieties of acne, or skin inflammation, which are caused by a combination of heredity, hormones, and bacteria—particularly a species known as *Propionibacterium acnes.* When a hair follicle becomes plugged by sebum, a fatty substance secreted by the sebaceous, or oil, glands, this forms what we know as a blackhead; a pimple, on the other hand, results when a bacterial infection, brought about by *P. acnes,* inflames the blackhead and turns it red. For this reason, antibiotics may sometimes cure acne or at least alleviate the worst symptoms.

Acne may seem like a life-and-death issue to a teenager, but it goes away eventually. On the other hand, toxic shock syndrome (TSS), caused by other bacteria at the surface of the skin—species of *Staphylococcus* and *Streptococcus*—can be extremely dangerous. The early stages of TSS are characterized by flulike symptoms, such as sudden fever, fatigue, diarrhea, and dizziness, but in a matter of a few hours or days the blood pressure drops dangerously, and a sunburn-like rash forms on the body. Circulatory problems arise as a result of low blood pressure, and some extremities, such as the fingers and toes, are deprived of blood as the body tries to shunt blood to vital organs. If the syndrome is severe enough, gangrene may develop in the fingers and toes.

In 1980, several women in the United States died from TSS, and several others were diagnosed with the condition. As researchers discovered, all of them had been menstruating and using high-absorbency tampons. It appears that such tampons provide an environment in which TSS-causing bacteria can grow, and this led to recommendations that women use lower-absorbency tampons if possible, and change them every two to four hours. Since these guidelines were instituted, the incidence of toxic shock has dropped significantly, to between 1 and 17 cases per 100,000 menstruating women.

Many bacteria produce toxins, poisonous substances that have effects in specific areas of the body. An example is *Clostridium tetani,* responsible for the disease known as *tetanus,* in which one's muscles become paralyzed. A related bacterium, *C. botulinum,* releases a toxin that causes the most severe form of food poisoning, botulism. Salmonella poisoning comes from another genus, *Salmonella,* which includes *S. typhi,* the cause of typhoid fever.

VIRAL INFECTIONS

With viruses, as we have noted, there is no need even to discuss "good" kinds, because there is no such thing—*all* viruses are harmful, and most are killers. The particular strains of virus that attack animals have introduced the world to a variety of ailments, ranging from the common cold to AIDS and some types of cancer. Other diseases related to viral infections are hepatitis, chicken pox, smallpox, polio, measles, and rabies.

One reason why physicians and scientists have never found a cure for the common cold is that it can be caused by any one of about 200 viruses, including rhinoviruses, adenoviruses, influenza viruses, parainfluenza viruses, syncytial viruses, echoviruses, and coxsackie viruses. Each has its own characteristics, its favored method of transmission, and its own developmental period. These viruses can be transmitted from one person to another by sneezing on the person, shaking hands, or handling an object previously touched by the infected person. Surprisingly, some more direct forms of contact with an infected person, as in kissing, seldom spread viruses.

A group of viruses called the *orthomyxoviruses* transmit influenza, an illness usually characterized by fever, muscle aches, fatigue, and upper respiratory obstruction and inflammation. The most common complication of influenza is pneumonia, a disease of the lungs that may be viral or bacterial. The viral form of pneumonia that goes hand in hand with influenza can be very severe, with a high mortality (death) rate; by contrast, bacterial pneumonia, which typically appears five to ten days after the onset of flu, can be treated with antibiotics.

THE EVER ELUSIVE VIRUS. Viruses are tricky. Because their generations are very short and their structures extremely simple, they are constantly mutating (altering their DNA and hence their heritable traits) and thus becoming less susceptible to vaccines. This is the reason why flu vaccine has to be prepared anew each year to target the current strains, and even then the vaccine is far less than universally effective. On the other hand, vaccination has a high rate of success for strains of virus that undergo little mutation—for example, the smallpox virus.

One particularly elusive type of virus is known as a *retrovirus,* which reverses the normal process by which living organisms produce proteins. Ordinarily, DNA in the cell's nucleus carries directions for the production of new protein. Coded messages in the DNA molecules are copied into RNA molecules, which direct the manufacture of new protein. In retroviruses, that process is reversed, with viral RNA used to make new viral DNA, which then is incorporated into host cell DNA, where it is used to direct the manufacture of new viral protein. Among the diseases caused by retroviruses is AIDS, discussed in Infectious Diseases and The Immune System.

FIGHTING THE INVISIBLE WAR

Every day of our lives, we are at war with microorganisms, both individually and as a species. It is a war that has lasted for several million years, with billions of lives in the balance, yet it is an invisible war. Up until a few centuries ago, in fact, we had no idea what we were fighting. Before the advent of germ theory, the most scientific theories of disease blamed them either on an imbalance of "humors" (blood, phlegm, yellow bile, and green bile), or on inhaling bad air. These were the most advanced ideas, the ones held by men of learning; most of the populace, by contrast, believed that disease was caused by evil spirits, cast upon individuals or populations by an angry God as punishment for disobedience.

TWO CHILDREN ARE CONFINED TO IRON LUNGS AS THE RESULT OF INFECTION WITH POLIOVIRUS; SIX OF THE CHILDREN IN THIS ONE FAMILY WERE STRICKEN WITH THE VIRUS. AN EPIDEMIC DISEASE CAN AFFECT A LARGE PROPORTION OF A POPULATION, AS HAPPENED IN THE CASE OF POLIO IN THE MIDDLE OF THE TWENTIETH CENTURY. *(© Bettmann/Corbis. Reproduced by permission.)*

Personal hygiene and public health were completely foreign concepts: not only did people bathe infrequently, but they also thought nothing of throwing trash—including rotting food and even human excrement—into the city streets. This image of trash in the streets may call to mind a city of medieval Western Europe, a place and time widely known for its filth, squalor, and ignorance. Yet such an image also describes Athens during the fifth century B.C., when human imagination, wisdom, and appreciation for beauty reached perhaps their highest points in all of history. In the Athens of Socrates, Herodotus, Hippocrates, and Sophocles, the streets were piled with trash and crawling with vermin. In fact, this lack of concern for cleanliness contributed directly to the end of the Greek golden age, sometimes known as the Age of Pericles, after Athens's great leader (495–425 B.C.)—who died in a great plague that swept the germ-ridden city.

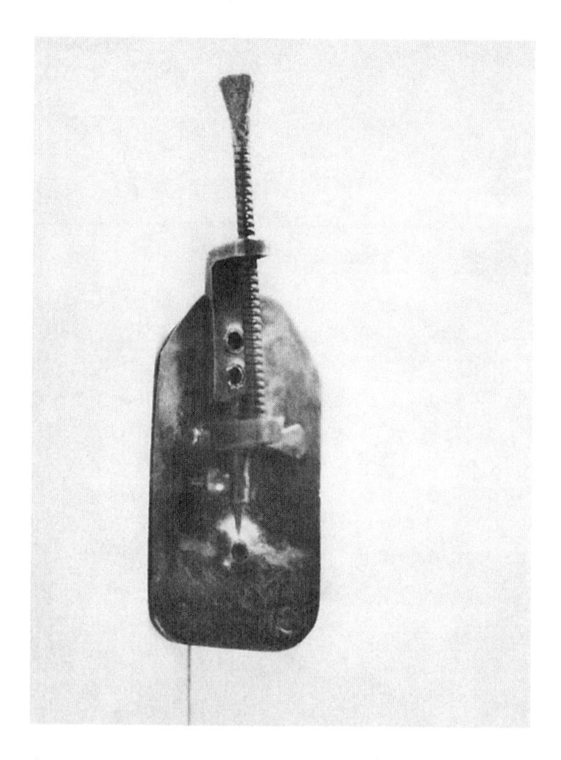

ANTON VAN LEEUWENHOEK'S PROTOTYPE MICROSCOPE. THE INVENTION OF THE MICROSCOPE MADE IT POSSIBLE TO SEE BACTERIA AND OTHER MICROORGANISMS, WHICH VAN LEEUWENHOEK, THE FIRST HUMAN BEING TO OBSERVE THEM, DUBBED ANIMALCULES, OR "TINY ANIMALS." *(© Bettmann/Corbis. Reproduced by permission.)*

BACTERIOLOGY AND ANTISEPSIS. The first inkling of any etiology other than that of imbalanced humors and demons was the work of the Italian physician Girolamo Fracastoro (*ca.* 1483–1553), who put forth the theory that disease is caused by particles so small they are almost imperceptible. The invention of the microscope in 1590 made it possible to glimpse those particles, which Holland's Anton van Leeuwenhoek (1632–1723)—the first human being to observe bacteria and other microorganisms—dubbed *animalcules,* or "tiny animals." The German scholar Athanasius Kircher (1601–1680) also observed "tiny worms" in the blood and pus of plague victims and theorized that they were the source of the infection. This was the first theory that dealt with microbial agents as infectious organisms.

In 1848 Ignaz P. Semmelweis (1818–1865), a Hungarian physician working in German hospitals, came up with a novel idea: after examining the bodies of women who had died of puerperal (childbed) fever, he suggested that doctors should wash their hands in a solution of chlorinated lime water before touching a pregnant patient. Semmelweis's idea resulted in a drastic reduction of puerperal fever cases, but his colleagues denounced his outlandish notion as a useless and foolish waste of time. Six years later, in 1854, modern epidemiology was born when the English physician John Snow (1813–1858) determined that the source of a cholera epidemic in London could be traced to the contaminated water of the Broad Street pump. After he ordered the pump closed, the epidemic ebbed—and still many physicians refused to believe that invisible organisms could spread disease.

GERM THEORY. A major turning point came just three years later, in 1857, when the great French chemist and microbiologist Louis Pasteur (1822–1895) discovered that heating beer and wine to a certain temperature killed bacteria that caused these liquids to spoil or turn into vinegar. Thus was born the process of pasteurization, still used today to purify such foods as milk, because, as Pasteur observed, "There are similarities between the diseases of animals or man and the diseases of beer and wine." Pasteur also dealt the final blow to spontaneous generation, a centuries-old belief that living organisms could originate from nonliving matter. As he showed in 1861, microorganisms present in the air can contaminate solutions that seem sterile.

Then, in 1876, the German physician Robert Koch (1843–1910) proved what Kircher had postulated two centuries earlier: that bacteria can cause diseases. Koch showed that the bacterium *Bacillus anthracis* was the source of anthrax in cattle and sheep and generalized the methodology he had used in that situation to form a specific set of guidelines for determining the cause of infectious diseases. Known as Koch's postulates, these guidelines define a truly infectious agent as one that can be isolated from an infected animal, cultured in a laboratory setting, introduced into a healthy animal to produce the same infection as in the first animal, and isolated again from the second animal. These ideas formed the basis of research into bacterial diseases and are still dominant in the sciences devoted to the study of disease.

Koch's postulates helped usher in what has been called the golden era of medical bacteriology. Between 1879 and 1889 German microbiologists isolated the organisms that cause cholera, typhoid fever, diphtheria, pneumonia, tetanus, meningitis, and gonorrhea as well the *Staphylococcus* and *Streptococcus* organisms. Even as

KEY TERMS

ANTIBIOTIC: A substance produced by or derived from a microorganism, which in diluted form is capable of killing or at least inhibiting the action of another microorganism. Antibiotics are not usually effective against viruses.

ANTISEPSIS: The practice of inhibiting the growth and multiplication of microorganisms, generally by ensuring the cleanliness of the environment.

BACTERIOLOGY: An area of the biological sciences concerned with bacteria, including their importance in medicine, industry, and agriculture.

DNA: Deoxyribonucleic acid, a molecule in all cells, and many viruses, containing genetic codes for inheritance.

ENDOGENOUS: A term for an infection that occurs when the body's resistance is lowered. Compare with *exogenous.*

EPIDEMIC: Affecting or potentially affecting a large proportion of a population (*adj.*) or an epidemic disease (*n.*)

EPIDEMIOLOGY: An area of the medical sciences devoted to the study of disease, including its incidence, distribution, and control within a population.

ETIOLOGY: A branch of medical study concerned with the causes and origins of disease; also, a general term referring to all the causes of a particular disease or condition.

EXOGENOUS: A term for an infection that originates outside the body. Compare with *endogenous.*

GENE: A unit of information about a particular heritable (capable of being inherited) trait that is passed from parent to offspring, stored in DNA molecules called *chromosomes.*

GERM THEORY: A theory in medicine, widely accepted today, that infections, contagious diseases, and other conditions are caused by the actions of microorganisms.

IMMUNITY: The condition of being able to resist a particular disease, particularly through means that prevent the growth and development of pathogens or counteract their effects.

IMMUNOLOGY: The study of the immune system, immunity, and immune responses.

INFECTION: A state or condition in which parasitic organisms attach themselves to the body or to the inside of the body of another organism, producing contamination and disease in the host.

INOCULATION: The prevention of a disease by the introduction to the body, in small quantities, of the virus or other microorganism that causes the disease.

MUTATION: Alteration in the physical structure of an organism's DNA, resulting in a genetic change that can be inherited.

PANDEMIC: Affecting an extremely high proportion of a population over a wide geographic area (*adj.*) or a disease of pandemic proportions (*n.*)

PARASITE: A general term for any organism that depends on another organism for support, which it receives at the expense of the other organism.

PARASITOLOGY: A biological discipline devoted to the study of parasites, primarily those among the animal and protist

KEY TERMS CONTINUED

kingdoms. Parasitic bacteria, fungi, and viruses usually are studied within the context of infectious diseases.

PATHOGEN: A disease-carrying parasite, typically a microorganism.

PATHOLOGY: The study of the essential nature of diseases.

PUBLIC HEALTH: A set of policies and methods for protecting and improving the health of a community through efforts that include disease prevention, health education, and sanitation.

RNA: Ribonucleic acid, the molecule translated from DNA in the cell nucleus, the control center of the cell, that directs protein synthesis in the cytoplasm, or the space between cells.

VACCINE: A preparation containing microorganisms, usually either weakened or dead, which is administered as a means of increasing immunity to the disease caused by those microorganisms.

VECTOR: An organism, such as an insect, that transmits a pathogen to the body of a host.

Koch's work was influencing the development of the germ theory, the influence of the English physician Joseph Lister (1827–1912) was being felt in operating rooms. Building on the work of both Semmelweis and Pasteur, Lister—for whom the well-known antiseptic mouthwash Listerine was named—began soaking surgical dressings in carbolic acid, or phenol, to prevent postoperative infection.

ANTIBIOTICS. Whereas antisepsis was the great battleground of the invisible war during the nineteenth century, in the twentieth century the most important struggle concerned the development of antibiotics. The first effective medications to fight bacterial infection in humans were sulfa drugs, developed in the 1930s. They work by blocking the growth and multiplication of bacteria and were initially effective against a broad range of bacteria, but many strains of bacteria have evolved resistance to them. Today, sulfa drugs are used most commonly in the treatment of urinary tract infections and for preventing infection of burn wounds.

The importance of sulfa drugs was eclipsed by that of penicillin, first discovered in 1928 by the British bacteriologist Alexander Fleming (1881–1955). Working in his laboratory, Fleming noticed that a mold that had fallen accidentally into a bacterial culture killed the bacteria. Having identified the mold as the fungus *Penicillium notatum,* Fleming made a juice with it that he called penicillin. He administered it to laboratory mice and discovered that it killed bacteria in the mice without harming healthy body cells.

It would be more than a decade before the development of a form of penicillin that could be synthesized easily. This drug arrived on the scene in 1941—just in time for the years of heaviest fighting in World War II—and after the war pharmaceutical companies began to manufacture numerous varieties of antibiotic. By the last decade of the twentieth century, however, a new problem emerged: bacteria were becoming resistant to antibiotics. This has been the case with medications used to treat conditions ranging from children's ear infections to tuberculosis.

An example is amoxicillin, a penicillin derivative developed in the late twentieth century. Many pediatricians found it a better treatment than penicillin for ear infections, because it did not tend to cause allergic reactions sometimes associated with the other antibiotic. However, by the late 1990s evidence surfaced indicating that certain types of bacteria had developed a protein that rendered amoxicillin ineffective against ear infections. Critics of amoxicillin (or of antibiotic treatments in general) maintained that widespread prescription of the antibiotic actually helped create that situation, because the bacteria developed the protein mutation defensively.

Because of these and similar concerns associated with antibiotics, doctors have begun taking measures toward controlling the spread of antibiotic-resistant diseases, for instance by prescribing antibiotics only when absolutely necessary. Research into newer types and combinations of drugs is ongoing, as is research regarding the development of vaccines to prevent bacterial infections.

WHERE TO LEARN MORE

Biddle, Wayne. *A Field Guide to Germs.* New York: Henry Holt, 1995.

The Big Picture Book of Viruses. Tulane University (Web site). <http://www.tulane.edu/~dmsander/Big_Virology/BVHomePage.html>.

Cells Alive! (Web site). <http://www.cellsalive.com/>.

Centers for Disease Control and Prevention (Web site). <http://www.cdc.gov/>.

Infection Index. Spencer S. Eccles Health Sciences Library, University of Utah (Web site). <http://medlib.med.utah.edu/WebPath/INFEHTML/INFECIDX.html>.

"Oral Health Topic: Infection Control." American Dental Association (Web site). <http://www.ada.org/public/topics/infection.html>.

The Race Against Lethal Microbes: Learning to Outwit the Shifty Bacteria, Viruses, and Parasites That Cause Infectious Diseases. Chevy Chase, MD: Howard Hughes Medical Institute, 1996.

Virtual Museum of Bacteria. Bacteria Information from the Foundation for Bacteriology (Web site). <http://www.bacteriamuseum.org/>.

Weinberg, Winkler G. *No Germs Allowed!: How to Avoid Infectious Diseases at Home and on the Road.* New Brunswick, NJ: Rutgers University Press, 1996.

BRAIN AND BODY

CHEMORECEPTION
BIOLOGICAL RHYTHMS

CHEMORECEPTION

CONCEPT

Chemoreception is a physiological process whereby organisms respond to chemical stimuli. Humans and most higher animals have two principal classes of chemoreceptors: taste (gustatory receptors), and smell (olfactory receptors). Though our sense of smell assists us in distinguishing among tastes, the gustatory and olfactory receptors are different in many respects—not only in their locations but also in terms of their chemical and neurological makeup. Capabilities of taste and smell vary widely among people, as a function of genetics, age, and even personal habits. Likewise, culture influences attitudes toward taste and smell. As for the animal kingdom, certain creatures are gifted with exceedingly acute senses, particularly where smell is concerned, but for some invertebrates, such as worms, there is really little distinction between taste and smell. Among the most interesting aspects of chemoreception in animals is the use of smell for communication, particularly through the release of special chemicals called pheromones. As to whether pheromones, which function as sex attractants, play a role in human interaction, many scientists remain skeptical.

HOW IT WORKS

The Senses

The term *sense,* which also may be called sensory reception and sensory perception, refers to the means by which an organism (usually an animal) receives signals regarding physical or chemical changes or both in its environment. Sensory reception and perception entail the translation of these signals, which represent changes in the matter or energy of the environment, into processes within the body and brain. For example, if you eat a cookie, your sense of taste translates the chemical data from that cookie into the sensation of sweetness, a sensation your brain most likely perceives as a pleasing one.

In everyday language, people are accustomed to speaking of five senses possessed by humans and, to a greater or lesser degree, by other animals: sight, touch, smell, hearing, and taste. The reality is rather more complex. For one thing, there are not really just five senses; rather, there are at least five others—and there may be more, depending on just how one defines and classifies the senses. These other senses include the kinesthetic sense, or the discernment of motion; the sensation of temperature, or distinguishing relative heat and cold; the awareness of pressure; the sense of equilibrium or balance; and the perception of pain.

All of these senses involve a response to stimuli, which may be defined as any phenomenon (that is, an observable fact or event, such as an environmental change) that directly influences the activity or growth of a living organism. Each of these senses has a biological component as well as a physical or a chemical one. In the next few paragraphs, we briefly discuss what this means, first by considering the nervous system in general terms and then by looking at the physical and chemical receptors that transmit data through that system.

THE NERVOUS SYSTEM. All sensation is biological in the sense that an organism not only must be living to experience it but also must have a functioning nervous system. The latter is a network, found in the bodies of all verte-

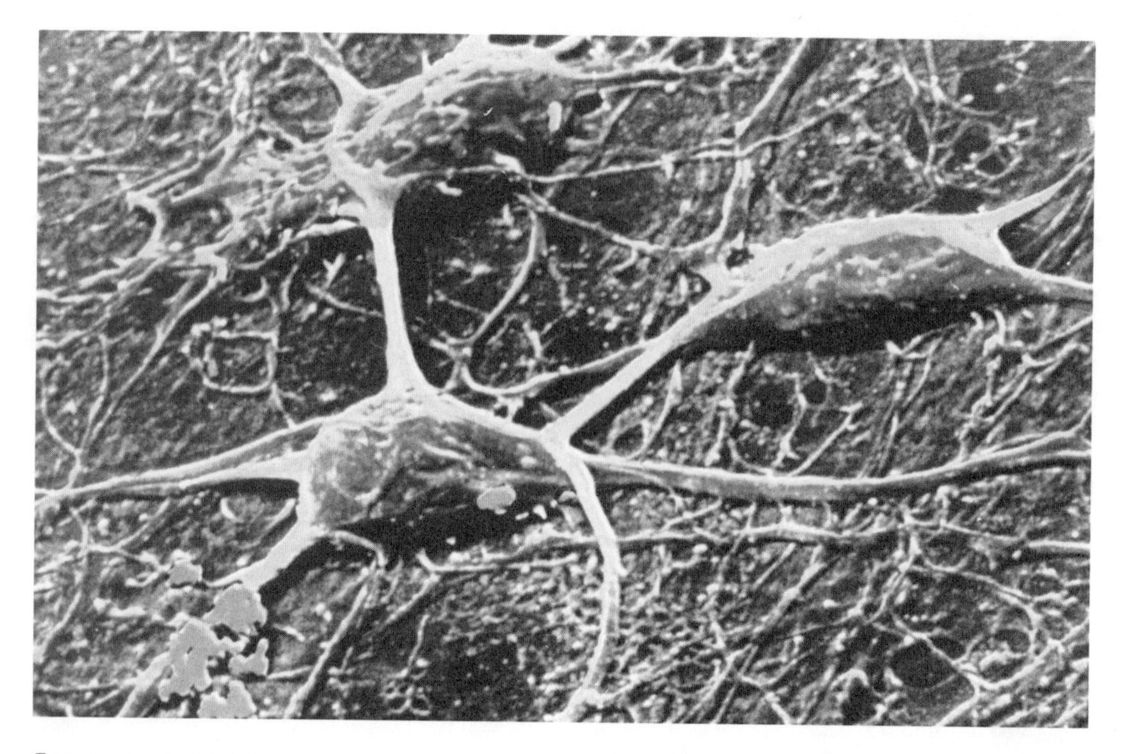

CHEMORECEPTION IS THE MEANS BY WHICH AN ORGANISM RECEIVES SIGNALS REGARDING CHEMICAL CHANGES IN ITS ENVIRONMENT, TRANSLATING THEM INTO PROCESSES WITHIN THE BODY AND BRAIN. NERVE CELLS, SUCH AS THE ONES SHOWN HERE, MAKE UP A NETWORK OF PRIMARY RECEPTORS THAT RECEIVE AND INTERPRET STIMULI AND TRANSMIT MESSAGES BASED ON THESE SENSORY DATA TO THE BRAIN. *(Photograph by Secchi-Lecague/Roussel-UCLAF/CNRI/Science Photo Library. National Audubon Society Collection/Photo Researchers, Inc. Reproduced by permission.)*

brates (animals with internal skeletons), whose purpose is to receive and interpret stimuli and to transmit impulses. Parts of the nervous system include the brain, spinal cord, nerves, and other components.

To be experienced through the senses, all data must be transmitted to the brain through the nervous system. This happens through the conversion and transmission of physical or chemical information, which takes place in sensory nerves known as receptors. A receptor is a structure in the nervous system that receives specific stimuli and is affected in such a way that it sends particular messages to the brain. The brain interprets these messages as sensations corresponding to the stimuli. Primary receptors are those that directly convert stimuli to electronic signals, which they send to the brain. Neurons, or nerve cells, serve as primary receptors. In addition, there are secondary receptors, which simply transmit signals between neurons. Secondary receptors induce a response in an adjoining neuron, thus sending the signal down the line toward the brain.

PHYSICAL AND CHEMICAL RECEPTORS. As we have noted, the changes in the outside environment that the brain interprets as sensation are either physical or chemical in origin. By *physical,* we mean the types of data and phenomena that are studied in the science of physics: matter, energy, and the interactions between them. Likewise, *chemical,* refers to objects of study in the scientific realm of chemistry: elements, compounds, and mixtures.

In the human body there are sensory nerves devoted to the interpretation of physical or chemical data. Among the receptors of physical data are photoreceptors, which respond to light and therefore play a part in the sense of sight; thermoreceptors, which respond to temperature and are concerned with the sense of heat and cold; and nociceptors, or pain receptors. In addition, there are mechanoreceptors, which respond to the mechanical properties of matter and are involved in the senses of touch, hearing, and equilibrium.

As for chemical data, they are interpreted through chemoreceptors, which govern the senses of smell and touch. For this reason, chemoreceptors sometimes are referred to as the chemical senses. Our focus in the present context, of course, is on *chemoreception,* the term for the physical process whereby organisms—not just humans, but virtually all animals—respond to chemical stimuli.

DISTINGUISHING GUSTATION AND OLFACTION

We tend to associate taste and smell, and indeed there is some relation between them, but there are also numerous distinctions. Not everyone with an acute sense of smell, for instance, has as finely honed a sense of taste. Conversely, people with poor senses of smell do not necessarily suffer a corresponding impairment in their taste buds. The neurological structures that have to do with taste and smell, at least in higher animals, are not the same. Whereas gustatory sensation travels via secondary receptors called epithelial cells, which pass messages to adjacent neurons, olfactory signals enter the nervous system through primary receptors.

The principal sensory organs of gustation and olfaction in a human being are, respectively, the taste buds on the tongue and the olfactory receptors located within the olfactory epithelium in the nasal cavity. They are designed to respond to two entirely different types of chemical materials. Taste buds are best at receiving sensory data from water-soluble chemicals, or those that dissolve in water, while the olfactory receptors are most attuned to vapors that are water-insoluble.

WATER AND OIL. Almost everyone who has ever eaten a hot, spicy dish has tried to "put out the fire" in their mouths by drinking water, only to discover to their dismay that water only seemed to make the problem worse. On the other hand, milk is usually quite effective. The reason is that most spicy-hot substances tend to be oily, and therefore they do not readily form intermolecular bonds with water. On the other hand, milk, though it is largely composed of water, also contains oily fat particles. Thanks to the chemistry of water- and oil-based substances, it is true (as the old saying goes) that "oil and water don't mix."

In water, hydrogen and oxygen have significantly different levels of electronegativity, or the relative ability of an atom to attract valence electrons, which are used in chemical bonding. Therefore, a water molecule tends to be electrically bipolar, with all the negative charges on the end or side where the oxygen atom is located and all the positive charges on the end or side of the hydrogen molecules. In petroleum and most other oily substances, however, the molecules typically include some combination of carbon and hydrogen, which have similar electronegativity values. For that reason, the electric charges are distributed more or less evenly throughout the molecule. As a result, oily substances form very loose intermolecular bonds, and they tend not to bond with water-based substances.

As for taste buds and olfactory membranes, it is fitting that the taste buds would be more receptive to water-based foods and liquids, since most foods (including jalapeño peppers) contain at least some water. Obviously, if we can taste spicy-hot substances, then there must be some receptivity to oil-based foods. Part of this receptivity involves the olfactory membranes, which, as we have noted, are more receptive to oily materials.

ANIMALS AND CHEMORECEPTION

Just as taste and smell are sharply distinguished in humans (despite the fact that smell aids us in the process of tasting), the same is true of most vertebrates. In the case of invertebrates, such as worms, however, there is much less differentiation between gustatory and olfactory receptors. These animals, in fact, may have only one chemical sense, with only slight differences between what scientists call distance chemoreception, more or less the same as smell in vertebrates, and contact chemoreception, which corresponds to vertebrates' sense of taste.

Such analogies can be made because distance chemoreceptors appear to respond to non-water-soluble substances in the same way that the olfactory receptors in vertebrates do, while contact chemoreceptors are more responsive to water-soluble chemicals. As with many other animals, these senses are linked to numerous behaviors—not just feeding—in invertebrates. Distance chemoreception enables invertebrates to sense the presence of chemicals that pose a danger, signaling the need to move away, while contact chemoreception assists the invertebrate in determining when to mate and when to lay eggs.

THE MANY FUNCTIONS OF CHEMORECEPTION. Terrestrial, or land-based, animals whose skins secrete mucus

TERRESTRIAL ANIMALS WHOSE SKINS SECRETE MUCUS, LIKE THE SNAIL, HAVE WHAT SCIENTISTS CALL THE COMMON CHEMICAL SENSE, WHICH MAKES THEM SENSITIVE TO THE PRESENCE OF FOREIGN CHEMICALS ON THE SURFACE OF THEIR BODIES. *(JLM Visuals. Reproduced by permission.)*

(e.g., snails and slugs) as well as aquatic animals have what scientists call the *common chemical sense,* which makes them sensitive to the presence of foreign chemicals anywhere on the surface of their bodies. Even humans and other animals whose skin does not secrete mucus across its entire surface have an evolutionary remnant of this sense. Thus, the mucous membranes in the eyes, mouth, nose, and genitals respond to chemical irritants.

This is yet another example of the fact that, while chemoreception in animals is associated most readily with tasting and smelling, it is linked to myriad other functions as well. Some insects may use chemoreception to detect the presence of moisture, and many animals apply it for a variety of purposes. Such purposes include selection and courtship of a mate as well as the identification of friends or foes. A dogs' sense of smell tells it everything it needs to know about a new animal or person it encounters; similarly, a cat may identify another cat by sniffing its rectum. Smell also can be used to mark territory, which is why dogs and cats mark their "turf" by urinating. Both chemical senses, particularly smell, are important mediums of communication in the animal kingdom.

MECHANISMS OF TASTE AND SMELL

Scientists in the nineteenth century believed that human tongues have receptors for four basic tastes: sweet, sour, salty, and bitter. More current research, however, shows that the taste receptors on the tongue are more complicated than was previously thought. Still, it does seem to be the case that specific kinds of taste buds are clustered in certain areas. Taste buds, so named because they look like plant buds when viewed under a microscope, cover the tongue and, to a lesser extent, can be found on the cheek, throat, and roof of the mouth. As we shall see, however, some people have a greater concentration of taste buds than do others. In the mouth, saliva breaks down the chemical components of substances, which travel through the pores in the papillae (small protuberances on the surface of the tongue) to reach the taste buds themselves.

When specific proteins in food bind to receptors on the taste buds, these receptors send messages to the cerebral cortex, a surface layer on the brain that coordinates sensory information. And though taste buds in certain regions of the tongue have an affinity for particular flavors, as we discuss later, the intricacies of the neural and chemical networks tend to suggest that nothing is clear-cut about this highly complex biological process.

OLFACTION: A DIRECT SENSE. Before modern times, several of the leading theories of vision maintained that the eye actually interacts with the objects it sees. Now we know that our eyes simply receive light reflected off those objects. By contrast, smelling is a direct experience, because we inhale microscopic portions of substances that have evaporated and make their way into the nasal cavity, where they chemically interact with sense receptors. Cells in the nose detect odors through receptor proteins on the cell surface, which bind to odor-carrying molecules. A specific odorant docks with an olfactory receptor protein in much the same way that a key fits into a lock. This, in turn, excites the nerve cell, causing it to send a signal to the brain.

Although there are many tens of thousands of odor-carrying molecule types in the world, meaning that there are as many different smells, there are only hundreds (or at most about 1,000) different types of olfactory receptors in even the most sensitive animal species. This finding has led scientists to speculate that not every receptor recognizes a unique odorant molecule; rather, similar odorants can bind to the same receptor. Another way to put this, in light of the lock and key analogy, is that a few loose-fitting odorant "keys" of roughly similar structure can enter the same receptor "lock."

Just as we sense smells directly, the olfactory sense is also direct in the way that signals are transmitted to the brain. Vision, by contrast, puts into play several steps: a receptor cell detects light and passes the signal to a nerve cell, which passes it on to another nerve cell in the central nervous system, which then relays it to the visual center of the brain. In olfaction, on the other hand, the olfactory nerve cells perform all these functions.

In most animals these cells take scent messages directly to the nerve cells of the olfactory bulb in the brain. With insects and other invertebrates whose brains are relatively simple, functioning primarily as clearinghouses for sensations, the olfactory nerves send signals to the olfactory ganglia, a mass of nerve tissue that connects nerve cells external to the brain and spinal cord. In higher animals, such as humans, olfactory signals go to the olfactory cortex, a structure in the brain where higher functions, such as memory and emotion, are coordinated with the sense of smell. Hence, as many people have observed, the sense of smell is linked strongly with long-term memory in a way that such senses as sight and touch are not.

REAL-LIFE APPLICATIONS

Taste Buds

Humans have only about 10,000 taste buds, whereas rabbits have 17,000 and cows some 25,000. This seems more than a little ironic, since humans enjoy by far the most varied diet. Both of the other animals are herbivores, meaning that they do not eat meat, nor are they accustomed to sweets and the many other varieties of taste in the diet of the average well-fed American. If anything, cows, with about 50% more taste buds than rabbits, eat a diet even more plain than that of their furry, fleet-footed fellow mammals.

Though our tasting equipment (that is, the chemoreceptors for taste in our tongues) may be much less sophisticated than that of cows or rabbits, the number of tastes our palate can recognize is as varied as a spectrum of color swatches at a paint store. Despite such variation, there are only a few basic tastes, most notably, the ones that once were thought to constitute primary tastes analogous to the primary colors: sweet, sour, salty, and bitter.

THE FOUR "BASIC TASTES." When you lick an ice cream cone, you may notice that you are experiencing the sweetness of it primarily at the tip of your tongue. This is not simply because you are licking it with the tip but also because there is a heavier concentration of sweetness receptivity in that area. On the other hand, if you eat a sour gumball, you experience the taste most notably on the sides of your tongue, where receptivity to sour tastes is strongest. Reception of salty tastes takes place near the front of the tongue, just behind the tip. As for bitter tastes, the focal point of receptivity appears to be near the back of the tongue. The latter may be a highly useful adaptive mechanism we have developed along the way, since many poisons are bitter, and the gagging reflex takes place near the back of the mouth. Not all bitter tastes are revolting, however: olives, which many people love, are bitter as well, as is coffee when it has no cream or sugar to alter its flavor.

As we noted earlier, the organization of taste receptors on the tongue is not quite as simple as once was believed. For example, though our receptivity to sweetness seems to take place primarily at the tip, to a lesser extent we taste sweetness and other flavors all over the tongue. Furthermore, such flavors as sweet, sour, bitter, and salty are not the sum total of "taste" as we experience it; not only smell but also texture affects the taste of substances. Genetic and cultural factors also influence a person's unique tastes and may explain why one person loves sweets while another cannot get enough of tangy tastes.

TASTE AND OTHER SENSES

Taste does not work alone; on the contrary, our sense of the smell, texture, and temperatures of foods affects our overall perception of its flavor and in some cases its desirability. When food is in the mouth, it produces a scent, which enters the nose through the nasopharynx, an opening that links the mouth and the nose. Because we experience smell more directly and our noses are more sensitive to olfactory sensations than our taste buds are to gustatory ones, people often experience the flavor of food first by its smell. This greatly affects our perception of what we eat.

For example, while many people like blue cheese, many others despise it, and this probably has more to do with its smell than with its taste. While its taste is rather tangy, it is not quite as "radical" as the aroma of this cheese, which has been compared to everything from old socks to vomit. Other cheeses, such as gorgonzola and, particularly, Limburger, are even more pungent and therefore have more than their share of detractors—but again, the smell is more extreme than the taste.

On a more pleasant note, anyone who has ever enjoyed a good steak or a hamburger cooked on an open grill will attest to the fact that a great deal of that enjoyment comes from the aromas of cooking. Usually it is the smell of a delectable food item, which we detect long before we taste it, that causes our salivary glands to begin operating, preparing us for the process of consuming and digesting the dish. Additionally, different types of cooking have particular smells and tastes associated with them, which people may find more or less appealing. For example, scrambled eggs cooked over an open campfire are likely (all other things being equal) to be more appealing to most people than eggs cooked in a skillet on an electric stove. But if the campfire is fueled with burning dung instead of wood, most Americans would choose the stove. In sparsely forested parts of the third world, however, animal dung is a principal form of fuel, and without it people might have to eat meals raw.

TEXTURE AND TEMPERATURE. Texture and temperature may have less impact on taste than does smell, but these are still significant factors. Take the example of three plates of French fries. One is limp and soggy in consistency, and another is so crispy that the French fry crunches like potato chips. The third, however, is just a bit crispy on the outside and just a bit soft on the inside. Most people, though certainly not all, would judge the third plate of French fry the most delectable—purely on the basis of texture.

By the same token, many people are less than enthusiastic about boiled okra, owing to its slimy consistency, whereas fried okra is more appealing to most Americans, since it lacks that gooey texture. Likewise, temperature plays more of a role than one might think. Many people, for example, find cold coffee unappealing, though others have a fondness for iced coffee, at least if it has milk and sweetener. Similarly, many Americans enjoy the combination of cold ice cream and hot pie or cobbler, partly because the contrast of temperatures adds to the overall flavor.

DIFFERENT PEOPLE, DIFFERENT SENSATIONS

Not all people "taste" the same—that is, not all people have the same sense of taste or the same level of acuity for distinguishing different flavors. One person may have 10–1,100 taste buds per square inch (6.45 cm^2) on the tongue, indicating a huge range of sensitivities with regard to gustatory data. Research also has shown that women, on average, have more taste buds than men, proving what many a woman has long asserted—that women have "better taste" than men.

Though it does appear that the average female has a more acute sense of gustation than the typical male, taste buds are not the only biological factor involved in recognizing flavors. For example, the amount of saliva a person naturally generates and the amount of salt that appears in one's saliva play a major role in determining an individual's response to salty foods. A person whose mouth generates less saliva is more sensitive to the salt in foods, whereas a person prone to generating a greater quantity of saliva is less likely to taste the salt that has been added to a dish. That person is therefore more likely to add salt.

Ability to smell also varies from person to person, though it appears that a less acute sense of smell may be a sign not merely of fewer olfactory receptors but also of an actual olfactory disorders. Just as some people may be color-blind, it appears that others are "smell-blind." And just as being color-blind can have very serious consequences, for instance by causing a color-blind driver to miss a red light, much the same is true with an olfactory disorder. To a greater extent than one might immediately guess, smell serves a protective function. For example, without a sense of smell, one cannot tell if food is spoiled, unless, of course, it has reached such a state of putrefaction that it shows visible signs. For a person with an ordinary sense of smell, however, rotten food sends a signal through the olfactory receptors, which may cause a gag reflex when smelling food that has spoiled.

AGE AND SENSE OF TASTE. It is an experience familiar to many people, and it goes something like this. Let us say that in his boyhood, a man enjoyed a particular brand of candy, of which he could never get enough. Left to his own devices, he probably would have eaten so much that he would have become sick. The fact that this never happened had more to do with his parents—and the fact that his allowance money had to go for other things as well—than it did with his own natural sense of restraint. So he dreamed of the day when he became a grown-up, when he could eat whatever he wanted.

Eventually, he forgets this dream amid the many distractions of adolescence, but then one day many years later, as a grown man, he happens to see this particular item of candy in a store, and all his childhood memories come back to him. He buys several pieces, thrilled that he can enjoy in complete freedom what was once a rare treat. He can barely wait to get into his car, open the first piece, put it in his mouth—and then he recoils in disgust, thinking, *How did I ever enjoy that?* Disappointed, he throws away the rest of the candy.

The candy, of course, has not changed, but the man—and his taste buds—have. As we mature, so do our taste buds, and their numbers increase, leading to greater sophistication of taste. Children tend to like very basic tastes, particularly sweet and sour, and respond much less favorably to the subtlety in more complex dishes. This fact, combined with an increased awareness of health issues as one ages, explains why an adult might relish a broccoli casserole but find cotton candy so sweet as to be repugnant, whereas a child's reaction would probably be just the opposite.

Just as taste buds mature along with the people who own them, they also age. Every 3-10 days, on average, our taste buds regenerate themselves, replacing old ones that have been worn out by foods that are too hot, too cold, or otherwise too taxing to the chemoreceptors in our tongues. But as people grow older, their taste buds replace themselves less frequently, and therefore their sense of taste becomes less finely tuned. An older person may require much more sweetness or spice to taste a particular food.

CHEMORECEPTION IMPAIRMENT AND DISORDERS

Sense of smell also deteriorates with age; as we noted earlier, this can pose dangers, because a person depends on the sense of smell for protection more than one might imagine. For example, in addition to the inability to detect spoiled food, an elderly person would be far less likely to smell smoke if a building were on fire. Older people also are less likely to be cognizant of olfactory data that send messages concerning unpleasant smells of a less critical nature—body odor, for example.

Many of the problems of gustation and olfaction suffered by the elderly are reflected, at a much younger age, in the bodies of smokers. In addition to its many other negative effects, smoking deadens taste buds and desensitizes the olfactory receptors. It is not uncommon to see a heavy smoker salting pizza or some other food that for people with ordinarily functioning taste buds would not seem to require any salt. As for olfactory sensation, a smoker becomes accustomed to the reek of stale smoke and ashes.

On a more temporary basis, many people find their senses of taste and smell impaired by illness. A person with a cold or flu, even at its final stages, usually has enough congestion that the senses of both smell and taste are limited, if not almost nonexistent. In this instance, the lack

TASTE AND SMELL ARE NOT PURELY BIOLOGICAL, BUT ALSO REFLECT CULTURAL FACTORS. AS LATE AS THE 1970S, MOST AMERICANS WOULD HAVE SAID THAT THE IDEA OF EATING RAW FISH WAS REVOLTING. TODAY, AMERICA'S CITIES BRISTLE WITH JAPANESE RESTAURANTS THAT SERVE SUSHI, SHOWING THAT CULTURAL TASTES CAN CHANGE. *(© Japack Company/Corbis. Reproduced by permission.)*

of ability to taste serves to illustrate the strong link between gustation and olfaction: the taste buds themselves are working fine, but the lack of smell, resulting from congestion, hinders the brain's ability to process flavor.

TASTE AND SMELL DISORDERS. In addition to people whose olfactory and gustatory senses are impaired by age, illness, or smoking, between two million and four million Americans suffer from some sort of taste or smell disorder. The inability to taste or smell not only robs an individual of certain sensory pleasures, it also can be dangerous to one's mental health. Some psychiatrists believe that the lack of taste and smell can have a profoundly negative effect on a person's quality of life, leading to depression or other psychological problems.

Whereas impairments of smell and taste brought on by cold, flu, various viral and bacterial infections, and even allergies are usually temporary, some other illness-related taste and smell disorders are more long term. Such is the case with neurological disorders due to brain injury or diseases such as Parkinson's or Alzheimer's (conditions marked by tremors and mental deterioration, respectively). These conditions can cause more permanent damage to the intricate neural networks that process tastes and smells.

Drugs such as lithium, used to treat bipolar disorder (what used to be called manic depression) also may cause taste and smell disorders. This occurs because certain drugs (lithium is just one example among many) inhibit the action of certain enzymes, affect the body's metabolism, and interfere with the neural networks and receptors involved in tasting and smelling. Exposure to such environmental toxins as lead, mercury, insecticides, and solvents (e.g., paint thinner) also can damage taste buds and sensory cells in the nose or brain.

CULTURE AND CHEMORECEPTION

One of the favorite delicacies in the Philippines is known as *dinuguan,* or pork cooked in pork blood. Chances are that a visitor from England or northern Europe, when told the constituents of the dish, would feel right at home; an American, however, most likely would try to think of an excuse to pass up this Filipino delight. To most Americans, it would seem that eating dinuguan, or the many varieties of blood pudding or blood sausage common in England and Scandinavia, is simply "gross"—that it is objectively and unquestionably disgusting. But in the 1960s or even the 1970s, most Americans, even in large cities, would have said that the idea of eating raw fish was revolting. Today, however, America's cities and suburbs bristle with Japanese restaurants that serve sushi, an indication of the fact that cultural tastes can change.

As is discussed in Parasites and Parasitology, it should be noted that improperly cooked pork and fish (especially raw fish) are quite likely to serve as hosts for disease-carrying worms, so one should exercise care before sitting down to a plate of dinuguan or sushi. But then again, many Americans like their steaks on the rare side, and undercooked beef certainly has its share of pathogens (disease-carrying parasites) as well. It seems that it is not health concerns that explain our cultural double standards about certain food items.

Of course, Americans are not the only people with quirky standards regarding tastes and smells; in fact, every culture has its idiosyncrasies. In China it is not considered at all offensive for one's breath to smell heavily of onions; on the other hand, the smell of dairy products, virtually nonexistent in Chinese cuisine, is considered highly offensive. What does all of this prove? Only that taste and smell are not purely biological but also reflect cultural factors.

During the nineteenth century, many European scientists embraced a racist theory concerning the olfactory capabilities of different peoples around the world. According to this highly unscientific "theory," non-Europeans were more primitive than Europeans and therefore closer to animals, which meant that they had a stronger sense of smell. Completing the loop of this circular logic, subscribers to this nonsense maintained that because non-Europeans had a stronger sense of smell, it proved they were more primitive than Europeans!

HUMANS, ANIMALS, AND SMELL

By the early twentieth century, physiologists had begun to explore much more scientific ideas concerning olfactory and gustatory abilities in humans—abilities that, needless to say, are not a function of race or ethnicity. Only one assumption of the old-fashioned European scientists was correct: that animals have a stronger sense of smell than do humans. The human nose is capable of detecting odors so faint that their proportion of the surrounding air is in the range of only a few parts per trillion. Many researchers are beginning to wonder whether smell does not play a greater role in human behavior and biology than previously was believed. For example, research has shown that only a few days after the birth of her baby, a human mother can smell the difference between a vest worn by her baby and one worn by another.

Nevertheless, the fact remains that the olfactory abilities of many animals are far beyond those of humans. This is a fact that hardly needs scientific verification, since most of us have observed dogs' reliance on their strong sense of smell. This explains why police use dogs to detect illegal drugs and explosives and to track runaway prisoners or the bodies of murder victims. Dogs are not the only animals gifted with acute senses of smell, which aid them in finding their way to specific targets. Salmon, for example, manage to find their way back to the streams where they were hatched, guided by their sense of smell. (For more about animals' navigational abilities, see Migration and Navigation.)

Most vertebrates other than humans have many more olfactory nerve cells in a proportionately larger olfactory epithelium, and this probably gives them much more sensitivity to odors. In addition, most land vertebrates have a specialized scent organ in the roof of their mouths called the vomeronasal organ, which gives them far more sensitivity to odors than humans have.

PHEROMONES. Some animals are known particularly for the odors they excrete, especially when it is an animal such as a skunk or stinkbug that puts off a repellent odor as a defense mechanism. But animals also send out much more subtle smells known as pheromones. Chemical substances produced and secreted by animals, pheromones serve as stimuli for behavioral responses on the part of other animals of the same species. Pheromones are common among insects as well as many vertebrates, but they are nonexistent among bird species.

Among so-called "social insects" such as bees and ants, pheromones play a particularly strong role. The queen honeybee gives off what is called the queen substance, a pheromone that acts to prevent the development of ovaries among workers, which are biologically unproductive females. Pheromones are vital for communication among

KEY TERMS

CHEMORECEPTION: A physiological process whereby organisms respond to chemical stimuli.

GUSTATORY: Of, or involving, the sense of taste (*gustation*).

INVERTEBRATE: An animal without an internal skeleton.

NERVOUS SYSTEM: A network, found in the bodies of all vertebrates, whose purpose is to receive and interpret stimuli and to transmit impulses. Parts of the nervous system include the brain, spinal cord, and nerves.

OLFACTORY: Of, or involving, the sense of smell (*olfaction*).

PHEROMONES: Chemical substances produced and secreted by animals, which serve as stimuli for one or more behavioral responses on the part of other animals of the same species.

RECEPTOR: A structure in the nervous system that receives specific stimuli and is affected in such a way that it sends particular messages to the brain. These messages are interpreted by the brain as sensations corresponding to the stimuli.

SENSE: The means by which an organism (usually an animal) receives signals regarding physical or chemical changes or both in its environment. Sometimes called *sensory reception* or *sensory perception.*

STIMULUS: Any phenomenon (i.e., an observable fact or event, such as an environmental change) that directly influences the activity or growth of a living organism.

VERTEBRATE: An animal with an internal skeleton.

social insects, which have little or no sense of sight. Ants, bees, and wasps send out smells to alarm others of danger, and ants may create a path of pheromones to guide others to a food source.

The function for which pheromones are most widely known, however, is as a sex attractant. Male musk deer are noted for their excretion of musk, which, as the result of to its pleasant and powerful smell, is often an ingredient in perfume manufacture. Though musk is a sex attractant, it is not a pheromone, which is a much less obvious scent and which, as we have noted, likely has an effect *only* on animals of the same species.

Experiments have shown that male mice who lack a gene for a pheromone receptor are likely to attempt to mate with males, simply because they cannot tell the difference. With humans, of course, it is easy to tell the difference between males and females on sight, but do humans also respond to pheromones? The fact that these chemicals theoretically could induce mating behavior led cologne and perfume makers long ago to embrace the idea of human pheromones, whose presence in a manufactured scent obviously would be a boon to many a single man or woman. Despite the enthusiastic claims of perfume manufacturers, however, many scientists have yet to be convinced that pheromones play a significant role for humans.

Regarding the fact that the human anatomy includes the vestige of a vomeronasal organ, the olfactory researcher Charles Wysocki told Lee Bowman in an article published on the National Library of Medicine Web site, "It's like the appendix—it's there, but it doesn't seem to do anything." As for scent makers' promises that pheromones will help users attract partners, Wysocki said, "Sure the claims are out there. ... 'All you have to do is put this on and you'll score.' But there's nothing in the published biomedical literature [to indicate] that we have any kind of pheromone that draws a partner." Research does suggest that people give off chemical messages that correspond to certain moods, but it is a long way from this to the idea of a spray-on aphrodisiac.

WHERE TO LEARN MORE

Ackerman, Diane. *A Natural History of the Senses.* New York: Vintage Books, 1991.

Bowman, Lee. "A Nose for Romance?" U.S. National Library of Medicine/National Institutes of Health (Web site). <http://www.nlm.nih.gov/medlineplus/news/fullstory_6125.html>.

Chemical of the Week—Chemoreception: The Chemistry of Odors. Science Is Fun/University of Wisconsin–Madison (Web site). <http://scifun.chem.wisc.edu/CHEMWEEK/Odors/chemorec.html>.

Chemoreception Links. Leffingwell & Associates (Web site). <http://www.leffingwell.com/links5.htm>.

The ChemoReception Web (Web site). <http://www.csa.com/crw/home.html>.

Evans, David H. *The Physiology of Fishes.* Boca Raton, FL: CRC Press, 1998.

Finger, Thomas E., Wayne L. Silver, and Diego Restrepo. *The Neurobiology of Taste and Smell.* New York: Wiley-Liss, 2000.

Monell Chemical Senses Center (Web site). <http://www.monell.org/>.

Pybus, David, and Charles Sell. *The Chemistry of Fragrances.* Cambridge, England: Royal Society of Chemistry, 1999.

Rivlin, Robert, and Gravelle, Karen. *Deciphering the Senses: The Expanding World of Human Perception.* New York: Simon and Schuster, 1984.

Whitfield, Philip, and D. M. Stoddart. *Hearing, Taste and Smell: Pathways of Perception.* Tarrytown, NY: Torstar Books, 1984.

BIOLOGICAL RHYTHMS

CONCEPT

People frequently talk about body clocks, a term that refers to the patterns of energy and exhaustion, functioning and resting, and wakefulness and sleep that characterize everyday life. In fact, the concept of the body clock, or circadian rhythm, is part of a larger picture of biological cycles, such as menstruation in mammalian females. Such cycles, which assume a variety of forms in a wide range of organisms, are known as biological rhythms. These rhythms may be defined as processes that occur periodically in an organism in conjunction with and often in response to periodic changes in environmental conditions—for example, a change in the amount of available light. Not all aspects of the body clock are part of day-to-day experience, and this is fortunate, since these interruptions in the healthy flow of biological rhythms can threaten the well-being of the human organism. Among these challenges to the ordered working of bodily "clocks" are jet lag, seasonal affective disorder (SAD), and other disorders linked to a range of causes, including drug use.

HOW IT WORKS

UNDERSTANDING BIOLOGICAL RHYTHMS

Among the many varieties of biological rhythm, the most well known are those relating to sleep and wakefulness, which are part of the circadian rhythm that we discuss later in this essay. Circadian, or daily, cycles are only one type of biological rhythm. Some rhythms take place on a cycle shorter than the length of a day, while others are based on a monthly or even an annual pattern.

Nor do all cycles involve sleep and wakefulness: menstruation, for instance, is a monthly cycle related to the sloughing off of the lining of the uterus, a reproductive organ found in most female mammals. Another biological rhythm is the beating of the heart, which, of course, takes place at very short intervals. Nonetheless, the circadian rhythm is the most universal of biological cycles, and it is the focus of our attention in this essay.

BIOLOGICAL CLOCKS. In discussing the operation of biological rhythms, the term biological clock often is used. A biological clock is any sort of mechanism internal to an organism that governs its biological rhythms. One such mechanism, which we examine in the next section, is the pineal gland. Internal clocks operate independently of the environment but also are affected by changes in environmental conditions.

Examples of such alterations of conditions include a decrease (or increase) in the hours of available light due to a change of seasons or a change in time alteration due to rapid travel from west to east or north to south. In the latter instance, a condition known as jet lag—increasingly familiar to humans since the advent of regular air travel in the mid–twentieth century—may ensue.

THE PINEAL GLAND

Governing human biological cycles—the "computer" that operates our biological clocks—is the pineal gland, a cone-shaped structure about the size of a pea located deep inside the brain. At one time, the great French philosopher and mathematician René Descartes (1596–1650) held that the pineal gland was actually the seat of the soul. Though it might seem absurd now that a respected thinker would seriously attempt to locate the soul in space, as though it were a physical object, Descartes's claim resulted from hours of painstaking dissection conducted on animals.

In searching for the human soul, Descartes sought that ineffable quality described some fifteen centuries earlier by the Roman emperor and philosopher Marcus Aurelius (121–180), who wrote, "This being of mine, whatever it really is, consists of a little flesh, a little breath, and the part which governs." As it turns out, the pineal gland *is,* in a sense, "the part which governs": it may not be the home of the soul (which, in any case, is not a question for science), but it does govern human circadian rhythms and thus has a powerful effect on the manner in which we experience the world.

MELATONIN. The pineal gland secretes two hormones (molecules that send signals to the body), melatonin and serotonin. During the late 1990s, melatonin became a popular over-the-counter treatment for persons afflicted with sleep disorders, because it is believed that the hormone is associated with healthful sleep. Scientists do not fully understand the role that melatonin plays in the body, although it appears that it regulates a number of diurnal, or daily, events.

In addition, melatonin seems to serve the function of controlling fat production, which is one reason why good sleep is associated not only with a healthy lifestyle but also with a healthy physique. Many health specialists maintain that for adults there is a close link between a "spare tire" (that is, fat accumulation around the waist) and stress, lack of sleep, and low melatonin levels.

Among the many roles melatonin plays in the body is its job of regulating glucose levels in the blood, which, in turn, serve to govern the production of growth hormone, or somatotropin. Growth hormone is associated with the development of lean body mass, as opposed to fat, which is why athletes involved in the Olympics and other major sporting competitions sometimes have illegally "doped" with it as a means of increasing strength. It is not surprising, then, to learn that children—who clearly need and use more growth hormone and who also need more hours of sleep than adults—also have higher melatonin levels.

SEROTONIN. Melatonin is not the only important hormone that is both secreted by the pineal gland and critical to the regulation of the body clock. Complementary to melatonin is serotonin, which is as important to waking functions as melatonin is to sleepiness. Like melatonin, serotonin serves several functions, including the regulation of attention.

Serotonin is among the substances responsible for the ability of a human with a healthily functioning brain to filter out background noise and sensory data. Thanks in part to serotonin, you are able to read this book without having your attention diverted by other sensory data around you: the voice of someone talking nearby, the sunlight or a bird singing outside, the hum of a light or a fan in the room.

By contrast, a person under the influence of the drug LSD (lysergic acid diethylamide) is not able to make those automatic filtering adjustments facilitated by serotonin. Instead, he or she is at the mercy of seemingly random intrusions of outside stimuli, such as the color of paint on a wall or the sound of music playing in the background. The secret of LSD's powerful hallucinatory effect can be attributed in part to the fact that it apparently mimics the chemistry of serotonin in the brain, "tricking" the brain into accepting the LSD itself as serotonin.

With regard to body clocks and biological rhythms, serotonin plays an even more vital governing role than does melatonin, since melatonin, in fact, is created by the chemical conversion of serotonin. On regular daily cycles the body converts serotonin to melatonin, thus influencing the organism to undergo a period of sleep. Then, as the sleeping period approaches its end, the body converts melatonin back into serotonin.

REAL-LIFE APPLICATIONS

CIRCADIAN RHYTHMS

The term *circadian* derives from the Latin *circa* ("about") and *dies* ("day"), and, indeed, it takes "about" a day for the body to undergo its entire cycle of serotonin-melatonin conversions. In fact, the cycle takes almost exactly 25 hours. Why 25 hours and not 24? This is a fascinating and perplexing question.

It would be reasonable to assume that natural selection favors those organisms whose body clocks correspond to the regular cycles of Earth's rotation on its axis, which governs the length of a day—or, more specifically, a solar day. Yet the length of the human daily cycle has been confirmed in countless experiments, for instance, with subjects in an environment such as a cave, where levels of illumination are kept constant for weeks on end. In each such case, the subject's body clock adopts a 25-hour cycle.

POSSIBLE EXPLANATIONS FOR THE 25-HOUR CYCLE. One might suggest that the length of the cycle has something to do with the fact that Earth's rate of rotation has changed, as indeed it has. But the speed of the planet's rotation has *slowed,* because—like everything else in the universe—it is gradually losing energy. (This is a result of the second law of thermodynamics.)

About 650 million years ago, long before humans or even dinosaurs appeared on the scene, Earth revolved on its axis about 400 times in the interval required to revolve around the Sun. This means that there were 400 days in a year. By the time *Homo sapiens* emerged as a species about two million years ago, days were considerably longer, though still shorter than they are now. This only means that the 25-hour human body clock would have been even less compatible with the length of a day in the distant past of our species.

One possible explanation of the 25-hour body clock is the length of the lunar day, or the amount of time it takes for the Moon to reappear in a given spot over the sky of Earth. In contrast to the 24-hour solar day, the lunar day lasts for 24 hours and 50 minutes—very close in length to the natural human cycle. Still, the exact relationships between the Moon's cycles and those of the human body have not been established fully: the idea that lunar cycles have an effect on menstruation, for instance, appears to be more rumor than fact.

PEAKS AND TROUGHS. On the other hand, circadian rhythms do mirror the patterns of the Moon's gravitational pull on Earth, which results in a high and low tide each day. Likewise, the human circadian rhythm has its highs and lows, or peaks and troughs. In the circadian trough, which occurs about 4:00 A.M., body temperature is at its lowest, whereas at the peak, around 4:00 P.M., it reaches a high. A person may experience a lag in energy after lunchtime, but usually by about 4:00 in the afternoon, energy picks up—a result of the fact that the body has entered a peak time in its cycle.

This fact, by the way, points up the great wisdom of a practice common in Spanish-speaking countries and some other parts of the world: siestas. The siesta devotes one of the least productive parts of the day, the post-lunch lag, to rest, so that a person is equipped with energy for the rest of the afternoon and early evening—at precisely the time when energy is at a high. To compensate for the time "lost" on napping, many such societies maintain a later schedule, with offices closing in the early evening rather than late afternoon and with evening meals served at about 9:00 P.M.

Note that even though our body clocks run on a 25-hour day, they readily adjust to the 24-hour world in which we live. As long as a person is exposed to regular cycles of day and night, the pineal gland automatically adapts to the length of a 24-hour solar day. If a person has been living in a sunless cave, with no exposure to daylight for a length of time, it would take about three weeks for the pineal gland to reset itself, but thereafter it would track with Earth time consistently.

The adjustment of the body clock is not simply a matter of sending signals for sleep and wakefulness. In fact, the pineal gland is at the center of a complex information network that controls sleep cycles, body temperature, and stress-fighting hormones. Hence the link that we noted earlier between body temperature and circadian rhythms: just as the body reaches its lowest temperature in the circadian trough, it also enters a period of extremely deep sleep.

REGULATING THE BODY CLOCK. Tied in with these sleep patterns are many other bodily functions. For example, bodybuilders and others who work out with weights experience their greatest benefits not when lifting (which, in fact, tears muscles down rather than building them up) but when resting—and particularly when sleeping—after having worked out earlier in the day. Likewise, deep sleep is associated with growth, as we have noted. Furthermore, it appears that dreaming may be essential to the well-being of the psyche, providing an opportunity for the brain to "clean out" the signals and data it has been receiving for the preceding 16 hours of wakefulness.

Given these and other important functions associated with deep sleep, it follows that the maintenance of the body clock is of great importance to the health of the human organism. For-

VIEW OF THE MIDNIGHT SUN IN LYNGENFJORD, NORWAY. THE "BODY CLOCK" CAN BE DISRUPTED BY CHANGES IN THE AMOUNT OF AVAILABLE LIGHT, SUCH AS OCCUR IN REGIONS OF THE EXTREME NORTH THAT UNDERGO PERIODS OF ALMOST CONSTANT DAYLIGHT FROM MID-MAY TO LATE JULY. *(© Bettmann/Corbis. Reproduced by permission.)*

tunately, animals' brains are programmed to make adjustments of the body clock so as to accommodate the daily cycles of light and dark. We have discussed the means by which the human brain achieves this accommodation, but it is not the only animal brain thus equipped. "Bird brains" (quite literally) are similarly able to make an adjustment: whereas humans have a natural 25-hour clock, birds run on a 23-hour circadian cycle, but their pineal glands likewise assist them in adapting to the 24-hour solar day.

The brains of birds, humans, and other animals respond to environmental features known collectively as *zeitgebers* (German for "time givers"), which aid in the adjustment to the solar schedule. The most obvious example is the change from day to night, but there are other zeitgebers of which we are less aware in our ordinary experience. For example, Earth's magnetic field goes through its own 24-hour cycle, which subtly influences our biological rhythms.

INTERFERING WITH THE BODY CLOCK

In modern life humans often interfere with their own body clocks, either deliberately and directly or indirectly and by accident. On the one hand, a person may drink coffee to stay awake at night, but he or she also may experience a sleep disorder as a result of some other situation, which may or may not be the result of purposeful action. Examples of sleep disorders that are the by-product of other activities include jet lag as well as the malfunctioning of the body clock that often stems from recreational drug use.

The causes for interference with a person's body clock may be outside that person's control to one degree or another. Working at night, for instance, is a condition that almost never suits a human being, no matter how much a person may insist that he or she is a "night person." Nevertheless, a person may be required by circumstances, such as schedule, economic necessity, or job availability, to take a night job. Another example of interference with the body clock would be narcolepsy (a condition characterized by brief attacks of deep sleep) or some other condition that is either congenital (something with which a person is born) or symptomatic (a symptom of some other condition rather than a condition in and of itself).

WHITE NIGHTS. At least one example of human experience involving interference with the body clock relates to conditions *completely* outside people's control. This is the situation of the "white nights" or "midnight sun," whereby regions in the extreme north—Russia, Alaska, and Scandinavia—undergo periods of almost constant daylight from mid-May to late July. (These are matched by a much less pleasant phenomenon: near constant darkness from mid-November to late January.)

During those times people often line their windows with dark material to make it easier to go to sleep in a world where the Sun is nearly as bright at 3:00 A.M. as it is at 3:00 P.M. The situation is even more pronounced in Antarctica, where researchers and adventurers may find themselves much closer to the South Pole than people in Saint Petersburg, Anchorage, or Oslo are to the North Pole.

In Antarctica the human population is much higher in the summer, a period that coincides with the depth of winter in the Northern Hemisphere, and scientists or mountaineers trekking through remote regions may be forced to sleep in tents that keep out the cold but let in the light. Usually, however, the rugged conditions of life near the South Pole involve such exertions that by nighttime people are ready to sleep, light or no light.

SOME SLEEP DISORDERS. Few people ever get to experience the white nights, but almost everyone has suffered through a temporary bout of insomnia—a condition known specifically as *transient insomnia.* An unfortunate few suffer from chronic insomnia or some other sleep disorder. Insomnia, the inability to go to sleep or to stay asleep, is one of the two most common sleep disorders, the other being hypersomnia, or excessive daytime sleepiness.

Transient forms of insomnia are usually treatable with short-term prescription drugs, but more serious conditions qualify as actual disorders and may require long-term treatment. These disorders may have as their cause drug use (either prescription or illegal) as well as medical or psychological problems. Among the most common of these more specialized disorders is apnea, the regular cessation of breathing whose most noticeable symptom is snoring.

Apnea, which affects a large portion of the United States population, is a potentially very serious condition that can bring about suffocation or even death. More often its effects are less dramatic, however, and manifest in hypersomnia, which is a result of lost sleep due to the fact that the sufferer actually is waking up numerous times throughout the night.

At the other extreme from apnea, in terms of prevalence among the population, is Kleine-Levin syndrome, which typically affects males in their late teens or twenties. The syndrome may bring about dramatic symptoms that range from excessive sleepiness, overeating, and irritability to abnormal behavior, hallucinations, and even loss of sexual inhibitions. Added to this strange mix is the fact that Kleine-Levin syndrome typically disappears after the person reaches the age of 40.

JET LAG. There are numerous classes of sleep disorders, among them circadian rhythm disorders—those related to jet lag or work schedules. As we have seen, the pineal gland can adjust easily from a natural 25-hour cycle to a 24-hour one, but it can do so only gradually, and it cannot readily adapt to sudden changes of schedule, such as those brought about by air travel.

Jet lag is a physiological and psychological condition in humans that typically includes fatigue and irritability; it usually follows a long flight through several time zones and probably results from disruption of circadian rhythms. The name is fitting, since jet lag is associated almost exclusively with jets: traveling great distances by ship, even at the speeds of modern craft, allows the body at least some time to adjust.

Older modes of travel were too slow to involve jet lag; for this reason, the phenomenon is a relatively recent one. The only people who manage to experience jet lag without riding in a jet are those traveling in even faster craft—that is, astronauts. An astronaut orbiting Earth in a space shuttle experiences rapid shifts from day to night; if manned vessels ever go out into deep space, scientists will face a new problem: assisting the adjustment of circadian cycles to that sunless realm.

On a much more ordinary level, there is the jet lag of people who travel from the United States East Coast to Europe or between the East Coast and West Coast of the United States. The worst kinds of jet lag occur when a person flies from west to east across six or more time zones: anyone who flies to Europe from the East Coast is likely to spend much of the first day after arrival sleeping rather than sightseeing. Thereafter, it may take up to ten days (usually as long as or longer than most European vacations) for the body to adjust fully.

By contrast, someone who has flown from the East Coast to the West Coast feels unexpected energy. The reason is that when it is 6:00 A.M. in the Pacific time zone, it is 9:00 A.M. in the eastern time zone, to which a person's body clock (in this particular scenario) is still adapted. Therefore, at 6:00 in the morning, the newly arrived traveler will feel as good as he or she would normally feel at 9:00 A.M. back east. Conversely, at 9:00 P.M. in the west, it is midnight in the east. This means that the traveler is likely to feel tired long before his or her ordinary bedtime.

There are steps one can take to avoid, or at least minimize the effects of, jet lag. One is to ensure a regular sleep schedule prior to traveling, so as to minimize the effects of sleep deprivation, if the latter does occur. It is even better if one can, in the days prior to leaving, adopt a schedule adjusted to the new time zone. For example, if one were traveling from the East Coast to California, one would start going to bed three hours earlier, and rising three hours earlier as well. Changing eating habits in the days prior to departure may also help. Some experts on the subject recommend a four-day period in which one alternates heavy eating (days one and three) and very light eating (days two and four.) It is believed that high-protein breakfasts stimulate the active, waking cycle, while high-carbohydrate evening meals stimulate the resting cycle; conversely, depriving the liver of carbohydrates may prepare the body clock to reset itself.

ON THE NIGHT SHIFT. At least the body does adjust to jet lag; on the other hand, it may never become accustomed to working a night shift. If you stay up all night studying for a test, you will find that around 4:00 A.M. you hit a "lull" when you feel sleepy—and because of the lowered temperature at the circadian trough, you also feel cold. You might assume that this situation would improve if you worked regularly at night, but the evidence suggests that it does not.

As long as a person lives in a sunlit world of 24-hour solar days, the body clock remains adapted to that schedule, and this will be true whether the person is at home and in bed or at work behind a desk or counter during the hours of night. In other words, the person always will hit the circadian trough about 4:00 A.M. This is one of the reasons why most people find the idea of working at night so unattractive, even though it is clear that in our modern society some night-shift positions are essential.

People who have offices in their homes may find it beneficial to work at late hours, when the phone is not ringing and the world is quiet, but the "extra time" gained by working at night ultimately is counterbalanced by the body's reaction to changes in its biological rhythms. Such is also the case with night-shift workers, who never really adjust to their schedules even after years on the job.

There *is* such a thing as a "night person," or someone with a chronic condition known as *delayed sleep phase syndrome.* A person with this syndrome is apt to feel most alert in the late evening and night, with a corresponding lag of energy in the late mornings and afternoons. Even so, given the role of sunlight in governing the body clock, the condition does not really lend itself to regular night work but rather merely causes a person to experience problems adapting to the schedule maintained by most of society. One possible means of dealing with this problem is to go to bed three hours later than would be normal for an ordinary 9-to-5 schedule, and wake up three hours later as well; unfortunately, that is not practical for most people. Another treatment applied with success is exposure of a person to artificial, high-intensity, full-spectrum light, which augments the effect of sunlight, between the hours of 7:00 and 9:00 A.M.

COLONIZING THE NIGHT? In this vein it is interesting to note that some of the optimistic predictions made in 1987 by Murray Melbin in his fascinating book *Night as Frontier: Colonizing the World After Dark* have not come to pass. Melbin, who explains circadian rhythms and the body clock in a highly readable and understandable fashion, makes a brilliant analysis of the means by which industrialized societies have extended their daily schedules into the nighttime hours. Thus, to use his analogy, such societies have "colonized" the night.

Until the invention in 1879 of the first successful incandescent lamp by the American inventor Thomas Edison (1847–1931), activity at night was limited. Torches, crude lamps, and candles in ancient times; metal lamps in the Middle Ages; and the various oil-burning lamps that applied the glass lantern chimney devised in 1490 by the Italian scientist and artist Leonardo

UNTIL THOMAS EDISON INVENTED THE FIRST SUCCESSFUL INCANDESCENT LAMP IN 1879, ACTIVITY AT NIGHT WAS LIMITED. WITH ELECTRIC LIGHTING, INDUSTRIALIZED SOCIETIES HAVE BEEN ABLE TO "COLONIZE" THE NIGHT, EXTENDING DAYTIME ACTIVITIES INTO THE NIGHTTIME HOURS. *(© Bettmann/Corbis. Reproduced by permission.)*

da Vinci (1452–1519) all made it possible for a person to read at night and to perform other limited functions. After their introduction in the nineteenth century, street lamps in London, the first of their kind, made the streets safe for walking at late hours, but travel, large gatherings, and outdoor work after dark remained difficult before the advent of electric light.

Since 1879 the Western world has indeed "colonized" the night with all-night eateries, roads that are never free of traffic, and round-the-clock entertainment on radio, TV, and now the Internet. There are even hardware stores open all night in some major cities. Certainly today there are more gas stations, restaurants, television programs, and customer-service telephone lines that operate 24 hours than there were in 1987, when Melbin wrote his book, but it is unlikely that Americans will ever fully "colonize" the night in the thoroughgoing fashion that their ancestors colonized the New World. An example of the limits to night colonization is in air travel.

Before the events of September 11, 2001, when terrorists crashed hijacked planes into the World Trade Center in New York City and the Pentagon in Washington, D.C., the burden on America's airports had become almost unbearable. The concourses of Hartsfield International in Atlanta, Georgia, the world's busiest airport, were a nonstop melee of people, luggage, and noise, as travelers fought to change flights or pick up their bags. One obvious solution to the problem would have been to adopt a round-the-clock airport schedule, with flights regularly leaving at 3:00 or 4:00 in the morning.

No airport rushed to enact such a measure, however, and after September 11 heightened security concerns made it unlikely that any facility would adopt a 24-hour schedule, with the additional security threats it entailed. For a time at least, the volume of air traffic decreased dramatically, but even as it climbed back up in the months after the terrorist attacks, airports continued to operate on their ordinary schedules. The reason appears to be the difficulty of persuading people to adjust to a late-night schedule—that is, finding enough people willing to fly in the middle of the night and enough baggage handlers and ticket agents willing to service them. There are, it seems, limits to the extent to which nighttime can be colonized.

OTHER EXAMPLES OF BIOLOGICAL RHYTHMS

Although the circadian rhythms of sleep and wakefulness are particularly important examples of biological cycles, they are far from the only ones. Not all rhythms, in fact, are circadian. Some are ultradian, meaning that they occur more than once a day. Examples include the cycles of taking in fluid and forming urine as well as cell-division cycles and cycles related to hormones and the endocrine glands that release them. For instance, the pituitary gland in the brain of a normal male mammal secretes hormones about every one to two hours during the day.

The overall cycle of sleeping and waking is circadian, but there is an ultradian cycle within sleep as the brain moves from drowsiness to REM (rapid eye movement, or dream, sleep) to dozing, then to light and deep sleep, and finally to slow-wave sleep. Over the course of the night, this cycle, which lasts about 90 minutes, repeats itself several times. Among the functions affected by

A ROCKY MOUNTAIN GOAT SHEDS ITS THICK WINTER FUR. THE SHEDDING OF FUR, SKIN, OR ANTLERS IS ONE EXAMPLE OF A CIRCANNUAL CYCLE, WHICH TAKES A YEAR TO COMPLETE. *(© W. Wayne Lockwood, M.D./Corbis. Reproduced by permission.)*

this cycle are heart rate and breathing, which slow down in deep sleep. Additionally, heartbeat and respiration are themselves ultradian cycles of very short duration.

MENSTRUATION AND OTHER INFRADIAN CYCLES. In contrast to the ultra-quick ultradian cycles of the beating heart and the lungs' intake and outflow of oxygen, there are much longer infradian, or monthly, cycles. By far the most common is menstruation, which begins when a female mammal reaches a state of physical maturity and continues on a monthly basis until she is no longer able to conceive offspring.

When she becomes pregnant, the menstrual cycle shuts down and, in some cases, does not resume until several months after delivery of the offspring. Assuming she is in good health, the human female will experience fairly regular menstrual periods at intervals of 28 days. Among human females, it has long been known that the menstrual cycles of women who live or work in close proximity to one another tend to come into alignment. For example, college girls on the same floor in a dormitory are likely to share menstrual cycles.

The reasons for this alignment of menstrual cycles are not completely understood. Nor is the cause of the 28-day cycle evident. If it were the result of the Moon's cycles, all women on Earth would have menstrual cycles that last 29.5 days, which is how long it takes the Moon to travel around Earth. Furthermore, if there were a clear connection between the Moon and menstruation, the periods of all menstruating females on Earth would be aligned with the Moon's phases. Neither of these, of course, is the case.

CIRCANNUAL CYCLES. Longer still than infradian cycles, circannual cycles, as their name suggests, take a year to complete. Among them is the cycle of dormancy and activity marked by the hibernation of certain species in the winter. There are also certain times of the year when animals shed things—fur, skin, antlers, or simply pounds. Likewise, at some points in the year animals gain weight.

People are affected strongly by the seasonal changes associated with the circannual cycle. There is almost no person who lives in a temperate zone (that is, one with four seasons) who is not capable of calling strong emotions to mind when imagining the sensations associated with winter, spring, summer, or fall. Some sensations,

KEY TERMS

BIOLOGICAL CLOCK: A mechanism within an organism (for example, the pineal gland in the human brain) that governs biological rhythms.

BIOLOGICAL RHYTHMS: Processes that occur periodically in an organism in conjunction with and often in response to periodic changes in environmental conditions.

CHRONOBIOLOGY: A subdiscipline of biology devoted to the study of biological rhythms.

CIRCADIAN RHYTHM: A biological cycle that takes place over the course of approximately a day. In humans circadian rhythms run on a cycle of approximately 25 hours and govern states of sleep and wakefulness as well as core body temperature and other biological functions.

HORMONE: Molecules produced by living cells, which send signals to spots remote from their point of origin and which induce specific effects on the activities of other cells.

INFRADIAN RHYTHM: A biological cycle that takes place over the course of a month.

JET LAG: A physiological and psychological condition in humans that typically includes fatigue and irritability; it usually follows from a long flight through several time zones and probably results from disruption of circadian rhythms.

MENOPAUSE: The point at which menstrual cycles cease, a time that typically corresponds to the cessation of the female's reproductive abilities.

MENSTRUATION: Sloughing off of the lining of the uterus, which occurs monthly in nonpregnant females who have not reached menopause (the point at which menstrual cycles cease) and which manifests as a discharge of blood.

PINEAL GLAND: A small, usually cone-shaped portion of the brain, often located between the two lobes, that plays a principal role in governing the release of certain hormones, including those associated with human circadian rhythms.

ULTRADIAN RHYTHM: A biological cycle that takes place over the course of less than a day. Compare with *circadian rhythm.*

however, are better than others, and though there can be negative associations with spring or summer, by far the season most likely to induce ill effects in humans is winter.

The thirteen weeks between the winter solstice in late December and the vernal equinox in late March have such a powerful impact on the human psyche that scientists have identified a mental condition associated with it. It is SAD, or seasonal affective disorder, which seems to be related to the shortened days (and thus, ultimately, to the altered circadian rhythm) in wintertime.

As we have noted, the body responds to the onset of night and sleep by the release of melatonin, but when darkness lasts longer than normal, melatonin secretions become much more pronounced than they would be under ordinary conditions. The result of this hormone imbalance can be depression, which may be compounded by other conditions associated with winter. Among these conditions is "cabin fever," or restlessness brought about by lengthy confinement indoors. An effective treatment for SAD is exposure to intense bright light.

STUDYING BIOLOGICAL RHYTHMS

Treatment of SAD is just one example of the issues confronted by scientists working in the realm of chronobiology, a subdiscipline devoted to the study of biological rhythms. Naturally, a particularly significant area of chronobiological study is devoted to sleep research. The latter is a relatively new field of medicine stimulated by the discovery of REM sleep in 1953. In addition to studying such disorders as sleep apnea, sleep researchers are concerned with such issues as the effects of sleep deprivation and the impact on circadian rhythms brought about by isolation from sunlight.

Note that the scientific study of biological rhythms has nothing to do with "biorhythms," a fad that peaked in the 1970s but still has its adherents today. Biorhythms are akin to astrology in their emphasis on the moment of a person's birth, and though biorhythms have a bit more scientific basis than astrology, that in itself is not saying much. As we have seen, biological rhythms do govern much of human life, but the study of these rhythms does not offer special insight into the fate or future of a person—one of the principal claims made by adherents of biorhythms. As with all pseudosciences, belief in biorhythms is maintained by emphasizing those examples that seem to correlate with the theory and ignoring or explaining away the many facts that contradict it.

An example of scientific research in chronobiology and related fields is the work of the psychologist Stephany Biello at Glasgow University in Scotland, who in June 2000 announced findings linking the drug, ecstasy, to long-term damage to the body clock. As with LSD and many another drug, ecstasy plays havoc with serotonin and may exert such a negative impact on the pathways of serotonin release in the pineal gland that it permanently alters the brain's ability to manufacture that vital hormone. Thus the drug, which induces a sense of euphoria in users, can induce serious sleep and mood disorders as well as severe depression.

WHERE TO LEARN MORE

Biological Rhythms (Web site). <http://faculty.washington.edu/chudler/clock.html>.

Center for Biological Timing (Web site). <http://www.cbt.virginia.edu/>.

Circadian Rhythms (Web site). <http://www.bio.warwick.ac.uk/millar/circad.html>.

"Ecstasy 'Ruins Body Clock.'" British Broadcasting Corporation (Web site). <http://news.bbc.co.uk/hi/english/health/newsid_803000/803633.stm>.

Hughes, Martin. *Bodyclock: The Effects of Time on Human Health.* New York: Facts on File, 1989.

Melbin, Murray. *Night as Frontier: Colonizing the World After Dark.* New York: Free Press, 1987.

Orlock, Carol. *Inner Time: The Science of Body Clocks and What Makes Us Tick.* Secaucus, NJ: Carol Publishing Group, 1993.

Rose, Kenneth Jon. *The Body in Time.* New York: John Wiley and Sons, 1988.

Sleep Disorders Information (Web site). <http://www.stanford.edu/~dement/sleepinfo.html>.

Waterhouse, J. M., D. S. Waters, and M. E. Waterhouse. *Your Body Clock.* New York: Oxford University Press, 1990.

Winfree, Arthur T. *The Timing of Biological Clocks.* New York: Scientific American Library, 1987.

LEARNING AND BEHAVIOR

BEHAVIOR
INSTINCT AND LEARNING
MIGRATION AND NAVIGATION

BEHAVIOR

CONCEPT

In biology the term behavior refers to the means by which living things respond to their environments. At first glance, this might seem to encompass only animal behavior, but, in fact, plants display observable behavior patterns as well. One of the principal manifestations of plant behavior is tropism, a response to a stimulus that acts in a particular direction, thus encouraging growth either toward or away from that stimulus. Behavior in plants is primarily a matter of response to stimuli, which may be any one of a variety of influences that derive either from inside or outside the organism. Response to stimuli is automatic, and even humans are capable of making these types of programmed responses. In most cases, behaviors in organisms are designed to ensure their survival. Such is the case, for instance, with the complex of behaviors known as territoriality, whereby animals defend what they perceive to be their own.

HOW IT WORKS

STIMULUS AND RESPONSE

A stimulus is any phenomenon that directly influences the activity or growth of a living organism. *Phenomenon,* meaning any observable fact or event, is a broad term and appropriately so, since stimuli can be of so many varieties. Chemicals, heat, light, pressure, and gravity all can serve as stimuli, as indeed can any environmental change. Nor are environmental changes limited to the organism's external environment. In some cases its internal environment can act as a stimulus, as when an animal reaches the age of courtship and mating and responds automatically to changes in its body.

All creatures, even humans, are capable of automatic responses to stimuli. When a person inhales dust, pepper, or something to which he or she is allergic, a sneeze follows. The person may suppress the sneeze (which is not a good practice, since it puts a strain on blood vessels in the head), but this does not stop the body from responding automatically to the irritating stimulus by initiating a sneeze. Similarly, plants respond automatically to light and other stimuli in a range of behaviors known collectively as *tropisms,* which we explore later in this essay.

INNATE AND LEARNED BEHAVIOR. Not all responses to stimuli are automatic, however. Certainly not all behavior on the part of higher animals is automatic, though, as we have noted, even humans are capable of *some* automatic responses. In general, behavior can be categorized as either innate (inborn) or learned, but the distinction is frequently unclear. In many cases it is safe to say that behavior present at birth is innate, but this does not mean that behavior that manifests later in life is learned. (Later in this essay we look at an example of this behavior as it relates to chickens and pecking.)

Behavior is considered innate when it is present and complete without any experience whereby it was learned. At the age of about four weeks, human babies, even blind ones, smile spontaneously at a pleasing stimulus. Like all innate behavior, babies' smiling is stereotyped, or always the same, and therefore quite predictable. Plants, protista (single-cell organisms), and animals that lack a well-developed nervous system rely on innate behavior. Higher animals, on the other hand, use both innate and learned behavior. A fish is born knowing how to swim, whereas a human or a giraffe must learn how to walk.

ETHOLOGY

Ethology is the study of animal behavior, including its mechanisms and evolution. The science dates back to the British naturalist Charles Darwin (1809–1882), who applied it in his research concerning evolution by means of natural selection (see Evolution). Darwin presented many examples to illustrate the fact that, in addition to other characteristics of an organism, such as its morphologic features or shape, behavior is an adaptation to environmental demands and can increase the chances of species survival.

The true foundations of ethology, however, lie in the work of two men during the period between 1930 and 1950: the Austrian zoologist Konrad Lorenz (1903–1989) and the Dutch ethologist Nikolaas Tinbergen (1907–1988). Together with the Austrian zoologist Karl von Frisch (1886–1982), most noted for his study of bee communication and sensory perception, the two men shared the 1973 Nobel Prize in physiology or medicine.

Lorenz and Tinbergen, who together are credited as founders of scientific ethology, contributed individually to the discipline and, during the mid–twentieth century, worked together on a theory that animals develop formalized, rigid sequences of action in response to specific stimuli. According to Lorenz and Tinbergen, animals show fixed-action patterns (FAPs) of behavior which are strong responses to particular stimuli. Later in this essay, we look at examples of FAPs in action. In addition, Lorenz put forward the highly influential theory of imprinting, discussed briefly in this essay and in more detail elsewhere (see Instinct and Learning).

BEHAVIORISM AND CONDITIONING

The development of ethology by Lorenz and Tinbergen occurred against the backdrop of the rise of the behaviorist school in the realms of philosophy, psychology, and the biological sciences. This school of thought had its roots in the late nineteenth century, with the writings of a number of philosophers and psychologists as well as practical scientists, such as the Russian physiologist Ivan Pavlov (1849–1936). Pavlov showed that an animal can be trained to respond to a particular stimulus even when that stimulus is removed, so long as the stimulus has been associated with a secondary one.

Pavlov began his now famous set of experiments by placing powdered meat in a dog's mouth and observing that saliva flowed into the mouth as a reflex reaction to the introduction of the meat. He then began ringing a bell before he gave the dog its food. After doing this several times, he discovered that the dog salivated merely at the sound of the bell. Many experiments of this type demonstrated that an innate behavior can be modified, and thus was born the scientific concept of conditioning, or learning by association with particular stimuli.

The variety of conditioning applied by Pavlov, known as *classical conditioning,* calls for pairing a stimulus that elicits a specific response with one that does not, until the second stimulus elicits a response like the first. Classical conditioning is contrasted with operant conditioning, which involves administering or withholding reinforcements (that is, rewards) based on the performance of a targeted response.

OPERANT CONDITIONING. During operant conditioning, a random behavior is rewarded and subsequently retained by an animal. According to operant conditioning theory, if we want to train a dog to sit on command, all we have to do is wait until the dog sits and then say, "Sit," and give the dog a biscuit. After a few repetitions, the dog will sit on command because the reward apparently reinforces the behavior and fosters its repetition.

Human parents apply operant conditioning when they admonish their offspring with such phrases as "You can't watch TV until you've cleaned your room." Likewise, young chimpanzees learn through a form of operant conditioning. By observing their parents, young chimps learn how to strip a twig and then use it to pick up termites (a tasty treat to a chimpanzee) from rotten logs. Their behavior thus is rewarded, an example of the way that operant conditioning enables animals to add new, noninherited forms of behavior to their range of skills.

Though the theory of operant conditioning goes back to the work of the American psychologist Edward L. Thorndike (1874–1949), by far its most famous proponent was another American psychologist, B. F. Skinner (1904–1990). In applying operant conditioning to human beings,

Skinner and his followers took the theory to extremes, maintaining that humans have no ideas of their own, only conditioned responses to stimuli. Love, courage, faith, and all the other emotions and attitudes that people hold in high esteem are, according to this school of thought, simply a matter of learned responses, rather like a parrot making human-like sounds to earn treats. This extreme form of behaviorism is no longer held in high regard within the scientific or medical communities.

REAL-LIFE APPLICATIONS

BEHAVIOR IN PLANTS

As noted earlier, the term *behavior* would seem at first glance to apply only to animals and not to plants. Certainly the majority of attention in behavioral studies, outside the realm of humans, is devoted to ethology, but plants are not without their observable behavioral characteristics. These features primarily manifest in the form of tropism, a response to a stimulus that acts in a particular direction, thus encouraging growth either toward or away from that stimulus. Tropism primarily affects members of the plant kingdom, though it has been observed in algae and fungi as well.

Though the word tropism itself may be unfamiliar to most people, the phenomenon itself is not. There are plenty of opportunities in daily life to observe the response of plants to energy, substances, or forms of stimulation. For example, perhaps you have noticed the way that trees or flowers grow toward sunlight, even bending in their growth if it is necessary to reach the energy source. Similarly, plants in a parched region are likely to develop roots directed laterally toward a water source.

Among the various forms of tropism are phototropism (response to light), geotropism (response to gravity), chemotropism (response to particular chemical substances), hydrotropism (response to water), thigmotropism (response to mechanical stimulation), traumatropism (response to wounds), and galvanotropism or electrotropism (response to electric current). Most of these types involve growth toward a stimulus, a phenomenon known as positive growth, or orthotropism. Plants tend to grow toward light or water, for instance. On the other hand, some kinds of stimuli tend to evoke diatropism, or growth away from the stimulus. Such is bound to be the case, for instance, with traumatropism and electrotropism.

ONE EXAMPLE OF INNATE ANIMAL BEHAVIOR IS THE REFLEX, A SIMPLE, INBORN, AUTOMATIC RESPONSE TO A STIMULUS BY A PART OF AN ORGANISM'S BODY. SUCH A MECHANISM IS AT WORK, FOR INSTANCE, WHEN JELLYFISH WITHDRAW THEIR TENTACLES. *(© Henry Horenstein/Corbis. Reproduced by permission.)*

Tropism, along with movement due to changes in water content, is one of the two principal forms of innate behavior on the part of plants. In general, stems and leaves experience positive phototropism, as they grow in the direction of a light source, the Sun. At the same time, roots exhibit positive gravitropism, or growth toward the gravitational force of Earth, as well as positive hydrotropism, since they grow toward water sources below ground. On the other hand, a plant may move in a specific way regardless of the direction of the stimulus. Such movements are temporary, reversible, and result from changes in the water pressure inside the plant.

ANIMAL BEHAVIOR

An excellent example of an innate animal behavior, and one in which humans also take part, is

IN CONTRAST TO SIMPLE FIXED-ACTION PATTERNS OF BEHAVIOR, OR FAPS, ARE COMPLEX PROGRAMMED BEHAVIOR PATTERNS, WHICH COMPRISE SEVERAL STEPS AND ARE MUCH MORE COMPLICATED. ONE TYPE OF COMPLEX BEHAVIOR IS THE BUILDING OF DAMS BY BEAVERS. *(© Harry Engels/Nas. Photo Researchers. Reproduced by permission.)*

the reflex. A reflex is a simple, inborn, automatic response to a stimulus by a part of an organism's body. The simplest model of reflex action involves a receptor and sensory neuron and an effector organ. Such a mechanism is at work, for instance, when certain varieties of coelenterate (a phylum that includes jellyfish) withdraw their tentacles.

More complex reflexes require processing interneurons between the sensory and motor neurons as well as specialized receptors. These neurons send signals across the body, or to various parts of the body, as, for example, when food in the mouth stimulates the salivary glands to produce saliva or when a hand is pulled away rapidly from a hot object.

Reflexes help animals respond quickly to a stimulus, thus protecting them from harm. By contrast, learned behavior results from experience and enables animals to adjust to new situations. If an animal exhibits a behavior at birth, it is a near certainty that it is innate and not learned. Sometimes later in life, however, a behavior may appear to be learned when, in fact, it is a form of innate behavior that has undergone improvement as the organism matures.

For example, chickens become more adept at pecking as they get older, but this does not mean that pecking is a learned behavior; on the contrary, it is innate. The improvement in pecking aim is not the result of learning and correction of errors but rather is due to a natural maturing of muscles and eyes and the coordination between them.

FAPS. In studying fixed-action patterns of behavior, or FAPs, Lorenz and Tinbergen observed numerous interesting phenomena. Male stickleback fish, for example, recognize potential competition—other breeding stickleback males—by the red stripe on their underside and thus engage in the FAP of attacking anything red on sight. Tinbergen discovered that jealous stickleback males were so attuned to the red stripe that they tried to attack passing British mail trucks, which were red, when they could see them through the glass of their tanks. Tinbergen termed the red stripe a behavioral releaser, or a simple stimulus that brings about a FAP.

Once a FAP is initiated, it continues to completion even if circumstances change. If an egg rolls out of a goose's nest, the goose stretches her neck until the underside of her bill touches the egg. Then she rolls the egg back to the nest. If someone takes the egg away while she is reaching for it, the goose goes through the motions anyway, even without an egg. Not all animal behavior is quite so predictable, however. In contrast to FAPs are complex programmed behavior patterns, which comprise several steps and are much more complicated. Birds making nests or beavers building dams are examples of complex programmed behavior.

IMPRINTING. As we noted earlier, Lorenz initiated the study of a learning pattern that came to be known as *imprinting.* Witnessed frequently in birds, imprinting is the learning of a behavior at a critical period early in life, such that the behavior becomes permanent. The very young bird or other organism is like wet concrete, into which any pattern can be etched; once the concrete has dried, the pattern is set.

Newly hatched geese are able to walk. This is something they learn the moment they are hatched, and they do so by following their parents. But how, Lorenz wondered, do young geese distinguish their parents from all other objects in their environments? He discovered that if he

ROCKY MOUNTAIN BIGHORN RAMS MEET HEAD TO HEAD AT THE BOUNDARIES OF THEIR RESPECTIVE TERRITORIES. USING THEIR HORNS, THESE RAMS WILL STRONGLY DEFEND THEIR TERRITORIES AGAINST INVADERS. *(© W. Perry Conway/Corbis. Reproduced by permission.)*

removed the parents from view the first day after the goslings hatched and if he walked in front of the young geese at that point, they would follow him. This tactic did not work if he waited until the third day after hatching, however.

Lorenz concluded that during a critical period following birth, the goslings follow their parents' movement and learn enough about their parents to recognize them. But since he also had determined that young geese follow any moving object, he reasoned that they first identify their parents by their movement, which acts as a releaser for parental imprinting. (Imprinting is discussed further in Instinct and Learning.)

INTERACTIVE BEHAVIOR

Much of an animal's behavior (this is true of the human animal as well) takes place in interaction with others. This interaction may include rudimentary forms of communication, such as bee dances, studied by Lorenz and Tinbergen's colleague Frisch. As he showed in perhaps the most important research of his career, bees communicate information about food supplies, including their direction and the distance to them, by means of two different varieties of "dance," or rhythmic movement. One is a circling dance, which informs the other bees that food is near (about 250 ft., or 75 m, from the hive), and the other is a wagging dance, which conveys the fact that food is farther away.

There are numerous other forms of communication using one or more sense organs. Birds hear each other sing, a dog sees and hears the spit and hiss of a cornered cat, and ants lay down scent signals, or pheromones, to mark a trail that leads to food. This is only one level of interactive behavior, however. Quite a different variety of interaction is courtship, discussed in Reproduction. Other forms of interactive behavior include the establishment of an animal's territory, a subject we discuss at the conclusion of this essay.

LIFE IN COMMUNITIES. Interactive behavior comes into play when animals live in close proximity to one another. Certainly there are benefits to group life for those species that practice it: the group helps protect individuals from predators and, through cooperation and division of labor, ensures that all are fed and sheltered. In order to be workable, however, a society must have a hierarchy. Thus, in a situation quite removed from the human ideals of freedom and democracy, insect and animal societies are ones in which every creature knows its place and sticks to it.

Bees, ants, and termites live in complex communities in which some individuals are responsible for finding food, others defend the colony, and still others watch over the offspring. In such a highly organized society, a dominance hierarchy or ranking system helps preserve peace and discipline. Chickens, for example, have a pecking order from the most dominant to the most submissive. Each individual knows its place in the order and does not challenge individuals of higher rank. This, again, is quite unlike humans, who at least occasionally step out of line and challenge bullies; by contrast, that never happens with chickens (fittingly enough).

Territoriality

Almost everyone has seen a dog "mark its territory" by urinating on a patch of ground or has watched a cat arch its back in fury at an intruder to what it perceives as its territory. In so doing, these household pets are participating in a form of behavior that cuts across the entire animal kingdom: territoriality, or the behavior by which an animal lays claim to and defends an area against others of its species and occasionally against members of other species as well.

The physical size of the territory defended is extremely varied. It might be only slightly larger than the animal itself or it might be the size of a small United States county. The population of the territory might consist of the animal itself, the animal and its mate, an entire family, or an entire herd or swarm. Time is another variable: some animals maintain a particular territory year-round, using it as an ongoing source of food and shelter. Others establish a territory only at certain times of the year, when they need to do so for the purposes of attracting a mate, breeding, or raising a family.

Territorial behavior offers several advantages to the territorial animal. An animal that has a "home ground" can react quickly to dangerous situations without having to seek hiding places or defensible ground. By placing potential competitors at spaced intervals, territoriality also prevents the depletion of an area's natural resources and may even slow down the spread of disease. Furthermore, territorial behavior exposes weaker animals (which are unable to defend their territory) to attacks by predators and thus assists the process of natural selection in building a stronger, healthier population.

EXAMPLES OF TERRITORIES. A territory established only for a single night, for the sole purpose of providing the animal or animals with a place to rest, is known as a roost. Even within the roost, there may be a battle for territory, since not all spots are created equal. Because roosting spots near the interior are the safest, they are the most highly prized.

Another type of specialized territory is the lek, used by various bird and mammal species during the breeding season. Leks are the "singles bars" of the animal world: here animals engage in behavior known as *lekking*, in which they display their breeding ability in the hope of attracting a mate. Not surprisingly, leks are among the most strongly defended of all territories, since holding a good lek increases the chances of attracting a mate. Like the singles-only communities that they mimic, leks are no place for families: generally of little use for feeding or bringing up young, the lek usually is abandoned by the animal once it attracts a mate or mates.

THREATENING DISPLAYS. An animal has to be prepared to defend its territory by fighting off invaders, but naturally it is preferable to avoid actual fighting if a mere display of strength will suffice. Fighting, after all, uses up energy and can result in injury or even death. Instead, animals rely on various threats, through vocalizations, smells, or visual displays.

The songs of birds, the drumming of woodpeckers, and the loud calls of monkeys may seem innocuous to humans, but they are all warnings that carry for long distances, advertising to potential intruders that someone else's territory is being approached. As noted earlier, many animals, such as dogs, rely on smells to mark their territories, spraying urine, leaving droppings, or rubbing scent glands around the territories' borders. Thus, an approaching animal will be warned off the territory without ever encountering the territory's defender. Or, if the invader is unfortunate enough to have trespassed on a skunk's territory, it may get a big blast of scent when it is too late to retreat.

Suppose an animal ignores these warnings, or suppose, for one reason or another, that two animals meet nose to nose at the boundaries of their respective territories. Usually there follows a threatening visual display, often involving exaggeration of the animals' sizes by the fluffing up of feathers or fur. The animals may show off their

KEY TERMS

CONDITIONING: Learning by association with particular stimuli. There are two varieties of conditioning: classical conditioning, which involves pairing a stimulus that elicits a specific response with one that does not until the second stimulus elicits a response like the first, and operant conditioning, which involves administering or withholding reinforcements (i.e., rewards) based on the performance of a targeted response.

ETHOLOGY: The study of animal behavior, including its mechanisms and evolution.

FAPS: Fixed-action patterns of behavior, or strong responses on the part of an animal to particular stimuli.

IMPRINTING: The learning of a behavior at a critical period early in life, such that the behavior becomes permanent.

INNATE: A term to describe behaviors that are present and complete within the individual and which require no experience to learn them. For example, fish have an innate ability to swim, whereas humans must learn how to walk.

NATURAL SELECTION: The process whereby some organisms thrive and others perish, depending on their degree of adaptation to a particular environment.

REFLEX: An inborn, automatic response to a stimulus by a part of an organism's body.

STIMULUS: Any phenomenon (for example, an environmental change) that directly influences the activity or growth of a living organism.

TERRITORIALITY: The behavior by which an animal lays claim to and defends an area against others of its species and occasionally against members of other species as well.

TROPISM: A response to a stimulus that acts in a particular direction, thus encouraging growth either toward or away from that stimulus.

weapons, whether claws or fangs or other devices. Or the two creatures may go through all the motions of fighting without ever actually touching, a behavior known as *ritual fighting.*

FIGHTING. The degree to which a creature engages in these displays of bravado helps define its territory. If the creature perceives that it is at the center of its own territory and is being attacked on home ground, it will go into as threatening a mode as it can muster. If, on the other hand, the animal is at the edge of its territorial boundaries, it will be much more halfhearted in its efforts at intimidation. As with humans, few animals want to fight when there is nothing really at stake. Also like humans, animals many times may seem to be spoiling for a fight without actually fighting, such that when a fight does break out, it is an aberration. This typically happens only in overcrowded conditions, when resources are scarce—again, not unlike the situation with humans.

Late in his career, Lorenz devoted himself to studying human fighting behavior. In *Das sogenannte Böse* (*On Aggression,* 1963), he maintained that fighting and warlike behavior are innate to human beings but that they can be unlearned through a process whereby humans' basic needs are met in less violent ways. Just as fighting in animal communities has its benefits, Lorenz maintained, inasmuch as it helps keep competitors separated and enables the larger group to hold on to territory, so fighting among humans might be directed toward more useful means. As discussed in Biological Communities, it is possible that sports and business competition in the human community provides a more peaceful outlet for warlike instincts.

WHERE TO LEARN MORE

Animal Behavior Resources on the Internet. Nebraska Behavioral Biology Group (Web site). <http://cricket.unl.edu/Internet.html>.

Applied Ethology (Web site). <http://www.usask.ca/wcvm/herdmed/applied-ethology/>.

Dugatkin, Lee Alan. *Cheating Monkeys and Citizen Bees: The Nature of Cooperation in Animals and Humans.* New York: Free Press, 1999.

Ethology: Animal Behavior (Web site). <http://www.nua-tech.com/paddy/ethology.shtml>.

"Growth Movements, Turgor Movements, and Circadian Rhythmics." Department of Biology, University of Hamburg *(Germany)* (Web site). <http://www.biologie.uni-hamburg.de/b-online/e32/32c.htm>.

Hart, J. W. *Light and Plant Growth.* Boston: Unwin Hyman, 1988.

Hauser, Marc D. *Wild Minds: What Animals Really Think.* New York: Henry Holt, 2000.

Hinde, Robert A. *Individuals, Relationships, and Culture: Links Between Ethology and the Social Sciences.* New York: Cambridge University Press, 1987.

Immelmann, Klaus, and Colin Beer. *A Dictionary of Ethology.* Cambridge, MA: Harvard University Press, 1989.

Tropisms (Web site). <http://www.ultranet.com/~jkimball/BiologyPages/T/Tropisms.html>.

INSTINCT AND LEARNING

CONCEPT

Among the most fascinating areas in the biological sciences is ethology, or the study of animal behavior—in particular, the areas of ethology that deal with instinct and learning. Instinct is a stereotyped, or largely unvarying, behavior that is typical of a particular species. An instinctive behavior does not have to be learned; rather, it is fully functional the first time it is performed. On the other hand, learning, in an ethological context, is the alteration of behavior as the result of experience. Clearly, the distinction between instinct and learning revolves around the question of whether an animal, in responding to a specific situation, is acting on the basis of experience or instead is guided by instincts "hardwired" within its brain. The difference would seem to be a simple one, but nothing is simple in the study of instinct and learning. Plenty of gray area exists between pure instinct and genuine learning, and within that gray area is a fascinating concept known as *imprinting,* or the learning of a behavior at a critical period early in life, such that the behavior becomes permanent.

HOW IT WORKS

INSTINCT

The founding fathers of ethology (the study of animal behavior, including its mechanisms and evolution) were the German zoologist Konrad Lorenz (1903–1989) and the Dutch ethologist Nikolaas Tinbergen (1907–1988). In addition to their separate contributions to the field, which we discuss later, the two scientists together developed a theory that animals employ formalized, rigid sequences of action in response to specific stimuli. These sequences they called fixed-action patterns of behavior, or FAPs.

In studying male stickleback fish, Tinbergen discovered an excellent example of a FAP. Males of that species recognize potential competitors, in the form of other stickleback males capable of breeding, by the red stripe on their underside. Tinbergen termed the red stripe a behavioral releaser, or a simple stimulus that brings about a FAP. (A stimulus is any observable fact or event that directly influences the activity or growth of a living organism.)

This particular FAP compels the male stickleback to attack anything red, even when it is not a competitor. Thus, as Tinbergen observed, jealous stickleback males actually would try to attack red British mail trucks when they could see them through the glass of their tanks. Clearly, then, a FAP such as the response to the red stripe is not something that an animal thinks through; rather, it is automatic, almost as though the animal were being acted upon, instead of acting in its own right.

There have been stories and scientific studies of animals seemingly acting above and beyond the call of duty by raising the offspring of other individuals—even those of another species. Tinbergen, for example, observed and photographed a cardinal feeding baby minnows at water's edge. Such behavior seems altruistic, touching, and even inspiring, yet for all the significance we might be inclined to place on it from our human perspective, it is nothing but a FAP. Most likely the bird had lost its own offspring, and was simply acting on a parental instinct, which caused it to respond to the sight of open, upturned, hungry mouths. It so happened that the open mouths were not those of its offspring at all, but of fish, yet this made no difference in the behavior of the cardinal, acting as it was on an inborn, stereotypic pattern of response.

SURVIVAL-FRIENDLY, CONTROLLED BEHAVIOR. This is one characteristic of instinct, a concept virtually synonymous with that of the FAP: the creature acting on instinct is not thinking about what it does but is behaving almost as though it were controlled by some outside force. The use of the "outside force" concept perhaps makes instinct sound like something mystical when it is not. On the contrary, instinctive behavior is simply a survival mechanism in the brains of each member of a particular species, developed through countless generations of natural selection.

If a particular behavior helps foster the survival of a species, natural selection favors it. In other words, those individuals of a species who possess the tendency toward a particular survival-friendly behavior are the ones that survive and pass on their genes to others, while those that do not possess this tendency do not. The survival-friendly instinct may be geared toward the survival of the individual, its offspring, or others.

Closely tied to instinct is the innate animal behavior known as a reflex: a simple, inborn, automatic response to a stimulus by a part of an organism's body. Reflexes help animals (including humans) respond quickly to a stimulus, thus protecting them from harm. Again, the animal does not think about what it is doing. If you touch a hot stove or receive an electric shock, you do not *decide* to pull your hand back: you whip your hand away from the painful stimulus faster than you can blink. As with instinct, a reflex is a survival-friendly behavior over which the animal (in this case, you) has little control.

INNATE VERSUS LEARNED BEHAVIOR

Instinct is innate, meaning that instinctive behaviors and responses are present and complete within the individual at birth. In other words, the individual does not have to undergo any experience to acquire such behaviors. For example, fish have an innate ability to swim, whereas most mammals must learn how to walk. It is fairly easy to identify innate behavior when an animal exhibits it at birth, but in some cases innate behavior manifests only later in life.

In such situations improvements in the creatures' ability to perform an innate behavior may seem to indicate that the animal is learning, when, in fact, another process is at work. For example, chickens exhibit the innate tendency toward pecking as a way of establishing and maintaining a dominance hierarchy. This is the "pecking order," to which people often refer in everyday speech as a metaphor for various hierarchies in human experience, such as those at a workplace. Though pecking is innate, chickens' ability to perform it actually improves as they grow older.

Older chickens display a better aim when pecking than do younger ones, but this does not mean that they have learned from experience. Indeed, one clue that pecking is innate in older chickens is the fact that they uniformly improve in their ability to peck. On the other hand, if they were simply learning, with practice, how to peck more accurately, one could expect that some chickens would exhibit more dramatic improvement than others, in the same way that some humans play basketball (or sing or write poetry) better than others. In fact, what has happened is that the ability to perform an innate behavior simply has improved as a result of growth: as the chickens' eyes and muscles mature, their aim improves, but this has nothing to do with experience per se.

IMPRINTING

This is one example of the ways in which instinctive and learned behavior can become confused, though in the pecking example there is really no gray area; rather, what is actually an innate behavior merely *seems* to be a learned one. Yet there truly is a great deal of gray area between instinct and learning. Many behaviors that at first glance might appear purely instinctive can be shown to have an experiential component—that is, an aspect of the behavior has been modified through experience or learning.

A fascinating example of how instinct and learning can be blurred or combined is imprinting, or the learning of a behavior at a critical period early in life, such that the behavior becomes permanent. Lorenz, who first developed the theory of imprinting, noted that newly hatched geese learn to walk by following their parents, but he wondered how they distinguished their parents from all other objects in their envi-

Instinct and learning can be blurred in the case of imprinting, or the learning of a behavior at a critical period early in life, such that the behavior becomes permanent. Newly hatched geese learn to walk by following their parents. If the goslings are removed from their parents, they will imprint instead on another moving object, even a human being. *(© Galen Rowell/Corbis. Reproduced by permission.)*

ronments. He discovered that if he removed the parents from view the first day after the goslings hatched and if he walked in front of the young geese at that point, they would follow him.

From these experiments, Lorenz concluded two things. First of all, during a critical period following birth, goslings follow their parents' movements and learn enough about their parents to recognize them. This critical period is short: if he walked in front of three-day old geese, they were already too old to imprint on him. Second, Lorenz determined that the parents' movement must be a behavioral releaser for imprinting; thus, if the tiny goslings happened to fix on another moving object, they would imprint on that one.

We can see in this example that imprinting has both innnate, or instinctive, *and* learned components. The tendency to imprint is innate, but if the act of imprinting itself were likewise innate, then it would be the same for all individuals within a species, and this is not the case. The vast majority of goslings will imprint on adult geese, of course, but in an experimental setting (or through some freak accident in nature), it is possible that some other object will come between the offspring and its parents, causing the offspring to imprint on it.

Once imprinting is complete, it brings about numerous consequences that are more or less automatic or inevitable. Yet for all their automatic, inevitable qualities, these consequences are not the result of innate behaviors or instincts but of learned behaviors. Thus, learning is destiny, at least to an extent. The individual's mind, at an early stage, is like wet concrete into which virtually any impression can be made. Once the individual has imprinted on something or someone, however, the concrete begins to set, and it hardens around the impressions made in it.

REAL-LIFE APPLICATIONS

Instinct at Work

When a kangaroo rat hears a rattling sound, even if that sound comes from a drum or a fan or some otherwise harmless contraption, it performs a lightning-quick escape jump maneuver. Why? Because its brain interprets the rattling sound as that of a rattlesnake ready to strike, and it acts automatically. In other words, the kangaroo rat does not reason out a course of action; it simply moves. Nor is it acting on experience in any way: even if it has never encountered a rattlesnake, it will respond in exactly the same way.

Virtually all instinctive behavior, as we noted earlier, is geared toward preserving the life of the individual or of its offspring, and thus it serves to preserve the life of the species as well. Such is the case with the rat acting to protect itself from the snake, and so, too, with a mother when her young are threatened. People do not normally think of ducks as intimidating creatures, and in most cases they are not, but try stepping anywhere near a mother's ducklings, and watch her response. She will start to hiss, spit, and waddle forward menacingly in such a way as to terrify any would-be intruder. If the outsider is still foolhardy enough to press forward, the duck or goose will readily become a squawking, flapping, biting, pecking army of one. This behavior does not vary from mother to mother but is the same in all instances of mother ducks who perceive a threat to their ducklings—a hallmark of an instinctive action.

AUTOMATIC BEHAVIORS AND RELEASERS. There is an automatic, almost robotlike character to many an example of instinctive behavior. If an egg rolls out of a goose's nest, the goose stretches her neck until the underside of her bill touches the egg—an action that, like all instinctive ones, clearly is geared toward the survival of her offspring. Suppose someone takes the egg away while the goose is reaching for it: she continues to go through the motion of stretching to retrieve the egg. This may not seem very "smart," but, of course, instinct has nothing to do with intelligence.

Likewise, a spider preparing to lay its eggs spins a silk cocoon in a particular way, always the same and without any regard for outside factors. She begins by building a base plate, then constructs the walls of her cocoon before laying her eggs within it and sealing it with a lid. So rigid are these behavioral patterns that they cannot be altered, even if the spider needs to do so. (It is hard to imagine a spider "wanting" to do something, a term that implies decision-making abilities and a degree of self-awareness common only among higher mammals, particularly humans.)

If the spider is moved physically after she has built the base plate, she nonetheless will set about spinning walls and depositing her eggs, even though there is no base plate to hold them. The eggs, therefore, will fall out of the bottom of the incomplete web, but the spider will continue working as before, building the lid for the top. Fortunately for the spider, she has several cocoons. If she is returned to her original completed base plate as she prepares to spin the next cocoon, she will not use the base plate she already has spun; rather, like a robot, she will start from the very beginning, spinning a new base plate as though the original were not there.

Many of these automatic behaviors are triggered by a releaser. For example, a bright red spot on the bill of an adult gull serves as a releaser to its offspring, which peck on the parent's bill to obtain food. For a female rat in heat, rubbing of her hindquarters acts as a releaser for an instinctive behavior pattern known as *lordosis.* In lordosis the female flexes her front legs, lowers her torso, raises her rump, and moves the tail to one side. This posture, in turn, acts as a releaser for a male rat, who initiates copulation—yet another example of instinctive behavior as life-preserving. In this case, however, what is being preserved is not the life of an individual but of the entire species, since intercourse yields offspring.

CHALLENGING SITUATIONS OF INSTINCT AND IMPRINTING

Instincts are ingrained so deeply that some animals possess what appears to be an instinct for exploiting, or taking advantage of, the instinctive behaviors of other animals. We typically think of parasites as microbes, or at least as no larger than insects, but there are species of bird regarded as parasites, inasmuch as they take advantage of other species. When people speak metaphorically of another person as a "parasite," what they really mean is that the person in question exploits the good behavior of others. A human "parasite" never has to pay for dinner at a restaurant, for instance, because he or she can count on good, decent people always to pick up the tab. Likewise, parasitic bird species, such as the North American cowbird or the European cuckoo, can rely on other species' instinctive tendency to do what all animals (and humans) should do: take care of their own offspring.

Avian parasites lay their eggs in the nest of an unwitting host and then leave, "knowing" instinctively (which is not really the same thing as knowing in the way that humans think of it) that their victims will take care of the eggs. Cowbirds and cuckoos have developed the practice of laying their eggs in the nests of birds that are smaller than they, meaning that their hatchlings will be larger than the victims' offspring. This only adds to the detriment caused by the interlopers: since the parasites' children are larger than the hosts', this triggers a more powerful release of the host parents' instinctive feeding behavior. In other words, the parasites' offspring get more food than the hosts'. As for the parasitic adults, they are long gone, having deposited their young in the care of the hosts.

ERRORS IN IMPRINTING. The situation of the parasitic bird species is one example of the "dark side" to instinct and learning mechanisms in animals. Other examples include the many possible errors that can occur in imprinting. Such errors arise when an animal either fails to receive an imprint from an appropriate parent figure or receives an imprint from a creature of another species.

In their efforts to establish and maintain their territories and attract females, male birds of a given species learn a particular song. This takes place at a critical period when, as a nestling, the bird hears the song of its father and from this exposure eventually develops its own mature song. The process is a lengthy one: the immature bird does not begin singing until the following spring, when it starts trying to match its own, juvenile song with the one it heard from its father during the critical period.

If during that early critical period, the nestling is prevented from hearing an adult song of its own species, it never will develop a species-typical song, and thus its very life and its ability to propagate (which, in turn, affects the well-being of the entire species) are threatened. The bird may hear the songs of other species, but this does it little good. It appears that there is a strongly instinctive aspect to what the bird can learn during the critical period and also that birds are highly selective toward songs produced by other members of their species. Therefore, it learns either the right song or no song at all.

WHY IMPRINTING IS CRUCIAL

Imprinting is crucial for an animal's development and not only because the act of imprinting helps the animal learn one or another important early function, such as walking. Far beyond what happens to the animal in the first minutes of its life, imprinting will affect its entire destiny both as an individual and as a member of its species. Suppose an animal has imprinted on a creature of another species: this will haunt it for the rest of its life, for instance, determining its choice of a mate and its courtship behavior.

Many species will avoid social contact with animals that are not similar to the one to which they have imprinted. In other words, a duckling that has imprinted properly on its duck parents will want to spend time with other ducks. From the larger perspective of nature and the continuing propagation of life-forms, this is a good thing, because it helps prevent attempts to breed between different species. If, however, through some accident (or as the result of exposure to artificial conditions, such as those in an experiment), an animal has imprinted on an individual of a different species, that animal will attempt to court a member of that other species later in life—usually with disastrous results.

TARZAN THE APE MAN. In discussing the critical nature of imprinting, we could use all sorts of animal examples, some of which we have discussed. More compelling, however, is the example of a man, albeit a fictional one, who lived among the apes and attempted to become one of them. That man, of course, is Tarzan, creation of the American novelist Edgar Rice Burroughs (1875–1950). First appearing in *Tarzan of the Apes* (1914) and immortalized in countless novels, movies, comic books, and cartoons, Tarzan is the epitome of the noble savage, or the man at one with nature.

The son of an English lord, the boy is left alone as an infant in Africa after his parents die and is raised by an ape who has lost her own child. The apes name him Tarzan, which, in their language, means "white skin," and he grows up as one of them. Later, he discovers his parents' cabin and the books left there, teaches himself to read, learns English, and begins to uncover the truth of his background. As a grown man, he meets and falls in love with Jane Porter, a beautiful young American who, like his parents when he was infant, was marooned nearby after a shipboard mutiny. Jane helps expose him to civilization, and Tarzan eventually travels to France, the United States, and other far-off lands.

One can find in this astonishing tale antecedents in other rough-and-ready characters from American literature, most notably Buck, the

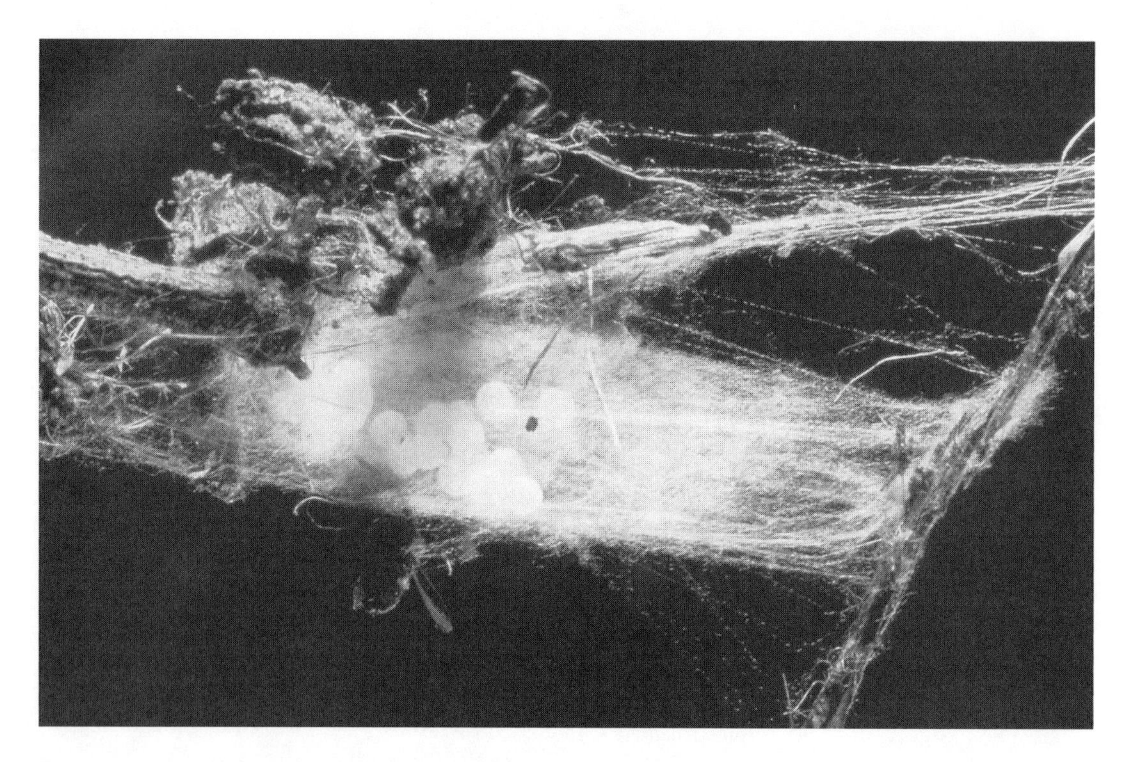

THERE IS AN AUTOMATIC CHARACTER TO MANY EXAMPLES OF INSTINCTIVE BEHAVIOR. A SPIDER SPINS A COCOON IN A PARTICULAR WAY, BEGINNING WITH A BASE PLATE AND THEN CONSTRUCTING WALLS BEFORE LAYING EGGS AND SEALING THE COCOON WITH A LID. IF THE SPIDER IS MOVED AFTER SHE HAS BUILT THE BASE PLATE, SHE WILL CONTINUE TO SPIN WALLS AND DEPOSIT HER EGGS, LEAVING THEM TO FALL OUT THE BOTTOM OF THE INCOMPLETE WEB. *(© Science Pictures Limited/Corbis. Reproduced by permission.)*

dog turned wolf in Jack London's (1876–1916) novel *The Call of the Wild* (1903). Buck is a particularly interesting example from the standpoint of biological study, because London, an advocate of Darwin's theory of evolution by natural selection, was attempting to show the common evolutionary thread linking dog and wolf and, by inference, ape and human.

While Burroughs is not so didactic as London, he had a purpose as well, an aim opposite to that of London: to show that a human, thrust by circumstances beyond his control into a world of animals, would assert his humanness. There may even have been a class element to the Tarzan scenario, inasmuch as the young Tarzan is the offspring of nobility and ultimately proves his noble lineage by rising above his circumstances. (This, too, is a theme quite far removed from the heart of London, who was a socialist.)

All literary analysis aside, is there any truth, from the standpoint of what we know about imprinting, instinct, and learning, to Burroughs's portrayal of Tarzan? The answer is a resounding "no." Assuming that a boy could be raised by apes, he would become socialized as an ape. Appropriately enough, in one story Burroughs shows Tarzan falling in love with a female ape, thinking that he is of the same species. The ape, for her part, recognizes the difference between them and rejects the human in favor of a male ape.

A real-life Tarzan would be human only in a biological sense; otherwise, he would lack virtually all human characteristics. Not only would he be foul smelling (from our perspective) and covered with hair, but he also would be entirely ignorant of human speech or even human thought processes, which have to be learned. Most preposterous of all is the idea that he would find books and recognize them as a mode of communication, let alone teach himself to read them. Suppose that he *had* been raised as a human but that he had never seen a book or even so much as a written word. Even if he were the most brilliant human being who had ever lived, he would treat the books the way an ape would—as mere objects.

COMPARING HUMANS AND ANIMALS

The case of Tarzan provides an appropriate place to close this discussion, with a few words on humans and their place in the context of the larger concepts of instinct, imprinting, and learning. Do humans possess instincts? It appears that we do, inasmuch as we are prone to certain automatic, innate responses. For example, human babies are capable of smiling at the age of four weeks, even though they do not know what a smile "means." On the other hand, humans are not nearly as inclined to instinctive behavior as animals; in general, with higher mammals and especially humans, instinctive characteristics diminish in favor of the capacity to learn.

Our methods of learning are quite different from those of lower animals, who "learn" in ways that do not involve conscious thought. For instance, a snail will pull its head back into its shell when touched, but when it is touched repeatedly with no subsequent harm, the withdrawal response ceases. The snail has experienced habituation, a type of behavior in which an animal develops a tendency to ignore a stimulus that is repeated over and over. Apparently, the snail's nervous system has "learned" that the stimulus is not threatening and so stops the reflex.

This is a far cry from learning as humans experience it, particularly as we move past the first few weeks and months of life. Even from the earliest moments of a human's existence—that is, even in the womb—a human is self-aware in a way that few, if any, animals other than mammals are. Furthermore, the self-awareness of a human from about the age of two years old is far beyond any concept of self possessed by even the highest forms of mammal. From what we can discern, a house cat or even an ape does not experience any thought along the lines of *Who am I?* or *What is my place in the order of things?* (Of course, almost everyone who has had a pet has at one time or another believed that a cat or dog was embroiled in such types of philosophical inquiry, but this may be merely anthropomorphism—ascribing human qualities to animals.) By contrast, a two-year-old human already is forming complex judgments about his or her role with regard to Mommy, Daddy, siblings, other relatives, and household pets.

IMPRINTING IN HUMANS. Whereas humans have fewer instinctive responses than most animals and are far more capable of learning than any other creature, in the area of imprinting we are not so different from cows or even birds. We have discussed the importance of imprinting in birds, but it should be noted that

KEY TERMS

BEHAVIORAL RELEASER: A simple stimulus that brings about a fixed-action pattern of behavior, or FAP.

ETHOLOGY: The study of animal behavior, including its mechanisms and evolution.

FAPS: Fixed-action patterns of behavior, or strong responses on the part of an animal to particular stimuli. *FAP* is virtually synonymous with *instinct.*

IMPRINTING: The learning of a behavior at a critical period early in life, such that the behavior becomes permanent.

INNATE: A term to describe behaviors that are present and complete within the individual and which require no experience to learn them. For example, fish have an innate ability to swim, whereas humans must learn how to walk.

INSTINCT: A stereotyped, or largely unvarying, behavior that is typical of a particular species. An instinctive behavior does not have to be learned; rather, it is fully functional the first time it is performed.

LEARNING: The alteration of behavior as the result of experience.

NATURAL SELECTION: The process whereby some organisms thrive and others perish, depending on their degree of adaptation to a particular environment.

REFLEX: An inborn, automatic response to a stimulus by a part of an organism's body.

RELEASER: See *Behavioral releaser.*

STIMULUS: Any phenomenon (that is, an observable fact or event, such as an environmental change) that directly influences the activity or growth of a living organism.

imprinting is at least as critical among hoofed mammals, such as cattle or sheep, since they tend to congregate in large herds wherein a young animal could be separated easily from its mother. Likewise, with humans imprinting of some sort is critical.

Humans do not imprint as rigidly as geese do, but it is clear enough that some type of imprinting takes place in the mind of a baby. It is conceivable that Tarzan, having spent a year around his human parents, might retain a few human qualities—but only because it was the *first* year of his life, when the mind is by definition most impressionable. By the same token, a child that had spent six or seven years around humans would be socialized so thoroughly as a human that he or she probably would retain this human quality even if placed among apes.

From the time of the Greeks, humans have understood that part of what makes us human is contact with other humans. Therefore, an infant separated from its mother for a prolonged period during its first year of life may experience serious mental retardation; irreparable damage and even death may result from a separation of several months. There are all too many terrible stories of people who, as a result of neglect, abuse, or mere misfortune, have been forced to grow up in some form of isolation and have been stunted as a result. Usually, the outcome of this isolation is not nearly as attractive as the noble savage portrayed in literature or in such movies as *Nell* (1994), in which Jodie Foster plays a young woman who has grown up with little exposure to other humans.

WHERE TO LEARN MORE

Animal Behaviour. University of Plymouth Department of Psychology (Web site). <http://salmon.psy.plym.ac.uk/year1/animbeha.htm>.

Animal Cognition and Learning. Tufts University (Web site). <http://www.pigeon.psy.tufts.edu/psych26/default.htm>.

Dr. P's Dog Training. University of Wisconsin—Stevens Point (Web site). <http://www.uwsp.edu/psych/dog/lib-sci.htm>.

Gould, James L., and Carol Grant Gould. *The Animal Mind.* New York: Scientific American Library, 1994.

Instinct (Web site). <http://www.a2zpsychology.com/a2z%20guide/instinct.htm>.

Milne, Lorus Johnson, and Margery Joan Greene Milne. *The Behavior and Learning of Animal Babies.* Chester, CT: Globe Pequot Press, 1989.

Rogers, Lesley J., and Gisela T. Kaplan. *Songs, Roars, and Rituals: Communication in Birds, Mammals, and Other Animals.* Cambridge, MA: Harvard University Press, 2000.

Tinbergen, Niko. *The Study of Instinct.* New York: Clarendon Press, 1989.

Topoff, Howard R. *The Natural History Reader in Animal Behavior.* New York: Columbia University Press, 1987.

"Where's My Mommy? Imprinting in the Wild and in Operation Migration—Journey North Whooping Cranes." Annenberg/Corporation for Public Broadcasting (Web site). <http://www.learner.org/jnorth/tm/crane/ImprintingProtocol.html>.

MIGRATION AND NAVIGATION

CONCEPT

Among the most intriguing aspects of animal behavior and perception is the tendency to migrate long distances, coupled with the navigational ability that makes this possible. Most such migration is seasonal, a primary example being birds' proverbial flight south for the winter. Sometimes, however, animals widely separated from their home environments nonetheless manage to find their way home. This fact has long fascinated humans, as reflected in a number of true and fictional stories on the subject that have circulated over the years. For example, *The Incredible Journey,* a 1963 Disney film remade in 1993, is a fictional tale, but there are numerous true stories of dogs and cats making their way home to their masters across thousands of miles. How do animals do this? Scientists do not fully understand the answers, but theories regarding animal navigation abound. In any case, there is no question that animals possess navigational abilities unavailable to humans, for example, echolocation, used by bats, whales, and dolphins for local navigation, requires an ability to hear sounds far beyond the range of the human ear.

HOW IT WORKS

REASONS FOR MIGRATION

Why do animals migrate? Seasonal temperature changes, of course, are a factor, as in the well-known instance of birds flying south for the winter. But to justify enduring the dangers and hardships of long-distance migration, there must be an underlying cost-benefit equation whereby the benefits of migration outweigh the costs. Or, to put it another way, the physical "cost" of migrating must be less than the cost of staying home.

Human beings would perform such calculations rationally, of course, by thinking through the options and weighing them. Animals, on the other hand, rely on instinct—a word that, like many terms in science, has a somewhat different meaning within the scientific community than it does for the world at large. People tend to think of instinct as a matter of "just knowing" something, as in "I just know he/she is a good/bad person," but, in fact, instinct seems to have nothing to do with "knowing" at all.

On the contrary, instinct can be defined as a stereotyped (that is, largely unvarying) behavior that is typical of a particular species. Instinctive behavior does not have to be learned; rather, it is fully functional the first time it is performed. Though animals do exhibit some problem-solving ability, when a bird flies south for the winter, it has not thought that process through in any way. Instead, it is on "autopilot." This may seem almost magical, but it probably just reflects the processes of natural selection (see Evolution): for a particular bird species, those individuals "hard-wired" with a tendency to fly south were those that survived harsh winters, and therefore this tendency became favored in the gene pool.

Just as circumstances in the creature's home environment present a compelling need for migration, so there are other circumstances in the wintering environment that force the animal to leave as spring approaches. The wintering environment, after all, has its own native species, which most likely remain in the area even as the influx of visitors from up north arrives. Competition for food and shelter thus can become rather intense. Over time, this increased competition creates a situation in which it is advantageous for the migrating creature to return home.

TYPES OF MIGRATION

The idea of birds flying south en masse for an entire winter represents only one of four different types of migration: complete, as opposed to partial, differential, or irruptive migration. Complete migration involves the movement of all individuals within a population away from their breeding grounds at the conclusion of the breeding season. Usually this entails migration to a wintering site that may be thousands of miles away.

Some species practice partial migration, whereby some individuals remain at the breeding ground year-round, while others migrate. Others employ differential migration, in which all members of the population migrate, but for periods of time and over distances that vary as a function of age or sex. For example, herring gulls migrate for increasingly shorter distances the older they get, and male American kestrels remain at the breeding grounds longer than females. Even when the male birds do set out on their journeys, they do not travel as far as their female counterparts. Finally, there is irruptive migration, whereby certain species do not migrate at all during some years but may do so during other years. The likelihood of migration seems to be tied to climate and resource availability: the colder the winter and the more scarce the food, the more likely migration will occur in species prone to irruptive migratory behavior.

DIRECTIONS OF MIGRATION

Though southward migration is the most widely known form of migratory behavior, not all migration is from the north to the south. Actually, this type of movement is more properly called *latitudinal migration,* since it also takes place in the Southern Hemisphere, where, of course, it is from south to north. (Also, winter in those latitudes occurs at the same time as summer in the Northern Hemisphere.) There is far less habitable land below the equator than above it, however, so latitudinal migration in the Southern Hemisphere is not nearly as significant as it is at northerly latitudes.

There are, in fact, species of bird, such as the prairie falcon (*Falco mexicanus*), that travel longitudinally, or from east to west. This type of movement probably is related to seasonal changes in the location, availability, and choice of prey. Nor does all migration involve movements across Earth's surface; there is also elevational migration, which entails a change of altitude or depth beneath the sea.

Animals that live on mountains, for instance, may take part in elevational migration, moving to lower elevations in winter just as other species move to lower latitudes. For zooplankton, tiny animals that float on the waters of the open ocean, migration is a matter of changing depths in the water. During the summertime, when populations of zooplankton are large, these organisms live on the surface and feed on the plant life there. During the cold months, however, zooplankton migrate to depths of about 3,300 ft. (1 km) and do not feed at all.

REAL-LIFE APPLICATIONS

THE PROCESS OF MIGRATION

When animals migrate, they move along more or less the same corridors or paths each year. For North American birds migrating south for the winter, one of the most commonly used "flyways" is across the Gulf of Mexico, a journey of 500-680 mi. (800–1,000 km). To make it across the open waters of the gulf, birds have to store up fat, on which they can live for some time while out of sight of food sources.

Migrating birds are not like the proverbial parent (usually a father) who will not let the children stop to go to the bathroom on a long road trip. As they make their way south, birds stop regularly to rest and eat, sometimes for days at a time. These stops are particularly frequent and long just before crossing a large expanse of water. Only when the bird has stored sufficient quantities of body fat does it resume the journey.

DAY AND NIGHT TRAVEL. In North America it is common to see flocks of birds apparently flying south during the daytime in the autumn months. Yet most migrating bird species travel at night. The fewer species that travel by day tend to follow paths that are slower and less direct than those of nocturnal migrants. The reason for this difference has to do with the differences in feeding opportunities.

BIRDS ARE NOT THE ONLY CREATURES CAPABLE OF GREAT NAVIGATIONAL FEATS. SALMON COME BACK FROM THE OCEANS AND FIGHT THEIR WAY UPSTREAM TO SPAWN IN THE VERY SAME SPOT AT WHICH THEY WERE HATCHED. *(© Ralph A. Clevenger/Corbis. Reproduced by permission.)*

Nocturnal migrants have the entire day in which to rest, forage for food, and eat, thus building up reserves of energy for the nonstop flight that night. Daytime migrants, on the other hand, have to forage at the same time that they are traveling. For this reason, they tend to stick close to the coastlines, which offer abundant supplies of insect life. This slows them down but offers a dependable food supply.

FEATS OF NAVIGATION

In making their journeys, some creatures display navigational skills that would put such great mariners as Ferdinand Magellan and James Cook to shame. For example, the arctic tern (*Sterna paradisaea*) is a complete migrant in every sense of the word. Not only does the tern engage in complete migration, as defined earlier, but it also literally crosses the globe from pole to pole. In traveling from the Arctic to the Antarctic and back again in a single year, the arctic term completes a round-trip migration of more than 21,750 mi. (35,000 km). This distance, nearly the circumference of the planet, is the longest regular migratory path of any animal.

A specimen of another bird species, the Manx shearwater, was taken via airplane to Boston from its home on an island off the coast of Wales. Released experimentally in Boston, the bird took only 13 days to return to its point of origin, a flight of some 3,050 mi. (4,880 km). In another experiment, an albatross from Midway Island, deep in the Pacific, was released in the Philippines and made its way to its home area, a distance of 4,120 mi. (6,592 km), in only 32 days. There are also many examples of swallows finding their way across great distances. For instance, swallows that winter in southern Africa still manage to get back to their homes in northern Europe each spring. Then there are the swallows of San Juan Capistrano, southwest of Los Angeles, which leave every year on October 23 and return like clockwork on March 19.

Birds are not the only creatures capable of such great navigational feats. Monarch butterflies (*Danas plexippus*), which are born in Canada or the northern United States, winter each year in southern California, just as they have done for countless years. Then, when winter is over, they make their way back to their home regions. Likewise, salmon come back from the oceans and fight their way upstream to spawn in the very same spot at which they were hatched.

HOW DO THEY DO IT?

Based on these and other examples, one is left wondering, how do they do it? Numerous observations and theories have been put forward to answer this question. Salmon, for instance, seem to distinguish their home streams on the basis of smell, whereas some birds appear to use visual signals, primarily the position of the Sun or star patterns. Auditory cues and the sensitivity of migratory species to these cues often have been advanced as a key to migration behavior. Finally, one theory, which we discuss shortly, holds that sensitivity to Earth's magnetic field provides long-distance travelers with the navigational aid they need.

Several experiments have been performed on a variety of creature well known for its navigational abilities: the homing pigeon, which often can return across many hundreds of miles to its home. While they are undergoing training by humans, these pigeons are released from a series of sites, each just a little farther from the birds' home area. This training seems to make them accustomed to traveling long distances and to finding their way back. Thus, once trained, a pigeon released 100 mi. (63 km) or more from its home will begin flying in the correct direction within a few minutes.

BIOLOGICAL CLOCKS AND NAVIGATION. Several theories regarding pigeons' homing skills cite internal or biological clocks (see Biological Rhythms). One such theory, which is no longer accepted widely, held that the pigeon's perception of the Sun's position in the sky, combined with its biological clock, helped it navigate. Experiments have not proved this to be the case, however.

In one such trial pigeons were kept in a laboratory from which they could see the Sun for only very limited periods of time each day. After an extended period, the pigeons, with their eyes covered, were taken away about 40 mi. (64 km) to the south and released, so that the moment of release was their first unobstructed view of the Sun in weeks. Assuming the theory was correct, the pigeons would have been disoriented, but, in fact, they quickly took stock of their position and began flying north.

"MAGNETIC MAPS." One intriguing theory of animal navigation holds that creatures carry in their brains "magnetic maps," or strong sensitivities to Earth's magnetic field, that assist them in finding their way across distances. As reported in *National Geographic Today* on-line (October 12, 2001), research on loggerhead turtles has shown that hatchlings are sensitive to the strength and direction of Earth's magnetic field and apparently use this in their migratory navigation.

A FASCINATING EXAMPLE OF ANIMAL NAVIGATION IS ECHOLOCATION, WHICH IS NOT NECESSARILY TIED TO MIGRATION. RATHER, IT PROVIDES A MEANS OF LOCAL NAVIGATION—AND ESCAPE FROM PREDATORS—FOR CREATURES THAT LACK THE ABILITY TO SEE IN THEIR ENVIRONMENTS, SUCH AS THIS WHALE IN THE CLOUDY REALM OF THE OCEAN. *(© Terry Whittaker. Photo Researchers. Reproduced by permission.)*

By rigging up harnesses attached to electronic tracking units and outfitting the turtles with these devices, researchers were able to follow the course of their migration. Because these are, after all, turtles (though seaborne rather than terrestrial or land-based), migration is no speedy affair. To complete the trip, an 8,000-mi. (12,900-km) circuit from south Florida around the Saragasso Sea in the north Atlantic and back again, takes 5–10 years.

In the experiment, baby loggerheads were tagged with tracking systems just after they came out of their underground nests on the eastern coast of Florida. Still babies, they would normally begin a journey across the Atlantic, past the Canary and Cape Verde islands on the west coast and Africa and then back to Florida. But instead of going on this journey, the turtles in the experiment were placed in a large circular water tank surrounded by an electric coil capable of generating specific magnetic fields.

By turns, the research team exposed the animals to fields simulating those in three key spots along the route: northern Florida, the area off the coast of Portugal, and the region near the Cape Verde Islands. In each case, turtles responded to these magnetic stimuli by turning in the appropriate direction—for instance, south when they perceived that they were in the magnetic field equivalent to that of Portugal and west in the field resembling that of Cape Verde.

The research team, which published its results in the distinguished journal *Science,* concluded that the turtles were hardwired with magnetic sensitivities. The team leader Kenneth Lohmann, a biologist at the University of North Carolina in Chapel Hill, told *National Geographic Today,* "These turtles have never been exposed to water, yet they were able to process magnetic information and change their swimming direction accordingly. It seems they inherited some sort of magnetic map."

ECHOLOCATION

One final, fascinating example of animal navigation is echolocation, which differs from most of the navigational behaviors we have discussed so far in that this one is not necessarily tied to migration. Rather, echolocation provides a means of local navigation for creatures that lack the ability to see in their environments: bats flying through caves and dolphins, porpoises, and

KEY TERMS

BIOLOGICAL CLOCK: A mechanism within an organism (for example, the pineal gland in the human brain) that governs biological rhythms.

ECHOLOCATION: The use of sound waves, which are reflected back to the emitter, as a means of navigating.

INSTINCT: A stereotyped, or largely unvarying, behavior that is typical of a particular species. An instinctive behavior does not have to be learned; rather, it is fully functional the first time it is performed.

MIDDLE EAR: A small cavity that transmits sound waves, via a network of tiny bones, from the eardrum, which lies between it and the outer ear. Bats use the middle ear to separate transmission and reception signals in echolocation.

MIGRATION: A pattern of movement, usually regular and seasonal, whereby animals travel (typically guided by instinct) to specific locations.

NATURAL SELECTION: The process whereby some organisms thrive and others perish, depending on their degree of adaptation to a particular environment.

toothed whales swimming beneath the ocean's surface.

The frequency of a sound is related not to volume but to pitch: the higher the note, the higher the frequency. The frequency is measured in Hertz, or cycles per second. Human beings are capable of hearing sounds between 20 Hz and 20,000 Hz, whereas cats can hear sounds up to 32,000 Hz and dogs up to 46,000 Hz. This is why these creatures can hear dog whistles and other sounds inaudible to humans. Even the hearing ability of these household pets pales compared with that of bats and oceangoing mammals, which can hear tones up to 150,000 Hz.

Echolocation represents an evolutionary triumph in the form of adaptation to environments—the dark, nocturnal world of the bat and the cloudy realm of sea mammals. In the distant past, bats that hunted for insects during the day would have been at a disadvantage compared with birds, which are nimble of movement and extremely sharp-eyed in spotting their insect prey. Whales, porpoises, and dolphins were in an even worse situation with regard to sharks. Sharks, known for a finely tuned sense of smell, were not only competitors for prey but, as tertiary consumers (see Food Webs), they were and are also potential predators of sea mammals.

IN BATS. Only certain kinds of bats use echolocation. These are the carnivorous varieties that live on frogs, fish, and insects, as opposed to the herbivorous eaters of fruit and nectar. These carnivorous bats fly through the darkness, emitting extremely high frequency sounds and receiving the echoes from these sounds. Contained in the echo is a whole database of information regarding the object off which it has reflected: its distance, direction, size, surface texture, and even material composition.

Interestingly, the volume of these sounds is so great—as high as 100 decibels (dB)—that if people were able to hear them, the noises would be almost literally ear-splitting. (The upper range of safety is 120 dB.) As it is, a person would hear only clicks or chirps. Scientists have long wondered how the bats can both emit these sounds, which would be deafening to the bat, and hear them at the same time. Experimental evidence indicates that when the bat emits a sound, the middle ear (that is, the middle portion of the ear's interior) adjusts in such a way as to momentarily deafen the bat. A split second later, the bat's inner ear readjusts so as to permit it to hear the echo from the previous sound.

IN OCEANGOING MAMMALS. It is true, as the tagline for the 1979 film *Alien* threatened, that "In space no one can hear you scream." Another way of putting this is that sound requires a material medium through which to travel, and the more dense the medium, the more efficiently it moves. Thus, sound moves more effectively through water than through air, and for this reason echolocation is even more suited to the deep-sea environment than it is to the world above ground.

As with bats, undersea mammals send out sounds and then listen for the echoes. Owing to their heightened sense of hearing, these creatures obtain far more information from sounds than a human would. Sound, in fact, does for them what sight would do for a human, providing a detailed, three-dimensional image of their surroundings, but the images it offers are even more precise because sound waves are less subject to interference and diffraction than light waves. (For example, sound waves can simply go around a building, whereas the light waves that hit one side of a building are not visible from the other side.)

Toothed whales, dolphins, and porpoises, all of which normally would be disadvantaged by their weak senses of sight and smell, use more or less the same means to navigate by echolocation. A fatty deposit in the mammal's head helps it direct the sound; then an area of the lower jaw called the *acoustic window* receives reflected noises. A fatty organ in the middle ear transmits vibrations from the echo, which are translated into neural impulses that go to the brain.

As with the bat, the toothed whale has special structures in its head that help it hear even as it sends out noises. Some species have bony insulating structures, which separate the portion of the head where sounds are received from that part where sounds are generated. Likewise, structures in the middle ear assist the whale in distinguishing whether sounds come from the left or the right, thus facilitating the whale's navigation.

WHERE TO LEARN MORE

Caras, Roger A. *The Endless Migrations.* New York: Dutton, 1985.

Journey North: A Global Study of Wildlife Migration (Web site). <http://www.learner.org/jnorth/>.

McDonnell, Janet. *Animal Migration.* Elgin, IL: Child's World, 1989.

Monarchs and Migration (Web site). <http://www.smm.org/sln/monarchs/>.

Neuroethology: Echolocation in the Bat (Web site). <http://soma.npa.uiuc.edu/courses/physl490b/models/bat_echolocation/bat_echolocation.html>.

Penny, Malcolm. *Animal Migration.* Illus. Vanda Baginska. New York: Bookwright Press, 1987.

Trivedi, Bijal P. "'Magnetic Map' Found to Guide Animal Migration" National Geographic Today (Web site). <http://news.nationalgeographic.com/news/2001/10/1012_TVanimalnavigation.html>.

Waterman, Talbot H. *Animal Migration.* New York: Scientific American Library, 1989.

THE BIOSPHERE AND ECOSYSTEMS

THE BIOSPHERE
ECOSYSTEMS AND ECOLOGY
BIOMES

THE BIOSPHERE

CONCEPT

The biosphere is simply "life on Earth"—the sum total, that is, of all living things on Earth. Yet the whole is more than the sum of the parts: not only is the biosphere an integrated system whose many components fit together in complex ways, but it also works, in turn, in concert with the other major earth systems. The latter include the geosphere, hydrosphere, and atmosphere, through which circulate the chemical elements and compounds essential to life. Among these elements is carbon, a part of all living things, which also cycles through the nonliving realms of soil, water, and air—just one of many vital biogeochemical cycles. As for the compounds on which life depends, none is more important than water, which, though it is the focal point of the hydrosphere, passes through the various earth systems as well. Organisms participate in the hydrologic cycle by providing moisture to the air through the process of transpiration, and they likewise benefit from the downward movement of moisture in the form of precipitation. These and many other interactions make it easy to see why scientists speak of Earth as a system—and why some go even further and call it a living thing.

HOW IT WORKS

EARTH SYSTEMS

Chances are that the mention of the word *system* calls to mind something mechanical or electrical, produced by humans: for example, a heating and cooling system. This application of the word is close to the scientific meaning, but in the sciences *system* identifies a wider range of examples. In scientific terms a system is any set of interactions that can be set apart mentally from the rest of the universe for the purposes of study, observation, and measurement. Thus, virtually by definition, a system is something in which the various parts fit together harmoniously, as though they were designed—or adapted over millions of years—to do so.

Anything outside the system is known as the *environment,* and numerous qualifying terms identify the level of interaction between the system and its environment. An isolated system is one so completely sealed off from its environment that neither matter nor energy passes through its boundaries. This is a merely theoretical construct, however, because, in practice, some matter *always* flows between system and environment. For example, regardless of how tightly a vault or other interior chamber is sealed, there is always room for matter at the microscopic or atomic level to pass through the barrier; moreover, energy, which in many forms does not require any material medium for its transmission, will pass through as well.

Earth itself is an approximation of a closed system, or a system in which, despite the sound of its name, exchanges of energy (but not matter) with the environment are possible. Earth absorbs electromagnetic energy from the Sun and returns that energy to space in a different form, but very little matter enters or departs Earth's system. Earth is not a perfectly closed system, however, since meteorites can enter the atmosphere and hydrogen can escape. Without the intrusion of meteorites, in fact, it is unlikely that life could exist on the planet, because these projectiles from space first brought water (and possibly even the carbon-based rudiments of life) to the planet Earth more than four billion years ago.

Nevertheless, Earth more closely resembles a closed system than it does an open system, or one that allows the full and free exchange of both matter and energy with its environment. The human circulatory system is an example of an open system, as are the various "spheres" of Earth (geosphere, hydrosphere, biosphere, and atmosphere) that we discuss later in this essay. The distribution of matter and energy within these earth systems may vary over time, but the total amount of energy and matter within the larger earth system is constant.

THE FOUR "SPHERES." On the other hand, the four subsystems, or "spheres," within the larger Earth system are very much open systems. Of these four subsystems, the name of only one, the atmosphere, is familiar from everyday life, whereas those of the other three sectors (geosphere, hydrosphere, and biosphere) may sound at first like scientific jargon. Yet each has a distinct identity and meaning, and each represents a part of Earth that is at once clearly defined and virtually inseparable from the rest of the planet.

The geosphere is the upper part of the planet's continental crust, the portion of the solid earth on which human beings live and which provides them with most of their food and natural resources. It is also the oldest, followed by the hydrosphere, which had its beginnings with several hundred millions years' worth of rains that took place about four billion years ago. Today the hydrosphere includes all water on Earth, except for water vapor in the atmosphere. The latter, incidentally, was probably the last of the four subsystems to take shape: though Earth in its early stages had a blanket of gases around it, there was no oxygen. (See Paleontology for more about the early atmosphere, oxygen, and early life.)

THE ATMOSPHERE. Today the atmosphere is 78% nitrogen, 21% oxygen, and 0.93% argon. The remaining 0.07% is made up of water vapor, carbon dioxide, ozone (a form of oxygen in which three oxygen atoms bond chemically), and noble gases. The noble gases, including argon and neon, are noted for their lack of reactivity, meaning that they are extremely resistant to chemical bonding with other elements.

Nitrogen also tends to be unreactive, and the reason for its abundance in the atmosphere lies in the fact that it never attempted to bond with other elements. Therefore, nitrogen, along with the noble gases, is simply "hanging in the air" (literally), left over from the time when volcanoes hurled it into the atmosphere several billion years ago. By contrast, oxygen (both in O_2 and O_3, or ozone, molecules) and the other elements in air are vital to life. Furthermore, oxygen is one of two elements, along with hydrogen, that goes into the formation of water.

OVERLAP BETWEEN SUBSYSTEMS. The present atmosphere would not exist without the biosphere. In order to put oxygen into the air, there had to be plants, which take in carbon dioxide and release oxygen in the process of photosynthesis. This resulted from an exceedingly complex series of evolutionary developments from anaerobic, or non-oxygen-breathing, single-cell life-forms to the appearance of algae. As plant life evolved, eventually it put more and more oxygen into the atmosphere, until the air became breathable for animal life. Thus, the atmosphere and biosphere have sustained one another.

Such overlap is typical and indeed inevitable where the open earth subsystems are concerned, and examples of this overlap are everywhere. For instance, plants (biosphere) grow in the ground (geosphere), but to survive they absorb water (hydrosphere) and carbon dioxide (atmosphere). Nor are plants merely absorbing: they also give back oxygen to the atmosphere, and by providing nutrition to animals, they contribute to the biosphere. At the same time, the many components of the picture just described are involved in complex biogeochemical cycles, which we look at later.

THE BIOSPHERE IN CONTEXT. The biosphere is, of course, integral to the functioning of earth systems. First of all, the present atmosphere, as we have noted, is the product of respiration on the part of plants, which receive carbon dioxide and produce oxygen. In addition, transpiration, a form of evaporation from living organisms (primarily plants), is a mechanism of fundamental importance for moving moisture from the hydrosphere through the biosphere to the atmosphere.

We examine transpiration later, within the larger context of evapotranspiration, along with another area in which the biosphere interacts closely with one or more of the other earth systems: soil. Though soil is part of the geosphere, its production and maintenance is an achievement of all spheres. The role of the biosphere in this instance is particularly important: the amount of decayed organic material (i.e., dead plants and animals) is critical to the quality of the soil for sustaining further life in the form of plants and other organisms that live underground.

As important as the biosphere is, it may be surprising to learn just what a small portion of the overall earth system it occupies. As living beings, we tend to have a bias in favor of the living world, but the overwhelming majority of the planet's mass and space is devoted to nonliving matter. The geosphere alone accounts for almost 82% of the combined mass of the four subsystems. (This is *not* the combined mass of Earth, which would be much larger; remember that the geosphere is only the extreme upper layer of Earth, and does not include the vast depths of the lower mantle and core.)

Of the remaining mass that makes up the four earth systems, the hydrosphere is a little more than 18%, the atmosphere less than 1%, and the biosphere a tiny 0.00008%. Note just how much greater the amount of mass is in the air, which we tend to think of as being weightless (though, of course, it is not), than in the biosphere. Even within the biosphere's almost infinitesimal fraction of total mass, the animal kingdom accounts for less than 2%, the remainder being devoted to other kingdoms: plants, fungi, monera (including bacteria), and protista, such as algae. (See Taxonomy and Species for more about the kingdoms of living things.) It need hardly be added that humans, in turn, are a very, very small portion of the animal world.

WATER AND THE HYDROLOGIC CYCLE

As any backyard horticulturist knows, plants need good soil *and* water. In the course of circulating throughout Earth, water makes its way through organisms in the biosphere as well as reservoirs housed within the geosphere. It also circulates continuously between the hydrosphere and the atmosphere. This movement, known as the hydrologic cycle, is driven by the twin processes of evaporation and transpiration.

The first of these processes, of course, is the means whereby liquid water is converted into a gaseous state and transported to the atmosphere, while the second one—a less familiar term—is the process by which plants lose water through their stomata, small openings on the undersides of leaves. Scientists usually speak of the two as a single phenomenon, evapotranspiration. The atmosphere is just one of several "compartments" in which water is stored within the larger environment. In fact, the atmosphere is the only major reservoir of water on Earth that is not considered part of the hydrosphere.

ACCOUNTING FOR EARTH'S WATER SUPPLY. The water that most of us see or experience is only a very small portion of the total. Actually, that statement should be qualified: the oceans, parts of which most people have seen, make up about 5.2% of Earth's total water supply. This may not sound like a large portion, but, in fact, the oceans are the second-largest water compartment on Earth. If the oceans are such a small portion yet rank second in abundance, two things are true: there must be *a lot* of water on Earth, and most of it must be in one place.

In fact, the vast majority of water on Earth is stored in aquifers, or underground rock formations, that hold 94.7% of the planet's water. Thus, deep groundwater and oceans account for 99.9% of the total. Glaciers and other forms of permanent and semipermanent ice take third place, with 0.065%. Another 0.03% appears in the form of shallow groundwater, the source of most local water supplies. Next are the inland surface waters, including such vast deposits as the Great Lakes and the Caspian Sea as well as the Mississippi-Missouri, Amazon, and Nile river systems and many more, which collectively make up just 0.003% of Earth's water.

ATMOSPHERIC MOISTURE AND WEATHER. That leaves only 0.002%, which is the proportion taken up by moisture in the atmosphere: clouds, mist, and fog, as well as rain, sleet, snow, and hail. While it may seem astounding that atmospheric moisture is such a small portion of the total, this fact says more about the vast amounts of water on Earth than it does about the small amount in the atmosphere. That "small" amount, after all, weighs 1.433×10^{13} tons (1.3×10^{13} tonnes), or 28,659,540,000,000,000 pounds (12,999,967,344,000,002 kg).

This moisture in the atmosphere is the source of all weather, which clearly has an effect on Earth's life-forms. (Weather is the condition of the atmosphere at a given time and in a given place, whereas climate is the pattern of weather in a particular area over an extended period of time.) On the one hand, rain is necessary to provide water to plants, and desert conditions can sustain only very specific life-forms; on the other hand, storms, icy precipitation, and flooding can be deadly.

BIOGEOCHEMICAL CYCLES

Water is not the only substance that circulates through the various earth systems. So, too, do six other substances or, rather, chemical elements. These elements are composed of a single type of atom, meaning that they cannot be broken down chemically to make a simpler substance, as is the case with such compounds as water. The six elements that cycle throughout Earth's systems are hydrogen, oxygen, carbon, nitrogen, phosphorus, and sulfur. The two following lists provide rankings for their abundance. The first shows their ranking and share in the entire known mass of the planet, including the crust, living matter, the oceans, and atmosphere. The second list shows their relative abundance and ranking in the human body.

Abundance of Selected Elements on Earth (Ranking and Percentage):

- 1. Oxygen (49.2%)
- 9. Hydrogen (0.87%)
- 12. Phosphorus (0.11%)
- 14. Carbon (0.08%)
- 15. Sulfur (0.06%)
- 16. Nitrogen (0.03%)

Abundance of Selected Elements in the Human Body (Ranking and Percentage):

- 1. Oxygen (65%)
- 2. Carbon (18%)
- 3. Hydrogen (10%)
- 4. Nitrogen (3%)
- 6. Phosphorus (1%)
- 9. Sulfur (0.26%)

Note that the ranking of all these elements (with the exception of oxygen) is relatively low in the total known elemental mass of Earth, whereas their relative abundance is much, much higher within the human body. This is significant, given the fact that these elements are all essential to the lives of organisms. All six of these elements take part in biogeochemical cycles, a term used to refer to the changes that a particular element undergoes as it passes back and forth through the various earth systems and particularly between living and nonliving matter.

THE CARBON AND NITROGEN CYCLES. Carbon, for instance, is present in all living things and is integral to the scientific definition of the word *organic.* The latter term does not, as is popularly believed, refer only to living and formerly living things, their parts, and their products, such as sweat or urine. Organic refers to the presence of compounds containing carbon and hydrogen. The realm of organic substances encompasses not only the world of the living, the formerly living, and their parts and products, but also such substances as plastics that have never been living.

The carbon cycle itself involves movement between the worlds of the living and nonliving, the organic and inorganic. This highly complex biogeochemical cycle circulates carbon from soil and carbonate rocks (which are inorganic because they do not contain carbon-hydrogen compounds) to plants and hence to animals, which put carbon dioxide into the air. Likewise, the nitrogen cycle moves that element among all these reservoirs.

Because nitrogen is highly unreactive, the participation of microorganisms in the nitrogen cycle is critical to moving that element between the various earth systems. These organisms "fix" nitrogen, meaning that by processing the element through their bodies, they bring about a chemical reaction that makes nitrogen usable to plant life. Additionally, detritivores and decomposers in the soil are responsible for transforming nitrites and nitrates (compounds of nitrogen and oxygen) from the bodies of dead animals into elemental nitrogen that can be returned to the atmosphere.

REAL-LIFE APPLICATIONS

SOIL AND THE LIFE IN IT

The soil is a sort of anchor to the biosphere. It teems with life like few other areas within the earth system, and, indeed, there are more creatures—plant, animal, monera, protista, and fungi—living in the soil than above it. Minerals from weathered rock in the soil provide plants with the nutrients they need to grow, setting in motion the first of several steps whereby organisms take root in and contribute to the soil.

NUMEROUS ANT SPECIES PERFORM A POSITIVE FUNCTION FOR THE ENVIRONMENT. LIKE EARTHWORMS, THEY AERATE SOIL AND HELP BRING OXYGEN AND ORGANIC MATERIAL FROM THE SURFACE WHILE CIRCULATING SOILS FROM BELOW. *(© Ralph A. Clevenger/Corbis. Reproduced by permission.)*

Plants provide food to animals, which, when they die, likewise become one with the soil. So, too, do plants themselves. Organisms from both the plant and animal kingdoms leave behind material to feed such decomposers as bacteria and fungi, which, along with detritivores, are critical to the functioning of food webs (see entry). Detritivores, of which earthworms are a great example, are much more complex organisms than the typically single-cell decomposers.

We cannot see bacteria, but almost anyone who has ever dug in the dirt has discovered another type of organism: the Annelida phylum of the animal kingdom, which includes all segmented worms, among them, the earthworm. (Incidentally, the leech family also falls within phylum Annelida.) These slimy creatures at first might seem disgusting, but without the appropriately named earthworm, our world could not exist as it does.

Detritivores consume the remains of plant and animal life, which usually contain enzymes and proteins far too complex to benefit the soil in their original state. By feeding on organic remains, detritivores cycle these complex chemicals through their internal systems, thus causing the substances to undergo chemical reactions that result in the breakdown of their components. As a result, simple and usable nutrients are made available to the soil.

Earthworms, which are visible and relatively large, are not the only worms at work in the soil. There are also the colorless creatures of the phylum Nematoda, or roundworms. Included in this phylum are hookworms and pinworms, which can be extremely detrimental to the body when they live inside it as parasites. (See Parasites and Parasitology.) This is one good reason (among many) not to eat dirt. But nematodes in the soil, most of them only slightly larger than microorganisms, perform the vital function of processing organic material by feeding on dead plants. Even in a soil situation, however, some nematodes are parasites that live off the roots of such crops as corn or cotton.

In addition to earthworms, ants and other creatures are also significant inhabitants of the soil. Like earthworms, ants aerate soil and help bring oxygen and organic material from the surface while circulating soils from below. Among the larger creatures that call the soil home are moles, which live off earthworms, grubs (insect larvae), and the roots of plants. By burrowing under the ground, they help to loosen the soil, making it more porous and thus receptive to both moisture and air. Other large burrowing creatures include mice, ground squirrels, and, in some areas, even prairie dogs.

SOILS AND THE ENVIRONMENTS THEY HELP CREATE

Five different factors determine the quality of soil: parent material (the decayed organisms and weathered rock that make it up), climate, the presence of living organisms, topography (the shape of the land, including prominent natural features), and the passage of time. These factors influence the ability of the soil to sustain life.

For example, in a desert, a place that obviously has a smaller abundance and complexity of life-forms (see Biological Communities), the soil itself is lacking in this life-sustaining quality. Desert soil, in fact, is usually referred to as *immature* soil. Healthy soil normally has a deep A horizon, the area in which decayed organic material appears and atop which humus sits. In general, the deeper the A horizon, the better the soil. Immature soil, on the other hand, has a very thin A horizon and no B horizon, which is the subsoil that typically separates the A horizon from a layer of weathered material that rests even lower, at the C horizon just above bedrock.

In deserts, by definition, the water supply is very limited, and only those species that require very little water—for example, the varieties of cactus that grow in the American Southwest—are able to survive. But lack of water is not the only problem. Desert subsoils often contain heavy deposits of salts, and when rain or irrigation adds water to the topsoil, these salts rise. Thus, watering desert topsoil actually can make it a worse environment for growth.

TROPICAL RAIN FORESTS. The soil in rain forests has just the opposite problem of desert soil: instead of being immature, it has gone beyond maturity and reached old age, a point at which plant growth and water percolation (the downward movement of water through the soil) have removed most of its nutrients. In environments located near the equator, whether these regions be desert or rain forest, soils tend to be "old." This helps to explain the fact that equatorial regions are usually low in agricultural productivity, despite the fact that they enjoy an otherwise favorable climate.

Given the poor quality of equatorial soil, one might wonder how it is that some of the most lush rain forests are located in the equatorial regions of Africa and South America. The answer is that a rain forest has such a wide assortment of life-forms that there is never a shortage of decayed organic material to "feed" the soil. Due to the sheer breadth and scope of the rain forest's ecological diversity, there are bound to always be plants and animals dying, replenishing what would otherwise be poor soil. The rapid rate of decay common in warm, moist regions (which are extremely hospitable to bacteria and other microbes) further supports the process of renewing minerals in the ground.

The fact that decayed organic material is critical to the life of soil helps to explain why many environmentalists have long been concerned about the destruction of tropical rain forests. For example, in Brazil, vast portions of the Amazon rain forest have been clear-cut, subjected to an extremely destructive form of slash-and-burn agriculture that is motivated by modern economic concerns—the desire of the country's leaders to create jobs and exports—but which resembles practices applied (albeit on a much smaller scale) by premodern peoples in Central and South America.

By removing the heavy jungle canopy of tall trees, clear-cutting exposes the ground to the heat of the Sun and the pounding of monsoon rains. Sun and rain, thanks to the removal of this protection, fall directly on the ground, parching it in the first instance and eroding it in the second. Furthermore, when trees and other vegetation are removed, the animal life that these plants supported disappears as well, and this has a direct impact on the soil by removing organisms whose waste products and bodies eventually would have decayed and enriched it.

WHAT MAKES GOOD SOIL? The soil of the Brazilian rain forest may be old and weak, but thankfully there are places in the world where the soil tells an entirely different story. Instead of being nutrient-poor, this soil is nutrient-rich, and instead of being red, such soil is a deep, rich black. Sometimes regions of good and bad soil exist in close proximity, as in ancient Egypt, where the Nile made possible a narrow strip of extraordinarily productive land, running the length of the country, flanked by deserts. The latter the Egyptians called "the red land" because of its nutrient-poor soil, whereas their own fertile region was "the black land."

Just as Egypt, after its annexation by the Romans in 30 B.C., became known as the breadbasket of the Roman Empire, there are other "black lands" that serve as the breadbaskets of today's world. Unlike Egypt, however, most are located in temperate rather than equatorial regions. (See Biomes for more about temperate regions.) Examples include the midwestern United States, western Canada, and southern Russia, regions characterized by vast plains of fertile black soil. Below this rich topsoil is a thick subsoil that helps hold in moisture and nutrients.

Rivers such as the Nile, the Mississippi-Missouri, or the Volga in Russia make possible the richest variety of soil on Earth, alluvial soil, a youngish sediment of sand, silt, and clay. A river pulls soil along with it as it flows, and with this comes nutrients from the regions through which the river has passed. The river then deposits these nutrients in the alluvial soil at the delta, the area where it enters a larger body, usually of salt water. (A delta is so named because, as it widens in the region near the sea, the shape of the river is like that of the Greek letter delta, or Δ.)

Because they are depositories for accumulated alluvial soil, delta regions such as Mississippi and Louisiana in the United States (where the mighty Mississippi-Missouri, the largest river system in North America, empties into the Gulf of Mexico), are exceedingly fertile. The same is true in the Volga and Nile deltas, the deltas of the Danube and other major European rivers, and those of lesser-known rivers in Canada or Australia.

FERTILIZER. It is also possible to artificially improve soil that is not in a river delta, or that has not otherwise been blessed by nature. The most significant way to achieve this is by using fertilizer, which augments the nutrients in the soil itself. As noted earlier, nitrogen is highly nonreactive, meaning that it tends not to bond chemically with other substances. However, because it is a necessary component of biogeochemical cycles, it is critical that nitrogen be introduced to the soil in such a way that it becomes useful, and typically this is done by combining it with a highly reactive element: oxygen.

Fertilizers may contain nitrogen in the form of a nitrate, which is a compound of nitrogen and oxygen, or they may include a nitrogen–hydrogen compound. The latter may be ammonia (NH_3) or ammonium (NH_4^+). Note that ammonia is much more than the product with which it is most readily associated: a household cleaner. In fact, ammonia is an extremely abundant substance, occurring naturally, for instance, in the atmospheres of Venus and other planets in our solar system. The importance of ammonia is reflected by the fact that it and water are the only two substances that chemists regularly refer to by their common names, as opposed to a scientific name such as carbon dioxide.

Desert tortoise and beavertail cactus in the Mojave Desert. Only those plant and animal species that can endure a limited water supply and immature soil with heavy deposits of salt in the lower layers can survive a desert environment. *(© D. Suzio. Reproduced by permission.)*

As for ammonium, its extra hydrogen atom makes it a substance that dissolves in water and is attracted to negatively charged surfaces of clays and organic matter in soil. Therefore, it tends to become stuck in one place rather than to move around, as nitrate does. Plants in acidic soils typically receive their nitrogen from ammonium, but in nonacidic soils, nitrate is typically the more useful form of fertilizer. The two fertilizers are also combined to form ammonium nitrate, which is powerful both as a fertilizer and as an explosive. (Ammonium nitrate was used both in the first World Trade Center bombing, in 1993, and in the even more devastating Oklahoma City bombing two years later.)

EROSION AND SOIL CONSERVATION

The mismanagement of agricultural lands, and/or the influence of natural forces, can produce devastating results, as illustrated by events during the years 1934 and 1935 in a region including Texas, Oklahoma, Kansas, and eastern Colorado. In just a few months, once-productive farmland turned into worthless fields of stubble and dust, good for virtually nothing. By the time it was over, the region had acquired a bitter nickname: the "dust bowl."

Ironically, in the years leading up to the early 1930s the future dust bowl farmlands had seemed remarkably productive. Farmers happily reaped abundant yields, year after year, not knowing that they were actually preparing the way for soil erosion on a grand scale. Farmers in the 1930s had long known about the principle of crop rotation as a means of giving the soil a rest and restoring its nutrients. But to be successful, crop rotation must include fallow years (i.e., no crops are planted), and must make use of crops that replenish the soil of nutrients.

Cotton and wheat are examples of crops that deplete nutrient content in the soil, and in fact wheat was the crop of choice in the future dust bowl. In some places, farmers alternated between wheat cultivation and livestock grazing on the same plot of land. The hooves of the cattle further damaged the soil, already weakened by raising wheat. The land was ready to become the site of a full-fledged natural disaster, and in the depths of the Great Depression, that disaster came in the form of high winds. These winds scattered vast quantities of soil from the Great Plains of the Midwest to the Atlantic seaboard, and acreage that once had rippled with wheat turned into desert-like wastelands.

The farmlands of the plains states have long since recovered from the dust bowl, and farming practices have changed considerably. Instead of alternating one year of wheat with a year of grazing livestock, farmers in the dust bowl region apply a three-year cycle of wheat, sorghum, and fallow land. They also have planted trees to serve as barriers against wind.

Years after the dust bowl, the American West could have again become the site of another disaster, had not farmers and agricultural officials learned from the mistakes of an earlier generation. During the 1970s, American farms enjoyed such a great surplus that farmers increasingly began to sell their crops to the Soviet Union, and farmers were encouraged to cultivate even marginal croplands to increase profits. This alarmed environmental activists, who called attention to the flow of nutrients from croplands into water resources. As a result of public concerns over these and related issues, Congress in 1977 passed the Soil and Water Resource Conservation Act, mandating measures to conserve or protect soil, water, and other resources on private farmlands and other properties.

LEACHING AND ITS EFFECT ON SOIL

Like erosion, leaching moves substances through soil, only in this case it is a downward movement. Leached water can carry all sorts of dissolved substances, ranging from nutrients to contaminants. The introduction of manufactured contaminants to the soil, and hence the water table, is of course a serious threat to the environment. On the other hand, where human waste and other, more natural forms of toxin are concerned, nature itself is able to achieve a certain amount of cleanup on its own.

In a septic tank system, used by people who are not connected to a municipal sewage system, anaerobic bacteria process wastes, removing a great deal of their toxic content in the tank itself. (These bacteria usually are not introduced artificially to the tank; they simply congregate in what is a natural environment for them.) The wastewater leaves the tank and passes through a drain field, in which the water leaches through layers of gravel and other filters that help remove more of its harmful content. In the drain field, the waste is subjected to aerobic decay by other forms of bacteria before it either filters through the drainpipes into the ground or is evaporated.

In addition to purifying water, leaching also passes nutrients to the depths of the A horizon and into the B horizon—something that is not always beneficial. In some ecosystems, leaching removes large amounts of dissolved nitrogen from the soil, and it becomes necessary to fertilize the soil with nitrate. However, soil often has difficulty binding to nitrate, which tends to leach easily, and this leads to an overabundance of nitrogen in the lower levels of the soil and groundwater. This is a condition known as nitrogen saturation, which can influence the eutrophication of waters (a topic discussed later) and cause the decline and death of trees with roots in an affected area of ground.

HOW DESERTS ARE FORMED

Let us now consider what one might call an *extreme* ecosystem: a desert. Of the world's deserts, by far the most impressive is the Sahara, which today spreads across some 3.5 million sq. mi. (9.06 million sq km), an area that is larger than the continental United States. Only about 780 acres (316 hectares) of it, a little more than 1 sq. mi. (2.6 sq km), is fertile. The rest is mostly stone and dry earth with scattered shrubs—and here and there the rolling sand dunes typically used to depict the Sahara in movies.

Just 8,000 years ago—the blink of an eye in terms of Earth's timescale—it was a region of flowing rivers and lush valleys. For thousands of years it served as a home to many cultures, some of them quite advanced, to judge from their artwork. Though they left behind an extraordinary record in the form of their rock-art paintings and carvings, which show an understanding of realistic representation that would not be matched until the time of the Greeks, the identity of the early Saharan peoples themselves remains largely a mystery.

EVOLUTION OF THE ANCIENT SAHARAN ECOSYSTEM. The phases in ancient Saharan cultures are identified by the names of the domesticated animals that dominated at given times, and collectively these names tell the story of the Saharan ecosystem's transformation from forest to grassland to desert. First was the Hunter period, from about 6000 to about 4000 B.C., when a Paleolithic, or Old Stone Age, people survived by hunting the many wild animals then available in the region. Next came the Herder period, from about 4000 to 1500 B.C. As their name suggests, these people maintained herds of animals and also practiced basic agriculture.

A high point of ancient Saharan civilization came with the Herder period, which, not surprisingly, also marked the high point of the local ecosystem in terms of its ability to sustain varied life-forms. Yet through a complex set of circumstances that scientists today do not understand fully, desertification—the slow transformation of ordinary lands to desert—had begun to set in. As the Sahara became drier and drier, the herds disappeared.

Eventually, the Egyptians began bringing in domesticated horses to cross the desert: hence the name of the Horse period (*ca.* 1500–*ca.* 600 B.C.), when the Sahara probably resembled the dry grasslands of western Texas or the sub-Saharan savanna in Africa today. By about 600 B.C., however, the climate had become so severe, the water supply so limited, and the ecosystem so depleted of supporting life-forms that not even horses could survive in the forbidding climate. There was only one mammal that could: the hardy, seemingly inexhaustible creature that gave its name to the Camel era, which continues to the present day.

CONTROLLING DESERTIFICATION. What happened to the Sahara? The answer is a complex one, as is the subject of desertification. Desertification does not always result in what people normally think of as a desert; rather, it is a process that contributes toward making a region more dry and arid, and because it is usually gradual, it can be reversed in some cases. Nor is it necessary for a society to undertake large-scale mechanized agricultural projects, such as those of the American dust bowl of the 1930s, to do long-term damage that can result in desertification. The Pueblan culture of what is now the southwestern United States depleted an already dry and vulnerable region after about A.D. 800 by removing its meager stands of mesquite trees.

And though human causes, either in the form of mismanagement or deliberate damage, certainly have contributed to desertification, sometimes nature itself is the driving force. Long-term changes in rainfall or general climate, as well as water erosion and wind erosion such as caused the dust bowl, can turn a region into a permanent desert. An ecosystem may survive short-term drought, but if soil is forced to go too long without proper moisture, it sets in motion a chain reaction in which plant life dwindles and, with it, animal life. Thus, the soil is denied the fresh organic material necessary to its continued sustenance, and a slow, steady process of decline begins.

THE PHOSPHORUS CYCLE

Having examined what can go wrong (and right) in soil, a key component of the biosphere, we will devote the remainder of this essay to two other aspects of the biosphere mentioned earlier: biogeochemical cycles and the hydrologic cycle. In the context of biogeochemical cycles, we will look at the phosphorus cycle, along with eutrophication—another instance of "what can go wrong."

Because of its high reactivity with oxygen, phosphorus is used in the production of safety matches, smoke bombs, and other incendiary devices. It is also important in various industrial applications and in fertilizers. In fact, ancient humans used phosphorus without knowing it when they fertilized their crops with animal bones. In the early 1800s chemists recognized that the critical component in the bones was the phosphorus, which plants use in photosynthesis—the biological conversion of energy from the Sun into chemical energy. With this discovery came the realization that phosphorus would make an even more effective fertilizer when treated with sulfuric acid, which makes it soluble, or capable of being dissolved, in water. This compound, known as superphosphate, can be produced from phosphate, a type of phosphorus-based mineral.

Microorganisms in the biosphere absorb insoluble phosphorus compounds and, through the action of acids within the microorganisms, turn them into soluble phosphates. These soluble phosphates then are absorbed by algae and other green plants, which are eaten by animals. When they die, the animals, in turn, release the phosphates back into the soil. As with all elements, the total amount of phosphorus on Earth stays constant, but the distribution of it does not. Some of the phosphorus passes from the geosphere into the biosphere, but the vast majority of it winds up in the ocean. It may find its way into sediments in shallow waters, in which case it continues to circulate, or it may be taken to the deep parts of the seas, in which case it is likely to be deposited for the long term.

Because fish absorb particles of phosphorus, some of it returns to dry land through the catching and consumption of seafood. Also, guano or dung from birds that live in an ocean environment (e.g., seagulls) returns portions of phosphorus to the terrestrial environment. Nevertheless, scientists believe that phosphorus is steadily being transferred to the ocean, from whence it is not likely to return. It is for this reason that phosphorus-based fertilizers are important, because they feed the soil with nutrients that would otherwise be steadily lost. However, phosphorus still ends up making its way through the waters, and this creates a serious problem in the form of eutrophication.

EUTROPHICATION. Eutrophication (from a Greek term meaning "well nourished") is a state of heightened biological productivity in a body of water. One of the leading causes of eutrophication is a high rate of nutrient input, in the form of phosphate or nitrate, a nitrogen-oxygen compound. As a result of soil erosion, fertilizers make their way into bodies of water, as does detergent runoff in wastewater. Excessive phosphates and nitrates stimulate growth in algae and other green plants, and when these plants die, they drift to the bottom of the lake or other body of water. There, decomposers consume the remains of the plants, and in the process of doing so, they also use oxygen that otherwise would be available to fish, mollusks, and other forms of life. As a result, those species die off, to be replaced by others that are more tolerant of lowered oxygen levels—for example, worms. Needless to say, the outcome of eutrophication is devastating to the lake's ecosystem.

Lake Erie—one of the Great Lakes on the border of the United States and Canada—became an extreme example of eutrophication in the 1960s. As a result of high phosphate concentrations, Erie's waters were choked with plant and algae growth. Fish were unable to live in the water, the beaches reeked with the smell of decaying algae, and Erie became widely known as a "dead" body of water. This situation led to the passage of new environmental standards and pollution controls by both the United States and Canada, whose governments acted to reduce the phosphate content in fertilizers and detergents drastically. Within a few decades, thanks to the new measures, the lake once again teemed with life. Thus, Lake Erie became an environmental success story.

EVAPOTRANSPIRATION

Many of the phenomena and processes we have described tie together the biosphere with other "spheres" of Earth. Such is the case with evapo-

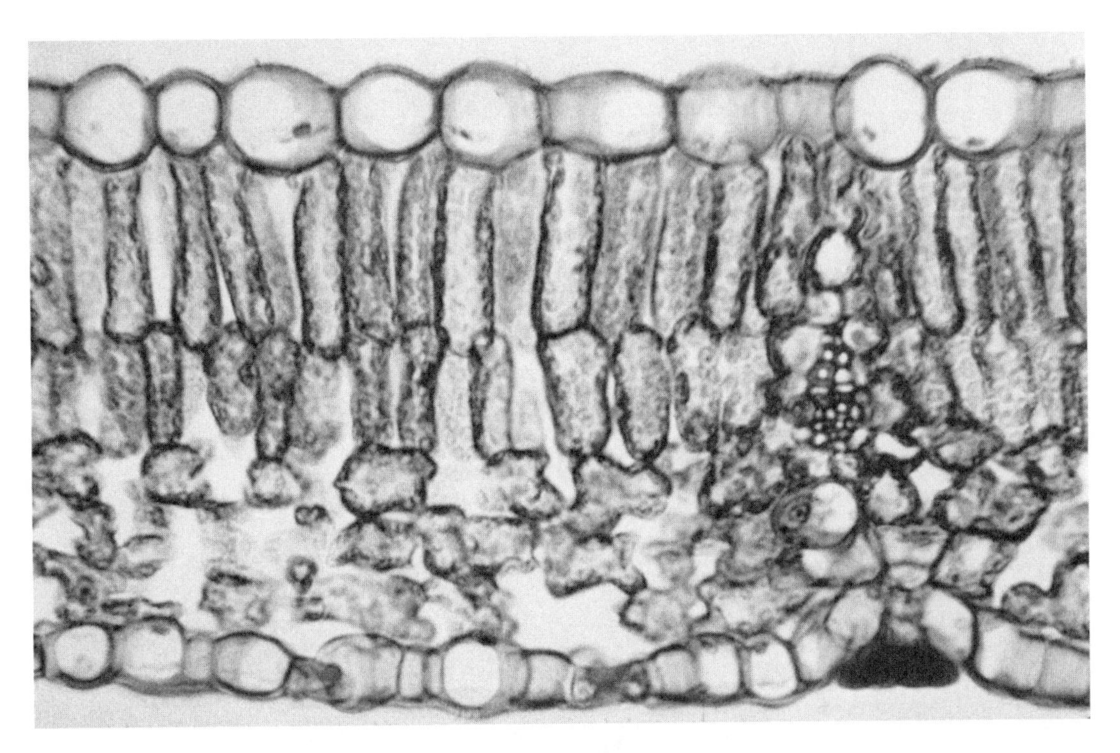

CROSS SECTION OF A LILAC LEAF. A KEY ELEMENT IN CIRCULATING LIFE-SUSTAINING MATERIALS AMONG THE VARIOUS EARTH SYSTEMS IS TRANSPIRATION, THE EVAPORATION OF MOISTURE FROM PLANTS. PLANTS LOSE THEIR WATER THROUGH MEMBRANES OF A TISSUE KNOWN AS SPONGY MESOPHYLL (SHOWN HERE), FOUND IN THE TINY CAVITIES THAT LIE BENEATH THE MICROSCOPIC LEAF PORES CALLED STOMATA. *(© Lester V. Bergman/Corbis. Reproduced by permission.)*

transpiration, the sum total of evaporation and transpiration. The second of these terms is less well-known than the first, but the words in fact refer to the same process. The only difference is that evaporation involves the upward movement of water from nonliving sources, while transpiration is the evaporation of moisture from living sources.

Transpiration is at least as important as evaporation when it comes to putting moisture into the atmosphere. It actually puts more water into the air than evaporation does: any large area of vegetation tends to transpire much larger quantities of moisture than an equivalent nonfoliated region, such as the surface of a lake or moist soil. Though animals can play a part in transpiration, plant transpiration has much greater environmental significance.

Water in plants is lost through moist membranes of a tissue known as spongy mesophyll, found in the tiny cavities that lie beneath the microscopic leaf pores called stomata. Stomata remain open most of the time, but when they need to be closed, guard cells around their borders push them shut. Because plants depend on stomata to "breathe" by pulling in carbon dioxide, they keep them open—just as a human's pores must remain open. Otherwise, the person would not take in enough oxygen, and would perish.

The fact that the stomata are exposed in order to receive carbon dioxide for the plant's photosynthesis also means that the stomata are open to allow the loss of moisture to the atmosphere. It can be said, then, that transpiration in plants—vital as it is to the functioning of our atmosphere—is actually an unavoidable consequence of photosynthesis, an unrelated process. (See Carbohydrates for more about photosynthesis.)

ANIMAL TRANSPIRATION. Transpiration in animals (including humans) takes place for much the same reason as it does with plants: as a by-product of breathing. Animals have to keep their moist respiratory surfaces, such as the lungs, open to the atmosphere. We may not think of our own breathing as transferring moisture to the air, but the presence of moisture in our lungs can be proved simply by breathing on a piece of glass and observing the misty cloud that lingers there.

Transpiration can cause animals to become dehydrated, but it also can be important in cooling down their bodies. When human bodies

KEY TERMS

A HORIZON: Topsoil, the uppermost of the three major soil horizons. This layer and the humus that lies above it house all the organic content in soil.

AEROBIC: Oxygen-breathing.

ANAEROBIC: Non-oxygen-breathing.

ATMOSPHERE: Earth's atmosphere is a blanket of gases that includes nitrogen (78%), oxygen (21%), argon (0.93%), and a combination of water vapor, carbon dioxide, ozone, and noble gases such as neon (0.07%). Most of these gases are contained in the troposphere, the lowest layer, which extends to about 10 mi. (16 km) above the planet's surface.

B HORIZON: Subsoil, beneath topsoil and above the C horizon. Though the B horizon contains no organic material, its presence is critical if the soil is to be suitable for sustaining a varied ecosystem.

BIOGEOCHEMICAL CYCLES: The changes that particular elements undergo as they pass back and forth through the various earth systems (e.g., the biosphere) and particularly between living and nonliving matter. The elements involved in biogeochemical cycles are hydrogen, oxygen, carbon, nitrogen, phosphorus, and sulfur.

BIOSPHERE: A combination of all living things on Earth—plants, animals, birds, marine life, insects, viruses, single-cell organisms, and so on—as well as all formerly living things that have not yet decomposed.

C HORIZON: The bottommost of the soil horizons, between subsoil and bedrock. The C horizon is made of regolith, or weathered rock.

CANOPY: The upper portion or layer of the trees in a forest. A forest with a closed canopy is one so dense with vegetation that the sky is not visible from the ground.

CLIMATE: The pattern of weather conditions in a particular region over an extended period. Compare with *weather.*

CLOSED SYSTEM: A system that permits the exchange of energy with its external environment but does not allow matter to pass between the environment and the system. Compare with *isolated system* on the one hand and *open system* on the other.

COMPOUND: A substance made up of atoms, chemically bonded to one another, of more than one chemical element.

DECOMPOSERS: Organisms that obtain their energy from the chemical breakdown of dead organisms as well as from animal and plant waste products. The principal forms of decomposer are bacteria and fungi.

DECOMPOSITION REACTION: A chemical reaction in which a compound is broken down into simpler compounds, or into its constituent elements. In the biosphere, this often is achieved through the help of detritivores and decomposers.

DETRITIVORES: Organisms that feed on waste matter, breaking organic material down into inorganic substances that then can become available to the biosphere in the form of nutrients for plants. Their function is similar to that of decomposers; however, unlike decomposers—which tend

KEY TERMS CONTINUED

to be bacteria or fungi—detritivores are relatively complex organisms, such as earthworms or maggots.

ECOLOGY: The study of the relationships between organisms and their environments.

ECOSYSTEM: A community of interdependent organisms along with the inorganic components of their environment.

ELEMENT: A substance made up of only one kind of atom. Unlike compounds, elements cannot be broken down chemically into other substances.

EUTROPHICATION: A state of heightened biological productivity in a body of water, which is typically detrimental to the ecosystem in which it takes place. Eutrophication can be caused by an excess of nitrogen or phosphorus, in the form of nitrates and phosphates, respectively.

EVAPORATION: The process whereby liquid water is converted into a gaseous state and transported to the atmosphere. When discussing the atmosphere and precipitation, usually evaporation is distinguished from transpiration. In this context, evaporation refers solely to the transfer of water from nonliving sources, such as the soil or the surface of a lake.

EVAPOTRANSPIRATION: The loss of water to the atmosphere via the combined (and related) processes of evaporation and transpiration.

FOOD WEB: A term describing the interaction of plants, herbivores, carnivores, omnivores, decomposers, and detritivores in an ecosystem. Each of these organisms consumes nutrients and passes it along to other organisms. Earth scientists typically prefer this name to *food chain,* an everyday term for a similar phenomenon. A food chain is a series of singular organisms in which each plant or animal depends on the organism that precedes it. Food chains rarely exist in nature.

GEOSPHERE: The upper part of Earth's continental crust, or that portion of the solid earth on which human beings live and which provides them with most of their food and natural resources.

HUMUS: Unincorporated, often partially decomposed plant residue that lies at the top of soil and eventually will decay fully to become part of it.

HYDROCARBON: Any organic chemical compound whose molecules are made up of nothing but carbon and hydrogen atoms.

HYDROLOGIC CYCLE: The continuous circulation of water throughout Earth and between various earth systems.

HYDROSPHERE: The entirety of Earth's water, excluding water vapor in the atmosphere but including all oceans, lakes, streams, groundwater, snow, and ice.

ISOLATED SYSTEM: A system that is separated so fully from the rest of the universe that it exchanges neither matter nor energy with its environment. This is an imaginary construct, since full isolation is impossible.

LEACHING: The removal of soil materials that are in solution, or dissolved in water.

MINERAL: A naturally occurring, typically inorganic substance with a specific

KEY TERMS CONTINUED

chemical composition and a crystalline structure. A crystalline structure is one in which the constituent parts have a simple and definite geometric arrangement that is repeated in all directions.

OPEN SYSTEM: A system that allows complete, or near complete, exchange of matter and energy with its environment.

ORGANIC: At one time, chemists used the term *organic* only in reference to living things. Now the word is applied to most compounds containing carbon, with the exception of carbonates (which are minerals) and oxides, such as carbon dioxide.

PATHOGEN: A disease-carrying parasite, usually a microorganism.

PHOTOSYNTHESIS: The biological conversion of light energy (that is, electromagnetic energy) from the Sun to chemical energy in plants.

SOLUBLE: Capable of being dissolved.

SYSTEM: Any set of interactions that can be set apart mentally from the rest of the universe for the purposes of study, observation, and measurement.

WEATHER: The condition of the atmosphere at a given time and place. Compare with *climate.*

become overheated, they produce perspiration, which cools the surface of the skin somewhat. If the air around us is too humid, however, then it already is largely saturated with water, and the perspiration has no place to evaporate. Therefore, instead of continuing to cool our bodies, the perspiration simply forms a sticky film on the skin. But assuming the air is capable of absorbing more moisture, the sweat will evaporate, cooling our bodies considerably.

EFFECTS OF HEAT AND COLD. When summer is at its height, air temperatures are warm and the trees are fully foliated (i.e., covered in leaves), and a high rate of transpiration occurs. So much water is pumped into the atmosphere through foliage that the rate of evapotranspiration typically exceeds the input of water to the local environment through rainfall. The result is that soil becomes dry, some streams cease to flow, and by late summer in extremely warm temperate areas, such as the southern United States, there is a great threat of drought and related problems, such as forest fires.

As trees drop their leaves in the autumn, transpiration rates decrease greatly. This makes it possible for the parched soil to become recharged by rainfall and for streams to flow again. Such is the case in a temperate region, which, by definition, is one that has the four seasons to which most people in the United States (outside Hawaii, Alaska, and extreme southern Florida and Texas) are accustomed. In a tropical region, by contrast, there is a "dry season," in which transpiration takes place, and a "rainy season," in which moisture from the atmosphere inundates the solid earth. This rainy season may be so intense that it produces floods.

WHERE TO LEARN MORE

Bial, Raymond. *A Handful of Dirt.* New York: Walker, 2000.

Biogeochemical Cycles (Web site). <http://www.bsi.vt.edu/chagedor/biol_4684/Cycles/cycles.html>.

Bocknek, Jonathan. *The Science of Soil.* Milwaukee, WI: Gareth Stevens, 1999.

Global Hydrology and Climate Center (Web site). <http://wwwghcc.msfc.nasa.gov/>.

Hancock, Paul L., and Brian J. Skinner. *The Oxford Companion to the Earth.* New York: Oxford University Press, 2000.

Kump, Lee R., James F. Kasting, and Robert G. Crane. *The Earth System.* Upper Saddle River, NJ: Prentice Hall, 2000.

Life and Biogeochemical Cycles (Web site). <http://essp.csumb.edu/esse/climate/climatebiogeo.html>.

Richardson, Joy. *The Water Cycle.* Illus. Linda Costello. New York: Franklin Watts, 1992.

Skinner, Brian J., Stephen C. Porter, and Daniel B. Botkin. *The Blue Planet: An Introduction to Earth System Science.* 2d ed. New York: John Wiley and Sons, 1999.

Smith, David. *The Water Cycle.* Illus. John Yates. New York: Thomson Learning, 1993.

ECOSYSTEMS AND ECOLOGY

CONCEPT

Composed of living organisms and the remains of living things as well as the nonliving materials in their surroundings, an ecosystem is a complete community. Its components include plants, animals, and microorganisms, both living and dead; soil, rocks, and minerals; water sources above and below ground; and the local atmosphere. An ecosystem can consist of an entire rain forest, geographically larger than many nations, or it can be as small as the body of an animal, which is likely to comprise far more microorganisms than there are people on Earth. Among the most significant areas of interest in the realm of ecosystems is ecology, or the study of the relationships between organisms and their environments. Forests, a broad category of ecosystem, provide a living laboratory in which to investigate the ways in which organisms interact with their environments. They also aptly illustrate the stresses placed on ecosystems by human activities.

HOW IT WORKS

THE BIOSPHERE AND FOOD WEBS

In the sciences, a system is any set of interactions that can be set apart mentally from the rest of the universe for the purposes of study, observation, and measurement. Modern earth scientists regard the planet as a massive ecosystem, the stage on which four extraordinarily complex earth systems interact. These systems are the atmosphere, the hydrosphere (all the planet's waters, except for moisture in the atmosphere), the geosphere (the soil and the extreme upper portion of the continental crust), and the biosphere.

The biosphere consists of all living things—microorganisms, plants, insects, birds, marine life, and all other forms of animals—as well as formerly living things that have not yet decomposed. (Decomposed remains of organic materials become part of the geosphere, specifically the soil.) Organisms in the biosphere, whether living or formerly living, are united by the interrelation of energy transfer that takes place through the food web.

A food web is similar to the more well known expression food chain. In scientific terms, however, a food chain is defined as a series of singular organisms in which each plant or animal depends on the organism that precedes or follows it. This rarely exists in nature; instead, the feeding relationships between organisms in the real world are much more complex and are best described as a web rather than a chain. (For more on the biosphere and earth systems, see The Biosphere.)

ENERGY FLOW AND NUTRIENTS

Energy is the ability of objects or systems to accomplish work. (The latter is defined as the exertion of force over a given distance: for example, a plant growing from the ground, an insect or bird flying, or a human or pack animal moving an object.) Food webs are built around energy transfer, or the flow of energy between organisms, which begins with plant life. Hence the importance of plants to ecosystems, as we illustrate later in discussing various types of forest, which are defined by their dominant varieties of tree.

Plants absorb energy in two ways. From the Sun they receive electromagnetic energy in the form of visible light and invisible infrared waves, which they convert to chemical energy by means of photosynthesis. In addition, plants take in nutrients from the soil, which contains energy in the forms of various chemical compounds. These compounds may be organic, which typically means that they came from living things, though, strictly speaking, organic refers to characteristic carbon-based chemical structures. Plants also receive inorganic compounds from minerals in the soil. (See Paleontology for an explanation of the scientific distinction between organic and inorganic.)

Contained in these minerals are six chemical elements essential to the sustenance of life on planet Earth: hydrogen, oxygen, carbon, nitrogen, phosphorus, and sulfur. These are the elements involved in biogeochemical cycles, through which they are circulated continually between the living and nonliving worlds—that is, between organisms on the one hand and the inorganic realms of rocks, minerals, water, and air on the other. (See The Biosphere for more about biogeochemical cycles.)

FROM PLANTS TO MEAT EATERS. As plants take up nutrients from the soil, they convert them into other forms. Eventually, the plants themselves become food either for herbivores (plant-eating organisms) or for omnivores (organisms that eat both plants and other animals), thus passing along these usable, energy-containing chemical compounds to other participants in the food web. It is likely that an herbivore will be eaten in turn either by an omnivore (for example, humans as well as a number of bird species and many others) or by a carnivore, an organism that eats only meat.

Carnivores and omnivores are not usually prey for other carnivores or omnivores, but this does happen in the case of what are known as tertiary consumers (see Food Webs). There is also the matter of cannibalism, discussed in Biological Communities. For the most part, however, the only creatures that eat carnivores and omnivores do so after these organisms have died or been killed.

DETRITIVORES AND DECOMPOSERS. An animal that obtains its energy in this way, from consuming the carcasses of carnivores and omnivores (as well as herbivores and perhaps even plants), is known as a detritivore. Large and notable examples include vultures and hyenas, though most detritivores—earthworms or maggots, for instance—are much smaller. Nonetheless, detritivores are relatively large, complex organisms compared with another variety of species that occupies a level or position in the food web that comes "after" carnivores and omnivores: decomposers, including bacteria and fungi.

This illustrates the reason why ecological sciences treat the expression food chain with disfavor. There is no such thing as "the top of the food chain," rather, there are simply stages, like a circular assembly line, with detritivores occupying a position between meat-eating animals and plants. Earthworms, in particular, help convert animal bodies into soil nutrients useful to the growth of plant life, yet even these and other detritivores must themselves eventually be converted to soil as well.

This "final" stage of conversion—that is, the last stop after the animal portion of the food web and before the cycle comes back around to ordinary plants—is occupied by decomposers, such as bacteria and fungi. These decomposers obtain their energy from the chemical breakdown of dead organisms as well as from animal and plant waste products. Like detritivores, they aid in decomposition, a chemical reaction in which a compound is broken down into simpler compounds, or into its constituent elements.

Often an element such as nitrogen appears in forms that are not readily usable to organisms, and therefore such elements (which may appear individually or in compounds) need to be processed chemically through the body of a decomposer or detritivore. This process brings about a chemical reaction in which the substance (whether an element or compound) is transformed into a more usable version. By processing these chemical compounds, decomposers and detritivores provide nutrients necessary to plant growth.

REAL-LIFE APPLICATIONS

FORESTS AND ECOLOGY

One easily understandable example of ecosystems and ecology in action is the forest. Virtually everyone has visited a forest at one time or another, and those who are enthusiasts for the great outdoors may spend a great deal of time in one. In the past, of course, people interacted with

A VARIETY OF ANGIOSPERM, MANGROVE TREES ARE FOUND IN LOW-LYING, MUDDY REGIONS NEAR SALTWATER, WHERE THE CLIMATE IS HUMID. A MANGROVE FOREST IS POOR IN SPECIES: ONLY ORGANISMS THAT CAN TOLERATE FLOODING AND HIGH SALT LEVELS ARE CAPABLE OF SURVIVING. *(© Wolfgang Kaehler/Corbis. Reproduced by permission.)*

forests not so much out of choice, and certainly not with recreation as the foremost aim in mind, but simply because they depended on the forest for survival. Not only did the forest provide hunters and food-gatherers with an abundance of wildlife and fruit, but trees provided material for building dwellings. It is no wonder, then, that many early human settlements tended to be in, or at the edges of, forests.

A forest is simply an ecosystem dominated by trees. There are many varieties of forest, however, because so many factors go into determining the character of a forest ecosystem. The fact that the forest *is* an ecosystem means that its qualities are defined by far more than just the varieties of trees, which are simply the most visible among many biological forms in the forest. Numerous abiotic, or nonbiological, factors also affect the characteristics of a forest as well. For instance, there is weather, defined as the condition of the atmosphere at a given time and place, and climate, the overall patterns of weather for extended periods.

These play a clear role, for instance, in defining the tropical rain forest, a place where constant rainfall ensures that there are always plenty of plants in flower. Because the trees and other species of vegetation do not all shed at the same time, the rain forest canopy—the upper layer of trees in the forest—remains rich in foliage year-round. Hence, the tropical rain forest is an example of an evergreen forest. Climate can determine the type of life forms capable of surviving in the forest ecosystem. This can be illustrated by referring to a forest almost perfectly opposite in character to a rain forest: the taiga, or boreal forest, that spans much of northern Eurasia.

The taiga is a deciduous forest, meaning that its trees shed their leaves seasonally; indeed, because of the very cold climate in taiga regions, where the temperature during winter is usually well below the freezing point, trees spend a great portion of the year bare. Rainfall is much, much lower than in a rain forest, of course: only about 10-20 in. (250–500 mm) per year, as compared with more than 70 in. (1,800 mm) for a typical rain forest. The dry, inhospitable climate of the taiga makes it a forbidding place for reptiles and amphibians, though the taiga is home to many endothermic (warm-blooded) creatures such as mammals or birds.

LATITUDE, ALTITUDE, AND FORESTS. Elevation or relief—that is, height above sea level—also determines the character of a forest, as do latitude (distance north or south of the equator) and topography, or the overall physical configuration of Earth's surface in a given area. Rain forests can exist anywhere, but by definition a *tropical* rain forest, such as those along the Amazon River in South America or the Congo River in Africa, must lie between the Tropic of Cancer in the north and the Tropic of Capricorn in the south.

Naturally, in the tropical rain forest, temperatures are high—typically, about 86°F (30°C) during the day, cooling down to about 68°F (20°C) at night. By contrast, there are much cooler rain forests in the temperate zones. An example is the Cherokee National Forest on the border between North Carolina and Tennessee, which, though located in the southeastern United States, is chilly even in the summer months.

Just as latitude affects a forest, so does altitude. Rain forests at relatively high elevations, such as the highlands of New Guinea, are known as montane forests. These forests, though they may be located at the same latitude as tropical rain forests—most montane forests are in eastern Brazil, southeastern Africa, northern Australia, and parts of southeast Asia—are much cooler. Lush by comparison to most non-rain forests, their vegetation is nonetheless much less dense than in a typical tropical rain forest.

In addition to its role in defining the overall character of the forest, differences in relative altitude or elevation resulting from the great height of trees in the rain forest also influence the formation of differing biological communities. For example, monkeys, flying squirrels, and other animals capable of swinging, gliding, or otherwise moving from tree to tree inhabit the canopy, which is rich in well-watered leaves and other food sources. These top-dwellers seldom even need to come down to the ground for anything. The rain forest floor, by contrast, is mostly bare, since the trees above shade it. On this level live creatures such as chimpanzees and gorillas, who feed off of low-lying plant forms. Other biological communities exist above or below the forest floor. (For more on these subjects, including the various types of forests, see Biomes. See The Biosphere for a discussion of soil quality in the rain forest.)

HUMANS AND FORESTS. Earlier we noted the fact that humans' early history kept them, like other primates, close to the forest. In modern times, a growing awareness of ecology, and of the distance that technology has placed between modern society and the forests, led to the movement for the establishment of national parks in general, and of national forests in particular.

The first of these preserves—places where commercial development is forbidden and commercial activity is limited—was Yellowstone National Park in Wyoming, established by the administration of President Ulysses S. Grant in 1872. Though Yellowstone contains enormous areas of forest land, the first national forest reserve (as national forests were called at the time) was Sequoia National Park, established in 1891. Home to some of the largest, most awe-inspiring trees in the world, Sequoia is part of a group of national forests and parks to the northeast of Bakersfield, California.

The United States Forest Service was actually founded earlier (1905) than the National Park Service (1916), a fact that illustrates the importance of pristine forests to maintaining a proper balance between humans and their environment. Since the establishment of the forest service under the aegis of the U.S. Department of Agriculture, the lands controlled by the forest service have grown to encompass about 191 million acres (77.3 million hectares), an area larger than Texas. During the same period, the U.S. example of national parks and forests has inspired nations around the world to create their own preserves.

Coupled with the rise of national parks and forests at the turn of the nineteenth century was a growing interest in conservation and management of environmental resources. This interest manifested across a broad spectrum, from environmentalists who urged that the forests be left in their original state to industrial foresters who view the forest as a resource that can be utilized. Both sides have their merits, and both have their complaints about the other. The close historical ties between conservationism and the science of forestry (the management of forest ecosystems for purposes such as harvesting timber), both of which had their origins during the nineteenth century, suggest that there is no inherent reason that the two sides should be in conflict. If anything, responsible forestry goes hand-in-hand with an attitude of conserving as many resources as is feasible.

BY EVOLVING BRIGHT COLORS, SCENTS, AND NECTAR, THE FLOWERS OF ANGIOSPERMS ATTRACT ANIMALS, WHICH TRAVEL FROM ONE FLOWER TO ANOTHER, MOVING POLLEN AS THEY GO. WHEREAS INSECTS AND ANIMALS POSE A THREAT TO GYMNOSPERMS, ANGIOSPERMS PUT BEES, BUTTERFLIES, HUMMINGBIRDS, AND OTHER FLOWER-SEEKING CREATURES TO WORK ASSISTING THEIR REPRODUCTIVE PROCESS. *(© Gallo Images/Corbis. Reproduced by permission.)*

COMPARING ANGIOSPERMS AND GYMNOSPERMS

Several times we have referred to angiosperms, a name that encompasses not just certain types of tree but also all plants that produce flowers during sexual reproduction. The name, which comes from Latin roots meaning "vessel seed," is a reference to the fact that the plant keeps its seeds in a vessel whose name, the ovary, emphasizes the sexual quality of the reproductive process it undergoes.

Angiosperms are a beautiful example of how a particular group of organisms can adapt to specific ecosystems and do so in a highly efficient manner, such that the evolutionary future heralds only greater dominance for these species. This is all the more interesting in light of the contrast between the success of the angiosperm and the rather less impressive results achieved by another broad category of sexually reproducing plant, one that formerly dominated Earth's forest: the gymnosperm.

Flowering plants evolved only about 130 million years ago, by which time gymnosperms (of which modern pines are an example) had long since evolved and proliferated. Yet in a relatively short period of time, angiosperms have become the dominant plants in the world today. About 80% of all living plant species are flowering plants, and based on the record of angiosperms and gymnosperms heretofore, it is likely that the world of 100 million years from now will be one in which the forests are typified by angiosperms. Gymnosperms, meanwhile, may well become a dying, if not a dead, breed.

POLLINATION BY GYMNOSPERMS. Gymnosperms reproduce sexually as well, but they do so by a much less efficient method than that of the angiosperm. Whereas the angiosperm keeps its seeds safely tucked away inside the ovary and coexists with its ecosystem most favorably by putting the insect and animal life to work, gymnosperm reproduction is an altogether less effective—and, indeed, less pleasant—process.

For starters, gymnosperms produce their seeds on the surface of leaflike structures, and this makes the seeds vulnerable to physical damage and drying as the wind whips the trees' branches back and forth. Furthermore, insects and other animals view gymnosperm seeds as a source of nutrition, as indeed they are. And in contrast to the angiosperm, which attracts bees and other creatures to it, gymnosperms package the male reproductive component in tiny pollen grains, which it releases into the wind.

Eventually, the grains make their way toward the female component of another individual within the same species, but the fact that they do is an example of the wonder inherent in life itself and not of the efficiency of gymnosperm reproduction. Gymnosperms shower their ecosystems with pollen, a fact familiar to anyone who lives in a place with a high gymnosperm population—and hence a high pollen count in the spring. In gymnosperm-heavy environments, yellowish dust forms on everything, and where humans interact with the natural world, this can create a great deal of discomfort in the form of hay fever and allergies. Meanwhile, cars, windowsills, mailboxes, and virtually every other available surface takes on a yellow film that usually is not relieved until a good rain falls or, more likely, pollination ends for the year.

Though pollen is unpleasant to humans, it should be noted that like all natural mechanisms, it benefits the overall ecosystem. Packed with energy, pollen grains contain large quantities of nitrogen, making them a major boost to the nutrient content in the soil. But it costs the gymnosperm a great deal, in terms of chemical and biological energy and material, to produce pollen grains, and the benefits are uncertain.

POLLINATION BY ANGIOSPERMS. If the gymnosperm and angiosperm varieties of pollination were compared to marketing campaigns, gymnosperm reproduction would involve the client (i.e., the gymnosperms themselves) investing maximum capital for minimal returns. In a very real sense, gymnosperm pollination is like the marketing of a company that bombards a neighborhood with leaflets, such that advertising rapidly becomes another form of trash simply to be swept up and thrown away.

By contrast, the "marketing" of angiosperms is like that of a company that uses carefully targeted, researched advertising, utilizing as many free means as possible for getting out word about itself. Just as a smart marketer sets in place the conditions to get consumers talking about a product—thus using advertising that is both free and extremely effective—the angiosperm enlists the aid of mobile organisms in its environment.

In addition, the angiosperm puts a great deal of its energy into producing reproductive structures, an effort that pays off bountifully. By evolving bright colors, scents, and nectar, the flowers of angiosperms attract animals, which travel from one flower to another, unintentionally moving pollen as they go. Thus, whereas insects and animals pose a threat to gymnosperms, angiosperms actually put bees, butterflies, hummingbirds, and other flower-seeking creatures to work assisting their reproductive process.

Because of this remarkably efficient system, animal-pollinated species of flowering plants do not need to produce as much pollen as gymnosperms. They can put their resources into other important functions instead, such as growth and greater seed production. In this way, the angiosperm solves its own problem of reproduction—and, as a side benefit, adds enormously to the world's beauty.

DEFORESTATION

Returning to the subject of forests in general, if a forest experiences significant disturbance, it may undergo deforestation. Despite the finality in the sound of the word, deforestation does not necessarily imply complete destruction of the forest. In fact, *deforestation* can describe any interruption in the ordinary progression of a forest's life, including clear-cut harvesting—even if the forest fully recovers.

Deforestation can occur naturally, as a result of changes in the soil and climate, but the most significant cases of deforestation over the past few thousand years have been the consequence of human activities. Usually deforestation is driven by the need to clear land to harvest trees for fuel or, in some cases, to obtain building materials in the form of lumber. Though deforestation has been a problem the world over, since the 1970s it has become an issue primarily in developing countries.

In developed nations such as the United States, environmental activism has raised public awareness concerning deforestation and led to curtailment of large-scale cutting in forests that are deemed important environmental habitats. By contrast, developing nations, such as Brazil, are cutting down their forests at an alarming rate. Generally, economics is the dominant factor, with the need for new agricultural land or the desire to obtain wood and other materials typically driving the deforestation process.

CONSEQUENCES OF DEFORESTATION. The deforestation of valuable reserves such as the Amazon rain forest is an environmental disaster in the making. As discussed in the essay The Biosphere, the soil in rain forests as a rule is "old," and leached of nutrients. Without the constant reintroduction of organic material from the plants and animals of the rain forests, it would be too poor to grow anything. Therefore, when nations cut down their own rain forest lands, they are in effect killing the golden goose to get at the egg: once the rain forest is gone, the land itself is worthless.

Deforestation has several other extremely serious consequences. From a biological standpoint, it greatly reduces biodiversity, or the range of species in the biota. In the case of tropical rain forests as well as old-growth forests (see Biological Communities), certain species cannot survive once the environmental structure has been rup-

A lone tree trunk stands in an area of deforested grassland in Maranhão, Brazil. The deforestation of valuable reserves such as the Amazon rain forest is an environmental disaster in the making, depleting and starving the soil, reducing biodiversity, and bringing about dangerous changes in atmospheric carbon content. *(© Barnabas Bosshart/Corbis. Reproduced by permission.)*

tured. From an environmental perspective, it leads to dangerous changes in the carbon content of the atmosphere, discussed later in this essay. In the case of old-growth forests or rain forests, deforestation removes an irreplaceable environmental asset that contributes to the planet's biodiversity—and to its oxygen supply.

Even from a human standpoint, deforestation takes an enormous toll. Economically, it depletes valuable forest resources. Furthermore, deforestation in many developing countries often is accompanied by the displacement of indigenous peoples, while still other political and social horrors may lurk in the shadows. For example, Brazil's forests are home to charcoal factories that amount to virtual slave-labor camps. Aboriginal peoples (i.e., "Indians") are lured from cities with promises of high income and benefits, only to arrive and find that the situation is quite different from what was advertised. Having paid the potential employer for transportation to the work site, however, they are unable to afford a return ticket and must labor to repay the cost.

The Greenhouse Effect

The most potentially serious aspect of cutting down forests may well be the greenhouse effect, which some scientists and activists believe is causing an overall warming of the planet. Today, thanks to the popularity of environmental causes among entertainment figures and on college campuses, terms such as "the greenhouse effect" and "global warming" are commonplace. However, these phrases are used so frequently, and sometimes so confusingly or misleadingly, that it is worthwhile to address their meaning briefly; then, we can conclude our discussion by looking at what impact the steady reduction in forest lands has had on the increasing release of greenhouse gases.

The greenhouse effect itself is not a consequence of any action on the part of human beings; rather, it is a part of life on Earth. In fact, without it, there could *be* no life on Earth. Though the planet receives an incredible amount of energy from the Sun, much of it is lost by being absorbed or reflected in the atmosphere or on the surface. So-called greenhouses gases such as carbon dioxide, however, help to trap this energy, keeping much more of the Sun's warmth

KEY TERMS

ANGIOSPERM: A type of plant that produces flowers during sexual reproduction.

ATMOSPHERE: Earth's atmosphere is a blanket of gases that includes nitrogen (78%), oxygen (21%), argon (0.93%), and a combination of water vapor, carbon dioxide, ozone, and noble gases such as neon (0.07%). Most of these gases are contained in the troposphere, the lowest layer, which extends to about 10 mi. (16 km) above the planet's surface.

BIODIVERSITY: The degree of variety among the species represented in a particular ecosystem.

BIOLOGICAL COMMUNITY: The living components of an ecosystem.

CANOPY: The upper portion or layer of the trees in a forest. A forest with a closed canopy is one so dense with vegetation that the sky is not visible from the ground.

CARNIVORE: A meat-eating organism, or an organism that eats *only* meat (as distinguished from an omnivore).

COMPOUND: A substance made up of atoms, chemically bonded to one another, of more than one chemical element.

CONIFER: A type of tree that produces cones bearing seeds.

DECIDUOUS: A term for a tree or other form of vegetation that sheds its leaves seasonally.

DECOMPOSERS: Organisms that obtain their energy from the chemical breakdown of dead organisms as well as from animal and plant waste products. The principal forms of decomposer are bacteria and fungi.

DECOMPOSITION REACTION: A chemical reaction in which a compound is broken down into simpler compounds or into its constituent elements. In the earth system, this often is achieved through the help of detritivores and decomposers.

DEFORESTATION: A term for any interruption in the ordinary progression of a forest's life.

DETRITIVORES: Organisms that feed on waste matter, breaking down organic material into inorganic substances that then can become available to the biosphere in the form of nutrients for plants. Their function is similar to that of decomposers; however, unlike decomposers—which tend to be bacteria or fungi—detritivores are relatively complex organisms, such as earthworms or maggots.

ECOLOGY: The study of the relationships between organisms and their environments.

ECOSYSTEM: A community of interdependent organisms along with the inorganic components of their environment.

ELEMENT: A substance made up of only one kind of atom. Unlike compounds, elements cannot be broken chemically into other substances.

ENERGY: The ability of an object (or in some cases a nonobject, such as a magnetic force field) to accomplish work.

ENERGY BUDGET: The total amount of energy available to a system or, more specifically, the difference between the

KEY TERMS CONTINUED

energy flowing into the system and the energy lost by it.

ENERGY TRANSFER: The flow of energy between organisms in a food web.

FOOD WEB: A term describing the interaction of plants, herbivores, carnivores, omnivores, decomposers, and detritivores in an ecosystem. Each of them consumes nutrients and passes it along to other organisms. Earth scientists typically prefer this name to *food chain,* an everyday term for a similar phenomenon. A food chain is a series of singular organisms in which each plant or animal depends on the organism that precedes it. Food chains rarely exist in nature.

FOREST: In general terms, a forest is simply any ecosystem dominated by tree-size woody plants. A number of other characteristics and parameters (for example, weather, altitude, and dominant species) further define types of forests, such as tropical rain forests.

GREENHOUSE EFFECT: Warming of the lower atmosphere and surface of Earth. This occurs because of the absorption of long-wavelength radiation from the planet's surface by certain radiatively active gases, such as carbon dioxide and water vapor, in the atmosphere. These gases are heated and ultimately re-radiate energy at an even longer wavelength to space. (Wavelength and energy levels are related inversely; hence, the longer the wavelength, the less the energy.)

GYMNOSPERM: A type of plant that reproduces sexually through the use of seeds that are exposed, not hidden in an ovary as with an angiosperm.

HERBIVORE: A plant-eating organism.

OMNIVORE: An organism that eats both plants and other animals.

ORGANIC: At one time chemists used the term *organic* only in reference to living things. Now the word is applied to most compounds containing carbon, with the exception of carbonates (which are minerals) and oxides, such as carbon dioxide.

PHOTOSYNTHESIS: The biological conversion of light energy (that is, electromagnetic energy) from the Sun to chemical energy in plants.

SYSTEM: Any set of interactions that can be set apart mentally from the rest of the universe for the purposes of study, observation, and measurement.

within Earth's atmosphere, much as a greenhouse helps trap heat. Without the greenhouse effect, Earth would be so cold that the oceans would freeze.

Obviously, then, the greenhouse effect is a good thing—but only if greenhouse gases are kept at certain levels. Earth, after all, is not the only planet in the solar system that experiences a greenhouse effect; there is also Venus, a hellish place where surface temperatures are as high as 932°F (500°C). To many environmentalists, there is a grave danger that Earth could be slowly going the way of Venus, building up greenhouse gases such that the temperature is slowly increasing. This is the phenomenon of global warming, which threatens to melt the polar ice caps and submerge much of Earth's land surface. At least, that is the opinion of environmentalists and others who subscribe to the idea that Earth is steadily warming as a result of human pollution and industrial activity.

There is a considerable body of scientific knowledge that challenges the environmentalist position on global warming and the greenhouse effect, but it is not our purpose here to judge the various positions. Rather, our concern is the link between forests and the increase of greenhouses gases in the atmosphere. Old-growth or mature forests of the type discussed in Succession and Climax contain vast amounts of carbon—the basis for all living things—and when these forests are cut down, that carbon has to go somewhere. Specifically, carbon, in the form of carbon dioxide, will be released into the atmosphere, increasing the amount of greenhouse gases there.

This release may occur quickly, as when wood is burned, or more slowly, if the timber from the forest is used over long periods of time—for instance, in the building of houses or other structures. Statistics suggest an alarming change in the amount of carbon in the forests as compared with that in the atmosphere: since about 1850, the amount of carbon stored in forests had dropped by about one-third, while the amount of carbon dioxide in the atmosphere has increased by a comparable factor. Thus, the effort to keep greenhouse gases at viable levels is inextricably tied to the movement to preserve forest ecosystems.

WHERE TO LEARN MORE

Ashworth, William, and Charles E. Little. *Encyclopedia of Environmental Studies.* New York: Facts on File, 2001.

Diamond, Jared M. *Guns, Germs, and Steel: The Fates of Human Societies.* New York: W. W. Norton, 1997.

The Ecological Society of America: Issues in Ecology (Web site). <http://esa.sdsc.edu/issues.htm>.

The Ecosystems Center, Marine Biological Laboratory, Woods Hole, Massachusetts (Web site). <http://ecosystems.mbl.edu/>.

Living Things: Habitats and Ecosystems (Web site). <http://www.fi.edu/tfi/units/life/habitat/habitat.html>.

Markley, O. W., and Walter R. McCuan. *21st Century Earth: Opposing Viewpoints.* San Diego, CA: Greenhaven Press, 1996.

Martin, Patricia A. *Woods and Forests.* Illus. Bob Italiano and Stephen Savage. New York: Franklin Watts, 2000.

Nebel, Bernard J., and Richard T. Wright. *Environmental Science: The Way the World Works.* Upper Saddle River, NJ: Prentice Hall, 2000.

Philander, S. George. *Is The Temperature Rising?: The Uncertain Science of Global Warming.* Princeton, NJ: Princeton University Press, 1998.

Rybolt, Thomas R., and Robert C. Mebane. *Environmental Experiments About Life.* Hillside, NJ: Enslow Publishers, 1993.

The State of the Nation's Ecosystems (Web site). <http://www.us-ecosystems.org/>.

Sustainable Ecosystems Institute (Web site). <http://www.sei.org/>.

BIOMES

CONCEPT

On a political map of the world, Earth is divided into countries, of which there are almost 200. But nature, of course, knows no national boundaries, and therefore the natural divisions of the planet are quite different from those agreed upon by humans. While continents are a useful concept to geographers and earth scientists, in the worlds of biology, ecology, and biogeography, the concept of a biome makes much more sense. There are more than a dozen basic terrestrial and aquatic biomes or ecosystems, including boreal coniferous forests, deserts, tundra, and underwater environments. Each is a distinct "world" unto itself, with characteristic forms of plant life as well as animal species that congregate around the plants for food or shelter or both. Combined with these features of the biological community are aspects of the inorganic realm that likewise define a biome, for instance, climate and the availability of water.

HOW IT WORKS

Ecosystems, Biomes, and Biological Communities

An ecosystem is a community of interdependent organisms along with the inorganic components of their environment, including water, soil, and air. Earth is the largest ecosystem, divided into biomes, large areas with similar climate and vegetation. A biome is a large ecosystem, extending over a wide geographic region, characterized by certain dominant life-forms—most notably, trees or the lack thereof. There are two basic varieties of biome: terrestrial, or land-based (of which there are six), and aquatic. The second of these types is divided further into marine and freshwater biomes.

Within a biome or ecosystem, the sum of all living organisms is referred to as the *biological community.* Sometimes the term *biota,* which refers to all flora and fauna (plants and animals) in a region, is used instead. Thus, biological community is a larger concept, since it includes microorganisms, which are vital to the functioning of the food web. The food web, which may be thought of as an interconnected network of food chains, is the means by which energy is transferred through a biological community. Without microorganisms known as decomposers, a key link in the food web would be missing. (See Food Webs for more on this subject.)

SUCCESSION AND CLIMAX. Over the course of time, ecosystems experience a process known as *succession,* the progressive replacement of one biological community by another. This is rather like the series of changes one might witness if one were to record the activity on a major city block over the space of a few decades, as stores come in and shut down and buildings are erected and demolished. In the case of biological succession, a process akin to natural selection (see Evolution) is occurring: the ecosystem becomes home, in turn, to a number of different biological communities until (in the absence of outside interference) the one that is most suited or adapted to local conditions finally takes root. (That is, until it is replaced, and the process of succession continues.)

This most suited or adapted biological community is described as a *climax community,* one that has reached a stable point as a result of ongoing succession. In such a situation, the community is at equilibrium with environmental conditions, and conditions are stable, such that the biota experiences little change thereafter. The most significant forms of climax vegetation are often the defining characteristics of terrestrial biomes. (See Succession and Climax for more about this subject.)

Defining Characteristics of a Biome

The boreal coniferous forest often is cited by biogeographers as a classic example of a biome, for a number of reasons. First, like most other terrestrial biomes, this one is defined by specific latitudinal positions: the term *boreal* means "northern," and these forests exist between 50 and 60 degrees north latitude. (Aside from the southernmost tip of South America and a few scattered islands, there is no significant landmass between 50 and 60 degrees south latitude.)

In North America the region between 50 and 60 degrees north latitude is the southerly band of Canadian provinces (Alberta and Saskatchewan, for example). The Eurasian equivalent of this region is a band encompassing the British Isles; an area of continental Europe that includes northern Germany, Poland, and southern Sweden; and a vast swath that spans the width of Russia from Saint Petersburg and Moscow in the west across the nation's wide expanse (10 time zones) to the Kamchatka peninsula north of Japan.

The boreal coniferous forest thus illustrates a key fact about biomes: they can occur in widely separated geographic regions as long as the environmental conditions are the same. In each of these locales average temperatures are low; summers are short, moist, and of moderate warmth; and winters are long, cold, and dry. Most precipitation is in the form of snow, and the A horizon of the soil, home of the organic material in which plants grow, is thin. Moreover, the soil is acidic and poor in nutrients. (See The Biosphere for more about soil.)

Most of the information conveyed in the preceding paragraph refers to the inorganic components of the boreal coniferous forest. (*Organic* does not necessarily mean "living," but it does refer to carbon-based chemical compounds other than carbonates, which are rocks, and carbon oxides, such as carbon dioxide.) As noted earlier, inorganic components of a biome include water and air, which in turn are involved in precipitation, weather, and climate. Although it does contain organic compounds from the decayed vegetable and animal matter that enriches it, soil, too, is largely inorganic, being formed from the weathering of rocks.

FLORA. Biomes are differentiated most clearly, however, on the basis of their organic components. The second term in the phrase *boreal coniferous* refers to a type of plant that produces cones containing seeds. Thus, the dominant plant life in the boreal coniferous forest includes evergreen conifers that can tolerate cold weather: pine, fir, and spruce.

The varieties that dominate may differ between geographic regions, however. The boreal coniferous forests of northeastern North America, for instance, are dominated by black spruce, while those in the northwest are characterized by stands of white spruce. In northeast Europe, Norway spruce is dominant, while species of pine and larch occupy the key positions in the forests of Siberia. Despite these differences in dominant species, the conditions are much the same, not only in terms of inorganic environment but also with regard to flora and fauna. In most boreal coniferous forests, the canopy or upper layer is so thick that it allows little light through. The result is that the understory, or lower layers of vegetation, is very limited.

FAUNA. As for animal life, species in the boreal coniferous forest include bear, moose, wolf, lynx, deer, weasels, rabbits, beavers, and chipmunks. With a few local variations, this roster of animal life is typical in most such biomes, whether in British Columbia or western Europe or Siberia.

A biome constitutes a complex network of interactions among plants, animals, and their surroundings, such that certain animals depend, either directly or indirectly, on certain plants for their sustenance. An obvious example is the beaver's use of coniferous tree limbs and even trunks for building shelter. Even more fundamental to the functioning of ecosystems is the role of plants as food.

Although few animals actually feed off the needles or bark of conifers, they do eat from these trees in more indirect ways. The woodpecker, for instance, consumes bugs that live in a tree's bark. Then there are the many insects that live off conifer seeds (see Ecosystems and Ecology for a discussion of conifer, or gymnosperm, reproduction), and these bugs, in turn, serve as food for birds, which are the prey of larger carnivores. Furthermore, though the understory in boreal coniferous forests is not dense, it provides enough vegetation to meet the needs of deer, rabbits, and other herbivores.

CLASSIFYING BIOMES

Earlier it was stated that there are "almost" 200 countries on Earth. It might seem strange that something like the number of countries could be so inexact, when it would seem to be a matter of very exact quantities, like the number of states in the United States. But defining sovereign nations is a bit more challenging. Obviously, the United States, Switzerland, and Japan are sovereign nations, but many another political entity exists in a gray area.

If the number of independent nations on Earth is so open to question, it would stand to reason that the number of basic biomes is as well. After all, nations typically are delineated by such things as borders, seats at the United Nations, currency, and so forth, whereas the boundaries between biomes are much less exact. Therefore, it would be futile to attempt to say exactly how many biomes there are on Earth, since the number varies according to interpretation.

TERRESTRIAL, AQUATIC, AND OTHER CATEGORIES. One of the more useful methods for classifying biomes is that of the American ecologist Eugene Pleasants Odum (1913–), introduced in his *Fundamentals of Ecology* (1953). The classification scheme that follows is based on that of Odum, who divided biomes into terrestrial and aquatic. In the present context, biomes have been grouped into five categories: forest, nonforest, freshwater, marine, and anthropogenic. The last of these categories refers to biomes strongly influenced by humans and their activities, though it should be noted that to some degree at least, human activities have influenced all of Earth's biomes. For example, many organisms carry in their fat cells trace amounts of human-manufactured contaminants, such as DDT. (See Food Webs for more on this subject.)

Biomes are organized here in such a way as to take into account their relative latitudes and corresponding climate. (Distinctions of latitude and climate are mostly relevant where terrestrial biomes are concerned.) As with biomes, there are many possible climate zones, particularly when rainfall patterns and other variables are considered. All climate zones, however, fall into one of three basic categories: tropical and subtropical, temperate, and polar and subpolar.

The first of these categories is a term comprising the region along the equator, extending north and south by about 30 degrees in either direction. In North America this would include southern Florida, Texas, and Louisiana. Temperate zones reach from about 30 to 60 degrees on either side of the equator, thus taking in most of the United States and southern Canada. Finally, subpolar and polar regions lie between 60 degrees and the poles, which are at 90 degrees.

REAL-LIFE APPLICATIONS

FOREST BIOMES

The term *forest,* as used in the realm of ecology, is one of those rare words that means the same thing within a scientific context as it does in the everyday world. An ecologist or biogeographer would define *forest* in more or less the same way that a nonscientist would: as any ecosystem dominated by tree-size woody plants. Of course, numerous other characteristics and parameters, such as weather, altitude, and dominant species, further characterize types of forests.

BOREAL CONIFEROUS FORESTS. Starting with the most northerly of forest biomes, there is the boreal coniferous forest, which we have discussed. Called *taiga* in Russia, boreal coniferous forests often are bordered on the north by tundra, discussed later in the context of nonforest ecosystems. An important subset of the boreal coniferous grouping is the montane forest, which also is dominated by conifers but which most often is found on mountains, at subalpine altitudes where the climate is cool and moist.

In addition to the dominant conifers, boreal coniferous forests also have important broadleafed angiosperms (plants that flower during sexual reproduction), including aspen, birch, poplar, and willow species. These forests are typically subject to periodic catastrophes, which result in at least partial destruction of the dominant trees within stands if not across a given forest as a whole. Among these catastrophic events are wildfires as well as defoliation by such pests as the spruce budworm.

TROPICAL GRASSLAND AND SAVANNA BIOMES, PRIMARILY IN AFRICA, ARE BEST CHARACTERIZED BY THE EXTRAORDINARILY ABUNDANT AND DIVERSE ANIMAL LIFE, WHICH INCLUDES SUCH LARGE MAMMALS AS THE CHEETAH AND THE GAZELLE. *(© Kevin Schafer/Corbis. Reproduced by permission.)*

TEMPERATE DECIDUOUS FORESTS. Moving farther away from the poles, the next major forest biome is that of the temperate deciduous forest. The average American, especially on the East Coast, is likely to be more familiar with the temperate deciduous forest than with any other biome. This type of forest develops in a climate that is relatively moist, with winters that are fairly cold. The larger grouping of temperate deciduous forests is divided into smaller categories depending on the relative amount of annual rainfall.

The term *deciduous* refers to a tree that sheds its leaves seasonally, and these forests are dominated by such trees, broad-leafed species that include ash, basswood, birch, cherry, chestnut, dogwood, elm, hickory, magnolia, maple, oak, and walnut, among others. (Note that many of these species are angiosperms. Likewise, most coniferous trees are gymnosperms, or plants that reproduce sexually through exposed seeds as opposed to seeds hidden in a flower.) Among the varieties of animal life are squirrels, rabbits, skunks, opossums, deer, bobcat, timber wolves, foxes, and black bears.

TEMPERATE RAIN FORESTS. Temperate rain forests are not necessarily farther from the poles than temperate deciduous forests, but they are subject to milder winters. For example, the temperate rain forests of Washington

State are north of many a temperate deciduous forest on the East Coast, but owing to differences in climate patterns, they are subject to milder winters than those typical of the deciduous forests to the east.

Characterized by abundant precipitation (most of it rain rather than snow, due to the milder temperatures), these systems are very moist—as the "rain forest" in their name implies. This, in turn, means that they are seldom subject to catastrophic wildfires, and therefore they often attain the climax stage of old-growth forests. In the temperate rain forest, coniferous trees are dominant, and many of these trees are extremely large and old. Among the tree species typical of this biome are Douglas fir, hemlock, cedar, redwood, spruce, and yellow cypress.

Tropical forests are discussed at some length in the essays Ecosystems and Ecology as well as The Biosphere. Among the two most basic varieties of this biome are semi-evergreen tropical forests and evergreen tropical rain forests. Most of the Amazon rain forest in South America, for instance, is an evergreen tropical forest, while surrounding biomes are semi-evergreen. Much the same is true of biomes in central and southern Africa, such as that surrounding the Congo River, with evergreen forests closest to the river and semi-evergreen ones in nearby areas.

The tropics, in general, are characterized not by the four seasons of the more temperate climate zones, but by a dry season and a wet season. The environment of a semi-evergreen tropical forest is one that is subject to great extremes of wet and dry, meaning that water is not available in abundance year-round. This means that most trees and shrubs in the biome are seasonally deciduous, shedding their leaves in anticipation of the drier season. In an evergreen tropical rain forest, on the other hand, rainfall is frequent and regular, so there is no seasonal drought. Deciduous trees may drop their leaves at various times of year, depending on the species, but with a wet climate and a wide range of trees, there is always something in bloom.

As with the temperate rain forest, the tropical variety experiences little in the way of wildfire or other catastrophic disturbances, and therefore an old-growth, climax community often develops in this biome. For this reason, tropical rain forests usually contain a wide diversity of trees, an enormous richness of species, and an extraordinary range of animals and microorganisms. Though biogeographers and ecologists often use the boreal coniferous forest as an example when examining biomes, those northerly forests are hardly examples of biological diversity. On the other hand, the tropical rain forest represents such diversity to its greatest extent.

NONFOREST BIOMES

Making an abrupt shift from the lush world of the tropical rain forest, let us look now at the tundra: a cold, treeless biome in the arctic and subarctic regions. (The Arctic Circle lies at approximately 66.5 degrees north latitude. Lands north of that line include northern Alaska and Canada, most of Greenland, extreme northern Scandinavia, and a northern strip of Russia and Siberia. The subarctic region, less clearly defined, comprises simply those lands that lie directly below the Arctic Circle.) Characterized by a short growing season, the tundra experiences very little precipitation in the form of liquid water. Yet the soil may well be marshy because temperatures are too low for significant evaporation and because the ground is usually frozen solid, preventing drainage.

In the most northerly tundras the dominant plants are small, hardy species that grow no more than 2–4 in. (5–10 cm) tall. In subarctic regions the dominant shrub species may grow as tall as 3.28 ft. (1 m), and the marshiest subarctic tundra may be home to sedge and cotton grass meadows. Among the larger forms of animal life on the tundra are the caribou and musk ox as well as the wolf, one of the larger predatory species.

GRASSLANDS AND CHAPARRAL. Temperate grasslands are known as prairies in North America and steppes in Eurasia, and these grasslands often are divided into smaller subgroups depending on the height of the dominant grasses. Fire, aided by the dry climate, acts as a curb to prevent the tall grass from giving way to larger trees and forests. In the United States, however, so much prairie has been converted to agricultural or other anthropogenic purposes that it constitutes an endangered biome.

Much further south are the tropical grassland and savanna biomes that appear primarily in Africa. Although they do have scattered trees and shrubs, these biomes are dominated by grasses and other plants. In any case, the plant life

ONE TYPE OF FRESHWATER BIOME IS THE SWAMP FOREST, WHICH MAY BE FLOODED EITHER SEASONALLY OR PERMANENTLY AND IS NOTABLE FOR ITS ANIMAL LIFE, INCLUDING HERON AND OTHER BIRD SPECIES. *(© S. R. Maglione. Photo Researchers. Reproduced by permission.)*

is not what best characterizes this biome in the minds of most people. Rather, it is the extraordinarily abundant and diverse animal life, which includes such large mammals as the rhinoceros, elephant, hippopotamus, buffalo, cheetah, gazelle and other antelope, wild dog, and hyena. Then, of course, there is the lion, "king of beasts," sometimes incorrectly portrayed (for instance, in many *Tarzan* movies) as a jungle creature.

A biome typical of coastal southern California, the chaparral is distinguished by what often is described as a Mediterranean climate: dry, rainy in winter, and prone to drought in summers. The characteristic plant in a chaparral region has thick, leathery leaves that help it preserve moisture during the dry seasons. As with most other nonforest biomes, wildfire is a major controlling factor.

DESERT. Deserts, discussed in more detail within The Biosphere, constitute a biome that may be temperate or tropical and which usually appears near the center of a continent. Such is particularly the case with the Gobi and Taklimakan deserts in northwestern and southwestern China, respectively; both deserts are located almost as far away from ocean as it is possible to be on Earth. Deserts also may occur in "rain shadows," areas separated from oceans by high mountains.

The unavailability of water is the chief defining feature of the desert, a biome that receives less than 9.9 in. (25 cm) of precipitation per year. Extremely dry deserts support virtually no plant productivity and therefore little, if any, animal life either. Such is the case, for instance, with the extraordinarily forbidding desert known as Rub' al Khali ("The Empty Quarter"), which occupies the lower third of the Arabian peninsula. On the other hand, less dry deserts may support relatively diverse plant life, as is the case, for instance, in Arizona.

FRESHWATER BIOMES

Now we make another abrupt shift, in this case from the dry desert to aquatic biomes, beginning with the freshwater variety. Among these are lentic biomes, which appear in the area of lakes and ponds—any place where water is still. (This would include even a vast body of water such as Lake Superior, which, though it looks like a sea from the edge, and experiences waves and heavy swells, is nonetheless a freshwater body where water is not flowing. Hence, it is by definition a lentic environment.) The water in these bodies may take a few days to flush; on the other hand, it may take centuries. By sitting for such long periods of time, the water may accumulate large amounts of nutrients, and this is one of the variables whereby subgroups of the lentic biome are classified.

In a lotic biome, such as that of a river or stream, water is flowing. A lotic biome may be as small as a babbling brook that runs for less than a mile, or as great as the Mississippi River, which drains much of the continental United States. Where lotic biomes are concerned, the greatest variables involve the strength of the flow, including its quantity, velocity, and seasonal variations. These characteristics influence other aspects of the ecosystem: for example, if water flows calmly and slowly, the bottom tends to gather silt. This provides a habitat for certain small species, such as the crustacean, known variously as a crayfish, crawfish, or crawdaddy.

A lotic environment is not fully self-sustaining, and therefore it may not qualify as a true biome. Though they support some plant life, these ecosystems do not have a full complement of autotrophs, or life-forms (usually plants) that depend only on the Sun and the atmosphere, rather than other organisms, for sustenance. Usually, the lotic ecosystem relies on the input of organic matter from the nearby terrestrial environment or from lakes upstream or both. These other biomes provide the lotic biome with the nutrients necessary to feed its fish and other aquatic species.

WETLANDS. Yet another freshwater biome is that of the wetlands. Just as rain forests were known as *jungles* until ecology and environmentalism entered the mainstream in the 1970s, so the term *wetlands* replaced a more blunt-sounding word: *swamp.* That word and several other old-fashioned ones are preserved in the terms for the four major wetland types: marsh, swamp forest, bog, and fen. Regardless of the name, this is a biome found in shallow waters, often in regions known for their pronounced seasonal variations of water depth.

The most biologically productive wetlands, marshes typically are dominated by relatively tall angiosperm varieties, such as the reed, cattail, and bulrush, as well as by floating flowers, such as the water lily and lotus. Swamp forests, such as the Okeefenokee on the Georgia-Florida border, are heavily populated with trees that include bald cypress and silver maple and may be flooded either seasonally or permanently. The Okeefenokee is notable for its animal life: not just heron and other bird species but also some of the more terrifying reptilian forms, including alligators and water moccasins, for which swamps are notorious.

Much less biologically productive than either marshes or swamps are bogs, which generally have acidic soil that supports only a limited range of vegetation. Characterized by a cool, wet climate (many of them are found in England), bogs often are dominated by sphagnum moss of one species or another. Later, when the sphagnum moss dies and the remains of several generations are compacted together with other plant debris, this becomes the basis for peat, which provides fuel for some homes in the British Isles and Europe.

Finally, there is the fen, another wetland found throughout the British Isles. Fens resemble bogs in several ways, including the fact that the local climate is usually cold (unlike the swamp forest, where the climate is generally hot). The fen has a better nutrient supply than the bog, however, and consequently the soil is less acidic, meaning that the biome as a whole is more productive.

MARINE BIOMES

The largest biome, geographically, is that of the open ocean, sometimes called a *pelagic oceanic biome.* Yet in terms of primary productivity—the first level in the food web—the ocean might as well be a desert. Tossed by waves and tides and heavily affected by the powerful salt content in its chemistry, the open ocean depends for its primary productivity not on plants but on phytoplankton, microscopic organisms that include a range of bacteria and algae.

Small crustaceans known as zooplankton eat the phytoplankton, only to be consumed, in turn, by small fish. Thus it goes up the trophic levels of the oceanic food web to the largest predators: bluefin tuna, sharks, squid, and whales. At the bottom of the ocean are other ecosystems, which depend on the slow rain of dead organic matter, or biomass, from the surface. Little is known about the deep-ocean biomes, but they appear to be diverse, if low in productivity (i.e., they have a relatively wide range of species but a small number of individuals).

CORAL REEFS APPEAR ONLY IN TROPICAL REGIONS. THE PRINCIPAL CHARACTERISTIC OF THIS BIOME IS ITS SUBSTRATE OF CALCIUM CARBONATE, FORMED FROM THE EXOSKELETONS OF DEAD CORAL POLYPS AND OTHER CREATURES. ON THIS STRUCTURE IS BUILT A HIGHLY PRODUCTIVE BIOME IN WHICH CORAL, ALGAE, FISH, AND INVERTEBRATES THRIVE. *(© Andrew Martinez. Photo Researchers. Reproduced by permission.)*

Upwelling regions are relatively deep, nutrient-rich areas that sustain highly productive biomes. Among the large variety of species included in such biomes are fish and shark, marine mammals, and birds such as gulls. Upwellings off the west coast of South America and in the Antarctic Ocean provide some of the human world's most abundant fisheries.

CLOSER TO SHORE. Closer to the major landmasses are the biomes of the continental shelves. There the water is warm compared with that of the open ocean, and the nutrient supply is relatively high. This flow of nutrients is fed partly by rivers that empty into the seas but also by the occasional rising of deeper, richer waters to the surface. Not surprisingly, then, phytoplankton and animal life are highly productive here, and continental shelf regions such as those of the Grand Banks off northeastern North America offer highly abundant and commercially important fisheries.

Another ocean biome closer to shore is that of the estuary, an ecosystem that is enclosed by land on several sides but is still open to the sea. Because they typically experience substantial inflows of river water from the nearby land, estuaries feature characteristics of both marine and freshwater biomes and offer highly productive ecosystems. Many a commercially important species of fish, shellfish, and crustacean makes its early home in an estuary before moving on to deeper waters after reaching maturity.

Seashores constitute a variety of oceanic biome or, indeed, several varieties. Environmental factors such as the intensity of wave motion determine the characteristics of the seashore biome, as do latitude. For example, temperate seashore ecosystems can develop kelp "forests." (See Succession and Climax for more about the interrelations of species in the kelp forest.) In other areas, where the bottoms are soft and covered with sand or mud, dominant species include mollusks, crustaceans, and marine worms.

As with seashores, coral reefs are those rare oceanic biomes affected by latitude; in fact, this type of biome appears only in tropical regions. The principal characteristic of the coral reef is its substrate of calcium carbonate, formed from the exoskeletons of dead coral polyps and other creatures. On this structure is built a highly productive biome in which coral, algae, fish, and invertebrates (animals without a backbone) thrive.

KEY TERMS

ABUNDANCE: A measure of the degree to which an ecosystem possesses large numbers of particular species. An abundant ecosystem may or may not have a wide array of different species. Compare with *complexity.*

ALPINE: A term that refers to a biogeographic zone that includes mountain slopes above the timberline.

ANGIOSPERM: A type of plant that produces flowers during sexual reproduction.

ANTHROPOGENIC: Influenced by human activity.

BIOGEOGRAPHY: The study of the geographic distribution of plants and animals, both today and over the course of extended periods.

BIOLOGICAL COMMUNITY: The living components of an ecosystem.

BIOME: A large ecosystem, characterized by its dominant life-forms. There are two basic varieties of biome: terrestrial, or land-based, and aquatic.

BIOTA: A combination of all flora and fauna (plant and animal life, respectively) in a region.

CANOPY: The upper portion or layer of the trees in a forest. A forest with a closed canopy is one so dense with vegetation that the sky is not visible from the ground.

CARNIVORE: A meat-eating organism, or an organism that eats *only* meat (as distinguished from an omnivore).

CLIMAX: A theoretical notion intended to describe a biological community that has reached a stable point as a result of ongoing succession. In such a situation, the community is at equilibrium with environmental conditions, and conditions are stable, such that the biota experiences little change thereafter.

COMPLEXITY: The range of ecological niches within a biological community. The degree of complexity is the number of different species that theoretically *could* exist in a given biota, as opposed to its diversity, or actual range of existing species.

CONIFER: A type of tree that produces cones bearing seeds.

DECIDUOUS: A term for a tree or other form of vegetation that sheds its leaves seasonally.

DECOMPOSERS: Organisms that obtain their energy from the chemical breakdown of dead organisms as well as from animal and plant waste products. The principal forms of decomposer are bacteria and fungi.

ANTHROPOGENIC BIOMES

Finally, there are anthropogenic biomes, such as the urban-industrial techno-ecosystem found in many a large metropolitan area. Such an ecosystem may include many species in addition to humans, but these—pets, houseplants, and the like—are not always native to the region, and probably would not flourish unless returned to their native biomes. New York City would be perhaps the ultimate example of an urban-industrial techno-ecosystem. Though it is far from a natural environment, it teems with life, from the oaks and elms in Central Park to the rats of the sewers, and from the pigeons that peck at crumbs on the sidewalks to the houseplants on the balconies and fire escapes of apartment buildings.

KEY TERMS CONTINUED

DETRITIVORES: Organisms that feed on waste matter, breaking organic material down into inorganic substances that then can become available to the biosphere in the form of nutrients for plants. Their function is similar to that of decomposers; however, unlike decomposers—which tend to be bacteria or fungi—detritivores are relatively complex organisms, such as earthworms or maggots.

DIVERSITY: A measure of the number of different species within a biological community.

ECOLOGY: The study of the relationships between organisms and their environments.

ECOSYSTEM: A community of interdependent organisms along with the inorganic components of their environment.

FAUNA: Animals.

FLORA: Plants.

FOOD WEB: A term describing the interaction of plants, herbivores, carnivores, omnivores, decomposers, and detritivores in an ecosystem. Each of these consumes nutrients and passes them along to other organisms (or, in the case of the decomposer food web, to the soil and environment). The food web may be thought of as a bundle or network of food chains, but since the latter rarely exist separately, scientists prefer the concept of a food web to that of a food chain.

FOREST: In general terms, a forest is simply any ecosystem dominated by tree-size woody plants. A number of other characteristics and parameters (for example, weather, altitude, and dominant species) further define types of forests, such as tropical rain forests.

GYMNOSPERM: A type of plant that reproduces sexually through the use of seeds that are exposed, not hidden in an ovary, as with an angiosperm.

HERBIVORE: A plant-eating organism.

OMNIVORE: An organism that eats both plants and other animals.

ORGANIC: At one time chemists used the term *organic* only in reference to living things. Now the word is applied to most compounds containing carbon, with the exception of carbonates (which are minerals), and oxides, such as carbon dioxide.

SUCCESSION: The progressive replacement of earlier biological communities with others over time.

TERRESTRIAL: Land-based.

UNDERSTORY: Layers of vegetation below the canopy in a forest.

Another anthropogenic biome is the rural techno-ecosystem, which is not as removed from human civilization as the "rural" in its name would imply. This type of biome appears in regions around transportation and transmission corridors, including highways, railways, canals, and aqueducts, as well as alongside power and telephone lines. Small towns are a characteristic area for such a biome, as are the regions around coal mines and other industrial plants devoted to the extraction, processing, or manufacture of products from natural resources. This biome usually supports a mixture of introduced species and native species, the latter being those varieties that can survive the disturbances, pollution, and other stresses associated with the human presence. For example, in the woods along an interstate highway, there are bound to be omnivorous creatures such as raccoons, which thrive on litter thrown out of passing cars. Such creatures must be agile enough to survive the threat of becoming "road kill," as well as other hazards associated with the environment. As with the city biome, the rural techno-ecosystem includes plenty of life-forms that have been introduced artificially, an example being the wildflowers planted on a median by state highway workers.

Agro-ecosystems are ecosystems that are managed and harvested for human use: farms, orchards, fisheries, commercial forests, and other agricultural concerns. Here the defining characteristic is the level of management, or the degree of anthropogenic influence. Very heavily managed agro-ecosystems involve the planting of non-native crop species and the introducing of non-native plants, often to the exclusion of native species. The "crop" may be a herd of animals, as when western ranchers introduced non-native cattle, and in the process killed off native predators such as coyotes. On the other hand, there are less heavily managed agro-ecosystems that do allow native wildlife species to thrive alongside those species introduced for commercial purposes.

WHERE TO LEARN MORE

Biomes of the World (Web site). <http://www.snowcrest.net/geography/slides/biomes/>.

Earth Floor: Biomes (Web site). <http://www.cotf.edu/ete/modules/msese/earthsysflr/biomes.html>.

Habitats/Biomes (Web site). <http://www.enchantedlearning.com/biomes/>.

Johnson, Rebecca L. *A Walk in the Boreal Forest.* Illus. Phyllis V. Saroff. Minneapolis: Carolrhoda Books, 2001.

———. *A Walk in the Desert.* Illus. Phyllis V. Saroff. Minneapolis: Carolrhoda Books, 2001.

——— *A Walk in the Rain Forest.* Illus. Phyllis V. Saroff. Minneapolis: Carolrhoda Books, 2001.

———. *A Walk in the Tundra.* Illus. Phyllis V. Saroff. Minneapolis: Carolrhoda Books, 2001.

Major Biomes of the World (Web site). <http://www.runet.edu/~swoodwar/CLASSES/GEOG235/biomes/main.html>.

The World's Biomes (Web site). <http://www.ucmp.berkeley.edu/glossary/gloss5/biome/>.

BIOLOGICAL COMMUNITIES

SYMBIOSIS
BIOLOGICAL COMMUNITIES
SUCCESSION AND CLIMAX

SYMBIOSIS

CONCEPT

Symbiosis is a biological relationship in which two species live in close proximity to each other and interact regularly in such a way as to benefit one or both of the organisms. When both partners benefit, this variety of symbiosis is known as *mutualism*. The name for a situation in which only one of the partners benefits is far more well known. Such an arrangement is known as *parasitism*, and a parasite is an organism that obtains nourishment or other life support from a host, usually without killing it. By their very nature, parasites are never beneficial, and sometimes they can be downright deadly. In addition to the extremes of mutualism and parasitism, there is a third variety of symbiosis, called *commensalism*. As with parasitism, in a relationship characterized by commensalism only one of the two organisms or species derives benefit, but in this case it manages to do so without causing harm to the host.

HOW IT WORKS

VARIETIES OF SYMBIOSIS

When two species—that is, at least two individuals representing two different species—live and interact closely in such a way that either or both species benefit, it is symbiosis. It is also possible for a symbiotic relationship to exist between two organisms of the same species. Organisms engaging in symbiotic relationships are called symbionts.

There are three basic types of symbiosis, differentiated as to how the benefits (and the detriments, if any) are distributed. These are commensalism, parasitism, and mutualism. In the first two varieties, only one of the two creatures benefits from the symbiotic relationship, and in both instances the creature who does not benefit—who provides a benefit to the other creature—is called the host. In commensalism the organism known as the commensal benefits from the host without the host's suffering any detriment. By contrast, in parasitism the parasite benefits at the expense of the host.

MUTUALISM: HUMAN AND DOG. Mutualism is distinguished from the other two types of symbiosis, because in this variety both creatures benefit. Thus, there is no host, and theoretically the partners are equal, though in practice one usually holds dominance over the other. An example of this inequality is the relationship between humans and dogs. In this relationship, both human and dog clearly benefit: the dog by receiving food, shelter, and care and the human by receiving protection and loving companionship—the last two being benefits the dog also receives from the human. Additionally, some dogs perform specific tasks, such as fetching slippers, assisting blind or disabled persons, or tracking prey for hunting or crime-solving purposes.

For all this exchange of benefits, one of the two animals, the human, clearly holds the upper hand. There might be exceptions in a few unusual circumstances, such as dog lovers who are so obsessive that they would buy food for their dogs before feeding themselves. Such exceptions, however, are rare indeed, and it can be said that in almost all cases the human is dominant.

OBLIGATE AND FACULTATIVE RELATIONSHIPS

Most forms of mutualism are facultative, meaning that the partners can live apart successfully.

Some relationships of mutualism are so close that the interacting species are unable to live without each other. A symbiotic relationship in which the partners, if separated, would be unable to continue living is known as an *obligate* relationship. In commensalism or parasitism, the relationship is usually obligate for the commensal or the parasite, since by definition they depend on the host. At the same time, and also by definition, the host is in a facultative relationship, since it does not need the commensal or parasite—indeed, in the case of the parasite, would be much better off without it. It is possible, however, for an organism to become so adjusted to the parasite attached to its body that the sudden removal of the parasite could cause at least a short-term shock to the system.

INQUILINISM

A special variety of commensalism is inquilinism, in which the commensal species makes use of the host's nest or habitat, without causing any inconvenience or detriment to the host. Inquilinism (the beneficiary is known as an inquiline) often occurs in an aquatic environment, though not always. In your own yard, which is your habitat or nest, there may be a bird nesting in a tree. Supposing you benefit from the bird, through the aesthetic enjoyment of its song or the pretty colors of its feathers—in this case the relationship could be said to be a mutualism. In any case, the bird still benefits more, inasmuch as it uses your habitat as a place of shelter.

The bird example is an extremely nonintrusive case of inquilinism; more often than not, however, a creature actually uses the literal nest of another species, which would be analogous to a bird nesting in your attic or even the inside of your house. This is where the analogy breaks down, of course, because such an arrangement would no longer be one of commensalism, since you would be suffering a number of deleterious effects, not the least of which would be bird droppings on the carpet.

Inquilinism sometimes is referred to as a cross between commensalism and parasitism and might be regarded as existing on a continuum between the two. Certainly, there are cases of a creature making use of another's habitat in a parasitic way. Such is the case with the North American cowbird and the European cuckoo, both of which leave their offspring in the nests of other birds to be raised by them. (See Instinct and Learning for a discussion of how these species exploit other birds' instinctive tendency to care for their young.)

REAL-LIFE APPLICATIONS

MYCORRHIZAE

One of the best examples of mutualism is known by the unusual name mycorrhiza, which is a "fungus root," or a fungus living in symbiosis with the roots of a vascular plant. (A vascular plant is any plant species containing a vascular system, which is a network of vessels for moving fluid through the body of the organism.) The relationship is a form of mutualism because, while the fungus benefits from access to carbohydrates, proteins, and other organic nutrients excreted by or contained in the roots of the host plant, the host plant benefits from an enhanced supply of inorganic nutrients, especially phosphorus, that come from the fungus.

The fungus carries out this function primarily by increasing the rate at which organic matter in the immediate vicinity of the plant root decomposes and by efficiently absorbing the inorganic nutrients that are liberated by this process-nutrients it shares with the plant. (The term *organic* refers to the presence of carbon and hydrogen together, which is characteristic not only of all living things but of many nonliving things as well.) The most important mineral nutrients that the fungus supplies to the plant are compounds containing either phosphorus or, to a lesser degree, nitrogen. (These elements are present in biogeochemical cycles—see The Biosphere.) So beneficial is the mycorrhizal mutualism that about 90% of all vascular plant families, including mustards and knotweeds (family *Brassicaceae* and *Polygonaceae,* respectively), enjoy some such relationship with fungi.

SOME EXAMPLES OF MYCORRHIZAE. Many mycorrhizal fungi in the *Basiodiomycete* group develop edible mushrooms, which are gathered by many people for use in gourmet cooking. Mushroom collectors have to be careful, of course, because some mycorrhizal fungi are deadly poisonous, as is the case with the death angel, or destroying angel—*Amanita virosa.*

Perhaps the most famous of the edible mushrooms produced by mycorrhizae are the many varieties known by the name truffle. Among these mushrooms is *Tuber melanosporum,* which is commonly mycorrhizal on various species of oak tree. The spore-bearing bodies of the truffle fungi develop underground and are usually brown or black and covered with warts. Truffle hunters require the help of truffle-sniffing pigs or dogs, but their work is definitely worth the trouble: good truffles command a handsome price, and particularly in France the truffle industry is big business. Given the lucrative nature of the undertaking, one might ask why people do not cultivate truffles rather than hunting for them. To create the necessary conditions for cultivation, however, so much effort is required that it is difficult to make a profit, even at the high prices charged for truffles. The soil composition must be just right, and under conditions of cultivation this takes about five years.

Orchids are an example of a plant in an obligate mutualism: they can thrive only in a mycorrhizal relationship. Tiny and dustlike, orchid seeds have virtually no stored energy to support the seedling when it germinates, or begins to grow. Only with the assistance of an appropriate mycorrhizal fungus can these seedlings begin developing. Until horticulturists discovered this fact, orchids were extremely difficult to propagate and grow in greenhouses; today, they are relatively easy to breed and cultivate.

THE IMPORTANCE OF MYCORRHIZAE. Some species of vascular plants do not contain chlorophyll, the chemical necessary for photosynthesis, or the conversion of light energy from the Sun into usable chemical energy in a plant. Such a plant is like a person missing a vital organ, and under normal circumstances, it would be impossible for the plant to survive. Yet the Indian pipe, or *Monotropa uniflora,* has managed to thrive despite the fact that it produces no chlorophyll; instead, it depends entirely on mycorrhizal fungus to supply it with the organic nutrients it needs. This obligate relationship is just one example of the critical role mycorrhizae perform in the lives of plants throughout the world.

Mycorrhizae are vital to plant nutrition, especially in places where the soil is poor in nutrients. Whereas many plant roots develop root hairs as a means of facilitating the extraction of water and nutrients from the soil, plant roots

ONE OF THE BEST EXAMPLES OF MUTUALISM IS THE MYCORRHIZA, A FUNGUS LIVING IN SYMBIOSIS WITH THE ROOTS OF A VASCULAR PLANT. MANY MYCORRHIZAL FUNGI DEVELOP EDIBLE MUSHROOMS, SUCH AS THE TRUFFLE, USED IN GOURMET COOKING. *(© Owen Frnaken/Corbis. Reproduced by permission.)*

that have a mycorrhizal fungus usually do not. Instead, these plants rely heavily on the fungus itself to absorb moisture and vital chemical elements from the ground. This means that it may be difficult or impossible for plants to survive if they are removed from an environment containing mycorrhizal fungus, a fact that indicates an obligate relationship.

Often, when species of trees and shrubs grown in a greenhouse are transplanted to a nonforested outdoor habitat, they exhibit signs of nutritional distress. This happens because the soils in such habitats do not have populations of appropriate species of mycorrhizal fungi to colonize the roots of the tree seedlings. If, however, seedlings are transplanted into a clear-cut area that was once a forest dominated by the same or closely related species of trees, the plants generally will do well. This happens because the clearcut former forest land typically still has a population of suitable mycorrhizal fungi.

Plants' dependence on mycorrhizal fungi may be so acute that the plants do not do well in the absence of such fungi, even when growing in soil that is apparently abundant in nutrients.

Although most mycorrhizal relationships are not so obligate, it is still of critical important to consider mycorrhizal fungi on a site before a natural ecosystem is converted into some sort of anthropogenic habitat (that is, an area dominated by humans—see Biomes). For example, almost all the tree species in tropical forests depend on mycorrhizae to supply them with nutrients from the soils, which are typically infertile. (See The Biosphere for more about the soil in rain forests.) If people clear and burn the forest to develop new agricultural lands, they leave the soil bereft of a key component. Even though some fungi will survive, they may not necessarily be the appropriate symbionts for the species of grasses and other crops that farmers will attempt to grow on the cleared land.

INTERKINGDOM AND INTRAKINGDOM PARTNERSHIPS

Mycorrhizae are just one example of the ways that mutualism brings into play interactions between widely separated species—in that particular case, between members of two entirely different kingdoms, those of plant and fungi. In some cases, mutualism may bring together an organism of a kingdom whose members are incapable of moving on their own (plants, fungi, or algae) with one whose members are mobile (animals or bacteria). An excellent example is the relationship between angiosperm plants and bees, which facilitate pollination for the plants (see Ecosystems and Ecology.)

Another plant-insect mutualism exists between a tropical ant (*Pseudomyrmex ferruginea*) and a shrub known as the *bull's horn acacia* (*Acacia cornigera*). The latter has evolved hollow thorns, which the ants use as protected nesting sites. The bull's horn acacia has the added benefit, from the ant's perspective, of exuding proteins at the tips of its leaflets, thus providing a handy source of nutrition. In return, the ants protect the acacia both from competition with other plants (by removing any encroaching foliage from the area) and from defoliating insects (by killing herbivorous, or plant-eating, insects and attacking larger herbivores, such as grazing mammals).

A much less dramatic, though biologically quite significant, example of interkingdom mutualism is the lichen. Lichen is the name for about 15,000 varieties, including some that are incorrectly called *mosses* (e.g., reindeer moss). Before the era of microscopy, botanists considered lichens to be single organisms, but they constitute an obligate mutualism between a fungus and an alga or a blue-green bacterium. The fungus benefits from access to photosynthetic products, while the alga or bacterium benefits from the relatively moist habitat that fungus provides as well as from enhanced access to inorganic nutrients.

BIG AND SMALL. In contrast to these cross-kingdom or interkingdom types of mutualism, there may be intrakingdom (within the same kingdom) symbiotic relationships between two very different types of animal. Often, mutualism joins forces in such a way that humans, observing these interactions, see in them object lessons, or stories illustrating the concept that the meek sometimes provide vital assistance to the mighty. One example of this is purely fictional, and it is a very old story indeed: Aesop's fable about the mouse and the lion.

In this tale a lion catches a mouse and is about to eat the little creature for a snack when the mouse pleads for its life; the lion, feeling particularly charitable that day, decides to spare it. Before leaving, the mouse promises one day to return the favor, and the lion chuckles at this offer, thinking that there is no way that a lowly mouse could ever save a fierce lion. Then one day the lion steps on a thorn and cannot extract it from his paw. He is in serious pain, yet the thorn is too small for him to remove with his teeth, and he suffers hopelessly—until the mouse arrives and ably extracts the thorn.

Many real-life examples of this strong-weak or big-small symbiosis exist, one of the more well-known versions being that between the African black rhinoceros (*Diceros bicornis*) and the oxpecker, or tickbird. The oxpecker, of the genus *Buphagus,* appears in two species, *B. africanus* and *B. erythrorhynchus.* It feeds off ticks, flies, and maggots that cling to the rhino's hide. Thus, this oddly matched pair often can be seen on the African savannas, the rhino benefiting from the pest-removal services of the oxpecker and the oxpecker enjoying the smorgasbord that the rhino's hide offers.

HUMANS AND OTHER SPECIES. Humans engage in a wide variety of symbiotic relationships with plants, animals, and bacteria. Bacteria may be parasitic on humans, but far

INTRAKINGDOM SYMBIOTIC RELATIONSHIPS CAN EXIST BETWEEN TWO VERY DIFFERENT TYPES OF ANIMAL. THE OXPECKER, OR TICKBIRD, FEEDS OFF THE TICKS, FLIES, AND MAGGOTS THAT CLING TO THE HIDE OF THE AFRICAN BLACK RHINOCEROS. THIS ODDLY MATCHED PAIR OFTEN CAN BE SEEN ON THE AFRICAN SAVANNAS. *(© Joe McDonald/Corbis. Reproduced by permission.)*

from all microorganisms are parasites: without the functioning of "good" bacteria in our intestines, we would not be able to process and eliminate food wastes properly. The relationship of humans to animals that provide a source of meat might be characterized as predation (i.e., the relationship of predator to prey), which is technically a form of symbiosis, though usually it is not considered in the same context. In any case, our relationship to the animals we have domesticated, which are raised on farms to provide food, is a mixture of predation and mutualism. For example, cows (*Bos taurus*) benefit by receiving food, veterinary services, and other forms of care and by protection from other predators, which might end the cows' lives in a much more unpleasant way than a rancher will.

All important agricultural plants exist in tight bonds of mutualism with humans, because human farmers have bred species so selectively that they require assistance in reproducing. For example, over time, agricultural corn, or maize (*Zea mays*), has been selected in such a way as to favor those varieties whose fruiting structure is enclosed in a leafy sheath that does not open and whose seeds do not separate easily from the supporting tissue. In other words, thanks to selective breeding, the corn that grows on farms is enclosed in a husk, and the kernels do not come off of the cob readily. Such corn may be desirable as a crop, but because of these characteristics, it is incapable of spreading its own seeds and thereby reproducing on its own. Obviously, agricultural corn is not on any endangered species list, the reason being that farmers continue to propagate the species through breeding and planting.

Another example of human-animal mutualism, to which we alluded earlier, is the relationship between people and their pets, most notably dogs (*Canis familiaris*) and house cats (*Felis catus*). Fed and kept safe in domestication, these animals benefit tremendously from their interaction with humans. Humans, in turn, gain from their pets' companionship, which might be regarded as a mutual benefit—at least in the case of dogs. (And even cats, though they pretend not to care much for their humans, have been known to indulge in at least a touch of sentimentality.) In addition, humans receive other services from pets: dogs protect against burglars, and cats eradicate rodents.

KEY TERMS

COMMENSALISM: A symbiotic relationship in which one organism, the commensal, benefits without causing any detriment to the other organism, the host.

FACULTATIVE: A term for a symbiotic relationship in which partners are capable of living apart.

HOST: The term for an organism that provides a benefit or benefits for another organism in a symbiotic relationship of commensalism or parasitism.

INQUILINISM: A type of symbiosis in which one species, the inquiline, makes use of a host's nest or habitat without causing any detriment to the host. Inquilinism is considered a variety of commensalism.

OBLIGATIVE: A term for a symbiotic relationship in which the partners, if they were separated, would be incapable of continuing to live.

PARASITISM: A symbiotic relationship in which one organism, the parasite, benefits at the expense of the other organism, the host.

SYMBIOSIS: A biological relationship in which (usually) two species live in close proximity to one another and interact regularly in such a way as to benefit one or both of the organisms. Symbiosis may exist between two or more individuals of the same species as well as between two or more individuals representing two different species. The three principal varieties of symbiosis are *mutualism, commensalism,* and *parasitism.*

Symbiosis Among Insects

Where insects and symbiosis are involved, perhaps the ideas that most readily come to mind are images of parasitism. Indeed, many parasites are insects, but insects often interact with other species in relationships of mutualism, such as those examples mentioned earlier (bees and angiosperms, ants and bull's horn acacia plants). Additionally, there are numerous cases of mutualism between insect species. One of the most intriguing is the arrangements that exists between ants and aphids, insects of the order *Homoptera,* which also are known as plant lice.

In discussing the ant-aphid mutualism, scientists often compare the aphids to cattle, with the ants acting as protectors and "ranchers." What aphids have that ants want is something called honeydew, a sweet substance containing surplus sugar from the aphid's diet that the aphid excretes through its anus. In return, ants protect aphid eggs during the winter and carry the newly hatched aphids to new host plants. The aphids feed on the leaves, and the ants receive a supply of honeydew.

In another mutualism involving a particular ant species, *Formica fusca,* two organisms appear to have evolved together in such a way that each benefits from the other, a phenomenon known as coadaptation. This particular mutualism involves the butterfly *Glaucopsyche lygdamus* when it is still a caterpillar, meaning that it is in the larval, or not yet fully developed, stage. Like the aphid, this creature, too, produces a sweet "honeydew" solution that the ants harvest as food. In return, the ants defend the caterpillar against parasitic wasps and flies.

WHEN MUTUALISM ALSO CAN BE PARASITISM. As the old saying goes, "One man's meat is another man's poison"—in other words, what is beneficial to one person may be harmful to another. So it is with symbiotic relationships, and often a creature that plays a helpful, mutualistic role in one relationship may be a harmful parasite in another interaction. Aphids, for instance, are parasitic to many a host plant, which experiences yellowing, stunting, mottling, browning, and curling of leaves as well as inhibiting of its ability to produce crops.

One particular butterfly group, *Heliconiinae* (a member of the *Nymphalidae,* largest of the butterfly families) furnishes another example of the fact that a mutualistic symbiont, in separate interaction, can serve as a parasite. Moreover, in this particular case the heliconius butterfly can be a mutualistic symbiont and parasite for the very same plant. Heliconius butterflies scatter the pollen from the flowers of passionflower vines (genus *Passiflora*), thus benefiting the plant, but their females also lay eggs on young *Passiflora* shoots, and the developing larva may eat the entire shoot. As an apparent adaptive response, several *Passiflora* species produce new shoots featuring a small structure that closely resembles a heliconius egg. A female butterfly that sees this "egg" will avoid laying her own egg there, and the shoot will be spared.

COMMENSALISM

Years ago a *National Geographic* article on the Indian city of Calcutta included a photograph that aptly illustrated the idea of commensalism, though in this case not between animals or plants but between people. The photograph showed a street vendor in a tiny wooden stall with a window, through which he sold his wares to passersby. It was a rainy day, and huddled beneath the window ledge (which also served as a countertop) was another vendor, protecting himself and his own tray of goods from the rain.

The photograph provided a stunning example, in microcosm, of the overpopulation problem both in Calcutta and in India as a whole—a level of crowding and of poverty far beyond the comprehension of the average American. At the same time it also offered a beautiful illustration of commensalism (though this was certainly not the purpose of including the picture with the article). The vendor sitting on the ground acted in the role of commensal to the relatively more fortunate vendor with the booth, who would be analogous to the host.

The relationship was apparently commensal, because the vendor on the ground received shelter from the other vendor's counter without the other vendor's suffering any detriment. If the vendor in the booth wanted to move elsewhere, and the vendor on the ground somehow prevented him from doing so, then the relationship would be one of parasitism. And, of course, if the vendor with the booth charged his less-fortunate neighbor rent, then the relationship would not be truly commensal, because the vendor on the ground would be paying for his shelter. To all appearances, however, the interaction between the two men was perfectly commensal.

COMMENSALISM IN NATURE. Plants that grow on the sides of other plants without being parasitic are known as *epiphytic plants.* Among these plants are certain species of orchids, ferns, and moss. By "standing on the shoulders of giants," these plants receive enormous ecological benefits: the height of their hosts gives them an opportunity to reach a higher level in the canopy (the upper layer of trees in the forest) than they would normally attain, and this provides them with much greater access to sunlight. At the same time, the hosts are not affected either negatively or positively by this relationship.

Another commensal relationship, known as *phoresy,* is a type of biological hitchhiking in which one organism receives access to transportation on the body of another animal, without the transporting animal being adversely affected by this arrangement. The burdock (*Arctium lappa*) is one of several North American plant species that produce fruit that adheres to fur and therefore is dispersed easily by the movement of mammals. The burdock is special from a human standpoint, however, inasmuch as the anatomical adaptation that makes possible its adhesion to fur provided designers with the model for that extremely useful innovation, Velcro.

As with the illustration of the street vendors in Calcutta, it is always possible that commensalism, through a slight alteration, may yield a relationship in which the host is affected negatively. There are instances in which individual animals may become loaded heavily with sticky fruit from the burdock (or other plants that employ a similar mechanism), thus causing their fur to mat excessively and perhaps resulting in significant detriment. This is not common, however, and usually this biological relationship is truly commensal. Furthermore, phoresy should not be confused with parasitic relationships in which a creature such as a tick attaches itself to the body of another organism for transport or other purposes. (For much more about parasitism, see Parasites and Parasitology.)

WHERE TO LEARN MORE

"Biology 160, Animal Behavior: Symbiosis and Social Parasitism." Department of Biology, University of California at Riverside (Web site). <http://www.biology.ucr.edu/Bio160/lecture25.html>.

Knutson, Roger M. *Furtive Fauna: A Field Guide to the Creatures Who Live on You.* New York: Penguin Books, 1992.

Lanner, Ronald M. *Made for Each Other: A Symbiosis of Birds and Pines.* New York: Oxford University Press, 1996.

Lembke, Janet. *Despicable Species: On Cowbirds, Kudzu, Hornworms, and Other Scourges.* New York: Lyons Press, 1999.

Margulis, Lynn. *Symbiotic Planet: A New Look at Evolution.* New York: Basic Books, 1998.

Mutualism and Commensalism. Neartica: The Natural World of North America (Web site). <http://www.nearctica.com/ecology/pops/symbiote.htm>.

"Parasites and Parasitism." University of Wales, Aberystwyth (Web site). <http://www.aber.ac.uk/parasitology/Edu/Para_ism/PaIsmTxt.html>.

Sapp, Jan. *Evolution by Association: A History of Symbiosis.* New York: Oxford University Press, 1994.

Symbiosis and Commensalism. The Sea Slug Forum (Web site). <http://www.seaslugforum.net/symbio.htm>.

Trager, William. *Living Together: The Biology of Animal Parasitism.* New York: Plenum Press, 1986.

BIOLOGICAL COMMUNITIES

CONCEPT

An ecosystem is a complete community of interdependent organisms as well as the inorganic components of their environment; by contrast, a biological community is just the living members of an ecosystem. Within the study of biological communities there are a great number of complexities involved in analyzing the relationships between species as well as the characteristics of specific communities. Yet many of the concepts applicable to biological communities as a whole also apply to human communities in particular, and this makes these ideas easier to understand. For example, the competitive urge that motivates humans to war (and to less destructive forms of strife in the business or sports worlds) may be linked to the larger phenomenon of biological competition. Indeed, much of the driving force behind the development of human societies, as it turns out, has been biological in nature.

HOW IT WORKS

THE BIOSPHERE, ECOSYSTEMS, AND ECOLOGY

An ecosystem is a community of interdependent organisms along with the inorganic components of their environment—air, water, and the mineral content of the soil—and a biome is an ecosystem, such as a tundra, that extends over a large area. All living organisms are part of a larger system of life-forms, which likewise interact with large systems of inorganic materials in the operation of a still larger system called *Earth.* (The concepts of ecosystem, biosphere, and biome are discussed, respectively, in Ecosystems and Ecology, The Biosphere, and Biomes.)

Note the importance of the distinctions between living and nonliving and organic and inorganic. Although in popular terms, organic means anything that is living as well as anything that was once living, along with their parts and products (bones, leaves, wood, sap, blood, urine, and so on), in fact, the scientific meaning of the word is both much broader and more targeted. A substance is organic if it contains carbon and hydrogen, and thus organic materials also include such items as plastics that have never been living.

The study of the relationship between living things and their environment, pioneered by the German zoologist Ernst Haeckel (1834–1919) and others, is called ecology. Though the world scientific community was initially slow to accept ecology as a subject of study, the discipline has gained increasing respect since the mid–twentieth century. This change is due to growing acceptance for the idea that all of life is interconnected and that the living world is tied to the nonliving, or inorganic, world. On the other hand, there is also the gathering awareness that certain aspects of industrial civilization may have a negative impact on the environment, an awareness that has spurred further interest in the study of ecology.

INTRODUCTION TO BIOLOGICAL COMMUNITIES

The term biological community refers to all the living components in an ecosystem. A slightly different concept is encompassed in the word biota, which refers to all flora and fauna, or plant and animal life, in a particular region.

For the biological community to survive and thrive, a balance must be maintained between consumption and production of resources. Nature provides for that balance in numerous ways, but beginning in the late twentieth century students of ecology in the industrialized world have become more and more concerned with the possible negative impact their own societies exert on Earth's biological communities and ecosystems.

It should be noted, however, that nature itself sometimes replaces biological communities in a process called succession. This process involves the progressive replacement of earlier biological communities with others over time. Coupled with succession is the idea of climax, a theoretical notion intended to describe a biological community that has reached a stable point as a result of ongoing succession. (See Succession and Climax for more about these subjects.)

NICHE. Whereas climax and succession apply to broad biological communities, a niche refers to the role a particular organism or species plays within that community. Though the concept of niche is abstract, it is unquestionable that each organism plays a vital role and that the totality of the biological community (and, indeed, the ecosystem) would suffer stress if a large enough group of organisms were removed from it. Furthermore, given the apparent interrelatedness of all components in a biological community, every species must have a niche—even human beings.

An interesting idea, and one that is somewhat similar to a niche, is that of an indicator species. This is a plant or animal that, by its presence, abundance, or chemical composition, demonstrates a particular aspect of the character or quality of the environment. Indicator species, for instance, can be plants that accumulate large concentrations of metals in their tissues, thus indicating a preponderance of metals in the soil. This metal, in turn, could indicate the presence of valuable deposits nearby, or it could serve as a sign that the soil is being contaminated. (See Food Webs for more about indicator species.)

Another concept closely tied to the concept of niche is that of symbiosis. The latter refers to a biological relationship in which (usually) two species live in close proximity to one another and interact regularly in such a way as to benefit one or both of the organisms. Symbiosis may exist between two or more individuals of the same species, as well as between two or more individuals representing two different species. The three principal varieties of symbiosis are mutualism, in which both participants benefit; commensalism, in which only one participant benefits, but at no expense to the other participant; and parasitism, in which one participant benefits at the expense of the other. These subjects are covered in much greater depth within the essays on Symbiosis and Parasites and Parasitology.

EVALUATING BIOLOGICAL COMMUNITIES

A few billion years ago, Earth's oceans and lands were populated with just a few varieties of single-cell organisms, but over time increasing differentiation of species led to the development of the much more complex ecosystems we know now. Such differentiation is essential, since the life forms in a particular region must adapt to that biome, whether it be forest or grassland, desert or aquatic environments, mountain setting or jungle.

Diversity is a measure of the number of different species within a biological community, while complexity is the number of niches within it. Put another way, the complexity of a community is the number of species that *could* exist in it. Abundance is the measure of populations within individual species; thus, if a biological community has large numbers of individuals, even if it is not diverse in species, it is still said to be abundant.

During its brief summer growing season, the arctic tundra has vast numbers of insects, migratory birds, and mammals, and thus its abundance is high, whereas its diversity is low. On the other hand, a rain forest might have several hundred or even a thousand different tree species, and an even larger number of insect species, in only a few hectares, but there may be only a few individuals representing each of those species in that area. Thus, the forest could have extremely high diversity but low abundance of any particular species. Needless to say, the rain forest is likely to have a much greater complexity than the tundra, meaning that it is theoretically likely to contain far more species.

Another way to evaluate ecosystems is in terms of productivity. Productivity refers to the amount of biomass—potentially burnable energy—produced by green plants as they capture sunlight and use its energy to create new organic compounds that can be consumed by local animal life. Once again, a forest, and particularly a rain forest, has a very high level of productivity, whereas a desert or tundra ecosystem does not.

FOOD WEBS. Food web (in contrast to the more popular, but less correct term, food chain) is the designation preferred by scientists to describe the means by which energy is transferred through a biological community. Within the food web are various stages, called trophic levels, that identify the position of various organisms in relation to the organisms they consume and the organisms that consume them.

Green plants that depend for their nourishment on photosynthesis, or the biological conversion of electromagnetic energy from the Sun into chemical energy, are primary producers. Herbivores, or plant-eating creatures, are primary consumers, whereas the animals that eat herbivores (whether carnivores or omnivores) are secondary consumers. The largest carnivores and herbivores are usually not prey for any other creatures, but when they die, they, too, will be consumed by detritivores, or scavengers, as well as decomposers, such as bacteria and fungi.

The second law of thermodynamics, one of several laws governing energy and the systems in which that energy is applied, holds that in each energy transfer some energy is lost. In the case of food webs, this means that much of the energy in each trophic level is unavailable to organisms at the next level. This, in turn, means that each successive trophic level generally has far fewer members than the prey on which they feed. While there might be thousands of primary producers in a particular community, there might be only a few top predators, including humans. (See Food Webs for more on these subjects.)

REAL-LIFE APPLICATIONS

COMPETITION

The word competition typically brings to mind images of a basketball court or football field, on one hand, or a Wall Street trading floor or boardroom, on the other. In biological terms, however, competition is the interaction among organisms of the same or different species vying for a common resource that appears in a limited supply relative to the demand. Put another way, this means that the capability of the environment to supply resources is smaller than the potential biological requirement for those resources.

Scarcity of resources relative to the need for them is one of the governing facts of human life, reflected in such common expressions as "There is no such thing as a free lunch." In fact, humans have spent most of their history in a modified form of biological competition, war. Furthermore, it may well be that our modern predilection for sports or business competition is simply a matter of transforming biological competition into a more refined form. In any case, competition prevails throughout the world of living things.

INTRASPECIFIC AND INTERSPECIFIC COMPETITION. For example, plants often compete for access to a limited supply of nutrients, water, sunlight, and space. Intraspecific competition occurs when individuals of the same species vie for access to essential resources (later we look at intraspecific competition between humans), or for mating partners, whereas interspecific competition takes place between different species.

Individuals of the same species have virtually identical resource requirements: for example, all humans need food, water, air, and some protection from the natural elements. For this reason, whenever populations of a species are crowded together, intraspecific competition is intense. This also has been illustrated by experiments involving laboratory mice, which become increasingly brutal to one another when confronted with severely diminished resources. When intraspecific competition occurs in dense populations, the result is a process known as self-thinning, characterized by the mortality (death) of those individuals less capable of surviving coupled with the survival of individuals that are more competitive. If this sounds like the "survival of the fittest," an idea associated with evolution that became the justification for a number of nefarious social movements and activities (see Evolution), it is no accident.

Ideas related to intraspecific competition influenced the English naturalist Charles Darwin (1809–1882) in developing his theory of evolution by means of natural selection. Intraspecific competition is an important regulator of population size and can make the species as a whole more fit by ensuring that only the hardiest individuals survive. Likewise, interspecific competition, or competition between species, plays a

A BIOME IS AN ECOSYSTEM, SUCH AS A TUNDRA, THAT EXTENDS OVER A LARGE AREA. IN THE ARCTIC TUNDRA, COMPETITION IS LOW, BUT CREATURES LIKE THIS ARCTIC FOX FACE SUCH POWERFUL ENVIRONMENTAL STRESSES FROM THE LOCAL CLIMATE THAT COMPETITION IS NOT THE MOST SIGNIFICANT FACTOR LIMITING POPULATIONS. *(Photo Researchers. Reproduced by permission.)*

strong role in shaping ecological communities. Furthermore, competition between species is also an agent of natural selection.

COMPETITIVE RELEASE. Environmental changes that affect a biological community may change the competitive relationships within it, leading to interesting results. During the early 1950s, a fungal pathogen (a disease-carrying parasite in the form of a fungus) known as chestnut blight, or *Endothia parasitica,* ended the dominance of the American chestnut in the eastern United States. Up to that time, the American chestnut (*Castanea dentata*) had been the leading species in the canopy, or uppermost layer, of the deciduous (prone to seasonal shedding of leaves) forests in the region. Thanks to the chestnut blight, it was as though the winner had been disqualified from a race, meaning that all the runners-up changed their standings.

By being relieved of the stresses associated with competition, other trees were allowed to become more successful and dominant in their habitat. They took advantage of this change to fill in the canopy gaps left by the demise of mature chestnut trees. In the same way, if a wildfire, storm, or other stress disturbs a mature forest, plants that previously have been suppressed by the higher-canopy trees find themselves with much greater access to such resources as light, moisture, and nutrients. As a result, they thrive.

Given the distasteful aspects of biological competition, such as the destruction of the "weak" in favor of the "strong" (actually, less adapted and more adapted are much more accurate terms in this context), one might wish for a situation in which no competition exists. Indeed, there are such situations in nature, but they are far from pleasant. In such biomes as the arctic tundra, for instance, competition is low, but this is not because all nature lives in happiness and harmony; instead, organisms face such powerful environmental stresses from the local climate that competition is not the most significant factor limiting populations.

THE TUNDRA BIOLOGICAL COMMUNITY. Creatures on the tundra face little stress from competition, but a great deal of stress in the form of very short growing seasons, thin soil, limited ground cover, low average temperatures and rainfall, high winds, and so on. If the density of individual plants of the tundra is decreased experimentally by thinning, the residual plants do not thrive as a result, as they might in a less harsh climate. In the case of the tundra, it is not competition that constrains their productivity, and therefore the reduction of potential competition does little to improve conditions for the organisms that survive.

It is interesting to observe what happens in a tundra environment if the intensity of environmental stress is artificially and experimentally alleviated by enclosing an area under a greenhouse and by fertilizing it with nutrients. Such experiments have been performed, and the results are fascinating: under these more favorable environmental conditions, competition actually increases, resulting in a biological community not unlike that of a more hospitable biome.

BIOLOGICAL COMMUNITIES AND CIVILIZATIONS

In his best-seller *Guns, Germs, and Steel: The Fate of Human Societies,* ethnobotanist Jared M. Diamond showed that local biological communities are among the leading determinants of the success or failure of human civilizations. The book had its beginnings, he wrote, during his many years of work with the native peoples of New Guinea. One day, a young man put a simple question to him: why do the societies of the West enjoy an abundance of material wealth and comforts, while those of New Guinea have so little?

The question may have been simple, but the answer was not obvious. As a scientist, Diamond refused to give an answer informed by the politics of the Left or Right, which might have blamed the problem, respectively, on western exploitation or on the failures of the New Guineans themselves. Instead, he approached it as a question of environment, and the result was his thought-provoking analysis, contained in *Guns, Germs, and Steel.*

FAVORABLE AND UNFAVORABLE ECOSYSTEMS. As Diamond showed, the places where agriculture was born were precisely those blessed with favorable climate, soil, and indigenous plant and animal life. Of course, it is no accident that civilization was born in the societies where agriculture first developed. Before a civilization can evolve, a society must become settled, and for that to happen, it must have agriculture.

Agriculture came into existence in four places during a period from about 8000 to 6000 B.C. In roughly chronological order, these locations were Mesopotamia, Egypt, India, and China. All were destined to emerge as civilizations, complete with written language, cities, and organized governments, between about 3000 and 2000 B.C.

In the New World, by contrast, agriculture appeared much later and in a much smaller way. The same was true of Africa and the Pacific Islands. In seeking to find the reasons why this happened, Diamond noted a number of factors, including geography. The agricultural areas of the Old World were stretched across a wide area at similar latitudes. This meant that the climates were not significantly different and would support agricultural exchanges, such as the spread of wheat and other crops from one region or ecosystem to another. By contrast, the landmasses of the New World or Africa have a much greater north-to-south distance than they do east to west, resulting in great differences of climate.

DIVERSITY OF SPECIES. Today such places as the American Midwest support abundant agriculture, and one might wonder why that was not the case in the centuries before Europeans arrived. The reason is simple but subtle, and it has nothing to do, as many Europeans and their descendants believed, with the cultural "superiority" of Europeans over Native Americans. The fact is that the native North American biological communities were far less diverse than their counterparts in the Old World. Peoples of the New World successfully domesticated corn and potatoes, because those were available to them. But they could not domesticate emmer wheat, the variety used for making bread, when they had no access to that species (it originated in Mesopotamia and spread throughout the Old World).

The New World also possessed few animals that could be domesticated either for food or for labor. Every single plant or animal that is a part of human life today had to be domesticated—adapted in such a way that it becomes useful and advantageous for humans—and the range of species capable of domestication is far from limitless. In fact, it is safe to say that all major species capable of being domesticated have been, thousands of years ago. The list of animals that can be domesticated is a short one, much shorter than the list of animals that can be tamed. A bear, for instance, can be captured, or raised from birth in captivity, but it is unlikely that humans would ever be able to breed bears in such a way that

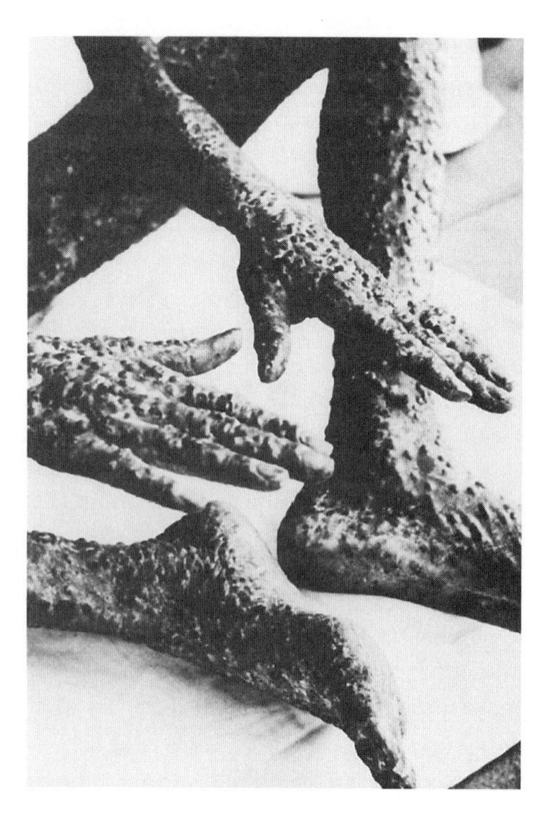

A SMALLPOX VICTIM SHOWS THE CHARACTERISTIC LESIONS ON HIS LIMBS. EUROPEANS' ADVANTAGE OVER THE NATIVE AMERICANS DERIVED FROM THE ECOLOGICAL COMPLEXITY OF THEIR BIOLOGICAL COMMUNITIES, EVEN DOWN TO THE WORLD OF MICROORGANISMS. THE NATIVE PEOPLES OF THE NEW WORLD HAD NO NATURAL RESISTANCE TO SMALLPOX OR A HOST OF OTHER DISEASES. *(© Corbis. Reproduced by permission.)*

their wild instincts all but disappeared and they became reliable, useful companions.

The animals that helped make possible the development of farms, villages, and ultimately empires in the Old World—cows and oxen, horses and donkeys, sheep and so on—were absent from the New World. (Actually, horses had once existed in the Americas, but they had been destroyed through overhunting, as discussed in the context of mass extinction within the Paleontology essay.) Many Indian tribes domesticated some types of birds and other creatures for food, but the only animal ever adapted for labor—by the most developed civilization of the pre-Columbian Americas, the Inca—was the llama, which is too small to carry heavy loads.

GREATER EXPOSURE TO MICROORGANISMS. The Europeans' advantage over the Native Americans derived ultimately from the ecological complexity of their biological communities compared with those of the Native Americans. This was also true of the "biological communities" they could not see, and of which people were unaware in 1500: the world of microorganisms, or the "germs" in the title of Diamond's book.

The native peoples of the New World had no natural resistance to smallpox or a host of other diseases, including measles, chicken pox, influenza, typhoid fever, and the bubonic plague. As with many plants and animals of the Old World, they simply had no exposure to these microorganisms. In the Old World, however, close contact with farm animals exposed humans to diseases, as did close contact with other people in crowded, filthy cities. This exposure, of course, killed off large numbers in such plagues as the Black Death (1347–1351), but those who survived tended to be much stronger and possessed vastly greater immunities. Therefore, the vast majority of Native American deaths that followed the European invasion were not a result of warfare, enslavement, or massacre of villages (though all of these occurred as well), but of infection. (See Infection and Infectious Diseases for more on these subjects.)

CANNIBALISM. Even the practice of cannibalism in such remote locations as the New Guinea highlands is, according to Diamond, a consequence of a relatively limited biological community. In the past, westerners assumed that only very "primitive" societies engaged in cannibalism. However, events have shown that, when faced with starvation, even people from European and European-influenced civilizations may consume human flesh in order to survive.

For instance, in 1846 members of the Donner party, making the journey west across North America, resorted to eating the bodies of those who had died in the perilous crossing. Much the same happened in the 1970s, when a plane carrying Uruguayan athletes crashed in the Andes, and the survivors lived off the flesh of those who had died. Though their upbringing and cultural norms may have told them that cannibalism was immoral or at the very least disgusting, their bodies told them that if they did not consume the only available food, they would die.

Whereas these circumstances were unusual and temporary, peoples in some parts of the world have been faced with a situation in which the only sources of protein provided within the biological community are ones whose consump-

KEY TERMS

ABUNDANCE: A measure of the degree to which an ecosystem possesses large numbers of particular species. An abundant ecosystem may or may not have a wide array of different species. Compare with *complexity*.

BIOENERGY: Energy derived from biological sources that are used directly as fuel (as opposed to food, which becomes fuel). Examples of bioenergy include wood or manure that can be burned. Usually, petrochemicals, such as petroleum or natural gas, though they are derived from the bodies of dead organisms, are treated separately from forms of bioenergy.

BIOLOGICAL COMMUNITY: The living components of an ecosystem.

BIOMAGNIFICATION: The increase in bioaccumulated contamination at higher levels of the food web. Biomagnification results from the fact that larger organisms consume larger quantities of food—and, hence, in the case of polluted materials, more toxins.

BIOMASS: Materials that are burned or processed to produce bioenergy.

BIOME: A large ecosystem, characterized by its dominant life-forms.

BIOSPHERE: A combination of all living things on Earth—plants, animals, birds, marine life, insects, viruses, single-cell organisms, and so on—as well as all formerly living things that have not yet decomposed.

BIOTA: A combination of all flora and fauna (plant and animal life, respectively) in a region.

CANOPY: The upper portion or layer of the trees in a forest. A forest with a closed canopy is one so dense with vegetation that the sky is not visible from the ground.

CARNIVORE: A meat-eating organism, or an organism that eats *only* meat (as distinguished from an omnivore).

CLIMAX: A theoretical notion intended to describe a biological community that has reached a stable point as a result of ongoing succession.

COMPLEXITY: The range of ecological niches within a biological community. The degree of complexity is the number of different species that theoretically *could* exist in a given biota, as opposed to its diversity, or actual range of existing species.

DECOMPOSERS: Organisms that obtain their energy from the chemical breakdown of dead organisms as well as from animal and plant waste products. The principal forms of decomposer are bacteria and fungi.

DECOMPOSITION REACTION: A chemical reaction in which a compound is broken down into simpler compounds, or into its constituent elements. In the biosphere, this often is achieved through the help of detritivores and decomposers.

DETRITIVORES: Organisms that feed on waste matter, breaking organic material down into inorganic substances that then can become available to the biosphere in the form of nutrients for plants. Their function is similar to that of decomposers; however, unlike decomposers—which tend to be bacteria or fungi—detritivores are

KEY TERMS CONTINUED

relatively complex organisms, such as earthworms or maggots.

DIVERSITY: A measure of the number of different species within a biological community.

ECOLOGY: The study of the relationships between organisms and their environments.

ECOSYSTEM: A community of interdependent organisms along with the inorganic components of their environment.

ENERGY TRANSFER: The flow of energy between organisms in a food web.

HERBIVORE: A plant-eating organism.

INDICATOR SPECIES: A plant or animal that, by its presence, abundance, or chemical composition, demonstrates a particular aspect of the character or quality of the environment.

NATURAL SELECTION: The process whereby some organisms thrive and others perish, depending on their degree of adaptation to a particular environment.

NICHE: A term referring to the role that a particular organism plays within its biological community.

OMNIVORE: An organism that eats both plants and other animals.

PATHOGEN: A disease-carrying parasite, usually a microorganism.

PHOTOSYNTHESIS: The biological conversion of light energy (that is, electromagnetic energy) from the Sun to chemical energy in plants.

PRIMARY PRODUCERS: Green plants that depend on photosynthesis for their nourishment.

PRODUCTIVITY: The amount of biomass produced by green plants in a given biome.

SUCCESSION: The progressive replacement of earlier biological communities with others over time.

TROPHIC LEVELS: Various stages within a food web. For instance, plants are on one trophic level, herbivores on another, and so on.

tion seems repugnant from the western viewpoint. Discussing the New Guinea highlands, Diamond noted that the area is virtually bereft of protein sources in the form of large, nonhuman mammals. Nor is bird life sufficient to support the local populace, and marine food sources are far away. For this reason, natives are prone not only to cannibalism but also to another culinary practice that most westerners find appalling: eating bugs, worms, grubs, caterpillars, and other creepy-crawly creatures.

WHERE TO LEARN MORE

Biota.org: The Digital Biology Project (Web site). <http://www.biota.org/>.

DeLong, J. Bradford. "Review of Diamond," *Guns, Germs, and Steel* (Web site). <http://econ161.berkeley.edu/Econ_Articles/Reviews/diamond_guns.html>.

"Designing a Report on the State of the Nation's Ecosystems." U.S. Geological Survey, Biological Resources Division (Web site). <http://www.us-ecosystems.org/index.html>.

Diamond, Jared M. *Guns, Germs, and Steel: The Fates of Human Societies.* New York: W. W. Norton, 1997.

Living Resources and Biological Communities (Web site). <http://www.chesapeakebay.net/ecointr5.htm>.

Miller, Kenton, and Laura Tangley. *Trees of Life: Saving Tropical Forests and Their Biological Wealth.* Boston: Beacon Press, 1991.

Nebel, Bernard J. *Environmental Science: The Way the World Works.* Englewood Cliffs, NJ: Prentice Hall, 1990.

NMITA: Neogene Marine Biota of Tropical America (Web site). <http://porites.geology.uiowa.edu/>.

Patent, Dorothy Hinshaw. *The Vanishing Feast: How Dwindling Genetic Diversity Threatens the World's Food Supply.* San Diego: Harcourt Brace, 1994.

Plant Communities of California (Web site). <http://encenter.org/habitat/habitatcontent.html>.

Quinn, John R. *Wildlife Survivors: The Flora and Fauna of Tomorrow.* Blue Ridge Summit, PA: TAB Books, 1994.

SUCCESSION AND CLIMAX

CONCEPT

Eventually almost everyone has the experience of watching an old neighborhood change. Sometimes we perceive that change for the better, sometimes for the worse, and the perception can have more to do with our individual desires or needs than it does with any qualities inherent in the change itself. For instance, one person might regard a new convenience store and gas station as an eyesore, while another might welcome it as a handy place to buy coffee, gasoline, or other items. Likewise, biological "neighborhoods" change, as when a complete or nearly complete community of living things replaces another. Once again, changes are not necessarily good or bad in any fundamental sense; rather, one community that happens to be better adapted to the changed environment replaces another. Sometimes a stress to the ecosystem brings about a change, such that life-forms that once were adapted to the local environment are no longer. Still, there appears to be a point when a community achieves near perfect adaptation to its environment, a stage in the levels of succession known as *climax.* This is the situation of old-growth forests, a fact that explains much about environmentalist opposition to logging in such situations.

HOW IT WORKS

BIOGEOGRAPHY

Elsewhere in this book, there is considerable material about ecosystems, or communities of interdependent organisms along with the inorganic components of their environment, as well as about biological communities, or the living components of an ecosystem. There are also discussions of biomes, or large ecosystems, and food webs, or the means by which energy transfer takes place across a biological community. Related to many of these ideas, as well as to succession and climax, is the realm of biogeography, or the study of the geographic distribution of plants and animals, both today and over the course of biological history.

Biogeography, which emerged in the nineteenth century amid efforts to explore and map the planet fully, draws on many fields. Among the areas that overlap with this interdisciplinary realm of study are the biological sciences of botany and zoology, the combined biological and earth sciences of oceanography and paleontology, as well as the earth sciences of geology and climatology. Not only do these disciplines contribute ideas to the growing field of biogeography, but they also make use of ideas developed by biogeographers. Biogeography is concerned with questions regarding local and regional variations in kinds and numbers of species and individuals. Among the issues addressed by biogeography are the reasons why particular species exist in particular areas, the physical and biotic (life-related) factors that influence the geographic range over which a species proliferates, changes in distribution of species over time, and so on.

Species interact by three basic means: competition for resources, such as space, sunlight, water, or food (see Biological Communities for more about competition); predation, or preying, upon one another (see Food Webs); and symbiosis. An example of the latter form of interaction, discussed elsewhere (see Symbiosis), occurs when an insect pollinates a plant while the plant provides the insect with nourishment, for instance, in the form of nectar. These interactions can and do affect the geographic distribution of species, and the presence or absence of a particular life-form may serve as a powerful control on the range of another organism.

Other significant concepts in the realm of biogeography are dispersal (the spread of a species from one region to another), and barriers (environmental factors that act to block dispersal). A species may extend its geographic range by gradually colonizing, or taking over, adjacent areas, or it may cross a barrier (for instance, a mountain range, an ocean, or a desert) and colonize the lands beyond. Later, we briefly examine the case of a bird that managed to do both.

SUCCESSION

Succession is the progressive replacement of earlier biological communities with others over time. It entails a process of ecological change, whereby new biotic communities replace old ones, culminating in a stable ecological system known as a climax community. In a climax community, climate, soil, and the characteristics of the local biota (the sum of all plants and animals) are all suited to one another.

At the beginning of the succession process, a preexisting ecosystem undergoes some sort of disturbance—for example, a forest fire. This is followed by recovery, succession, and (assuming there are no further significant disturbances) climax. If the environment has not been modified previously by biological processes, meaning that succession takes place on a bare substrate, such as a sand dune or a dry riverbed, it is known as primary succession. Primary succession also occurs when a previous biological community has been obliterated. Secondary succession takes place on a substrate that has been home to other life-forms and usually in the wake of disturbances that have not been so sweeping that they prevented the local vegetation from regenerating.

THE FACILITATION MODEL. Whether the conditions are those of primary or secondary succession, the outcome of the preceding disturbance is such that resources are now widely available, but there is little competition for them. One way of describing this situation is through what is known as the facilitation model, which identifies "pioneer species" as those life-forms most capable of establishing a presence on the site of the disturbance.

Pioneers modify a site by their presence, for instance, by regenerating the soil with organic material, thus making the area more attractive for invasion by other species. Eventually, new species move in, edging out the pioneers as they do so. This process may repeat itself several times, until the ecosystem reaches the climax stage, which we examine in greater depth a bit later in this essay. At the climax stage, there are few biological "openings" for further change, and change is only very slight and slow—at least until another disturbance arises and starts the process over again.

THE TOLERANCE MODEL. The tolerance model is another possible mechanism of succession. According to this concept, all species involved in succession are equally capable of establishing themselves on a recently disturbed site, but those capable of attaining a large population size quickly are most likely to become dominant. Unlike the facilitation model, the tolerance model does not depict earlier inhabitants as preparing the site biologically for new invader species; rather, this model is more akin to natural selection, discussed elsewhere (see Evolution).

According to the tolerance model, some species will prove themselves more tolerant of biological stresses that occur within the environment as succession proceeds. Among these stresses is competition, and those species less tolerant of competition may succeed earlier on, when there is little competition for resources. Later in the succession process, however, such species will be eliminated in favor of others more capable of competing.

THE INHIBITION MODEL. Yet another model of succession is the inhibition model, which, like the tolerance model, starts with the premise of an open situation at the outset: in other words, all species have equal opportunity to establish populations after a disturbance. In the inhibition model, however, some of the early species actually make the site less suitable for the development of other species. An example of this is when plants secrete toxins in the soil, thus inhibiting the establishment and growth of other species. Nevertheless, in time the inhibitory species die, thus creating opportunities that can be exploited by later successional species.

There is evidence to support all three models—facilitation, tolerance, and inhibition—but just as each has a great deal of basis in fact, none of the three fully depicts the dynamics of a suc-

CATTLE EGRET AND WILD HORSE IN SOUTHERN FRANCE. DURING THE NINETEENTH CENTURY, THE OLD WORLD CATTLE EGRET MANAGED TO CROSS THE ATLANTIC AND FOUND A BREEDING COLONY IN BRAZIL. SINCE THEN, IT HAS EXPANDED THE RANGE OF ITS HABITATS, SO THAT COLONIES CAN BE FOUND AS FAR NORTH AND EAST AS ONTARIO AND AS FAR SOUTH AND WEST AS SOUTHERN CHILE. *(© Hellio/Van Ingen. Photo Researchers. Reproduced by permission.)*

cessional environment. Put another way, each is fully right in one particular instance, but none are correct in all circumstances. Facilitation seems to work best for describing primary succession, whereas the more intense, vigorous environment of secondary succession is best pictured by tolerance or inhibition models. All of this shows that succession patterns tend to be idiosyncratic, owing to the many variables that determine their character.

CLIMAX

When a biological community reaches a position of stability and is in equilibrium with environmental conditions, it is said to have reached a state of climax. Often such communities are described as *old growth,* and in these situations change takes place slowly. Dominant species in a climax community are those that are highly tolerant of the biological stresses that come with competition. And well they might be, since by the climax stage, resources have been allocated almost completely among the dominant life-forms.

Despite its slow rate of change, the climax community is not a perfectly static or unchanging one, because microsuccession (succession on the scale of a single tree, or a stand of trees) is always taking place. In fact, frequent enough events of disturbance within small sections of the biological community may prevent climax from even occurring. Once the biota does achieve a state of equilibrium with the environment, however, it is likely that change will slow down considerably, bringing an end to the stages of succession. Climax remains a somewhat theoretical notion, and in practice it may be difficult to identify a climax community.

REAL-LIFE APPLICATIONS

COLONIZATION AND ISLAND BIOGEOGRAPHY

Earlier, in the context of biogeography, there was a reference to animals "colonizing." This may sound like the behavior of humans only, but other animals are also capable of colonizing. Nor are humans the only creatures that crossed the Atlantic Ocean from the Old World to colonize parts of the New World. During the nineteenth century, an Old World bird species known as the cattle egret managed to cross the Atlantic, perhaps driven by a storm, and founded a breeding colony in Brazil. Since then, it has expanded the range of its habitats, so that cattle egret colonies can be found as far north and east as Ontario and as far south and west as southern Chile.

THE BIOGEOGRAPHY OF ISOLATED BIOMES. Colonization is one example of the phenomena studied within the realm of biogeography. Other examples involve islands, and, indeed, island biogeography is a significant subdiscipline. The central idea of island biogeography, a discipline developed in 1967 by American biologists R. H. MacArthur (1930–1972) and Edward O. Wilson (1929–), is that for any landmass a certain number of species can coexist in a state of equilibrium. The larger the size of the landmass, the larger the number of species. Thus, reasonably enough, a large island should have a great number of species, whereas a small one should support only a few species.

These principles have helped in the study of other "island" ecosystems that are not necessarily on islands but rather in or on isolated lakes, mountain ranges surrounded by deserts, and patches of forest left behind by clear-cut logging. As a result of such investigations, loggers in the forests of the Pacific Northwest or the Amazon valley have been encouraged to leave behind larger stands of trees in closer proximity to one another.

This makes possible the survival of species at higher trophic levels (positions on the food web) and of those with very specialized requirements as to food or habitat. Examples of the latter species include Amazonian monkeys and the northern spotted owl, which we discuss later. Because studies in island biogeography have made land-use planners more aware of the barriers posed by clear-cut foresting, it has become common practice to establish "forest corridors." These long, thin lines of trees connecting sections of forest ensure that one section will not be isolated completely from another.

SUCCESSION IN ACTION

In discussing succession earlier, it was noted that a disturbance usually sets the succession process in motion. Examples of such disturbances can include seismic events (earthquakes, tidal waves, or volcanic eruptions) and weather events (hurricanes or tornadoes). Across larger geologic timescales, the movement of glaciers or even of plates in Earth's crust (see Paleontology for more about plate tectonics and its effect on environments) can set succession processes in motion.

There are also causes directly within the biosphere, or the realm of all life, that can bring about disturbances. Among them are wildfires as well as sudden infestations of insects that act to defoliate, or remove the leaf cover from, a mature forest. Quite a few disturbances can result from activities on the part of the biosphere's most complex species: *Homo sapiens.* Humans can cause ecological disturbances by plowing up ground, by harvesting trees from forests, by bulldozing land for construction purposes—even by causing explosions on a military reservation or battlefield.

Disturbances can take place on a grand scale or a small scale. It is theoretically possible for disturbances—even man-made ones—to wipe out forests as large as the Ardennes in northwestern Europe or the Amazon rain forest in South America. Fortunately, neither shelling in the Ardennes during the world wars nor logging in the Amazon valley in the late twentieth century managed to destroy those biomes, but it is quite conceivable that they could have. On the other hand, a disturbance can affect an individual lifeform, as when lightning strikes and kills a mature tree in a forest, creating a gap that will be filled through the growth of another tree—an example of microsuccession.

FORMS OF SUCCESSION. Once succession begins, it can take one of several courses. It may lead to the restoration of the ecosystem in a form similar to that which it took before the disturbance. Or, depending on environmental circumstances, a very different ecosystem may develop. For example, suppose that a forest fire has wiped out a biological community and secondary succession has begun. It is conceivable that this succession process will restore the forest to something approaching its former state. On the other hand, the wildfire itself may well have been a signal of a climate change, in this case, to a drier, warmer environment. In this instance, succession may bring about a community quite different from that which preceded the disturbance.

The "disturbance" itself actually may be the alleviation of a long-term environmental stress that has plagued the community. Suppose that a biological community has suffered from a local source of pollution, for instance, from a factory dumping toxins into the water supply. Suppose, too, that pressure from state or federal authorities finally forces a cleanup. How does this affect the biotic environment? In all likelihood, species that are sensitive to pollution (i.e., ones that normally could not survive in polluted conditions) would invade the area.

Removal of an environmental stress may not always be a matter of pollution and cleanup. For instance, a herd of cattle may be overgrazing a pasture, thus holding back the growth of plant species in the area. Imagine, then, that the cattle are moved elsewhere; as a result, new plant species will proliferate in the area, and, in all likelihood, the biological diversity of that particular ecosystem will increase.

PRIMARY SUCCESSION

As we noted earlier, primary succession occurs in an environment where there has never been a significant biological community or in the wake of

GLACIER BAY, ALASKA, IS AN EXAMPLE OF AN ECOSYSTEM THAT EXPERIENCED PRIMARY SUCCESSION IN THE WAKE OF DEGLACIATION; AS THE GLACIERS MELTED, VARIOUS PLANTS MOVED IN, EACH TAKING ITS TURN AS DOMINANT SPECIES. THE HABITAT NOW HAS REACHED MATURITY, AND ACCESS TO RESOURCES IS ALLOCATED AS FULLY AS IT CAN BE AMONG THE DOMINANT SPECIES. *(© Pat O'Hara/Corbis. Reproduced by permission.)*

disturbances that have been intense enough to wipe out all traces of a biological community. An example of primary succession in an area that has not possessed a biological community would be an abandoned paved parking lot. Eventually, the asphalt would give way to plant life, and given

ISRAELI FARMERS TRY TO WARD OFF A SWARM OF LOCUSTS. DEFOLIATION BROUGHT ABOUT BY INSECTS IS ONE EXAMPLE OF THE TYPE OF DISTURBANCE THAT MAY SERVE AS A PRECURSOR TO SECONDARY SUCCESSION. *(© Hulton-Deutsch Collection/Corbis. Reproduced by permission.)*

enough time, a wide-ranging biological community might develop around it.

These statements should be qualified in two ways, however. When it is said that an area has not maintained a significant biological community, this refers only to the recent or relatively recent past. In the case of the parking lot area, there probably have been countless biological communities in that spot over the ages, each replaced by the other in a process of primary succession. Also, by *significant biological community* we mean a biological community that exists *above* ground; even in the instance of the parking lot, there would be an extensive biological community underground. (See The Biosphere for more about life in the soil.)

GLACIER BAY. Glacier Bay, in southern Alaska, is an example of an ecosystem that experienced primary succession in the wake of deglaciation, or the melting of a glacier. The glaciers there have been melting for at least the past few hundred years, and as this melting began to occur, plants moved in. The first were mosses and lichens, flowering plants such as the river-beauty (*Epilobium latifolium*), and the mountain avens (*Dryas octopetala*), noted for their ability to "fix" or transform nitrogen into forms usable by the soil. (See The Biosphere for more about nitrogen fixing and biogeochemical cycles.)

These were the pioneer species, and over time they were replaced by larger plants, such as a short version of the willow. Later, taller shrubs, such as the alder (also a nitrogen-fixing species), dominated the area for about half a century. In time, Sitka spruce (*Picea sitchensis*), western hemlock (*Tsuga heterophylla*), and mountain hemlock (*T. mertensiana*) each had its turn as dominant plant species. With the last group, Glacier Bay reached climax, meaning that the dominant species are not those most tolerant of stresses associated with competition. The habitat thus has reached maturity, and access to resources is allocated as fully as it can be among the dominant species. Accompanying these changes have been changes in nonliving parts of the ecosystem as well, including the soil and its acidity.

SECONDARY SUCCESSION

When a disturbance has not been so intense or sweeping as to destroy all life within an ecosystem, regeneration may occur, bringing about secondary succession. But regeneration of existing species is not the only mechanism that makes secondary succession possible; invasions by new plant species typically augment the succession process. While much else changes in the environment of a secondary succession, the quality of the soil itself remains constant, as do other characteristics, such as climate.

Because it is rare for a disturbance to be powerful enough to obliterate all preexisting life-forms, secondary succession is much more common than primary succession. Examples of the type of disturbance that may serve as a precursor to secondary succession are windstorms, wildfires, and defoliation brought about by insects—provided, of course, that the destruction caused by these phenomena is less than total. The same is true of most disturbances associated with human activities, such as the abandonment of agricultural lands and the harvesting of forests by cutting down trees for lumber or pulp.

In a forest of mixed species in the eastern United States, the dominant trees are a mixture of angiosperms and coniferous species (respectively, plants that reproduce by producing flowers and those that reproduce by producing cones bearing seeds), and there are plant species capable or surviving under the canopy, or "roof," provided by these trees. Suppose that the forest has been clear-cut. This means that most or all of the large trees have been removed, but the entire biological community has not been wiped out, since loggers typically would not bother to cut down smaller plants that are not in their way.

As soon as the clear-cutting is over, regeneration begins. One form that this takes is the formation of new sprouts from the stumps of the old angiosperms. These sprouts are likely to grow rapidly and then experience a process of self-thinning, in which only the hardiest shoots survive. Within half a century, a given tree will have only one to three mature stems growing from its stump.

At the same time, other species regenerate seemingly from nowhere, though actually they are growing from a "seed bank" buried in the forest floor, where trees have dropped countless seeds over the generations. Species such as the pin cherry (*Prunus pennsylvanica*) and red raspberry (*Rubus strigosus*) are particularly adept at regeneration in this form. Therefore, these species are likely to feature prominently in the forest during the first several decades of secondary succession.

On the other hand, some tree species simply do not survive clear-cutting, or at least not in large numbers; if they are to obtain a stake in the secondary succession, they must do so by a process of re-invasion. Such often happens in the case of coniferous trees. Other species may also invade when they have not previously been a part of the habitat, yet they enter now because the temporary conditions of resource availability and limited competition make the prospect for invasion attractive. A great number of species, from alders and white birch to various species of grasses, fit into this last category.

Plants are not the only organisms involved in secondary succession. In a mature forest of the type described, the dominant bird forms probably include species of warblers, vireos, thrushes, woodpeckers, and flycatchers. When clear-cutting occurs, however, these birds are likely to be replaced by an entirely new avian community—one composed of birds more suited to the immature habitat that follows a disturbance. As time passes, however, and the forest regenerates fully, the bird species of the mature forest re-invade and resume dominance, a process that may well take three to four decades.

OLD-GROWTH FORESTS

Old-growth forests represent a climax ecosystem—one that has come to the end of its stages of succession. They are dominated by trees of advanced age (hence the name *old-growth*), and the physical structure of these ecosystems is extraordinarily complex. In some places, the forest canopy is dense and layered, whereas in others it has gaps. Tree sizes vary enormously, and the forest is littered with the remains of dead trees.

An old-growth forest, by definition, takes a long time to develop. Not only must it have been free from human disturbance, but it also must have been spared various natural disturbances of the kind that we have mentioned, disturbances that bring about the conditions for succession. Typically, then, most old-growth forests are rain forests in tropical and temperate environments, where they are unlikely to suffer such stresses as drought and wildfire. Among North American old-growth forests are those of the United States Pacific Northwest as well as in adjoining regions of southwestern Canada.

THE SPOTTED OWL. These old-growth forests of North America are home to a bird that became well known in the 1980s and 1990s to environmentalists and their critics: the

KEY TERMS

ABUNDANCE: A measure of the degree to which an ecosystem possesses large numbers of particular species. An abundant ecosystem may or may not have a wide array of different species. Compare with *complexity.*

ANGIOSPERM: A type of plant that produces flowers during sexual reproduction.

BIOGEOGRAPHY: The study of the geographic distribution of plants and animals, both today and over the course of extended periods.

BIOLOGICAL COMMUNITY: The living components of an ecosystem.

BIOME: A large ecosystem, characterized by its dominant life-forms.

BIOSPHERE: A combination of all living things on Earth—plants, animals, birds, marine life, insects, viruses, single-cell organisms, and so on—as well as all formerly living things that have not yet decomposed.

BIOTA: A combination of all flora and fauna (plant and animal life, respectively) in a region.

BIOTIC: Life-related.

CANOPY: The upper portion or layer of the trees in a forest. A forest with a closed canopy is one so dense with vegetation that the sky is not visible from the ground.

CLIMATE: The pattern of weather conditions in a particular region over an extended period. Compare with *weather.*

CLIMAX: A theoretical notion intended to describe a biological community that has reached a stable point as a result of ongoing succession. In such a situation, the community is at equilibrium with environmental conditions, and conditions are stable, such that the biota experiences little change thereafter.

COMPETITION: Interaction between organisms of the same or different species brought about by their need for a common resource that is available in quantities insufficient to meet the biological demand.

COMPLEXITY: The range of ecological niches within a biological community. The degree of complexity is the number of different species that *could* exist, in theory, in a given biota, as opposed to its diversity, or the actual range of existing species.

CONIFER: A type of tree that produces cones bearing seeds.

DIVERSITY: A measure of the number of different species within a biological community.

ECOSYSTEM: A community of interdependent organisms along with the inorganic components of their environment.

FOOD WEB: A term describing the interaction of plants, herbivores, carnivores, omnivores, decomposers, and detritivores in an ecosystem. Each of these organisms consumes nutrients and passes it along to other organisms. Earth scientists typically prefer this name to *food chain,* an everyday term for a similar phenomenon. A food chain is a series of singular organisms in which each plant or animal depends on the organism that precedes it. Food chains rarely exist in nature.

FOREST: In general terms, a forest is simply any ecosystem dominated by tree-

KEY TERMS CONTINUED

size woody plants. Numerous other characteristics and parameters (for example, weather, altitude, and dominant species) further define types of forests, such as tropical rain forests.

MICROSUCCESSION: Succession on a very small scale within a larger ecosystem or biological community. Microsuccession can occur at the level of a stand of trees or even a single tree.

NATURAL SELECTION: The process whereby some organisms thrive and others perish, depending on their degree of adaptation to a particular environment.

NICHE: A term referring to the role that a particular organism plays within its biological community.

OLD-GROWTH: An adjective for a climax community.

SUCCESSION: The progressive replacement of earlier biological communities with others over time. Succession, which can culminate in a climax community (see *climax*), is either primary, which occurs where there is no preexisting biological community (or no such community has survived), or secondary, in which a biological community regenerates in the wake of a disturbance, such as a forest fire.

TROPHIC LEVELS: Various stages within a food web. For instance, plants are on one trophic level, herbivores on another, and so on.

northern spotted owl, or *Strix occidentalis caurina.* A nonmigratory bird, the spotted owl has a breeding pattern such that it requires large tracts of old-growth, moist-to-wet conifer forest as its habitat. These are the spotted owl's environmental requirements, but given the potential economic value of old-growth forests in the region, the situation was bound to generate heated controversy as the needs of the spotted owl clashed with those of local humans.

On the one hand, environmentalists insisted that the spotted owl's existence would be threatened by logging, and, on the other, representatives of the logging industry and the local community maintained that prevention of logging in the old-growth forests would cost jobs and livelihoods. The question was not an easy one, pitting the interests of the environment against those of ordinary human beings. By the early 1990s the federal government had stepped in on the side of the environmentalists, having recognized the spotted owl as a threatened species under the terms of the U.S. Endangered Species Act of 1973. Even so, controversy over the spotted owl—and over the proper role of environmental, economic, and political concerns in such situations—continues.

CONTINUING CONTROVERSY. Another concern raised by the logging of old-growth forests has been the need to preserve dead trees, which provide a habitat for woodpeckers and other varieties of species. This concern, too, has brought about conflict with loggers, who find that dead wood gets in the way of their work. *Dead wood,* after all, is an expression for something or someone that is not performing a useful function (as in, "We're removing all the dead wood from the team"), and to loggers this literal dead wood is nothing more than a nuisance.

Unfortunately, the United States logging industry typically has not pursued a strategy of attempting to manage old-growth forests as a renewable resource, which these forests could be, given enough time. Instead, logging companies—interested in immediate profits and not much else—have tended to treat old-growth forests as though they were more like coal mines, home of a nonrenewable resource. In this "mining" model of tree harvesting, the forest is allowed to experience a process of succession such that a younger, second-growth forest emerges. Over time, this might become an old-growth forest, but the need to turn a quick profit means that the forest likely will be cut down before that time comes.

The average citizen, who typically has no vested interest in the side of either the loggers or the environmentalists, might well find good and bad on both sides of the issue. Certainly, the image of radical environmentalists chaining themselves to trees is as distasteful as the idea of loggers removing valuable natural resources. There is also a class dimension to the struggle, since a person deeply concerned about environmental issues is probably someone from an economic level above mere survival. This results in another distasteful image: of upper-middle-class and upper-class environmentalists inhibiting the livelihood of working-class loggers.

On the other hand, as we have already suggested, the logging companies themselves are big business and hardly representative of the working class. Largely as a result of pressure from environmentalists, these companies have attempted to develop more environmentally responsible logging schemes under the framework of what is called *new forestry.* These practices involve leaving a forest largely intact and removing only certain trees. Many environmentalists contend, however, that even the new forestry disturbs the essential character of old-growth forests.

WHERE TO LEARN MORE

Browne, E. J. *The Secular Ark: Studies in the History of Biogeography.* New Haven: Yale University Press, 1983.

Cox, C. Barry, and Peter D. Moore. *Biogeography: An Ecological and Evolutionary Approach.* Malden, MA: Blackwell Science, 2000.

The Eastern Old Growth Clearinghouse (Web site). <http://www.old-growth.org/>.

Environmental Biology—Grasslands (Web site). <http://www.marietta.edu/~biol/102/grasslnd.html>.

Forestry: Ecosystems: Forest Succession. Saskatchewan Interactive (Web site). <http://interactive.usask.ca/skinteractive/modules/forestry/ecosystems/forest_succession.html>.

Introduction to Biogeography and Ecology: Plant Succession. Fundamentals of Physical Geography <http://www.geog.ouc.bc.ca/physgeog/contents/9i.html>.

Old-Growth Forests in the United States Pacific Northwest (Web site). <http://www.wri.org/biodiv/b011-btl.html>.

Reed, Willow. *Succession: From Field to Forest.* Hillside, NJ: Enslow Publishers, 1991.

Succession (Web site). <http://www.cpluhna.nau.edu/Biota/succession.htm>.

GENERAL SUBJECT INDEX

Boldface type indicates main entry page numbers.
Italic type indicates photo and illustration page numbers.

A

Aardvarks, 222
Accidents, cause of death, 231
Acid reflux, 48
Acne, 286
Acquired characteristics, 165
Active sites (enzymes), 25
ADA (American Dietetic Association), 96
Adenosine diphosphate (ADP), 34
Adenosine triphosphate (ATP), 34, 56
Adrenaline, 267–268
Advanced Cell Technology (Worcester, MA), 123
Africa
  AIDS epidemic, 250
African black rhinoceros, 386, *387*
African trypanosomiasis, 276
Agriculture
  biomes, 380
  bred crop species, 387
  history, 395–396
  slash-and-burn, 350
  soil conservation, 352
  See also Crops
AIDS (Acquired immunodeficiency syndrome), 245, 250, 258–261, *259*
Air travel and jet lag, 310–311
Airports, 312
Alaska, *404*, 405
Albatross, 338
Albinism, *130*, 130–131
Alcohol
  cancer, 240
  fermentation process, 27–28
  use and abuse, 59
Alexis, son of Czar Nicholas II, *242*
Algorithms, 193
Alleles, 112
Allergies, 60, 264–268, *265*, 364–365
Allopatric species, 217
Alluvial soil, 351
Alpha-fetoprotein screening, 156
Alternative cancer treatment, 239
Altitude
  forest ecology, 362–363
Alzheimer's disease, *233*, 233–235
Amazon rain forest, 365, *366*
American chestnut trees, 394
American diet, 49–50, 79
  overweight Americans, 10, 37, 81, *82*, 85–86
American Dietetic Association (ADA), 96
American elm, 210
Amino acids, **11–17**
  enzymes, 24
  glossary, 16*t*
  proteins, 18–19, 79–80
Ammonia, in fertilizers, 351
Ammonium nitrate, 351–352
Amniocentesis, 156
Amoebic dysentery, 276
Amoxicillin, 290–291
Anabolism, 33, 34–35
Anaerobic respiration, 58–59
Anaphylactic shock, 265, 267–268
Ancestral record, evidence for evolution, 170–171
Anemia, 268–269
Anesthetics, in childbirth, 154
Angiosperms, 138, 140, 173–174, *362*, *364*, 364–365
Animals
  behavior, 321–326
  biomes, 375
  chemoreception, 297–298
  cloning, 123
  evidence of evolution, 169–171
  hibernation, 40–43
  vs. human intelligence, 167–168
  instinct and learning, 327–334
  kingdom animalia, 198, 205, 206
  mating rituals, 145–147
  migration, 338–341
  natural selection, 163–165

pregnancy and birth, 151–153
presence of proteins, 20–21, 23
respiration, 58
selective breeding, 128, 169
sense of smell, 303
shedding (fur or skin), 313, *313*
symbiotic relationships, 386–387
transpiration, 355, 358
See also Mammals; specific species, i.e. Bats

Annelids
asexual reproduction, *138*
detritivores, 349
respiration, 57, *57*

Anorexia nervosa, 38–39
Antacids, 48
Antarctica
midnight sun, 310

Anteaters, *223,* 223–224
Anthrax, *250,* 251–252
Anthropogenic biomes, 378–380
Anthropoidea, 220–221, 274–275
Antibiotics
amino acids in, 13
bacterial resistance to, 165, 290–291
discovery, 290

Antibodies, 122–123, 263–264
Antihistamines, 267
Antioxidants
Vitamin E, 91

Ants, *349,* 349–350, 386, 388
Apatosaurus, 184
Aphids, 388
Apis mellifera scutellata, 211
Apnea, 310
Aquatic animals
behavior, *321*
bioaccumulation, 72–73
echolocation, 340–341
endangered species, 208
evolutionary history, 195, 204–205, 221–222
food webs, 77, 376–377
migration, 336
mussels, 211
reproduction, 143–144
respiration, 57
taxonomy, *199*
See also Fish; Oceans

Aquatic biomes, 370
Aquifers, 347
Arachnida (class), 275
Arctic foxes, *394*
Arctic Ocean, 88
Arctic terns, 338
Aristotle, *192,* 196–197, 222
Arsenic, in human body, 78
Arteries. *See* Circulatory system
Arthropoda (phylum), 275, 279–282
See also Insects

Artichokes, 6
Artiodactyls, 223
Ascaris lumbricoides, 278
Ascorbic acid. *See* Vitamin C
Asexual reproduction, 135, 136–138, *138,* 285
See also Reproduction

Asteroids, cause mass extinction, 185–186, *186*
Astronauts, sleep cycles, 310
Atheism, 167
Athena, goddess, 138
Athens (ancient Greece), 287
Atherosclerosis, *36*
Atmosphere
dust following massive asteroid strike, 185
early Earth, 178
elements of, 346
role in biosphere, 347–348

Atomic bombs, radioactive fallout, 103, *104*
Atomic Energy Commission (AEC). *See* United States Atomic Energy Commission (AEC)
Atomic theory
proof of theories, 166

ATP (adenosine triphosphate), 34, 56
Attention (brain function), 307
Australia
marsupials, 219

Australopithecus, 215–216
Autosomes, 111–112
Autotrophs, 77, 87–88
Avery, Oswald, 111, 117

B

B cells, 255, 263–264
B vitamins, 92
Bacillus anthracis, 251
Bacteria
anaerobic, 58
and antibiotics, 165
decomposers, 361
in digestive system, *51,* 51–54
infection, 284–285, 286
infectious diseases, 245, 248, 250, 252, 264
origin of life, 179
ulcers, 48–49, *49*
See also Germs

Baking soda, 48
Balaena, 208
Banting, Frederick, 243
Barents, Willem, 88–89
Barrel sponge (animal), *199*
Bases, used as antacids, 48
Basketball, 261

Bats (animals)
echolocation, 340
order Chiroptera, 220
pentadactyl limb, *170*
pollinating plants, 141
BBC (British Broadcasting Corp.), 223
Bears, *40,* 89
Beauty (human perception), *146,* 147
Beavers, *322*
Beavertail cactus, *351*
Bedbugs, 281–282
Bees
behavior, 323–324
killer bees, 211, *211*
pheromones, 303–304
Behavior, **319–326**
glossary, 325*t*
See also Instinct; Learning and learned behavior
Behaviorism, 320–321
Berg, Paul, 119
Beriberi, 92, 93–95
Bertillon, Alphonse, 105
Best, Charles Herbert, 243
Bestiaries, *196,* 197
Beta-carotene, 90
Biello, Stephany, 315
Big bang theory, 177
Bile, human, 47
Bilirubin, 51
Bingeing, 39
See also Bulimia
Bioaccumulation, 71, 72–73, 76
Biodiversity, 365–366, 392
Biogeography, 400–401, 402–403
Biological communities, **391–399, 400–409**
glossary, 397–398*t*
See also Food webs
Biological rhythms, **306–315**
glossary, 314*t*
Biological warfare, 251–252
Biomagnification, 72–73
Biomes, **370–380**
glossary, 378–379*t*
Biopsy, diagnosing cancer, 239
Biorhythms (1970's fad), 315
Biosphere, **345–359,** 360
glossary, 356–358*t*
Biotechnology, 119
See also Genetic engineering
Birds
behavior, 322–323, 324, 327, 328–329, *329,* 330–331
biomes, *375*
circadian cycle, 309
colonization, 402, *402,* 406
endangered or extinct species, 208–209, *209,* 408
evolutionary history, 192
experiments in beriberi, 94–95
impact of DDT, 73
mating rituals, *145,* 146–147
migration, 336–338
parasites, 274, 330–331
pollinating plants, 140–141
respiration, 58
speciation, 217
symbiotic relationships, 384, 386, *387*
taxonomic keys, 193
training, 338
See also specific species, i.e. Chickens
Birth defects. *See* Congenital disorders
Biston betularia, 173
The Black Death (1347-1351), 247–248
Black-throated green warblers, 217
Blane, Sir Gilbert, 93
Blood
anemia, 268–269
hemoglobin, 19, 56
hemophilia, 115–116, 241–242
importance of bone marrow, *263*
importance of vitamin K, 91–92
infections, 276–277
lymphocytes, 263, 264
oxygen diffusion in lungs, 56–57
plasma, 47
proteins in, 21
sickle-cell anemia, *14,* 15–16
Blood vessels, *36*
Blue whales, 208
Body clock. *See* Biological rhythms
Bogs, 376
Bone marrow, 263, *263*
Boreal forest, 362, 371–372
Botulism, 252
Bowel movements, human, 50
Boyer, Herbert, 119
Brain, 168
Alzheimer's disease, *233,* 233–235
Creutzfeldt-Jakob disease, 232–233
neurological disorders, 302
pineal gland, 306–307
processing sensory data, 299
Bread, 28, *28*
Breast-feeding, 73
Breathing, 56–58
See also Respiration (human)
Breeding. *See* Selective breeding
British Broadcasting Corp. (BBC), 223
Bronchial disorders, 60
Buchner, Eduard, 25
Bulimia, *38,* 38–39
Bull's horn acacia, 386
Burdock (plant), 389

Burroughs, Edgar Rice, 331–332
Business application. *See* Industrial uses
Butterflies, 338, 388–389
Byzantine Empire, 230, 246

C

Cabbage, *26*
Cacti, *351*
Caesar, Julius, 157
Calcium carbonate, *377*
Calcium, importance in nutrition, 81, 91
Calvaria major, 209
Cambrian period, 179
Camels, 223
Camerarius, Rudolf Jakob, 138
Cancer, 230, 238–241
  lung cancer, 62
  mutation, 132
  treatment with designer proteins, 21
Candy bars, 10
Cannibalism, 396, 398
Carbohydrates, **3–10**, 44, *79,* 80, 81, 82
  glossary, 8–9*t*
Carbon
  in amino acids, 11–13
  in old-growth biological communities, 369
  origin of life, 178–179
  percentage of biosphere, 348
  percentage of human body mass, 78
Carbon dioxide, 28, *28,* 55–56, 369
Carcinogens, 238
Cardinals, 327
Cardiovascular disease, 230, 232
Carnivores
  dinosaurs, 184
  order Carnivora, 221
  place in food web, 361
Casal, Gaspar, 95
Catabolism, 33, 34–35
Catalysts
  enzymes, 24–25
  photosynthesis, 5
Cats
  human interaction with, 387
  sense of smell, 298
  taxonomy, 221
Cattle, 7–8, 123, 233
Cattle egrets, *402*
CCK (Cholecystokinin), 10
CDC. *See* United States Centers for Disease Control and Prevention (CDC)
Cellular biology
  amino acids, 14
  characteristics of bacteria, 284–285
  chromosomes and DNA, 99–100, 135
  metabolism, 34
  origins of life, 179
  photosynthesis, *4*
  starches, 7
  taxonomy, 198, 206
Cellular respiration, 56, *57,* 58–59
Cellulose, *6,* 7–8
Cenozoic era, 186
Centers for Disease Control. *See* United States Centers for Disease Control and Prevention (CDC)
Cereal foods, 81
Cervix, 153
Cesarean section, 156–157
Cetaceans, 221–222
Chaparral, 375
Chase, Martha, 111
Cheetahs, *373*
Chelicerata (subphylum), 275
Chemical bonding
  enzymes, 25
  lipids, 35
  peptide linkage, 13, 18
Chemical energy
  detritivores and decomposers, 361
  produced by photosynthesis, 4–5, 360–361
Chemical equations
  cellular respiration, 56
  photosynthesis, 5, 68
Chemical reactions
  catalysts, 24–25
  fertilizers, 351
  human digestion, 45–46
  peptide linkage, 13, 18–19
  spicy foods, 297
Chemoreception, **295–305**
  glossary, 304*t*
Chemotherapy, 239
Chestnut blight, 394
Chickens, 322, 324, 328
Childbirth, **151–157**
  glossary, 156*t*
  history, 153–155
Children
  mental development, 334
  sense of taste, 301
  vaccinations, 258
Chimney sweeps, 240, *240*
Chiropterans, 220
Chlorophyll, *4,* 5
Chloroplasts, *4*
Cholecystokinin (CCK), 10
Cholera, *247*
Cholesterol, *36,* 36–37
Chordata (phylum), 205

Chorionic villi, 155–156
Chromium, in human body, 78
Chromosomes, 99–100, *102,* 111–112, 117, 126, 135
Chronobiological study, 315
Circadian rhythms, 306, 307–312
Circannual cycles, 313
Circulatory system, 230, 232
    See also Blood
Cities (anthropogenic biomes), 378–379
Citrus fruits, 93
Cladistics, 192
Classification. *See* Taxonomy
Cleft palate, 128
Climate
    changes, 353
    classifying biomes, 372
Climax biological communities, 370–371, **400–409**
    glossary, 407–408*t*
Cloning, *121,* 122–123
    See also Genetic engineering
Clonorchis, 277
Closed systems, 345–346
Clostridium botulinum, 252
Coenzymes, 25–26
Cohen, Stanley, 119
Cold climate, 374, *394,* 394–395
Cold, common
    caused by virus, 60, 286–287
    sense of taste and smell, 301–302
Cold-blooded animals, 218
Colonization (movement of species), *402,* 402–403
Commensalism (symbiosis), 273, 383–384, 389
Common chemical sense, 298, *298*
Communities (animals), 323–324, 331
Competition (biological communities), 393–395
Complete migration, 336
Concentration camps, *121*
Conditioning (behavior), 320–321
Congenital disorders, 128–131, *129,* 155–156, 232
Coniferous forests, 371–372, 374
Contagious diseases. *See* Infectious diseases
Continental drift
    and evolution, 169, 180
    impact of massive asteroid, 185
Continental shelves, 377
Controversies
    cloning, 123
    DNA evidence, 108
    evolution, 165–169
    genetic engineering, 103–104, 121–122
    in vitro fertilization, 144
    logging, 408–409
Copper, in human body, 78
Coral reefs, 377, *377*
Corn, 82, 387
Courtship. *See* Mating rituals
Cousins, 115–116
Cowbirds, 274
Cowpox, 256–257, *257*
Cows. *See* Cattle
Creationism, 163, 168
Creutzfeldt, Hans Gerhard, 232–233
Creutzfeldt-Jakob disease, 128, 232–233
Crick, Francis, 111, 117
Criminal investigations. *See* Forensics and criminal investigation
Crops, 136, 209, 350–351, 352, 380
    See also Agriculture; Plants
Crosscurrent exchange, 58
Cystic fibrosis, 62, 128

D

Darwin, Charles, *162*
    Darwin's moth, 138
    ethology, 320
    introduces theory of evolution, 161, 169
    taxonomy, 197–198
Darwin, Erasmus, 167
Dating techniques
    amino acids, 17
    fossils, 172
    radiometric, 172
DDT (dichlorodiphenyltrichloroethane), *72,* 73
Deciduous trees, 362, 373
Decomposers, 68, 69, 349, 361
Decomposition, 68, 69
Deforestation, 365–366
Deinonychus, 184
Delayed sleep phase syndrome, 311
Deltas, fertile soil, 351
Deoxyribonucleic acid. *See* DNA (deoxyribonucleic acid)
Department of Energy. *See* United States Department of Energy (DOE)
Dermoptera, 220
Descartes, René, 306–307
Desert tortoises, *351*
Deserts
    biomes, 375
    formation of, 353
    soil, 350, *351*
Designer proteins, 21–22
Desmodus rotundus, 220
Detritivores, 68, 69, 221, 349, 361
Devonian period, 182, 183
Dextrose, 3–4
Diabetes mellitus, 242–243
Diamond, Jared M., 395
Diana, Princess of Wales, *38*
Diet. *See* American Diet; Fitness; Nutrition and nutrients

Differential migration, 336
Diffusion (respiration), 56–57, *57*
Digestion, human, **44–54**
 breakdown of amino acids, 14–15
 digestive system, 45–47, *46*, 276, 278, 285–286
 enzymes, 27
 glossary, 52–53*t*
 importance of cellulose, 8
 metabolism, 33–34
Dinosaurs, *183*, 183–185
Dinuguan (Filipino delicacy), 302–303
Diploid cells, 99
Disaccharides, 4
Diseases, human, **229–235**
 beriberi, 92, 93–95
 cancer, 21
 congenital disorders, 128–131, *129*, 155–156, 232
 digestive disorders, 48–50
 eating disorders, 38–40
 ethnic groups, 113, 127, 131
 genetic disorders, *113*, 113–116, 121, 127
 glossary, 234*t*
 Hartnup disease, 15
 infectious diseases, **244–252**, 396
 kwashiorkor, 85
 neurological disorders, 302
 noninfectious diseases, **236–243**
 parasites, 275–282
 pellagra, 15, 95
 respiratory disorders, 60, 62
 rickets, *90*, 91
 risk factors, 239–241
 scurvy, 93
 sickle-cell anemia, *14*, 15–16
 sleep disorders, 309–311
 social impact, 230, 246, 247–248
 unknown causes, 230, 232–235, *233*
 Vitamin A poisoning, 89–90
DNA (deoxyribonucleic acid)
 asexual reproduction, 135
 cancer, 238
 effect of viruses on, 285, 287
 forensics and criminal investigations, *20*, 21, 103, 104–105, *105*, 108
 genes, 100–101, 117–118, *118*, 119, 126, 164–165
 history, 111
 nanocomputer, 120
 phylogeny, 198
 synthesis of amino acids, 13
Doctors. *See* Physicians
Dodos (bird), 208–209, *209*
Dogs
 human interaction with, 383, 387
 Pavlov's dog, 320
 sense of smell, 298, 303
Dolly (cloned sheep), *122*, 123
Dolphins, 195, 204–205, 221–222, 340–341
Domesticated animals, 387, 395–396
Dominant genes, 112–113
Donkeys, 215
Double-helix model (DNA), 111, 117–118, *118*
 See also DNA (deoxyribonucleic acid)
Down syndrome, 126, 129, *129*
Dreissena polymorpha, 211
Drugs and medicines
 ecstasy, 315
 impact on taste and smell, 302
 insulin, 120, 242–243
 LSD (lysergic acid diethylamide), 307
 used in childbirth, 154
Dryja, Thaddeus R., 13
Ducks, 193, 330
Duodenum, 47, 48
Dust Bowl (1934-1935), 352
Dust mites, *265*
Dutch elm disease, 210
Dwarfism, 129

E

E. coli bacteria. *See Escherichia coli* bacteria
Ears
 bats, 340
 human ear infections, 290
 whales, 341
Earth
 biological communities, **391–399, 400–409**
 biomes, **370–380**
 Biosphere, **345–359**
 geologic time periods, 179
 geomagnetic field, 338–339
 greenhouse effect, 366, 368–369
 origin of life, 17, 177–178, 180–181, 217
 rotation, 308
 struck by massive asteroid, 185–186, *186*
Earthworms, 349
Eating disorders, *38*, 38–40
Eating habits, human, 39, 48, 49, 242, 311
Ebola virus, 250–251
Echolocation, *339*, 339–341
Ecology, **360–369**, 391
 glossary, 367–368*t*
Ecosystems, **360–369, 370–380, 391–399**
 glossary, 367–368*t*
 See also Food webs
Ecstasy (drug), 315
Edison, Thomas, *312*
Edwards, Robert G., 144
Egg cells, 100, 136
Eggs
 bird parasites, 274, 330–331

egg-laying mammals, 218
human, 143
Egypt, 350–351
Ehrlich, Paul, 255
Eijkman, Christiaan, 93–95
Elderly
Alzheimer's disease, 234
sense of taste, 301
Electromagnetic energy in photosynthesis, 4–5
Electronegativity, 297
Elements in the human body, 78
Elephantiasis, *278,* 279
Elephants, 200, 222
Elevation. *See* Altitude
Embryo, 152–153
Endangered species, 207–208, 223
See also Extinction
Endogenous infection, 283
Energy
human body, 10, 36, 308
metabolism, 33–34, 44, 80
moves from environment to system, 345, 360–361
produced in chemical reactions, 24
Sun's energy used by plants, 67–69
transfer, 69, 393
Entamoeba histolytica, 276
Enterobiasis, 278
Environment
definition, 345–346
Environmental concerns
forest conservation, 363, 408–409
soil conservation, 352
tropical rain forests, 350
Enzymes, 21, **24–30,** 37, 119
glossary, 29–30*t*
Eohippus (horse ancestor), 172
Epidemiology, 237
Epinephrine, 267–268
Epiphytic plants, 389
Erosion
Dust Bowl (1934-1935), 352
Escherichia coli bacteria, *51,* 285–286
Eskimo curlews, 208
Esophagus, 45
Estuaries, 377
Ethics, genetic engineering, 103–104, 122
Ethology, 320
Eugenics, 121–122
Europe
early humans, 162
Eutrophication, 354
Evapotranspiration, 346, 347, 354–355, *355*
Evergreen forests. *See* Coniferous forests
Evolution, **161–175**
amino acids dating, 17
choosing the ideal mate, 147–149
and geology, 169–170
glossary, 174–175*t*
history, 161, 169, 173
humans and food intake, *82*
mammals, 221–222
moving from water to land, 181–182
primates, 220
role of mutation, 101, 127
trees, 364–365
See also Natural selection; Phylogeny
Exhalation (human breathing), 55–56
Exogenous infection, 283
Explorers and exploration
Barents, Willem, 88–89
introduced species, 210, 395–396
Explosions and explosives
ammonium nitrate, 351–352
impact of massive asteroid, 185
Extinction
dodo bird and dodo tree, 208–209, *209*
mass extinctions, 181, 182–183, 185–186
Eyes
diseases, 279
heredity, 112–113

F

Facial characteristics, *129,* 129–130
Facilitation model (biological communities), 401
Facultative relationships (symbiosis), 273–274, 383–384
Fast food, 79, 81
Fat, human body, 10, 35–37, *36,* 88, 127, *146,* 147, 307
Fats and oils, 35–37, 44–45, 81, 88, 297
See also Lipids
Faunal dating, 172
FDA. *See* United States Food and Drug Administration (FDA)
Feces, human, 50–54
Felidae, 221
Female reproductive system, 143, 152–153
Fens, 376
Fermentation, 25, *26,* 27–30, *28*
Fertilization (sexual reproduction), 136, 143–144, *144*
Fertilizers
nitrogen-based, 351
phosphorus-based, 351
Fetus
development, 152–153, *155,* 155–156
similarities among animals, 170
Fiber
in human diet, 50
See also Cellulose
Field mice, 163–164

Fighting (behavior)
- defending territory, *323*
- displays, 324–326
- mating rituals, 145–146

Fingerprints, *20,* 105
First law of thermodynamics, in food webs, 69
Fischer, Emil, 25
Fish
- behavior, 322, 327–328
- bioaccumulation, 73, 76
- evolutionary history, 195
- food webs, *70,* 71
- introduced species, 210
- migraton, *337*
- origin of life, 181–182, 183
- respiration, 57
- speciation, 217
- sushi, *302,* 303
- *See also* Aquatic animals

Fission (asexual reproduction), 135, 285
Fitness
- exercise, 37
- ideals, 147
- weight loss programs, 37, 81

Fixed-action patterns, 322, *322,* 327–328
Flatulence, 54
Flavonoids, 141
Fleas, 282
Fleming, Alexander, 290
Flowers, 138, 140–141
- *See also* Plants

Flukes, 277
Folk taxonomy, 195–196
Food and Drug Administration. *See* United States Food and Drug Administration (FDA)
Food chains. *See* Food webs
Food webs, **67–76,** 360–361, 376–377, 393
- glossary, 74–76*t*

Foods
- bacteria, 285
- cultural attitudes, *302,* 302–303
- improper cooking and handling, 245, 251, 277–278
- nutrition labels, 79
- produced by lactic acid, 59–60
- spoilage, 27
- taste, 297, 299–301
- USDA food pyramid, 81
- *See also* Crops; Meat

Forceps, 153–154
Forensics and criminal investigation, *20,* 21, 103, 104–105, *105,* 108
Forestry. *See* Logging and forestry
Forests
- biological communities, 371–374, 403
- carbon content, 369
- deforestation, 365–366
- ecology, 361–363
- old-growth, 406, 408–409
- specialized climate, *362*

Fossils
- the fossil record, *171,* 171–172, 217
- mineralization, 176

Francisella tularensis, 252
Fried foods, 81
Frigate birds, *145*
Frisch, Karl von, 320, 323
Fruits and vegetables
- artichokes, 6
- cabbage, *26*
- healthy eating, 50, 81
- pineapples, *137*
- source of carbohydrates, 5–9
- source of protein, 23
- source of vitamin C, 93

Fungi
- decomposers, 361
- kingdom, 198
- mycorrhizae, 384–386

Fusion (sexual reproduction), 135–136

G

Galactose, 115
Galápagos Islands, 169
Gametes, 100, 136
Gasohol, 30
Gastric juices, human, 45–46
Gazelles, *373*
Geese, 322, 328–329, *329,* 330
Gender determination, 111, 112
Gender differences
- attitudes towards ideal mates, 145–149
- body fat, 37
- desirable characteristics, *146*
- eating disorders, 38
- genetic disorders, 115
- hemophilia, 241–242
- reproductive system, 142–143
- taste buds, 300–301

Genes, **99–109, 117–125**
- alleles, 112
- cancer, 238
- gene therapy, 103
- Human Genome Project, 103–104
- propensity to gain weight, 37, 127
- single-cell life-forms, 206
- speciation, 217

Genetic engineering, 102–103, **117–125**
- glossary, 124*t*

Genetic recombination, 101

Genetics, **99–109**
current research, 101–103, *102,* 118
in evolution, 164–165
glossary, 106–108*t*
history, 110–111
Genitals, human, 142–143, 238–239, 240
Genotype, 110, 112
Geosphere, 346
Germ (reproductive) cells, 99, 126–127
Germany, 81
Germs
bacteriology, 288, 290
disease-causing, 244, 283–284, 396
See also Bacteria; Viruses
Gessner, Konrad von, 197
Gestation, 152–153
Giardia lamblia, 276
Gills, 57
Glacier Bay, *404,* 405
Glaciers, 405
Glands, human, 45
diseases, 229–230, 242
pineal gland, 306–307
thymus gland, 263
Glossaries
amino acids, 16*t*
behavior, 325*t*
biological communities, 397–398*t*
biological rhythms, 314*t*
biosphere, 356–358*t*
carbohydrates, 8–9*t*
chemoreception, 304*t*
childbirth, 156*t*
climax biological communities, 407–408*t*
digestion, human, 52–53*t*
diseases, human, 234*t*
ecology, 367–368*t*
ecosystems, 367–368*t*
enzymes, 29–30*t*
evolution, 174–175*t*
food webs, 74–76*t*
genetic engineering, 124*t*
genetics, 106–108*t*
heredity, 114*t*
immunity and immunology, 260*t*
infectious diseases, 251*t*
instinct, 333*t*
learning and learned behavior, 333*t*
metabolism, 41–42*t*
migration, 340*t*
mutation, 131*t*
navigation, 340*t*
noninfectious diseases, 243*t*
parasites and parasitology, 280–281*t*
pregnancy, 156*t*
proteins, 22*t*
reproduction, 139–140*t*
respiration, human, 61–62*t*
sexual reproduction, 148–149*t*
species and speciation, 212–213*t,* 224–225*t*
succession (biological communities), 407–408*t*
symbiosis, 388*t*
taxonomy, 202–203*t*
vitamins, 94*t*
Glucose, 3–4, 56, 243
Glycogen, 44, *79,* 80
God and creationism, 163, 166
Goldberger, Joseph, 95
Gould, Stephen Jay, 177
Grand Canyon (U.S.), 113
Grasslands, 374–375
Great Lakes (North America), 211, 354
Greenhouse effect, 366, 368–369
Grimm, Jacob, 162
Gurdon, John B., 123
Gymnosperms, 138, 140, 173–174, *364,* 364–365

H

Habitats
See also Ecosystems
Habitats, human encroachment, 207–209
Habituation (behavior), 333
Haeckel, Ernst, 167, 197, 391
Haploid cells, 100
Harden, Sir Arthur, 25–26
Hares, 226
Hartnup disease, 15
Hay fever. *See* Allergies
Hazardous materials
anthrax scare, *250*
Health, human. *See* Human health
Health or organic foods, 86
Heart (human). *See* Circulatory system
Hemoglobin, 15–16, 19, 56, 170–171
Hemophilia, 115–116, 241–242, *242*
Hennig, Willi, 193
Herbivores
dinosaurs, 184
place in food web, 361
Heredity, 99, **110–116**
congenital disorders, 128–131, *129,* 232
disorders, *113,* 113–116, 121, 229, 235, 240, 241–242
glossary, 114*t*
history, 110–111
mutation, 126, 127
Herons, *375*
Hershey, Alfred, 111
Hibernation, *40,* 40–43
Hierarchical behavior (animals), 323–324

Highways, 380
Hinnies, 215
Histamines, 266–267
Historical geology, 177
HIV (human immunodeficiency virus), 259
Hominidae (family), 206
Homo (genus), 206
Homo sapiens, 206, 308
Homo sapiens neanderthalensis, 177
Homosexual community, 259
Hookworms, 278
Hormones
  amino acids in, 13–14
  biological rhythms, 307, 314–315
  insulin, 120, 242–243
  therapy, fighting cancer, 239
Horses, *402*
  the fossil record, 172–173
  mating with donkeys, 215
  species, 222
Hospitals, 154
Human behavior
  competition, 393
  fighting, 325–326
  learning, 331–334
  operant conditioning, 320–321
  symbiotic relationships, 383, 386–387
Human body
  elements in, 348
Human Genome Project, 103–104, 118, 120–121
Human health
  causes of death in U.S., 231, 236
  impact of nuclear radiation, 73
  impact of obesity, 39–40
  nutrition, 8–10
  presence of proteins, 20–23
  See also Diseases, human
Human history and development
  causing ecological disturbances, 403
  early civilizations, 186, 353, 395–396
  early migrations, 162–163
  encroaching on animal habitats, 207–209
  evolution of primates, 167–169, *177,* 220–221
  first *Homo sapiens,* 181, 308
Human immunodeficiency virus (HIV), 259
Human intelligence, 168, 333
Human sexuality, 145–149, *146*
Hume, David, 167
Huns, 163
Hunting, endangered species, 208
Huntington disease, 128
Hutton, James, 177
Hybridization (genetics), 110
Hydrocarbons, 178
Hydrogen
  in amino acids, 11–13
  percentage of biosphere, 348
  percentage of human body mass, 78
Hydrogen sulfide (intestinal gas), 54
Hydrogenation, 36
Hydrologic cycle, 347–348
Hydrolysis, 14
Hydrosphere, 346
Hypersomnia, 310
Hypothesis. *See* Theories and proofs
Hyracotherium (early horse ancestor), 172
Hyraxes, 222

I

Ileum, 47
Immune system, **262–269**
  AIDS, 245, 250, 258–261
  autoimmune diseases, 230, 242, 268–269
  fighting infectious diseases, 244–245
  glossary, 266–267*t*
Immunity and immunology, **255–261**
  glossary, 260*t*
Immunotherapy, 239
Imprinting (behavior), 322–323, 328–329, *329,* 333–334
In vitro fertilization, 144, *144*
Inbreeding, 115–116
Incest, 116
India
  early research in vaccines, 256
Indian pipe (plant), 385
Indicator species, 71–72, 392
Indigestion, 48
Indo-Europeans
  early human development, 162
Industrial melanism, 173
Industrial Revolution, 173, 240
Industrial uses
  genetic engineering, 119
  lactic acid in food production, 59–60
  nighttime activity, 312
Industrialized nations
  causes of death, 231
  deforestation, 365–366
Infection, **283–291**
  glossary, 289–290*t*
  history, 287–288, 290
Infectious diseases, **244–252**
  glossary, 251*t*
  parasites, 278
  plague, 230
  relation to cancer, 240
  See also Diseases, human
Influenza, 60
  caused by virus, 287

epidemic 1918-1920, 249–250
Infradian cycles, 313
Inhibition model (biological communities), 401–402
Innate behavior, 319–320, 322, 328
Inorganic substances. *See* Organic substances
Inquilinism, 384
Insecticides. *See* Pesticides
Insectivores, 219–220
Insects
ants, *349,* 386, 388–389
aphids, 388
bedbugs, 281–282
bees, 211, *211,* 303–304, 323–324
butterflies, 338, 388–389
class Insecta, 275
evolution, 165
fleas, 282
infectious diseases, 245
lice, 282
locusts, *405*
moths, 138, 173
parasites, 279–282
respiration, 57
symbiotic relationships, 386, 388–389
taxonomy, 200, 275
termites, 6, *6*
ticks, *279,* 282
Insomnia, 310
Instinct, **327–334,** 335
glossary, 333*t*
See also Behavior; Learning and learned behavior
Insulin, 120, 242–243
Intelligence, 168
Intelligent design theory, 168–169
Intestinal gas, 54
Intrinsic diseases, 229
Introduced species, 209–214
Inuit, *207*
Iridium, asteroid causes mass extinction, 185
Iron lung, 287
Irruptive migration, 336
Islam, 246
Island biogeography, 402–403
Isotopes, used in radiometric dating, 172

J

Jakob, Alfons Maria, 232–233
Jejunum, 47
Jellyfish, *321*
Jenner, Edward, 255–257, *257*
Jet lag, 310–311
Jews, *121*
Johnson, Earvin "Magic," 259–261
Junk food, 10, 79
Jurassic Park (movie), 184

K

Kangaroo rats, 329–330
Kangaroos, 219, *219*
Kaposi's sarcoma, 258, *259*
Kelp forests, 71
Ketoacidosis, 243
Kettlewell, Bernard, 173
Keystone species, 69–71, *70*
Kingdoms (taxonomy), 198–200
Kircher, Athanasius, 288
Kirchhoff, Gottlieb, 25
Kleine-Levin syndrome, 310
Kotler, Kerry, 108–109
Krebs cycle, 34–35
Krebs, Sir Hans Adolf, 34–35
Kudzu, 211–214, 274
Kwashiorkor, 85

L

Labels, food, 79
Labor (birth), 153
Laborers
diseases, 240, *240*
forced labor, 366
Lactic acid, 59
Lactose, 4
Lactose intolerance, 27
Lagomorphs, 224–226
Lake Erie (North America), 354
Lake Victoria (Africa), 210
Lakes
biomes, 375–376
eutrophication, 354
introduced species, 210–211
Lamarck, Jean Baptiste de, 165, 197
Languages, 162
Large intestine, 47
Lates niloticus, 210
Learning and learned behavior, 319–320, 322, **327–334**
glossary, 333*t*
See also Behavior; Instinct
Leewenhoek, Anton van, *288*
Left-hand or right-hand amino acids, 13, 17
Legal issues, teaching evolution, 169
Leks (animal territory), 324
Lemons, 93
Lemurs, 220
Lentic biomes, 375–376
Leprosy, 248–249, *249*

Libraries, analogy to taxonomy, 194
Lice, 282
Lichen, 386
Light, artificial, 311–312, *312*
    See also Sunlight
Lilac, *355*
Limes, 93
Lincoln, Abraham, 114–115
Lind, James, 93
Linnaeus, Carolus, 197, 205
Lipids, 44–45
    importance in nutrition, 80
    metabolism, 35
    in proteins, 19
Lister, Joseph, 290
Lithium, 302
Liver, human, *79,* 91–92
Loa loa, 279
Lobsters, *171*
Locusts, *405*
Loggerhead turtles, 338–339
Logging and forestry, 363, 365–366, 403, 406, 407–408
Lohmann, Kenneth, 339
Lordosis (behavior), 330
Lorenz, Konrad, 320, 322–323, 325–326, 327, 328–329
Lotic biomes, 375–376
Low-density lipoproteins, 36–37
Loxodonta africana, 200
Loxodonta cyclotisare, 200
LSD (lysergic acid diethylamide), 307
Lucy *(australopithecus),* 216
Luminol, 21
Lunar cycles, 308, 313
Lung cancer, 62
Lungs, 57–58, *59*
    See also Respiration (human)
Lupus (systemic lupus erythematosus), 268
Lymph nodes, 263
Lymphocytes, 255, 263, *263,* 264

M

MacArthur, R. H., 402
MacLeod, Colin Munro, 111
Macroscelideans, 224–226
Mad cow disease, 233
Madagascar, 138
Magnetic poles, 338–339
Magnetic resonance imaging, *237*
Magnus, Albertus, 196
Major histocompatibility complex, 262–263
Malaria, 249, 276–277
Male reproductive system, 142–143
Mallon, "Typhoid" Mary, 251
Malnutrition, 82–86, 89
Maltose, 4
Mammals
    class mammalia, 205, 218–226
    evolutionary history, 195, 204–205, 217–218
    respiration, 58
Management of forest resources, 363
Mangrove trees, *362*
Mantophasmatodea, 200
Manx shearwaters (bird), 338
Marfan syndrome, 114–115
Marrow (bone), 263, *263*
Marsupials, 218–219, *219*
Mating rituals
    animal, *145,* 145–147
    human, 145
    pheromones, 304
Mayr, Ernst, 206
McCarty, Maclyn, 111
Meat
    protein content, 23
    red meat in diet, 49–50
    undercooked meat, 277–278, 303
    See also Carnivores
Mechanist school, 167
Medical treatments and research
    bacteriology, 288, 290
    brain, 234
    cancer, 239
    childbirth, 155–157
    cloning cells, 123
    designer proteins, 21
    genetic engineering, 103, 120, 122–123
    immunology, 255–261
    in vitro fertilization, 144
    insulin, 120, 243
    use of amino acids, 15
Medicines. *See* Drugs and medicines
Meiosis, 100, 136
Melanin, 130–131, 173
Melatonin, 307, 314–315
Mendel, Gregor, 110–111
Menstruation, 286, 313
Merychippus (horse ancestor), 173
Mesozoic era, 183–184
Metabolic enzymes, 27
Metabolism, 33–43
    disorders, 37–38
    glossary, 41–42*t*
Metchnikoff, Élie, 255
Meteorites, 345
    cause mass extinction, 185–186, *186*
Methane, produced by ruminants, 7–8
Mettrie, Julien de La, 167
Mice, 123, 163–164, 226
Microscopes, *288*

Middle Ages
bestiaries, *196*
the Plague (1347-1351), 247–248
Midgets, 129
Midwives, 153–154
Miescher, Johann Friedrich, 111, 117
Migration
animal behavior, **335–341**
early humans, 162
glossary, 340*t*
Miller, Stanley, 179
Minerals
importance in nutrition, 80–81
serpentine, 71
Minnows, 327–328
Miscarriage, 152–153
Mitosis, 99–100, 135
Mobility (mammals), 198, 218
Mohave Desert, *351*
Molecular structure
amino acids, 11–13, *12*
proteins, 19
Monera, 198
Monism, 167
Monosaccharides, 3–4
Monotreme order, 218
Montagu, Lady Mary Wortley, 256
Montane forests, 363
Moon cycles, 308, 313
Morgan, Thomas Hunt, 111
Moss, 137
Mothers and babies, 73
Moths, 138, 173
Mountains, animal migration, 336
Mules, 215, *216*
Murder cases
Simpson, O. J., 105, 108
Mushrooms, 384–385, *385*
Mussels, 71, 211
Mutagens, 132
Mutation, **126–132**
DNA, 101
early genetics research, 111
glossary, 131*t*
importance in evolution, 164–165
Mutualism (symbiosis), 273, 383–386, 388
Mycobacterium leprae, 248
Mycorrhizae, 384–386

N

Nagasaki bombardment (1945), *104*
Naming conventions
amino acids, 12–13
binomial nomenclature, 206
geologic time periods, 179
germs, 283–284
vitamins, 88
worms, 274
Nanocomputers, 120
Narcolepsy, 309
National forests, 363
National Institutes of Health (NIH), 103, 120
Native Americans, 127, 130–131, 186, *207,* 396
Natural selection, 163–165, 204
animal migration, 335
behavior, 328, 393–394
choosing the ideal mate, 147–149
See also Evolution
Navigation
animal behavior, **335–341**
glossary, 340*t*
Nazis, *20, 121,* 121–122
Neanderthal man, *177*
Necator americanus, 278
Nematoda (phylum), 349
Nerve cells, *296,* 296–297, 299, 303
Nervous system, 295–297, *296*
See also Brain
New Guinea, 396, 398
New World. *See* Explorers and exploration
New York City, *247,* 378
Niacin, 15, 95
Niche (biological communities), 392
Nile perch, 210
Nitrogen
depletion by leaching soil, 352–353
percentage of biosphere, 346, 348
used in fertilizers, 351
Nocturnal activities
animal migration, 336–337
human, 311–312
Nomenclature. *See* Naming conventions; Taxonomy
Noninfectious diseases, **236–243**
glossary, 243*t*
See also Diseases, human
Northeast Passage, 88
Northern fur seal, *130*
Northern latitudes
midnight sun, 310
Norway, *309*
Novaya Zemlya (Russia), 88–89, *89*
Nuclear radiation, 73
Nuclear weapons, 103, *104*
Numerical taxonomy, 192–193
See also Taxonomy
Nursing mothers, 73
Nutrition and nutrients, 44–45, **77–86**
amino acids, 14–15
carbohydrates, 8–10
diseases, 230

fats, 37
glossary, 83–84*t*
"Nutrition Facts" label, 79
proteins, 22–23
vitamins, 95–96
See also American diet

O

Obesity, 37, 39–40, 81, 127
See also American diet
Obligate relationships (symbiosis), 273–274, 383–386
Obligatory taxonomy, 193–194, 205
See also Taxonomy
Obstetricians, 154
Occupational health
cancer, 240
nightshift, 309, 311
Oceania
marsupials, 219
Oceans
biomes, 376–377
food webs, 77, 376–377
mass extinctions, 182, 185
migrations, 336
percentage of Earth's water in, 347
phosphorus in, 354
Odum, Eugene Pleasants, 372
Oils. *See* Fats and oils
Old people. *See* Elderly
Old-growth biological communities, 369, 402, 406, 408–409
Oligosaccharides, 4
Omnivores, 361
On the Origin of the Species by Means of Natural Selection (Darwin), 169
Open systems, 345–346
Operant behavior, 320
Oranges, 93
Orchids, 138, 385
Orders (taxonomy), 200, 218–226
Ordovician period, 182
Organ transplants, 262–263
Organic and health foods, 86
Organic substances
distinguished from inorganic, 78, 176, 178, 348, 361, 371, 391
history, 178
Origin of life, 178–182
amino acids dating, 17
theories of evolution, 162
Origin of the universe, 177–178
Ornithischia, 184
Osteomalacia, 91
Overproduction and natural selection, 163
Oviparity, 151, *152*
Ovivparity, 151
Ovulation, 143
Owls, 406, 408
Oxidation, effect on human body, 91
Oxpecker, 386, *387*
Oxygen
absorbed in lungs, 55
early Earth, 178, 179
percentage of biosphere, 346, 348
percentage of human body mass, 78
produced in photosynthesis, 5, 58
used in fertilizers, 351
Oxytocin, 153

P

Paleontology, **176–188**
and geology, 176–177, 180
glossary, 187–188*t*
Pancreas, human, 47
Pangea and evolution, 180
Pangolins, *223*, 223–224
Parahippus (early horse ancestor), 173
Parasites and parasitology, **273–282**, 330–331, 383–384
glossary, 280–281*t*
Parthenogenesis, 137–138
The Parthenon, 138
Partial migration, 336
Passionflower, 389
Pasteur, Louis
pasteurization, 288
vaccinations, 256, 257
Pastuerlla pestis, 248
Patented research, genetic engineering, 121
Pathogens
cause infection, 285–286
glossary, 284
infectious diseases, 245
targeted by immune system, 255, 262
Pauling, Linus, *92*, 93
Pavlov, Ivan, 320
Pelagic ocean biomes, 376
Pellagra, 15, 95
Penicillin, 290
Pentadactyl limb, 170, *170*
Pepper moths, 173
Peptide linkage, 13, 18
Perissodactyls, 222
Permian period, 182–183
Pesticides
DDT, *72*, 73
Petroleum
fermentation, 28, 30

remains of dinosaurs, 185
Phenetics, 192–193
Phenotype, 110
Phenylalanine hydroxylase, 37
Phenylketonuria (PKU), 37
Pheromones, 303–304
Philosophy
theories and proofs, 167
Pholidota, *223,* 223–224
Phoresy (symbiosis), 389
Phosphorus
importance in nutrition, 91
percentage of biosphere, 348
phosphorus cycle, 354
plants and fungi, 384
Photosynthesis
creation of oxygen, 346
energy transfer, 360–361
micorrhizae, 385
production of carbohydrates, 3, *4,* 4–5
Phototropism, 321
Phylogeny, 191–192, 195, 197–198, 204, 216, 217
Phylum
chordata, 205
nematoda, 349
Physical fitness. *See* Fitness
Physicians, 153–154, 238–239
Phytoplankton, 77, 376
Pigeons, 338
Pima (Native American tribe), 127
Pineapples, *137*
Pinworms, 278
Placenta, 153
Plagues, 230, *231,* 246–248
Planets, origin, 178
Plants
asexual and sexual reproduction, 136–141
behavior, 321
biomes, 372–375
evapotranspiration, 346, 347, 354–355, *355*
evolution, 173–174
fermentation, 28
forensics, 108
genetic engineering, 119
hybridization, 110
introduced species, 210–214
kingdom plantae, 198
photosynthesis, 4–5, 58, 77, 87–88, 346, 360–361
protein content, 23
selective breeding, 128, 387
starches and cellulose, 6–8
symbiotic relationships, 384–386, 389
use of nitrogen, 348
vegetative propagation, 136
See also Fruits and Vegetables; Trees
Plasma (blood), 47
Plasmodium, 249, 276–277
Plate tectonics and evolution, 169, 180
Plato, teacher of Aristotle, 197
Pliohippus (horse ancestor), 173
Pneumonia, 60, 62, 287
Polar bears, 89
Polarity of molecules, 19
Poliomyelitis, 257–258, *287*
Pollen and pollination, 138–141, *364,* 364–365
Pollution
cancer-causing, 240–241
impact on biological communities, 403
Polymers, made of amino acids, 13, 18–19
Polysaccharides, 4
Pond biomes, 375–376
Pork, 277–278
Porpoises, 221, 340–341
Post-traumatic stress disorder, 232
Pott, Percivall, 240
Poverty and the poor
forced labor, 366
malnutrition, 85, 95
occupational health, 240
Precipitation
deserts, 375
rain, 358, 374
Pregnancy, **151–157**
glossary, 156*t*
Primates, 167–168
order primates, 205–206, 216, 220–221
Probability
amino acids in proteins, 18–19
base pairs in genes, 100, 118
Proboscideans, 222
Producers (food webs), 68
Proofs. *See* Theories and proofs
Propagation. *See* Reproduction; Vegetative propagation
Prosimii, 220–221
Proteins, **18–23,** 44
complete proteins, 15, 80
content of vegetables, 6
glossary, 22*t*
importance in nutrition, 10, 79–80, 81–82
made of amino acids, 13
synthesis, 100–101
Protista, 198
Protozoa, 275–277
Psychological disorders
Alzheimer's disease, *233,* 233–235
eating disorders, 38–40
mental retardation, 334
seasonal affective disorder, 314–315
Pygmies, *128*

Q

Quantum mechanics and evolutionary theory, 165–166

R

Rabbits, 226
Rabies, 257
Race (humans), 207
Radiation (electromagnetism)
  cancer treatments, 239
  cancer-causing, 240–241
  effects of exposure, 103, *104*
  nuclear, 73
Radiometric dating, 172
Rain. *See* Precipitation
Rain forests, 350, 362–363, 365, *366*, 373–374, 392–393
Rape cases, 108–109
Raphus cucullatus, 208–209
Rats, 226, 330
Reasoning ability, 168
Receptors (senses), 296–297, 298–299
Recessive genes, 112–113, 115
Red blood cells
  malaria, 276–277
  produced by bone marrow, *263*
Reflexes, *321*, 322
Religion
  creationism, 163
  and science, 167
Replication of DNA, 100–101
Reproduction, **135–141**
  glossary, 139–140*t*
  See also Asexual reproduction; Sexual reproduction
Reproductive system (human), 142–143
Respiration (human), **55–63**, *59*
  glossary, 61–62*t*
  respiratory system, 56–58, *59*
  See also Cellular respiration
Retroviruses, 287
Rheumatoid arthritis, 268
Rhinoceros, 386, *387*
Ribonucleic acid. *See* RNA (ribonucleic acid)
Rice, 95
Rickets, *90*, 91
Right whales, 208
Right-hand or left-hand amino acids, 13, 17
River blindness, 279
Rivers
  biomes, 376
  moving soil, 351
RNA (ribonucleic acid), 101, 287
Rocky Mountain bighorn sheep, *323*
Rodents, 224–226
Roman Empire, 230, 246
Roosts (animal territory), 324
Roundworms, 278
Royalty and hemophilia, 115–116, 241
Ruminants, 7–8, 53
Rural techno-ecosystems, 379–380
Russia, 241, *242*

S

Sabin, Albert, 257–258
Sahara desert, 353
Sailors, 93
Saliva, human, 45
Salk, Jonas, 257–258
Salmon, *337*
Salmonella typhosa, 251
Salts in soil, 350
San Blas Indians, 130–131
Sanitariums
  leprosy facilities, 248
Sanitary conditions, 287–288, 288, 290
Saturated fats, 36, 88
Sauerkraut, *26*
Saurischia, 184
Savannas, *373*, 374–375
Scandentia, 220
Schistosoma, 277
Schools, teaching evolution, 169
SCID (Severe combined immune deficiency syndrome). *See* Severe combined immune deficiency syndrome
Scientific theories. *See* Theories and proofs
Scrotum cancer, 240
Scuba diving, *199*
Scurvy, 93
Sea otters, 71
Seals, *130*
Seashores, 377
Seasonal affective disorder (SAD), 314
Seasons, 313–315
Second law of thermodynamics in food webs, 69, 393
Seeds and seed-bearing plants, 138
Selective breeding, 128, 169
Selenium, bioaccumulation, 71
Senses, 295–305
Septic tanks, 352
Sequoia National Park, 363
Serotonin, 307
Serpentine minerals, 71
Severe combined immune deficiency syndrome (SCID), 115
Sewage worms, 72

Sexual reproduction, 135–141, **142–150**
- basis of species, 206–207, 215–216
- glossary, 148–149*t*

Sexual revolution, 147–149
Sexuality. *See* Human sexuality
Sexually transmitted diseases, 240, 245, 258, 276
Shedding (fur or skin), 313, *313*
Sheep, *122,* 123, *323*
Shell shock, 232
Ships, introduced species, 210
Shores, sea, 377
Shrews, 219–220, 226
Sickle-cell anemia, *14,* 15–16
Siesta, 308
Simmelweis, Ignaz P., 288
Simple sugars. *See under* Sugars
Simpson, O.J., 105, *105,* 108
Sirenians, 221–222
Skin, 130, 244–245, 248, 262
Skinner, B. F., 320–321
Sleep, 40, 307, 308, 312–313
- disorders, 309–311, 315
- *See also* Hibernation

Small intestine, human, 47
Smallpox, 230–231, 252, 256, *257, 396*
Smell (olfaction), 297, 299, 301–304
Snails, *298,* 333
Sneath, Peter Henry Andrews, 193
Snow, John, 288
Social Darwinism, 164
Societal views
- AIDS, 259–261
- childbirth, 154
- evolution, 165–169
- in vitro fertilization, 147–149

Societies and human evolution, 162–163
Soil
- fungi, 385–386
- role in Biosphere, 346–347, 348–353

Sokal, Robert Reuven, 193
Solar system, origin, 177–178
Solubility of lipids, 35
Somatic (body) cells, 99, 126–127
Somatotropin, 307
Soot
- cause of cancer, 240, *240*
- industrial melanism, 173

Sound
- echolocation in animals, 339–341

South Pole
- midnight sun, 310

Species and speciation, *112,* 127, **204–214, 215–226**
- competition in communities, 393–395
- discovering new species, 194–195, 223
- glossary, 212–213*t,* 224–225*t*
- *Homo sapiens,* 206
- undiscovered species, 216

Sperm cells, 100, 132, 136, 142–143
Spiders, 330, *332*
Spoilage, foods, 27
Sponge (animal), *199*
Spongiform encephalopathy, 128
Spotted owls, 406, 408
Spruce trees, 371
Squirrels, *112,* 113
Staphylococcus, 286
Starches, 6–7, *7,* 25
Starfish, *70,* 71
Starvation
- children, 84–85
- dieting technique, 81

Steptoe, Patrick, 144
Steward, F.C., 123
Stickleback fish, 217, 322, 327
Stimuli (living organisms), 295, 296–297, 319, 320, 321, 327–328
Stomach, human, 46–47, 48, *49*
Strata (geology) and the fossil record, *171,* 171–172
Stream biomes, 376
Streptococcus, 286
Stress (psychology), 230, 232
Substrate, 25
Succession (biological communities), 370–371, 392, **400–409**
- glossary, 407–408*t*

Sucrose, 4
Sugars
- chemical reactions, 25
- metabolism, 34
- nutrition in carbohydrates, 9–10
- in proteins, 19
- simple sugars, 3–4

Suicide, 231
Sulfa drugs, 290
Sulfur, percentage of biosphere, 348
Summer, cause of dry conditions, 358
Sun, origin of the solar system, 177–178
Sunlight
- after massive asteroid strike, 185
- impact on sleep, 308, *309,* 309–310
- in photosynthesis, 4–5, 360–361
- phototropism in plants, 321
- vitamin D deficiency, 91

Surgery
- brain, 234
- fighting cancer, 239

Suriname toad, *152*
Survival of the fittest. *See* Evolution; Natural selection
Sushi, *302,* 303
Sutton, Walter S., 111
Swallows, 338

Swamps, *375,* 376
Symbiosis, **383–390,** 392
- dodo bird and dodo tree, 209
- glossary, 388*t*
- parasites, 273–274
- pollination, 140

Sympatric species, 217
Synthesis of proteins, 19, 21–22
Systema naturae (Linnaeus), 197
Systems (biology), 345–347

T

T cells, 255, 263–264
Table sugar, 4
Taiga. *See* Boreal forest
Tapeworms, 277–278
Tarzan (fictional character), 331–332
Taste buds, 298–299, 298–300
Taste (gustation), 297, 298–304
Taxonomy
- biology, **191–203,** 204–206, 215
- biomes, 372
- glossary, 202–203*t*
- history, *192, 196,* 196–198
- taxonomic keys, 193, 197
- worms and anthropods, 274–275
- *See also* Naming conventions

Teeth, aardvarks, 222
Temperate deciduous forests, 373
Temperate rain forests, 373–374
Temperature
- foods, impact on taste, 300

Termites, *6*
Terrestrial biomes, 370
Territory (animal behavior), 324–326
Terrorism (biological), 231, 251–252
Testosterone, 142
Texture
- foods, impact on taste, 300

Theories and proofs
- evolution, 166–167

Thermodynamics, laws of, 69
Thiamine (vitamin B1), 92, 95
Third world nations
- starvation, 84–85

Thymus gland, 263
Tickbirds, 386, *387*
Ticks, *279,* 282
Tidal waves, 185
Tinbergen, Nikolaas, 320, 322–323, 327
Toads, 123, *152*
Tobacco use
- cancer, 240
- sense of taste and smell, 301–302

Tolerance model (biological communities), 401
Tongue (human), 299–301
Tortoises, *351*
Toxic shock syndrome (TSS), 286
Toxins
- bioaccumulation, 72–73
- indicator species, 71–72
- targeted by immune system, 262

Trace elements in human body, 78
Tracheal respiration, 57
Transgenic animals or plants, 103
Transpiration, 354–355, 358
Travel
- jet lag, 310–311
- nighttime air travel, 312

Trees
- conifers, 371–372, 374
- dead trees, 408
- deciduous trees, 373
- dodo tree, 209
- dominate forest ecosystem, 362
- introduced species, 210
- micorrhizae, 385
- recovery after logging, 406
- specialized climate, *362*
- tree-dwellers, 363
- *See also* Forests

Triceratops, 183, 184
Trichinosis, 278
Trichomonas vaginalis, 276
Trophic levels in food webs, 68–69, 73
Tropics
- coral reefs, 377, *377*
- savannas, *373*
- tropical rain forests, 350, 362–363, 365, *366,* 374

Tropism, 321
Truffles, 385, *385*
Tryptophan, 15
Tuberculosis, 60, 248–249
Tubificid worms, 72
Tubulidentates, 222
Tumors, 238
Tundra, 374, 392–393, *394,* 394–395
Turtles, 338–339
Twins, genetic studies, 115
Typhoid fever, 251
"Typhoid" Mary Mallon, 251
Tyrannosaurus rex, 184

U

Ulcers, 48–49, *49*
Ulmus americana, 210
Ultradian rhythms, 312–313
Ultrasonics, medical usage, *155,* 155–156

Underwater diving, *199*
Uniramia (subphylum), 275
United Kingdom
  Creutzfeldt-Jakob disease, 233
  early vaccinations, 256–257
  royalty, 115–116, 241
United States Atomic Energy Commission (AEC), 103
United States Centers for Disease Control and Prevention (CDC), 258–259
United States Department of Agriculture (USDA), 81, 363
United States Department of Energy (DOE), 103, 120
United States Food and Drug Administration (FDA), 79
United States Forest Service, 363
Unsaturated fats. *See* Saturated fats
Upwelling regions (oceans), 377
Urine, of diabetics, 242
USDA (Department of Agriculture). *See* United States Department of Agriculture
Uterus, 152, 153

V

Vaccines and vaccination
  cancer, 239
  genetic engineering, 103
  history, 255–256
  infectious diseases, 249
  viruses, 287
Vampire bats, 220
Variola. *See* Smallpox
Vegetables. *See* Fruits and vegetables
Vegetarians, 23
Vegetative propagation, 136, *137*
Velociraptor, 184
Vestiges, 170
Viruses
  cellular activity, 285
  human immunodeficiency virus (HIV), 259–261
  infections, 286–287
  infectious diseases, 250–251
  respiratory disorders, 60
Vitamin A, 80, 88–90
Vitamin B1 (thiamine), 92, 95
Vitamin B2, 92
Vitamin B6, 92
Vitamin B12, 92, 268–269
Vitamin C, *92,* 93
Vitamin D, *90,* 90–91
Vitamin E, 91
Vitamin K, 91–92
Vitamins, 45, **87–96**
  glossary, 94*t*
  history, 93–95
  importance in nutrition, 80–81
  supplements, 95–96
Viviparity, 151–152
Vomiting, 39
Vries, Hugo De, 111

W

Wallace, Alfred Russel, 169
Warm-blooded animals, 218
Wastes, human, 50–54
"Watch analogy" (evolution), 161–162
Water
  biomes, 375–377
  cause of *giardiasis,* 276
  content of vegetables, 6
  early Earth, 178
  eutrophication, 354
  in healthy diet, 50
  in human digestion, 47, 88
  hydrologic cycle, 347–348
  molecular polarity, 19
  percentage of human body mass, 78
  pollution, 71–72
  in soil, 350, 352–353
  and spicy foods, 297
Watson, James D., 111, 117
Weather, 347–348, 358
Weight loss. *See* Fitness
Weightlifting, 308
Weizmann Institute, 120
Wetlands, 376
Whales
  echolocation, *339,* 340–341
  endangered species, *207,* 208
  taxonomy, 195, 204–205, 222
Wheat, 352
Wilson, Edward O., 402
Wings, birds, 192
Winter
  animal migration, 335–336
  impact on biological rhythms, 314–315
Wöhler, Friedrich, 178
Word origin
  kwashiorkor, 85
  metabolism, 33
  vitamins, 95
  See also Naming conventions
Work. *See* Laborers; Occupational health
World War II
  concentration camps, *121*
  Nagasaki Bombardment, 1945, *104*
Worms
  acorn worm, *138*
  detritivores, 349
  parasites, 275, 277–279

taxonomy, 274–275
Wuncheria bancrofti, 278, 279

Xenarthrans, 219

Yeast, 28, *28*
Yellowstone National Park, 363
Yersinia pestis, 248

Z

Zebra mussels, 211
Zeitgebers, 309
Zooplankton, 336
Zygotes, 100, 152